PSYCHOLOGY RESEARCH PROGRESS

EXPANDING HORIZONS OF THE MIND SCIENCE(S)

PSYCHOLOGY RESEARCH PROGRESS

Additional books in this series can be found on Nova's website
under the Series tab.

Additional e-books in this series can be found on Nova's website
under the e-books tab.

PSYCHOLOGY RESEARCH PROGRESS

EXPANDING HORIZONS OF THE MIND SCIENCE(S)

P. N. TANDON
R. C. TRIPATHI
AND
N. SRINIVASAN
EDITORS

New York

NOTICE TO THE READER

Library of Congress Cataloging-in-Publication Data

Expanding horizons of the mind science / editors, P.N. Tandon, R.C. Tripathi and N. Srinivasan.
    p. cm.
 Includes index.
 ISBN 978-1-62808-705-5 (softcover)
 1. Intellect. 2. Cognitive science. I. Tandon, P. N. (Prakash Narain), 1928- II. Tripathi, Rama Charan. III. Srinivasan, Narayanan.
 BF431.E937 2011
 153--dc22
            2011014543

Published by Nova Science Publishers, Inc. † New York

CONTENTS

FOREWORD

Science, as the human kind knows, is a result of thought processes of mind. In some sense, science is a product of mind. There have been questions as to whether science could be applied to understand the ways in mind itself works. In understanding the interconnection among brain-behavior and consciousness, mind science has remained a weak link. Ancient Indian civilization had understood the power of human consciousness and tools of yoga and meditation to control the behavior by regulation of thought processes of mind. With advances of technology in mind times, it is becoming possible to develop mind science employing rigorous tools and techniques. The Department of Science and Technology mounted a Cognitive Science Initiative as a part of the XI Plan Program me of the Science and Technology Sector.

The initiative has encouraged a group of scientists drawn from different disciplines in the country to self-assemble and establish a network. Some interesting opportunities for developing cognitive science in the country have been identified.

Under the Cognitive Science initiative, it was decided to capture the control status of research in the area of mind science and develop an aspirational framework in the form of an edited volume. Prof. Tandon is the lead editor of this publication which contains authoritative articles form 24 distinguished scientists in the country and abroad.

I congratulate the Task Force in Cognitive Science Initiative for proposing and promoting this publication on "Expanding Horizons in Mind Science(s)". I thank the authors who contributed to the volume. This presents the State-of-Art and provides a platform for proactively developing research in the area in the country.

Human mind is too complex a system to be understood though the principles of reductionism. Therefore modern science may need to prepare itself to be developed for understanding human cognition. This is a small beginning if modern Indian scientist to prepare themselves with the area of Mind Science.

If we understand the underlying mysteries of human thought processes better, may be it will be possible to add some happiness to the lives of people who with the advent of technology live longer. Can Cognitive Science aid in the Happiness Index of Nations?

T. Ramaswami
Secretary
DST, India

PREFACE

The journey of Mind Sciences in India has been through the peaks and troughs. It is generally accepted that no other cultural tradition except Indian has given so much attention to the matters of mind and consciousness. Yet, recognition to the sciences studying Mind like Psychology as a scientific discipline came in very late. There were only a handful of Universities in India, which had independent Departments of Psychology or Neuroscience at the time of India's independence. In the last few decades, Mind Sciences in the country have picked steam resulting in major discussions and interactions across disciplines like Psychology, Neuroscience and Computer Science.

A number of new centres have come up recently around the country with research focus on mental and psychological processes. The foremost among these is the new National Brain Research Centre (NBRC) at Manesar and a new Centre for Behavioural and Cognitive Sciences (CBCS) at the University of Allahabad, which has been established by the UGC. The Department of Science and Technology (DST) too have decided to add Cognitive Science as the fourth pillar of knowledge along with nano-, bio-, and information technology. For the past many years, DST has provided funds for the organization of an interdisciplinary Summer School 'From molecules to human behaviour' under the initiative taken by CBMR, Lucknow, IISc, Bangalore and G.B. Pant Social Science Institute, Allahabad. It also has taken initiative for developing Cognitive Science by providing research support to inter-disciplinary research on Mind and Behaviour in the Country.

In addition to its importance of understanding and realizing human potential, the advancement in Mind sciences is also important to address problems of mental health in the country. There is an alarming shortage of trained professionals to tackle rising mental health problems. The need for enhancing manpower in the Mind sciences has been felt by international institutions and national bodies like National Academy of Psychology (NAOP), India, Indian Academy of Neurosciences and Indian Association of Clinical Psychologists.

The present volume is a testimony to the interest in the study of mental processes and phenomena across levels and disciplines. The transdisciplinary interest that has emerged is evident in the contributions included in this volume and the disciplinary background of the contributors. The three editors of the volume represent different disciplines, namely, psychology, cognitive science, neuroscience and medicine, Contributors to this volume represent the disciplines of neuroscience, psychology, psychiatry, medicine, cognitive science, molecular biology, neurolinguistics and speech pathology. In this sense, this volume is the first of its kind in India and demonstrates the possible directions in which scientific disciplines could develop by transcending their narrow boundaries under the direction of the Department of Science and Technology, India.

The editors will like to express their gratitude to the Department of Science and Technology, India for their generous financial support and encouragement. We are particularly thankful to Dr. H. B. Singh played a key role at the Department of Science and Technology.

Finally, the authors of the various chapters who contributed to this volume deserve our special thanks. Although, the time that they had at their disposal was very short, they did their best to meet the deadlines set by us. It is our fond hope that multidisciplinarity reflected in this volume will grow not only within Mind sciences but in all other sciences in India and beyond.

P.N. Tandon
R.C. Tripathi
N. Srinivasan

Section 1. Introduction to the Mind Science(s)

In: Expanding Horizions of the Mind Science(s)   ISBN: 978-1-62808-705-5
Editors: P. N. Tandon, R. C. Tripathi and N. Srinivasan ©2013 Nova Science Publishers, Inc.

Chapter 1

NEW INTERDISCIPLINARY SPACES IN MIND SCIENCE(S)

R. C. Tripathi[1] and Narayanan Srinivasan[2]
[1]Department of Psychology, University of Allahabad, Allahabad, India
[2]Centre of Behavioural and Cognitive Sciences, University of Allahabad, Allahabad, India

ABSTRACT

This chapter examines the evolution of mind science(s) as a multifaceted scientific discipline in the last few decades. This has resulted due to the attention that has been paid to studying psychological phenomena by scientists from such diverse set of sciences as, biology, physics, computer science, and medicine apart from its sister disciplines, philosophy and the social sciences. The chapter discusses some frontier areas of research in mind science(s) which are at the traditional disciplinary boundaries of psychology. These are - the relationship between brain and mind, consciousness and free will, nature of cognition, social behaviour, culture and society. We also discuss problems in developing mind science(s) as a unified discipline and suggest some possible new directions which mind science(s) can take in continuing to evolve as a multidisciplinary 'soulful' science.

1. INTRODUCTION

Psychology belongs to that group of sciences which have a long past but a short history. In almost all the cultural traditions around the world, concepts which are equivalent of psyche, mind and behaviour have received a great deal of attention from philosophers and scholars; in the Indian cultural tradition, perhaps, more than the others. Yet, Psychology continues to struggle to create a space for itself and for recognition from scientists professing and practicing other scientific disciplines. As Carrel (1959) puts it, scientists have been fascinated by the beauty of the inert matter but have ignored focusing on the body and mind for a very long time. Psychology's problem has been largely due to the nature of phenomena it studies that are not amenable to easy study by scientific methodology that are widely used in the natural sciences. And the methods that are available do not permit us to study the multidimensional man as an indivisible whole. Yet, in a way no science can do without Psychology. So, is not geometry, not located in nature but in the human mind? That man is more than a concrete system composed of chemical substances and tissues, is being realised increasingly. This century promises to be a century of mind and consciousness, in which more than ever, the unanswered questions related to the humans, will be explored and hopefully answered.

Strange as it may sound, modern scientific psychology has a date of birth! No other Science can claim, even remotely, when it was born. Scientific Psychology, in that sense is not very old. It is widely accepted that it was born with the founding of the first Experimental Psychology Laboratory by Wilhelm Wundt in Germany in 1867. This was preceded by publication of a path breaking book in 1860 titled *Elements of Psychophysics* by Gustav Fechner. In this relatively short period of about 150 years, psychology has grown by leaps and bounds and has developed interfaces with almost all major sciences, be it Physics, Chemistry, Biology, Medicine and also with other Human Sciences.

Valsiner and Rosa (2007) note that Psychology has been involved in 'recurrent destruction efforts of its theoretical core, as well as to various efforts to eliminate the discipline by downward (to physiology, or genetics) and upward (to texts, cultural models) reductionism.'(p.3), yet, as they observe, it enjoys good health and excellent institutional grounding. Although, there is not a separate disciplinary category for Psychology under which a Nobel Prize is awarded, there are quite a few scientists who have worked on psychological phenomena in related disciplines who have been awarded the Nobel Prize. The first among these is Ivan Pavlov, who was awarded the Nobel in Physiology/Medicine in 1904 for his experiments on classical conditioning of salivation responses. The principle enunciated by Pavlov has stood the test of time and has served as the foundational principle for explaining how learning takes place and for acquisition of new behaviours. To this may be added, Nobel prizes awarded to Konrad Lorenz and N Tinbergen whose work on organization and elicitation of individual and social behaviour patterns was recognised in 1977. In 1978, Herbert Simon was awarded a Nobel for studying how cognitive limits influence decision making in organizations. Work of R W. Sperry on functional specialisation of cerebral hemispheres and David Hubel and Torsten Wiesel's research on information processing in the visual system were similarly recognised. One of the latest Nobel comes in Economics in 2002, awarded to Daniel Kahneman, a Professor in Psychology at Princeton University for his work on judgment and decision making under conditions of uncertainty.

To emphasize the point of emerging interfaces of Psychology with other sciences, we may point out that human brain, consciousness and thought processes have recently attracted the attention of Physicists and Information Scientists, like Penrose (1989), Pribram (1991) besides, of course, the biologists and neuroscientists. Their interest has led to the development of the concept of Quantum mind and also to the emergence of a new field called Quantum Brain Dynamics (QBD) which takes a field-theoretical approach to understand human consciousness. Nevertheless, phenomenal consciousness or qualia remains a hard problem for Physicists because qualities like taste and feelings cannot be explained using principles of Physics or expressed in mathematical terms. Ironically enough, Psychology still struggles with issues related to its definition, (Henriques, 2004), and questions are asked about its status as a scientific discipline within the hierarchy of sciences (Simonton, 2004). We turn to some of these questions in the next section.

2. ISSUES OF DEFINITION AND KNOWLEDGE INTEGRATION

The problem which gets raised relating to definitional issues of psychology pertains primarily to its focus and theoretical approaches used for integration of different psychological phenomena across levels. The different emphases have led to the development of various schools of psychology - Structuralism, Functionalism, Behaviourism, Cognitivism, etc., and, indeed, to the classification of psychology in terms of cultural traditions, such as, Indian Psychology (Rao et al., 2009) which focus more on the ways of minds and less on the ways of humans. Psychologists have been seeking integration of knowledge based on phylogeny as well as on human ontogeny but such approaches have generally not succeeded in finding convergence. In this context, Henriques's (2004) analysis appears correct when he says that the problem of Psychology emerges due to difficulties arising from seeking reconciliation between two different kinds of epistemologies, one focusing on developing laws of human behaviour where man is seen essentially as nothing more than a homo sapien and the second, where it seeks to engage in meaningful acts and also attempts to make sense of the acts of others within their socio-cultural settings. He suggests that Psychology can be properly defined in terms of its relationship to physical, biological and social sciences. This he does by using what he calls, the Tree of Knowledge System (ToK).

"The ToK System is a graphic depiction of the evolution of complexity from the Big Bang through the present. It offers a new vision of the nature of knowledge as consisting of four levels or dimensions of complexity (Matter, Life, Mind, and Culture) that correspond to the behavior of four classes of objects (material objects, organisms, animals, and humans), and four classes of science (physical, biological, psychological, and social)." (p.1209)

The ToK approach permits him to view the subject matter of Psychology as extending from simplest animal behaviours to human consciousness as forming one continuum and seeing humans as being both continuous with other animals and distinct from other animals at both individual and societal levels of analysis. In so doing, that is, in focusing on the study of mental processes and behaviours at different levels, it seeks to unravel biological bases of such processes and behaviours, by asking how brain and behaviour are related. Similar approaches and concerns have led to the development of emerging areas of Behavioural Neuroscience, Cognitive Neuroscience, Affective Neuroscience, Social Neuroscience, and Neuropsychology all of which look at the neural bases of psychological processes implicated in various behaviours and psychological disorders. Progress in techniques of brain mapping and brain imaging along with the application of such techniques as fMRI technology have contributed to the significant advancement of knowledge in these areas (Gazzaniga et al., 2002).

Although, the past few decades have seen emergence of new interest in studying behavioural phenomena by scientists from divergent disciplines as also emergence of new disciplinary areas, our understanding of human behaviour has remained at the descriptive level and there have been few theoretical advances that have been made. One of the reasons for this has been that scientists have confined themselves to working either at the genotypic level or at phenotypic levels which confines them to looking at phenomena which are relatively static. Panksepp et al. (2002) suggest scientists will be able to understand the ongoing activity of neurobiological systems a lot better if scientists were to look at information that exists between proximate genotypic level and the phenotypic level simultaneously. Perhaps, there is a need to revaluate the assumptions that are made by scientists which do not permit them to draw knowledge from research studies performed at other levels. For example, Panksepp et al. believe that if evolutionary psychologists were to accept that there are similarities in brain, behaviour and various basic mental processes across mammalian species, they would be able to link up with other scientific disciplines such as biochemistry, physiology, molecular genetics, and developmental biology along with scientists engaged in various kinds of neuro-scientific analysis of behaviour.

Bickle (1998) thinks that there is another kind of problem related to the levels. He has argued that scientists should not stop for looking at psychological phenomena at the neural levels and functional psychology now needs to move to molecular and cellular levels. He draws upon some fascinating experiments on memory, attention and consciousness which provide molecule to mind explanations to build his argument. Another important area which has emerged which sees human behaviour, basically, as evolved adaptations in specific cultural environments, is Evolutionary Psychology. Models of information processing have come to be applied to understand various cognitive processes such as of attention, perception, learning and memory, judgment and decision making, thought processes and also emotions and are studied under Cognitive Psychology and Cognitive Science.

One unique feature of the humans is their capacity for growth and development as a result of new experiences they have within the socio-cultural and physical environments of which they are a part. Human development and how it comes about is a major area of concern to psychologists as it has implications for developing new competencies and unfolding the untapped and undeveloped potential of the humans. Understanding the differences and uniqueness of humans and behaviour involving interaction with other humans in various settings, have been of abiding interest to psychologists.

Apart from these, there are a number of areas in which Psychology has come to be applied. Some of the more important areas are of Mental and Physical Health; Educational and Industrial Organizations; Human Factors; Forensic Psychology including Law and Justice, etc. In fact, integration of Psychology into health care systems in the United States of America and acceptance of psychologists in the Medicare Programme is seen as the final testimony to acceptance of Psychology as a scientific discipline. It is now realised, at least, in the United States by the public policy makers that psychological processes contribute to a large number of chronic illnesses, both physical and mental. APA's 2020 vision document sees psychologists playing a very significant role as primary health care providers of behavioural and psychological interventions. Although, psychology has been integrated with the medical practice in USA, it remains to be integrated with the academic curricula of medical education in spite of several recommendations which have been made in this connection. The scene in India, however, continues to be quite dismal in terms of integration of psychology and behavioural sciences in medical practice as well as in other sectors.

We may be asked what kind of psychological knowledge is being produced and what research trends can be observed in contemporary psychology. Although, the past few years have seen a rapid growth in researches as well as research grants in Cognitive Psychology and various sub-disciplines of psychology related to Neurosciences, Spear (2007) in a recent review of trends in Psychology does not find a major shift towards cognitive psychology or Neuroscience in U.S. universities as was reported earlier by Friman et al. (1993). Psychology continues to be dominated not by one approach or paradigm. In that sense one may call it a discipline which is not unified and as having multiple focuses and also employing multiple kinds of lenses. The debate over psychological universals and the possibility of coming up with universal laws that govern human behaviour still rages in Psychology (Kim, 1998) but it does so even in Physics (Hawking, 2009; Kauffman, 2009). As Staats (1981) points out, all sciences begin in disunity and move towards unification.

Valsiner (2009) argues in support of a unified discipline of Psychology but one which is not only culture-sensitive but also culture-inclusive. This, however, indicates a direction in which psychology may take in future. What is also questioned is whether the reductionist paradigm, which has so far ruled psychology and, indeed, all sciences, can ever deliver on such a promise. He points to what he calls have been three domains of oversight. First, he points out that, contrary to what our psyche is, a dynamic flow of psychological phenomena, be it of any kind, the data psychologists generate are for the most part, static representations of these states. Secondly, all psychological phenomena is embedded in a context where the higher order processes and contexts determines the lower order functioning. This hierarchy of levels and the manner in which they get connected is ignored. Lastly, according to him, scientists forget the context for the phenomenon which is studied using the predefined codes and discovering the phenomenon as it obtains in the 'other world' never takes place.

Elstrup (2009) attempts to provide an integrative model of psychology by fusing psycho-logic with socio-logic. He develops his model based on the relationship between intentional subject and subject-matter, like food, for example, arguing that the core of psychology is intentionality which permits integration across phylogenetic levels. Future directedness is used as another coordinate in the model. It is seen as operating at the base of relationships which develop and contribute to the making of human psyche. Tripathi and Sinha (2009) have critically evaluated the model and argue that development of integrative models should not be done by ignoring the complex and unique features of human and social systems. They suggest that integrative models of psychology need to deal simultaneously with information-theoretic and meaning making processes at the individual and collective levels and by asking how psychological and social processes related to being, the existential states, and becoming, the possible states, alternate in different contexts. The major problem which comes in the way of developing integrative models, as Stotz (2008) points out while developing a post genomic synthesis of nature and nurture, is the maintaining of false dichotomies, such as, innate-acquired, inherited-learned, gene-environment, biology-culture, and nature-nurture. These dichotomies need to be synthesized in a dialectical sense in order to develop a unified science of psychology.

Even when a unified science of Psychology becomes possible, Psychologists will still be called upon to set limits to their discovered scientific truths, more so than is true for physical and other sciences. After all the phenomena they are concerned with are not only far more complex but as pointed out above, are dynamic and continuously evolving, unlike physical or even other living systems. Clearly, Psychology, more than ever, is in need of a paradigm shift, not only in its approach, but also in its methods of inquiry. We are told that science answers 'how' questions but fails to answer 'why' questions satisfactorily (Kauffman, 2009). In the case of Psychology, the 'where' question, the question of human destiny and where we need to arrive, of how it can contribute towards making of minds that will nurture and contribute in building of 'ideal types' of social systems, is more important. It is just not enough to know the efficient causes even if one can find them. While these larger questions about explanation in the mind science(s) are yet to be addressed, Psychologists and Neuroscientists have been making steady progress in explaining mental processes. The predominant approach to explaining mental and neural function is the computational approach to mind based on notions of information processing.

3. MIND, INFORMATION PROCESSING, AND BRAIN

The computational view of mind characterizes the mind in terms of processes that operate on representations. The information processing approach has been the dominant paradigm in cognitive psychology for the past 50 years. David Marr (Marr, 1982) proposed that the mind could be explained in terms of computational, algorithmic, and implementational levels. The computational level is aimed at finding the computations performed by the system, which can be achieved through many different algorithms. The algorithmic level aims at identifying the algorithms used by the human mind, which can be implemented in multiple ways. The implementational level addresses those issues by studying the brain. In a related vein, Palmer and Kimchi (1986) characterized the information processing approach in terms of informational description, recursive decompositions and physical embodiment. According to this approach, mental events are described in terms of information (input and output) and the operation performed on the input. Systems are decomposed into events at a lower level with specification of relation between these events. The information carriers are representations and mental processes transform these representations.

The computational or information processing view of mind has formed the corner stone of theory development and empirical work in Psychology for the past fifty years. The approach has been successful in understanding many mental processes including perception, memory, language, decision making, attention, emotions and social cognition (Thagard 2005). Closely linked are developments in the computational modelling of mental processes using symbolic and connectionist approaches (Anderson et al., 2004; Rumelhart and McClelland, 1986; Simon and Newell, 1972; Sun, 2009).

Mind is usually conceived of as consisting of separate (possibly innate) structures that have a specific cognitive function that has developed through evolutionary processes. Different conceptions of modularity exist in the literature (Fodor, 1983). According to Fodor (1983), modules have certain properties that include domain specificity, information encapsulation, obligatory firing, fast, shallow outputs and a fixed neural architecture. Mostly low-level processes are modular and high-level processes like memory are not modular. Pylyshyn (1999) has emphasized that the most important aspect of a module is encapsulation i.e., processes inside a module are not subject to cognitive influences from outside the module. He argued that early visual perception is an example of such a module with the property of encapsulation.

In the context of cognitive neuroscience, this leads to the notion of localization of function i.e., identification of a neural structure that performs a specific, possibly unique function. Cosmides and Tooby (1997) have argued for modules not specific to perceptual or motor aspects but for specific adaptive functions like social understanding and tool use based on evolutionary arguments and have proposed evolutionary psychology as a way to study such processes. On a different note, Karmiloff-Smith (1994) has argued that modules or encapsulation of mental capacities occur through development and are not innate. It is experience and expertise that enables the formation of a module specialized for certain mental activities. A slightly different notion of modularity has been proposed by Jackendoff (2007).

A mental structure makes sense only for the part of the mental module or brain area processing that structure. Information is broadcast only through the interfaces that are available between levels i.e., interface between one level of a structure with a level of another structure. Each structure has its own format and hence is domain-specific. If a particular aspect in one structure does not have a direct correlate in another structure, then these structures are to some extent informationally encapsulated (Jackendoff, 2007).

What is the relationship between mind and brain? This has been often referred to as the mind-brain or mind-body problem. The developing sub-discipline interfacing psychology and neuroscience, cognitive neuroscience uses multiple techniques to investigate this problem including single cell electrophysiology, lesion studies, brain stimulation, electro-encephalography (EEG), magnetic encephalography (MEG), neuroimaging using positron emission tomography (PET) and functional magnetic resonance imaging (fMRI), and transcranial magnetic stimulation (Gazzaniga et al., 2002). The single cell electrophysiology is well represented by the work of Nobel Prize winning work by Hubel and Wiesel on the visual cortex primarily with monkeys (Hubel and Wiesel 2004). EEG measures brain electrical activity and provides important information about the time course of processes in the brain. Early localization studies before the advent of the hugely popular neuroimaging techniques tried to localize specific brain areas underlying specific cognitive mechanisms (Gazzaniga et al., 2002; Kolb and Wilshaw, 1996). These methodologies have been used to understand many cognitive processes and the brain areas involved in these processes (See Srivastava and Srinivasan, this volume for specific findings on attention and Pammi and Miyapuram, this volume for specific findings on decision making). An example of integration using multiple methodologies is the work on attention by Posner and his colleagues (Posner and Rothbart, 2007).

It should be noted that any cognitive task and involves the activation of large neural networks involving multiple areas in the brain. Some have criticized that neuroimaging is nothing but sophisticated phrenology and it would be difficult to localize functions to specific brain areas (Uttal, 2001). The brain activity underlying cognitive processes are distributed across many brain areas and a given brain area may underlie multiple functions (Barrett, 2009). Given these difficult issues, some argue that neuroimaging can provide a way to functionally decompose cognitive processes rather than just localization of functions (Bechtel and Richardson in press). Another development is the increasing use of computational models to form specific hypotheses about brain structure and function (O'Reilly and Munakata, 2000).

Barrett (2009) talking about the future of psychology has emphasized a distinction between observer-independent categories and observer-dependent categories. According to her, categories like "anger" are observer-dependent categories and would not map neatly to brain activations that are observer-independent categories. She proposes that complex psychological categories that are part of natural language enabling communication between humans might correspond to patterns in neural reference space. A momentary mental state might correspond to activity in a neural assembly. More importantly, she proposes an intermediate level of psychological primitives, which could map on to brain activity and more specifically to distributed neural networks. This would be critical for linking the mind and brain and it would be the job of future psychologists to come up with the appropriate psychological primitives (Barrett, 2009).

Recent approaches in psychology have emphasized the embodied and situated nature of psychological processes. The embodied approach focuses on the critical role of the body in understanding the mind since the body continuously affects and influences the mind in a substantial manner that cannot be reduced to neural activity (Anderson, 2003; Gallagher, 2005; Wilson, 2002). The embodied approach argues that other bodies play a critical role in addition to the special role by our own body in cognition. The related situated cognition approach focuses on the key role played by the environment in determining mind. According to this view, mind arises out of interactions of the body/brain with the environment (Brooks, 1991; Barnier et al., 2008; Hutchins, 1995; Smith and Semin, 2007). An example is the work by Edward Hutchins on the interaction between an agent and the artefacts created by humans (Hutchins, 1995). Hutchins focuses on the distributed nature of cognition and the way these artefacts reduce cognitive load and in a sense become part of the mind. Representations are neither in the human nor in the environment. These issues bring us to the discussion of behaviour in a given context and in the next section we discuss social behaviour in contexts.

4. SOCIAL BEHAVIOUR IN CONTEXT

If Psychology is recognised as a Social Science, it is because of its focus on studying behaviour and experiences of individuals which take place in social settings. Jones and Colman (1996) draw attention to the view that social psychology is the biochemistry of social sciences. It attempts to unravel what lies at the root of social behaviours asking how interpersonal interactions bind individuals together as part of groups and social systems and how social systems get the order they have at the same time. Individuals are what they are because of the influences of the cultural settings that contribute to the manner in which humans are socialised and acquire values and norms which serve as the foundation of the social behaviour which is displayed by them in diverse settings. Not only behaviour of individuals which takes place in the presence of others and with others gets influenced by the social settings but their thoughts, the meaning they give to the situations and actions of others and also their feelings are affected. It is important to note that individuals are influenced by those very settings which they have purposely created for themselves.

Does it follow from this that lower order organisms, that is organisms other than humans, have no 'social life' or are not part of any social system? This is not true. Gibbs et al. (2008) recently published an article in *Science* which supports the view that even organisms like bacteria, which do not have a nervous system, display behaviour which is indicative of social behaviour akin to self identity and social recognition. It is well accepted that different species display affiliation as well as aggression and carry notions of hierarchy and territoriality. This is very clearly seen in case of ants and bees. However, it should be noted that human social behaviour and the supportive systems are far more complex.

One important difference between the animal and human social behaviour may lie in the fact that in case of humans, it is supportive of life as well as health. Studies of social support have clearly brought out the positive role which it plays in recovery from ill health, such as, in the case of patients recovering from cardiac arterial bypass graft surgery (Schroder et al., 1997). It works through strengthening of cardiovascular system, the immune system and neuro-endocrine system (Uchino et al., 1996). But the question which still remains is whether social behaviour has any basis outside the social and cultural settings. Sociality is not unique to humans. It may be further pointed out that genes and regulatory sequences contribute to the organization and functioning of neural circuits and molecular pathways in the brain that support social behaviour (Robinson et al., 2008). It is found that social experience interacts with information in genome to modulate brain activities.

The nature of human social behaviour is also different both in terms of its diversity and the quality of relationship which they have with others of the same species and also in terms of the social and moral emotions which come in to play while they interact with each other. One clear distinction which is there between humans and other species is in their remembering of the past, both, individually and collectively, and the manner in which they privilege it for guiding their future social interactions. While elementary social processes like altruism and cooperation, competition and aggression may be found in lower order organisms, social behaviours associated in social cognition, self presentation, compassion and forgiveness, group decision making, leadership, normative behaviour, group conflicts, collective behaviour are not evident in the lower order organisms.

Broadly, there are three disciplinary influences which are clearly visible in approaches that seek to explain and understand social behaviour in a unified manner. The first approach draws from biological sciences and looks at genes, neural circuits in the brain, neurochemicals like serotonin and dopamine, and the neuropeptides such as Oxytocin, Vasopressin, etc., in the role they have in determining various kinds of social behaviour (Heinrichs and Domes, 2008). We place the work which makes use of Evolutionary Psychology, particularly, involving interpersonal relations, aggression and social dominance within the biological approach. This approach has been particularly helpful in understanding various kinds of social psychological disorders, such as, Asperger's Syndrome, Williams Syndrome and Autistic Spectrum Disorder and the like. Oliver Sacks' (1985) popular book 'The man who mistook his wife for a hat' has brought to attention to neurological bases of disorders like Korsakov's syndrome and visual agnosia. Nevertheless, the approach does not permit us to understand spontaneously occurring 'here and now' social behaviours and episodes.

The second approach is social cognitive and seeks to apply knowledge based on Cognitive Psychology and Cognitive Science (Wilson and Keil, 1999). The approach looks essentially at how various social episodes are understood and processed in terms of social schemas or mental structures which get created in our brains and represent the world around us. Social behaviour is seen as a result of activation or priming of these schemas. The approach has found application in a variety of areas but more particularly in development of theories related to interpersonal attraction, social identity, inter-group differentiation, prejudice, social and cognitive biases resulting from misattributions and regulation and control of deviant social behaviours. More recently, the cognitive and neuroscience approaches have been combined to form social neuroscience or social cognitive neuroscience.

The third one, which has formed the mainstream of social psychology from the time of August Compte, borrows concept and theories from sociology as well as Anthropology and looks at social behaviour as a characteristic behaviour within a given system's structure and processes (Myers, 2008). Thus, for example, feelings related to injustice and distribution are seen as having their location in psychological processes and also in social structures. There is also a fourth approach, pioneered by Kurt Lewin (1936; 1951) who used theoretical models borrowed from Physics and Mathematics in the form of Field Theory and Topology and applied them successfully to analyse behaviour in group settings. Lewinian approach flourished for many decades and had a major influence in the development of theories related to group dynamics and was responsible for the emergence of Experimental Social Psychology. It assumed that most social problems and issues facing us today, like inter-group and national conflicts, of environmental degradation, of unequal distribution, of ill health and poverty are due to the humans and, therefore require humans to be the part of solution. This approach had a great deal of impact in using social psychological knowledge for making of social policies and showed how effective social interventions could be designed to solve various social problems. Today, this approach appears to have fallen on the way side, although, it is also true that a lot of it survives in areas of Applied Social Psychology. It also has found integration with other approaches. The approaches mentioned above succeed only in a very limited way to provide integrative frameworks and that too at a certain level but not across levels. Social behaviour which occurs in diverse contexts, social, cultural and historical is definitely more complex than asocial, non-contextual behaviour. It therefore, requires more vectors and multidisciplinary frameworks to explain such behaviours. Such frameworks are likely to succeed only to the extent they achieve integration of levels, both horizontally and vertically, in terms of phylogenesis as well as ontogenesis.

5. CULTURE AND SOCIETY

The role of biological and cultural factors in explaining the mind has been debated through the centuries. Most such discourses take completely different paths and do not converge. In so doing, they simply follow the well accepted dichotomy between body/matter and mind. The role of biology traditionally has been seen as supporting rationalism i.e., that major aspects of the mind are innate and it is a result of our biological history. The role of culture traditionally has been seen as supporting empiricism i.e., that major aspects of the mind are not innate and it is a product of experience or culture. The work of Geertz and E. Durkheim falls within this category. Bickle's (1998) approach calling for ruthless reductionism and his call to move away to deeper levels to understand psychological functionalities, i.e., to cellular and molecular levels represents the former kind.

The entire debate hinges on the question whether human culture shapes human minds or it is human mind that shapes culture. Empirical support is offered in support of both propositions. We take two such studies. Psychologists have observed that there are differences in the manner in which individuals belonging to different cultures represent self and the manner in which they make distinction between individual-self versus others (Markus and Kitayama, 1991; Nisbett, 2003; Tripathi, 1988). In case of East Asian cultures, the self is seen to be interdependent and interconnected with other human beings where as in case of the Western cultures (Americans and Europeans), self is seen as independent of others and individual autonomy is emphasised. Following on these findings, Zhu et al. (2007) sought to examine whether such cultural differences show up in neural activations as well. In an fMRI experiment, they asked their participants to make judgments about self, about their mothers, and persons unrelated to them. They obtained significant differences in the activation of the medial prefrontal cortex (MPFC) region between participants located in the Western and Eastern cultures. Mother judgments in the East Asian participants showed enhanced MPFC activity which was not separate from the activity found in the same region for self judgments compared to the participants from the Western culture that showed low activation in this region.

While Zhu et al.'s study supports how culture mediates the ways in which information is processed, another study by Borg et al. (2003) has drawn attention to the way religious attitudes and spiritual acceptance may be modulated by serotonin system activity. They found an inverse relationship between the density of 5-HT$_{1A}$ receptors and spiritual acceptance. The researchers do sound a cautionary note that it is still too early to conclude that feeling religious can simply be a matter decided by the molecules.

There are, thus, two views which scientists are engaged in debating with respect to psychological phenomena as they obtain in different cultures. There is a group of scientists who argue in support of psychic unity holding that it is a given. Any differences which are found are only at the shallow levels and at the deeper level there is universality. This view is held by both biologists and cognitive scientists who make use of the computer metaphor where the brain is the hardware which is responsible for the software in terms of psychological processes. The other view is that psychological mechanisms are shared by the humans universally within cultures but, generally not across cultures. So cognitive and behavioural activities do not combine naturally where complex decisions are to be taken but the 'expertise' for decision making develops as a result of exposure to various decisions taken in different domains with in a culture. There are specific cognitive styles that develop within cultures (Berry, 1976; Sternberg and Grigorenko, 1997).

The dichotomy between biology and culture may be wholly unnecessary. While one overlooks the human phenomena, the other produces a fractured science. Valsiner (2009) argues in support of an integrative framework which makes it possible to unite the opposites with in the same whole in order to contribute towards the development of 'abstracted universal science'. Recent developments show that we may have already started moving in this direction. The last decade has seen the debunking of the nature-nurture distinction and launching of a number of journals which address and forge new joints of culture with various science disciplines. One such example is the recent launching in 2007 of the Journal of Social, Evolutionary and Cultural Psychology which seeks to integrate studies in these disciplinary domains. The proclaimed objective of the journal is to bridge sub-disciplines with in psychology and also develop bridges of psychology with other disciplines in order to gain holistic insights in to human behaviour, emotion, cognition and motivation. The convergence of culture and society with scientific disciplines is also reflected in a series of books which have been published based on conversations of psychologists, physicists, neuroscientists with the Dalai Lama under the Mind and Life series (Hayward and Varela, 1992; Goleman, 1997; Davidson and Harrington, 2002). As a result, there has been a sudden spurt in scientific studies of consciousness, a subject which scientists for a long time did not want to touch even with a barge-pole.

6. CONSCIOUSNESS AND FREE WILL

A core issue in the psychological science(s) today is consciousness and free will. Consciousness studies forms an integral part of current research in the psychological science(s). Consciousness research uses multiple approaches and studies phenomena ranging from states of consciousness (wakefulness, sleep, dreaming, meditative states, etc.) and awareness (of environment, body, and thoughts). Functionalist approaches to consciousness investigate the potential role of consciousness the functioning of the mind including information integration, global broadcasting, and better control of perception and behaviour, freewill and self (Baars, 1997; 2005; Seth et al., 2006; Jordan, 2003).

According to the global workspace theory, unconscious systems process information in parallel and these are made available for *conscious* access to all the processing systems through the global workspace or a theatre (Baars, 1997; Baars, 2005). The neuronal adequacy criterion proposes that at least 500 ms of electrical stimulation is required for conscious experience to occur (Libet, 2005). The cognitive neuroscience of consciousness has focused mostly on the neural correlates of consciousness (Koch, 2004). Block (2005) has proposed two types of awareness: primary awareness and access consciousness. It has been argued that primary awareness is due to recurrent activations in early perceptual areas and access consciousness is due to activations in higher centres like prefrontal cortex.

The concept for free will has important consequences for all types of human behaviour at both the individual level as well as society. Free will is also linked to the role played by consciousness in making decisions. Current approaches in Psychology agree that a significant amount of processing is unconscious. Libet (2005) argued that all decisions have an unconscious basis but consciousness can veto a decision since the conscious experience of deciding to act occurs roughly 200 ms before the action. Hence, there is sufficient time for consciousness to veto the action but the action itself is based on unconscious processes in the brain. It is possible that the conscious decision to veto itself might be due to other unconscious processes questioning the nature of free will. According to Wegner and Wheatley (1999), we can be only aware of a decision "consciously" after it has been made. While it does not play a causal role, it can affect future actions.

The issue of free will itself may dependent on the deterministic nature of psychological processes. While determinism is seen as an opposite to free will or volition, non-determinism also is not necessarily compatible with volition and hence is not of much use. A random decision maker has no free will even though it is non-deterministic. Irrespective of whether one thinks free will is compatible or incompatible with determinism, further studies in psychological science(s) cutting across many disciplines will play a critical role in understanding the nature of volition. This would also have important implications on all domains in which human decisions play a critical role including politics and justice. Advances in consciousness studies will also be critical given the potential role of consciousness in volition.

7. LOOKING AHEAD

This volume attempts to showcase developments related to some important research themes in mind science(s). Admittedly there are areas which have been left out, particularly, relating to professional psychology. The professional areas in which psychologists work are too numerous to be mentioned here. But some major areas in which psychologists are recognised as professionals for delivering psychological services need mention. Psychologists, besides working as researchers and teachers in the academic settings, work as Clinical and Counselling psychologists, as School and Community psychologists, Industrial and Organizational psychologists, and also as Engineering, Sports and Forensic

Psychologists. We needed to mention this because the future of psychology and mind science(s) will largely depend on how best mind and psychological science is able to combine with practice and the extent to which it is able to solve social problems and contribute towards building harmonious societies.

Reich (2008) advocates, in this context, that psychology should adopt the Pasteurian model where science and professional practice are integrated and develop in a seamless fashion. The knowledge so generated can then be used for promotion of human welfare. It is known that generation of knowledge in all scientific disciplines depends on what is demanded by the society and very often it is driven by economic and political considerations. That this is going to be the case in case of the future development of Psychology, is to state the obvious. Nevertheless, we still need to stretch our neck out and ask the direction which Psychology is likely to take in future. An exercise like this necessarily contains the wish list of the author/s about how they would like the discipline to grow. The exercise should still be useful because it can point to the possibilities which are seen. Perhaps, the exercise will also help in making available new choices to the researchers.

So, let us ask where do we stand with respect to mind science as of now, and where do we see it going. We are, of course, not the first one to raise this question which has been asked many times over and answered also. Solso (1999) asks a similar question related to mind and brain sciences and answers it in terms of three hypotheses which he proposed for this purpose. We feel the same three hypotheses can be made use of in our case also. These three hypotheses are called (i) vector extrapolation hypothesis (VEH), which holds that further developments are going to be in the direction that is already established, (ii) the contrarian reaction (CR) hypothesis, meaning that the established trend will either be modified or reversed, and, (iii) the multiple influence model (MIM) which suggests that multiple influences scientific, sociological, political, ethical, etc will determine the course a science is going to take eventually.

As can be guessed scientific developments of the last century, if assessed in retrospect, do not provide support for any one hypothesis but partial support to all of them, be it Physics or Psychology. Our hunch is that the development of mind sciences or of other sciences in the twenty first century is not going to turn out to be any different. We will find support for the existing paradigms, behaviourist and cognitive, that are in use till they are finally modified or replaced; contrarian reactions to the use of the kind of methods now used, for example, for studying brain structure and processes as well as human behaviour; and finally, the new kind of complex societies and humans which will evolve as they will be called upon to adapt to changes in the cyber space and the changes that may take place in regulating free flow of knowledge in emerging knowledge societies will require change in our theoretical models.

Let us speculate on some of the likely directions in which psychological sciences may develop in future:

(i)   The science of behavioural genetics is in a very nascent stage. Our expectation is that it will grow to establish clear links between genes and behaviour. Once these are established, new modes of assessment of human personality will be developed which will help in accurately diagnosing various kinds of psychological disorders.

(ii)   There will be more research to understand the cellular and molecular bases of human mind and behaviour. Chemical markers of various psychological processes will be deciphered. One expects integrative theoretical frameworks to emerge to synthesize knowledge generated across various sub-disciplines associated with neuroscience and across levels.

(iii) Psychology's interface with Evolutionary Biology will make it possible to identify accurately those behaviours which get modified as a result of past experiences and to know how adaptive responses actually develop, get codified and are retained.

(iv) Techniques will be developed to monitor changes that take place in mental processes through implanted chemical sensors and to possibly make inferences about the thoughts being thought and feelings being felt.

(v) Psychological techniques of inoculation will be developed to increase tolerance against life stresses, diseases and social and psychological traumas.

(vi) Integration of approaches and use of multiple methods will make it possible to study the dynamics of psychological processes such that one would be able to observe how a phenomenon emerges within a structure and accurately model the emergence of mental phenomena.

(vii) Mind scientists will pay more attention to understanding meta-level processes and phenomenon.

(viii) Mind scientists will come to assume that psychological processes are basically indivisible and are better understood in a 'wave' sense than in a 'particle' sense. They are better understood by assuming that the universe is holonomic.

(ix) Mind scientists will develop focus on complexity and develop non-linear models rather than simplistic linear models.

(x) Mind scientists will be able to explain the so-called easy problems of consciousness and may be even attempt to solve the hard problem of consciousness (Chalmers, 1996).

(xi) The grand model of mind science(s) will focus on how it can develop sentient and evolved beings rather than focus on developing humans that are efficient, competent and self-serving. It will be culture-inclusive and culture-sensitive.

To conclude, mind science(s) will be affected both by the development of artificial minds as well as further developments in understanding the brain (Page 48 in R. Solso 1999). Bechtel et al. (1998) have pointed out the existence of "simultaneous pulls downwards into the brain and outwards into the world". Integration of these multiple approaches will determine the success of the interdisciplinary mind science(s).

REFERENCES

Anderson, J.R., Bothell, D., Byrne, M.D., Douglass, S., Lebiere, C., and Qin, Y. (2004). An integrated theory of mind. *Psychological Review, 111*, 1036–1060.

Anderson, M.L. (2003). Embodied cognition: A field guide. *Artificial Intelligence, 149*, 91-130.

Barrett, L.F. (2009). The Future of Psychology: Connecting Mind to Brain. Perspectives in *Psychological Science, 4*, 326-339.

Baars, B.J. (1997). *In the Theater of Consciousness: The Workspace of the Mind.* Oxford, England: Oxford University Press.

Baars, B.J. (2005). Global workspace theory of consciousness: Toward a cognitive neuroscience of human experience. *Progress in Brain Research, 150*, 45-53.

Barnier, A. J., Sutton, J., Harris, C. B., and Wilson, R. A. (2008). A conceptual and empirical framework for the social distribution of cognition: The case of memory. *Cognitive Systems Research, 9*, 33-51.

Bechtel, W., and Richardson, R.C. (in press). Neuroimaging as a tool for functionally decomposing cognitive processes. In S. Hanson and M. Bunzl (Eds.), *Foundations of Functional Imaging*. Cambridge, MA: MIT Press.

Bechtel, W., and Graham, G. (Eds.). (1998). *A Companion to Cognitive Science*. Malden, MA: Blackwell.

Bickle, J. (1998). *Psychoneural reduction: The new wave*. Cambridge, MA: MIT Press.

Block, N. (2005). Two neural correlates of consciousness. *Trends in Cognitive Sciences, 9*, 46-52.

Boden, M.A. (2005). *Mind as Machine. Vol. 1 and 2*, Oxford University Press: London.

Borg, J., Andrec, B., Soderstrom, H., and Farde, L. (2003). The serotonin system and spiritual experiences. *American Journal of Psychiatry, 160*, 1965-1969.

Brooks, A. (1991). Intelligence without representation. *Artificial Intelligence, 47*, 139-160.

Carrel, A.(1959). *Man, the unknown*. Mumbai: Wilco Publishing House.

Chalmers, D.J. (1996). *The Conscious Mind*. New York: Oxford University Press.

Cosmides, L., and Tooby, J. (1997). Dissecting the computational architecture of social inference mechanisms. Characterizing human psychological adaptations. *Ciba Foundation Symposium*, 208 (pp. 132-161), New York: Wiley.

Davidson, R.J., and Harrington, A. (Eds,) (2002). *Vision of compassion*. NewYork: Oxford University Press.

Elstrup, O. (2009). The ways of human: Modelling the fundamentals of psychology and social relations. *Integrative Psychological and Behavioural Science, 43*, 267-300.

Fodor, J. A. (1983). *Modularity of Mind: An Essay on Faculty Psychology*. Cambridge: MIT Press.

Friman, P.C. Allen, K.D., Kerwin, M.L.E., and Larzelere, R. (1993). Changes in modern psychology, *American Psychologist, 48*, 658-664.

Gallagher, S. (2005). *How the Body Shapes the Mind?* Oxford: Oxford University Press.

Gazzaniga, M., Ivry, R.B., and Mangun, G.R. (2002). *Cognitive Neuroscience: The Biology of the Mind*. Cambridge: MIT press.

Gibbs, K.A. Urbanawski, M.L., and Greenberg, E.P.(2008). Genetic determinants of self identity and social recognition in bacteria, *Science, 321*, 256-259.

Gibson, J.J. (1979). *The Ecological Approach to Visual Perception*. Boston: Houghton Mifflin.

Goleman, D. (1997). Healing emotions: Conversation with the Dalai Lama on mindfulness, emotions and health. Boston: Shambhala.

Hawking, S.A. (2009). Godel and the end of physics. Dirac Centennial Celebrations. Cambridge, U.K. http//www.dampt.com.ac.uk/strings02/dirac/hawking.

Hayward, J. and Varela, F. (Eds.) (1992). *Gentle bridges: Dialogues between cognitive sciences and Buddhist traditions*. Boston: Shambhala.

Heirichs, M. and Domes, G. (2008). Neuropeptides and social behaviour: Effects of Oxytocin and vasopressin in humans. *Progress in Brain Research, 170*, 337-350.

Henriques, G. (2004). Psychology defined. *Journal of Clinical Psychology, 60*, 1207-1221.

Hubel, D.H., and Wiesel, T.N. (2004). *Brain and Visual Perception: The Story of a 25-Year Collaboration*. New York: Oxford University Press.

Hutchins, E. (1995). *Cognition in the Wild*. Cambridge: MIT Press.

Jackendoff, R. (2007). *Language, Consciousness, Culture: Essays on Mental Structure*. Cambridge: MIT Press.

Jones, E.E., and Colman, A.M. (1996). Social psychology. In A. Kuper and J. Kuper (Eds.) *The Social Science Encyclopaedia* (2nd Ed.). London: Routledge.

Karmiloff-Smith, A. (1994). Precis of beyond modularity: A developmental perspective on cognitive science. *Behavioral and Brain Sciences, 17*, 693-706.

Kauffman, S.A. (2009). Towards a post-reductionist science: The open universe. ArXiv: 0907.2492.

Kim, J. (1998). Problems of reduction. In Craig, E. (Ed.) *Routledge Encyclopedia of Philosophy*. London: Routledge.

Koch, C. (2004). *The Quest for Consciousness: A neurobiological Approach*. Roberts and Publishers.

Kolb, B., and Wilshaw, I.Q. (1996). *Fundamentals of Human Neuropsychology*. New York: W. H. Freeman.

Lewin, K. (1936). *Principles of Topological Psychology*. New York: McGraw Hill.

Lewin, K. (1951). Field theory in social science: Selected theoretical papers. D. Cartwright (ed.) New York: McGraw Hill.

Libet, B. (2004). *Mind Time: The Temporal Factor in Consciousness*. Harvard University Press.

Markus, H., and Kitayama, S. (1991), Culture and the self: Implications for cognitions, emotions, and motivation. *Psychological Review, 98*, 224-253.

Marr, D.C. (1982). *Vision: A Computational Investigation into the Human Representation and Processing of Visual Information*. San Francisco: W. H. Freeman.

Miller, G., Galanter, E., and Pribram, K. (1960). *Plans and the Structure of Behavior*. New York: Henry Holt.

Myers, D. (2008). *Social psychology* (9th Ed.). New York:McGraw Hill.

Newell, A. (1992). Précis of Unified theories of cognition. *Behavioral and Brain Sciences, 15*, 425-492.

Nisbett, R. (2003). *The Geography of Thought: How Asians and Westerners Think differently*. New York: The Free Press.

O'Reilly, R.C., and Munakata, Y. (2000). *Computational Explorations in Cognitive Neuroscience: Understanding the Mind by Simulating the Brain*. Cambridge: MIT Press.

Palmer, S. (1999). *Vision Science: Photons to phenomenology*. Cambridge: MIT Press.

Palmer, S., and Kimchi, R. (1986). The information processing approach to cognition. In T. J. Knapp and L. C. Robertson (Eds.), *Approaches to Cognition: Contrasts and Controversies*. Hillsdale, NJ: Erlbaum.

Pammi, V.S.C., and Miyapuram, K.P. (2011). Neuroeconomics of individual decision making at multiple levels: A review. In P. N. Tandon, R.C. Tripathi, and N. Srinivasan (Eds.), *Expanding Horizons of the Mind Science(s)*, New York: Nova Science Publications.

Panksepp, J., Moskel, J.R., Panksepp, J.B. and Krores, R.A. (2002). Comparative approaches in evolutionary psychology: Modular neuroscience meets the mind. *Neuroendocrinology Letters, 23*, 105-115.

Penrose, R. (1989). *The Emperor's New Mind*. Oxford: Oxford University Press.

Pribram, K. (1991). *Brain and Perception: Holonomy and Structure in Figural Processing.* New Jersey: Lawrence Erlbaum Associates.

Posner, M., and Rothbart, M. (2007). Research on attention networks as a model for integration of psychological science. *Annual Review of Psychology, 58,* 1-23.

Pylyshyn, Z.W. (1999). Is vision continuous with cognition? The case for cognitive impenetrability of visual perception. *Behavioral and Brain Sciences, 22,* 341-423.

Rao, K.R., Paranjpe, A., and Dalal, A.K. (2009). *The Handbook of Indian Psychology.* New Delhi: Cambridge University Press, India.

Reich, J.W. (2008). Integrating science and practice: Adopting the Pasteurian model. *Review of General Psychology, 12,* 365-377.

Robinson,G., Fernald, R. and Clayton, D. (2008). Genes and social behaviour. *Science, 322,* 896-900.

Rumelhart, D.E., and McClelland, J.E. (Eds.,). (1986). *Parallel Distributed Processing: Explorations in the Microstructure of Cognition. Volumes 1 and 2,* Cambridge: MIT Press.

Sacks, O. (1985). *The man who mistook his wife for a hat and clinical other tales.* New York: Touchstone.

Schroder, K.E., Schwarzer, R. and Endler, N.S. (1997). Predicting cardiac patients' quality of life. *Journal of Health Psychology, 2,* 231-244.

Scott Jordan, J. (2003). Emergence of self and other in perception and action: An event control approach. *Consciousness and Cognition, 12,* 633-646.

Seth, A.K., Izhikevich, E.M., Reeke, G.N., and Edelman, G.M. (2006). Theories and measures of consciousness: An extended framework. *Proceedings of National Academy of Sciences USA. 103,* 10799-10804.

Simonton, D.K. (2004). *Creativity in Science: Change, logic, genius, and zeitgeist.* Cambridge: Cambridge University Press.

Smith, E.R., and Semin, G.R. (2007). Situated social cognition. *Current Directions in Psychological Science, 16,* 132-135.

Spear, J.H. (2007). Prominent schools or other active specialities? A fresh look at some trends in psychology. *Review of General Psychology, 11(4),* 363-380.

Solso, R.L. (Ed.), (1999). Mind and Brain Sciences in the 21st Century, Cambridge: MIT Press.

Srivastava, P., and Srinivasan, N. (2011). Selective attention: Insights from multiple methodologies. In P. N. Tandon, R.C. Tripathi, and N. Srinivasan (Eds.), *Expanding Horizons of the Mind Science(s),* New York: Nova Science Publications.

Staats, A.W. (1981). Social behaviourism, unified theory, unified theory construction methods, and the zeitgeist of separatism. *American Psychologist, 36 ,*239-256.

Sternberg, R. J. and Grigorenko, R.L. (1997). Are cognitive styles still in style? *American Psychologist, 52,* 700-712.

Sun, R. (2009). Theoretical status of computational cognitive modeling. *Cognitive Systems Research, 10,* 124-140.

Thagard, P. (2005). *Mind: Introduction to Cognitive Science,* Cambridge, MA: MIT Press.

Tripathi, R.C. (1988). Aligning values with development in India. In D.Sinha and H.S.R. Kao (Eds.) *Social values and development: Asian perspectives* (pp. 314-332). New Delhi: Sage.

Tripahi, R.C. and Sinha,Y. (2009). The ways of 'being' and 'becoming' human. *Integrative Psychological and Behavioural Science, 43*, 311-323.

Uchino, B.N., Cacioppo, J.T., and Kiecalt-Glaser, J.K. (1996). The relationship between social support and physiological processes: A review with emphasis on underling mechanisms and implications for health. *Psychological Bulletin, 119*, 488-531.

Uttal, W. R. (2001). *The new phrenology: The limits of localizing cognitive processes in the brain.* Cambridge, MA: MIT Press.

Valsiner, J. (2009). Integrating psychology within the globalising world: a requiem to the post-modernist experiment with wissenschaft. *Integrative Psychological Behavioral Science, 43*, 1-21.

Valsiner, J. and Rosa, A. (Eds.) (2007). *The Cambridge handbook of socio-cultural psychology.* New York: Cambridge University Press.

Wegner, D.M., and Wheatley, T. (1999). Apparent mental causation: Sources of the experience of free will. *American Psychologist, 54*, 480-492.

Wilson, M. (2002). Six views of embodied cognition. *Psychonomic Bulletin and Review, 9*, 625–636.

Wilson, R.A. and Keil, F.C. (Eds) (1999). *The MIT Encyclopaedia of the Cognitive Sciences.* Cambridge, MA: MIT Press.

Zhu, Y., Zhang, L., Fan, J. And Han, S. (2007). Neural basis of cultural influence in self-representation. *Neuroimage, 34*, 1310-1316.

In: Expanding Horizions of the Mind Science(s)   ISBN: 978-1-62808-705-5
Editors: P. N. Tandon, R. C. Tripathi and N. Srinivasan ©2013 Nova Science Publishers, Inc.

Chapter 2

INFORMATION PROCESSING IN THE HUMAN BRAIN

P. N. Tandon

National Brain Research Centre, Manesar, India

"The Human brain is special both as an object and as a system – its connectivity,
dynamics, mode of functioning and relation to body and the world is like nothing else science
has yet encountered"

Gerald M. Edelman

ABSTRACT

To understand the human brain, a thorough understanding of its anatomy and
physiology is needed. The chapter discusses basics in brain anatomy and physiological
organization. In the recent decades, brain has been understood in terms of information
processing. The chapter discusses the basic principles of memory in the brain in the
context of the information processing approach. The chapter ends with a discussion on
the comparisons between the brain and computer.

1. INTRODUCTION

The human brain is often called a computer and several electronic devices commonly
used in information technology (IT) like computers, neural networks, artificial intelligence
equipment and robots, are believed to be modelled on the basis of our understanding of the
brain. It is also true that in recent years these artefacts have been invaluable in exploring and
understanding the structure and function of brain. And though some feature of the implements
used in IT may mimic the brain, the two have no more than superficial similarities. With
intense activity in both these fields, each is likely to benefit from the other.

2. THE BRAIN: ANATOMICAL ORGANIZATION

The brain of an adult human being weighs approximately 1300-1500 grams. It is enclosed in a rigid skull. Grossly it is divided into three major components – largest being the *Cerebrum* consisting of two cerebral hemispheres interconnected to each other. The *Cerebellum is* much smaller and situated below the cerebrum has distinct appearance and functions of its own. Situated in the centre of the brain, connecting the cerebrum and the cerebellum and extending outside the skull as the spinal cord, is the *Brain Stem* or the central core. Each one of these is further subdivided into well defined components. The brain receives its information from the world around us and our own body through its *sensory nerves* subserving the five major senses e.g. *vision, hearing, smell, touch and somesthetic sensations (touch, pain, muscle – joint sense).* Sensations from its *visceral organs* e.g. *heart, lung, intestines, urinary bladder* etc., are communicated by the autonomic nervous system. Brain executes its motor functions through its motor nerves and pathways.

3. BASIC BUILDING BLOCKS

The basic building blocks of the brain are the nerve cells (neurons), with their processes (axon and dendrites) and the supporting cells (the glia). Neurons are arranged in well organized columns covering the surface of the cerebrum and the cerebellum and constituting the so-called grey matter or the cortex. In addition they are seen as tightly packed clusters in the depth of the brain as distinct nuclei and structures like the *Thalamus, Hypothalamus, Striatum, Amygdala* etc., often collectively referred to as the *Basal Ganglia*. Out of the approximately 30 billion neurons in the brain, the cerebral cortex contains around 10 billion. To accommodate so many cells the cortex is folded to provide the characteristic appearance of the *gyri* and *sulci.*

The cerebral cortex with its supporting elements makes up approximately 80 percent of the brain's total volume. It is actually a recent development in the course of evolution. While a tripling of brain size from 400 to 1300 grams marks hominid brain evolution, the brain of Homo Sapiens has hardly changed in size, form or external morphology since the species first appeared less than hundred thousand years ago (Mountcastle, 1998). While the brain of the rodent or monkey has many structures identical to human brain, the latter is not a scaled up model of the former.

From the multi-potent progenitor cells, a well orchestrated process of development results in the migration of the specialized neurons to their final destination in a precisely organized columnar module, which characterize the cerebral cortex. Simultaneously these neurons develop precise connections with their genetically determined targets, which ultimately produce specific functional units. Each group is connected to only certain subsets of other groups to subserve specific functions e.g. vision, smell, somesthetic sensation, voluntary movement etc. These distinct cortical areas are known to be discrete cytoarchitecturally as described by Broadman and manifest a somatotopical (point to point) representation of the sensory and motor functions as revealed by Penfield (1958), Mountcastle (1986), Hubel and Wiesel (1968). The organization of the functional areas has been elaborated not only to particular region of the cortex but to the level of the individual cells within the columns playing a specific role. Thus at its higher level of anatomical organization it appears that the cortex is nicely divided into areas, regions and specialized parts that, in general are functionally segregated or specialized. We can trace the sensory pathways from the *peripheral end organs (sensor)* in the *skin, retina, cochlea, nasal mucosa, tongue etc.,* to specific and discrete regions of the brain, even specific columns or cells. Similarly we can trace the motor pathways from discrete regions of the cerebral cortex to the specific motor neurons in the spinal cord and onwards to their target muscles.

Each neuron has a cell body (the soma) and multiple processes to receive (the dendrites) and send messages (a single axon). The neuronal processes of one neuron in turn communicate with others not through physical contact but at a junction called the *Synapse,* which is a membrane bound cleft. On an average each neuron is connected to about one thousand other neurons. Some neurons are reported to have as many as 20,000 synapses. It is estimated that there are nearly one to ten trillion synapses. The neuronal process carrying a message to the synapse is called *Presynaptic* and the one receiving it is called *Postsynaptic.* During the last decade it has been established that in addition to the important role played by the neurons, the synapses and the circuits thus formed, other brain cells namely the glia also play important role in information processing (Tandon, 2007).

Over the years, it has become evident that the organization of local microcircuits, their connectivity to other cortical and subcortical structures so far more complex than so far believed. Many cognitive functions previously thought to be localized to a single cortical area have been shown to engage many and frequently widely separated areas of the cortex in a distributed rather than hierarchical arrangement. It is now established that in the case of vision there are as many as three dozen (or probably more) different areas in the monkey brain, each contributing to specific function e.g. the detection of the line orientation, the movement of objects, the colour etc., widely distributed in the occipital, parietal and temporal lobes. These functionally different segregated areas are no doubt reciprocally connected. The same is true for other functional as well. Thus brain activation related to voluntary movement has been found in at least fourteen areas including the primary motor area (Frackowiak, 1998). This is in control to the three, primary, secondary and supplementary motor areas, identified by Penfield nearly fifty years ago, on the basis of mild electrical stimulation of the cortex of conscious human subjects operations for epilepsy.

Before one gets an impression that all functional units in the brain are hard-wired as in a computer, it must be emphasized that at the synaptic level there is a remarkable degree of plasticity. Earlier believed to be a feature of early developmental stages of the brain, it is now unequivocally established that synaptic plasticity and modifiability are virtually universal properties of brains at all ages. No two brains are identical, not even those of the identical twins, at the finest level of microcircuitry. Neither is the same brain microstructure identical from moment to moment. Depending upon the environmental inputs and specific functional requirements at a given moment new synaptic connections are made and some existing ones dismantled. This led Gerald Edelman (1998) to comment, "Brain is 'special', that it is a gross mistake to think of it as an ordinary 'machine'. Each brain has uniquely marked in it the consequence of a developmental history and an experiential history". This plasticity of the neuronal circuitry would no doubt be a nightmare for anyone wanting to develop an artificial brain model.

4. Physiological Organization

A message generated or transmitted by a neuron travels along its axon as an action potential till it reaches a synapse, where through a cascade of chemical events it is passed on to the other neuron/neurons. Synaptic transmission depends upon the release of a transmitter substance form the presynaptic terminal into the synaptic cleft. This is turn modifies the membrane of the postsynaptic (recipient) neuron involving a variety of receptors, ion-channels, transporters and other molecules. More than fifty neurotransmitters have now been identified in our brain. In addition there are a host of other neuropeptides and hormones which are involved in the generation, propagation, transmission and modification of the electrical signals which form the basis of interneuronal communication. These diverse neurotransmitting molecules may exert strikingly different effects on the firing state of their target neurons. Moreover, the same neuron may synthesize and liberate upto ten chemical coding accessible to nerve cells is thus very large and complex (Changeux, 1993). The electrochemical repertoire involved in "cross talk" between the neurons is thus not only rich but subtle. Nothing like this has so far been attempted in any manmade machine. This is one of the many ways in which the brain differs from any computer so far developed.

The brain operates in terms of interactive groups of neurons, whose rate of firing is excited, inhibited or modified as a consequence of coded information transmitted to it either from a peripheral sensor or from another set of neurons within the neurons system. The resulting electrical activity at different levels of functional organization varies from millisecond to seconds. Thus an action potential, the basic unit of electrical signal across a synapse lasts a few milliseconds; an evoked potential, an expression of integrated electrical activity from a specific region of the cerebral cortex in response to a sensory stimulus, is measured in hundreds of milliseconds. The readiness potential that proceeds voluntary motor activity and recorded with electroencephalography (EEG) may last seconds. These constitute the temporal component of any spatial functional organization (Frackowiak, 1998). Recent experimental studies suggest that the brain encodes information, not just in the firing rates of individual neurons, but also in the patterns in which groups of neurons work together (Barinaga, 1998). Thus the repeated demonstration of consistent patterns of local brain activation during defined sensory, motor or cognitive activity now provide objective evidence of functional organization which can be measured non-invasively with the help of modern brain imaging techniques. As a result, it has now been possible to map even some of the mental activities, thought, memory, emotion on to specific regions of the brain (Hymann, 1998; Morris et al., 1998).

It must be realized that the brain does not function as a simple input-output system. Not only are the interconnections between the individual neurons and groups of neurons, between one functional unit and the other extremely complex but nonlinear. Practically at every level of its functional organization there are feedback loops which are capable of modifying or even blocking the input. Various functional units even in a single functional system e.g. vision or hearing are not in continuity or even contiguity but widely distributed. The grossly hard wired systems under genetic control (nature) are plastic and modifiable under influences of environment (nurture). While most of this plasticity resides at the level of synapses recent evidence reveals that it may involve other structures e.g. spines on the dendrites, change in circuitry, new circuit formation and even neurogenesis (new neuron formation) (Kempermann and Gage, 1999). Most functional units exhibit a great deal of redundancy. The system is endowed with the ability of selecting signal from noise at various levels of organization, categorize patterns out of a multiplicity of signals (see later). It has the ability to learn, memorize, recall and even create novel outputs thus endowing it with qualities of creativity and intuition. Functioning as an integrated system, not just isolated units, it has the capability of abstracting individual elements of an input (say vision), distributing these to different sets of specific neurons (concerned with shape, size colour, movement) for storing these, for future recall and when required, reconstituting these to the original composite image. Integration results jointly from the patterns of connectivity and the dynamic properties of the interacting neurons. This is specially so in respect to the cognitive functions, earlier thought to be localized to a single cortical area, but now shown to engage many and frequently widely separated areas of the cortex (Mountcastle, 1998).

5. INFORMATION PROCESSING

Based on knowledge of the structural and functional organization described above, it is now possible to have a better understanding of the information processing function of the brain. The classical model of information processing in human brain is depicted in Figure 1.

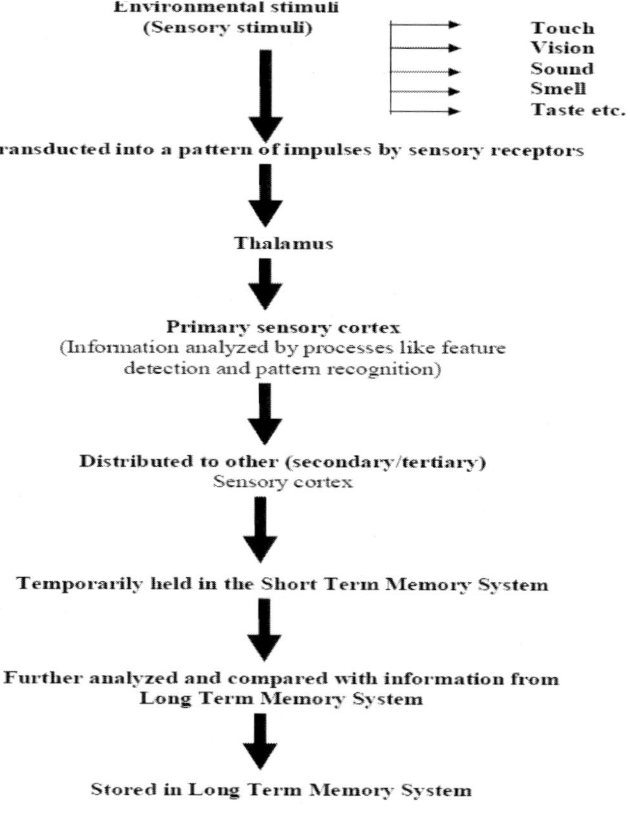

Figure 1. A classical information processing model in the human brain.

Utilizing this paradigm one may look at the example of vision, one of the most thoroughly investigated system. A basic organizing principle of the visual system (and to a great extent other sensory system) is that a relatively large number of specialized cells at each stage supply information to a smaller number of cells at the next stage, which it turn have their own specialized function. Up to a certain stage of organization the system functions in an hierarchical fashion but ultimately it becomes a distributed system. Thus any visual stimulus would pass though the retinal ganglion cells (the rods and cones), along their axons (in the optic nerve) spatially reorganized in the optic chiasma, carried to specific layers of neurons in the lateral geniculate body (a part of the thalamus – the principal sensory ganglion) to be projected to the primary visual cortex (VI) in the occipital lobe, reflecting a precise spatial representation of the visual field and further distributed to other visual areas – V2, 3, 4........, for processing information related to location, visual form, colour, orientation and motion of the object.

It must be appreciated that unlike a laboratory experiment where these visual signals can be studied with minimal or no distraction or intermingling with other sensory inputs, in day to day life these are embedded in diverse stimuli emanating from the environment outside and individuals own body. Hence the information generated by a visual stimulus (and for that matter any sensory stimulus) would of necessity be integrated and coordinated with a flood of information impinging on ones consciousness. Much of the information would need blocking or elimination at various stages of information transmission, analysis and storage. On the other hand, there are some inputs which require special attention, prompt voluntary response or imprinting in memory for future needs. It is obvious that in addition to the neural circuitry that serves the five primary senses, the human brain has numerous other systems for making sense of external stimuli and regulating the body's ability to function in the world (Ackerman, 1992).

A large amount of information has accumulated in recent years about the mechanisms involved in information processing of individual sensory/motor functions but very little is known as to how these are integrated as a unified perceptual experience or volitional activity which are the essence of human behaviour. There is much speculation on this subject, some comparing it to a giant computer engaged in a complex computation and others believing it to be in the nature of alterations in the ongoing intrinsic electrical oscillations resulting from thalamus-cortical reverberating projections. However, it must be recognized that this aspect of information processing in the brain remains an area of ignorance. This has recently been succinctly summarized by Vernon Mountcastle, an ardent investigator of this subject thus, "Many functions attributed to cerebral cortex like thresholding, amplification, feature convergence and new future construction, distribution, coincidence, detection, synchronization, long term storage and retrieval etc., are not yet fully understood at the level of neuronal circuit operations" (Mountcastle, 1998).

6. INFORMATION STORAGE OR MEMORY

Utilizing a host of new technologies including EEG and event related potential (ERP), functional magnetic resonance imaging (fMRI) and magneto-encephalography (MEG), Positron emission tomography (PET) and supplementing these with biochemical and molecular investigations in cell cultures, brain slices, laboratory animals and more recently conscious, cooperative human volunteers performing learning and memory tasks, unprecedented information has been gathered regarding the neuronal, neurochemical, molecular and biophysical substrates of memory. The formation of memory not only involves multiple anatomical regions but a complex set of cellular events, involving a variety of neurotransmitters, second messenger pathways, post-translation modification of proteins in the cytoplasm and regulation of gene expression in the nucleus. It would be impossible to even summarize this vast knowledge in this brief review. However, a few salient features are mentioned.

Memory entails sensory perception, encoding, storage, consolidation, stabilization and retrieval. The earlier prevailing concept that memory is a single entity that could be traced to a single structure or location has proved to be false. There is enough evidence to suggest that memory consists of multiple components subserved by many different regions of the brain based upon a distributed network of neurons. Thus the neural substrate of elements like short term or working memory and long term memory or episodic or semantic memory; explicit, implicit or associative memory, have been localized to diverse regions of the brain including the prefrontal cortex, the temporal cortex and hippocampus, the limbic system, medial and rostral part of the thalamus, fornix and mammilary bodies, dorsal mesencephalon and cerebellum (Frackowiak, 1994; Gabrieli et al., 1997; Goldman–Rakic, 1992; Thompson, 1986; Ungerleider, 1995; Wickelgren, 1997).

Let us take the example of working memory which is a part of the so-called short term memory and operates over a few seconds. A combination of moment – to – movement awareness and instant retrieval of archived information constitutes "Working-Memory". Like an erasable blackboard it allows us to hold briefly in our mind information – whether it be words, numbers, figures or a map of our surroundings – essential for comprehensive reason and planning (Wickelgren, 1997).

Micro-electrode recording form individual neurons in the prefrontal cortex revealed some cells showed heightened electrical activity when information was presented, where as others became active when the animal was recalling the information. A third set of neurons responded most strongly when the animals began their motor response. Goldman Rakic (1992) thus postulates that prefrontal cortex functions as an intermediate between memory and action. Furthermore, the prefrontal cortex, with its elaborate network of reciprocal connections with major sensory, limbic and pre-motor areas of the cerebral cortex, is dedicated to spatial information processing. She believed that the prefrontal cortex is divided into multiple memory domains with each specialized for encoding a different type of information such as the location of objects, the features of objects (color, size, shape) and in humans semantic and mathematical knowledge. Dopamine is considered one of the most important chemical implicated in the cognitive process, which subserves working memory, (William and Goldman-Rakic, 1995). On the other hand on the basis of neuroimaging data Flectcher and Rugg (1997) demonstrated that effective encoding in episodic memory was associated with enhanced activity in left prefrontal cortex, where as retrieval was accompanied by the enhancement of predominantly right side prefrontal activity. Thus the prefrontal cortex appears to be essential for working-memory duties, holding the relevant information on-line, performing complex processing functions, as well as for planning and attention. However, this is neither the only site for working memory nor the primary site for long-term memory.

It has long been known that the hippocampus and the surrounding temporal lobe structures are involved in memory functions. Bilateral hippocampal lesions in humans result in amnesia (Milner, 1958; Scoville and Correll, 1973). It is interesting to note that such lesions impair the capacity to record the daily current conscious experience without any disturbance of reasoning, attention of concentration. Gentle electrical stimulation in epileptic patients reactivated the past record if the "stream of consciousness" (Penfield and Perot, 1963). Functional brain imaging in human provides unambiguous demonstration that the medial temporal lobe is active at the time of verbal memory retrieval (Haxby, 1996; Nuberg et al., 1996). Gabrieli et al. (1997) observed that encoding tasks yielded increased signals for successfully remembered information in an anterior medial temporal-lobe memory system are active during distinct memory processes. Such information regarding the precise role of other regions of the brain play in memory function is rapidly accumulating but will not be discussed here any further (Helmuth, 1999; Wood et al., 1999).

As mentioned earlier, synapses play a pivotal role in neuronal communication, information transmission and probably also in information storage. It is now generally accepted that long-lasting activity – dependent changes in synaptic strength as observed in case of long term potentiation (LTP) are of fundamental importance for the development of neuronal circuitry and for information storage in mammalian brain (Nicoll and Malenka, 1995). Sensory experiences leave their imprint on the brain by altering the effectiveness of synapses on neurons. Based on the strength of the stimulus some synapses on a neurons grow stronger and other grow weaker and the pattern of synaptic changes represents a memory of experience, consolidation requires protein synthesis, which make the temporary changes permanent (Bear, 1996; 1997). However, it must be emphasized that synaptic change though essential is not identical to memory. Amongst the various hypotheses, the one currently finding favour is that memory results from the selective matching that occurs between ongoing neural activity in a particular region and the signals from the world, the body and the brain itself. Signals from the world or other parts of the brain act to select particular circuits from the myriads of circuits available in a given brain area. The synaptic alterations that ensure affect the future responses of the brain to similar or different signals. Thus in this view, a memory is dynamically generated from the activity of certain selected subsets of circuits (Edelman, 1998). Reverberating thalamo-cortical circuits seem to play a significant role in this process. There are other hypotheses too, but for the present there is a need for more empirical data and modeling studies to fully understand the process.

The above account may suggest the brain to be a modular system comprised of regions, area, circuits or even individual neurons dedicated to specific functions organized in an hierarchical fashion. This picture no doubt emerges as a consequence of reductionist approach commonly used in investigating complex systems. However, recent empirical evidence, specially based on modern brain mapping techniques reveals parallel, distributed and interactive processing, coding through neuronal assemblies that are not specialized for only one type of information, existence of non hierarchical and non modular processing. It is this type of organization that provides for the dynamic holistic aspect of many of the higher mental functions. Thus Edelman (1998) has suggested a comprehensive view of effective brain function to arise both the combined action of local segregated parts having different functions and from the global integration of these parts mediated by what he calls the process of reentry.

In summary today we have great deal of knowledge about the individual neurons, the way they connect to each other to form functional units. A large number of neurochemical they use to communicate with each other have been identified. The molecular events associated with neural communication have been unravelled. However we still do not fully understand how this neural machinery is integrated to subserve higher mental activity.

7. BRAIN AND COMPUTER

The brain is often called a computer and a computer is often compared to the brain. Therefore, an account of brain structure and function would be incomplete without referring to the significance each had for the future understanding and development of both these fields. It would be generally accepted that the human brain is enormously more complex than any electronic computer at present. The plastic microcircuitry of the brain, capable of responding to the inputs from the environment and needs of the ever changing repertoire of behavioural responses (output) is vastly different from the hard wired architecture and virtually preprogrammed outputs of a computer. While the brain predominantly depends on distributed parallel computing, most computers utilize serial processing. In contrast to the neurons in the brain that can take on any one of the series of values over a continuum, the transistors in digital circuits function on a limited 0 or 1 response. The ability of the brain microcircuitry to both structural and functional plasticity and use of diverse chemicals to do so is undoubtedly its most complex characteristic.

In some respect the existing generation of computers is "superior" to human brain. Their capability to carry out massive and complex calculations at speeds greatly faster and more precise is one of these. The computers can also have remarkable "memory". Scientists have already developed or soon to develop computers that can learn for themselves, generalize from particular examples, draw particular examples from general principles or even evolve new solutions to problems they have never faced including playing chess as good, if not better, than the best living player. However, it is true to say that so far the computers man has invented are something quite different from the brain. No doubt current developments in Neurosciences on one hand and those in Computer Science (including Artificial Intelligence and Neural Networks) have benefited both in a many ways. Consequently, whole new fields of Computational Neuroscience on one hand and Neuroinformatics on the other hand have emerged. However, as Francis Crick (1984) concluded, the hope of understanding the properties of the brain with the help of computer algorithms from neural network have not materialized. No doubt there are great optimists like Ereck De Solla Price of Yale University who "expect computerized artificial intelligence to team up with the human brain to change the very pattern of human thought. Attempts continue to me made in large number of laboratories all over the world to understand the brain to the extent that one could make humanoid robots solve tasks typically solved by the human brain by essentially the same principles. Kwato (2008) and his colleagues in collaboration with several others in the USA, UK etc. have been attempting to "Create the brain for over a decade. Nothing nearing the activities that even a toddler can do has, as yet emerged.

Therefore, as of today we have a long way to go to fully understand or artificially model and recreate even simple functions like sensing the nuances of environmental inputs or mimicking the smooth and graceful activities so effortlessly carried out even by a toddler, leave aside an accomplished pianist or a Kathak dancer. The mechanisms underlying our thoughts, emotions, creative urges, leave aside solving the riddle of brain-mind relationship remain unresolved. Nevertheless there is hope that with closer interaction between neuroscience and information science the path to this understanding will become easier and more rewarding.

REFERENCES

Ackerman, S. (1992). *Discovering the Brain*, National Academy Press, Washington DC.

Barinaga, M. (1998). Listening in on the brain. *Science, 280*, 376-378.

Bear, M.F. (1996). A synaptic basis of memory storage in the cerebral cortex. *Proceedings of National Academy of Sciences USA, 93*, 13453- 13459.

Bear, M.F. (1997). How do memories leave their mark? *Nature, 385*, 481-482.

Changeux, J. (1993). Chemical signaling in the brain. *Scientific American*, 58-62.

Crick, F. (1984). Memory and molecular turnover. *Nature, 312*, 101.

Edelman, G. (1984). Building a picture of the brain. *Daedalus, 127*, 37-69.

Fletcher, P.C., Frith C.D., and Rugg, M.D. (1997). The functional neuroanatomy of episodic memory. *Trends in Neurosciences, 20*, 213-218.

Frackowiak, R.S.J. (1994). Functional mapping of verbal memory and language. *Trends in Neurosciences, 17*, 109-115.

Frackowiak, R.S.J. (1998). The functional architecture of the brain. *Daedalus, 127*, 37-69.

Gabrieli, J.D.E., Brewer, D.J.E., and Glover, G.H. (1997). Separate neural basis of two fundamental memory processes in the human medial temporal lobe. *Science, 276*, 264-266.

Goldman-Rakic, P.S. (1992). Working memory and the mind. *Scientific American*, 73-79.

Helmuth , L. (1999). New role found for the hippocampus. *Science, 285*, 133-134.

Hubel, D.H. and Wiesel, T.N. (1968). Receptive field and functional architecture of monkey striate cortex. *Journal of Physiology (London), 195*, 215-243.

Hymann, S.E. (1998). A new image for fear and emotion. *Nature, 393*, 417-418.

Kempermann, G. and Gage, F.H. (1999). New nerve cells for the adult brain, *Scientific American*, 48-53.

Kawato, M. (2008). From "understanding the brain by creating the brain" towards manipulature neuroscience, *Philosophical Transactions of the Royal Society B, 363*, 2201-2214.

Milner, B. (1958). Psychological defects produced by temporal lobe excision. *Research Publications – Association of Nervous and Mental System Disease, 36*, 244-257.

Morris, J.S., Ohman, A. and Dolan, R.J. (1998). Conscious and unconscious emotional learning in the human amygdale. *Nature, 393*, 467-470.

Mountcastle, V.B. (1986). *Memory and Brain*. Oxford University Press, New York.

Mountcastle, V.B. (1998). Brain science at the century's ebb. *Daedalus, 127*, 1-36.

Nicoll, R.A., and Malenka, R.C. (1995). Contrasting properties of two forms of long-term potentiation in the hippocampus. *Nature, 377*, 115-118.

Nyberg, L., McIntosh, A.R., Houle, S., Nilsson, L. G., and Tulving, E. (1996). Activation of medical temporal structures during episodic memory retrieval. *Nature, 380*, 669-670.

Penfield, W. (1958). *The Excitable Cortex in Conscious Man*. The 5th Sherrington Lecture, Liverpool University Press.

Penfield, W. and Perot, P. (1963). The brain's record of auditory and visual experience. A final summary and discussion. *Brain, 86*, 595-696.

Seoville, W.B. and Correll, R.E. (1973). Memory and the temporal lobe. A review for the clinicians. *Acta Neurochirurg (Wien), 28*, 251-258.

Tandon, P.N. (2007). Brain Cells – recently unveiled secretes. Their clinical significance. *Neurology India, 55*, 322-327.

Thompson, R.F. (1986). The neurobiology of learning and memory. *Science, 233*, 941-947.

Ungerleider, L.G. (1995). Functional brain imaging studies of cortical mechanisms for memory. *Science, 270*, 769-775.

Wickelgren, I. (1978). Getting a grasp of working memory. *Science, 275*, 1580-1582.

Willam, G.V., Goldman-Rackic, P.S. (1995). Modulation of memory fields by dopamine D1 receptors in prefrontal cortex. *Nature, 376*, 572-575.

Wood, E.R., Dudchenko, P.A. and Eichenbaum, H. (1999) The global record of memory in hippocampal neural activity. *Nature, 397*, 613-616.

In: Expanding Horizions of the Mind Science(s)   ISBN: 978-1-62808-705-5
Editors: P. N. Tandon, R. C. Tripathi and N. Srinivasan ©2013 Nova Science Publishers, Inc.

Chapter 3

NEUROIMAGING STUDIES OF HUMAN COGNITION: MEASUREMENT TECHNIQUES AND RECENT DEVELOPMENTS

P. Raghunathan and P.K. Roy

National Brain Research Centre, Manesar, India

ABSTRACT

This article reviews the basic techniques of PET, fMRI, EEG and MEG and their applications for mapping human cognition. The emphasis is on the underlying principle that neural activity generates electromagnetic signal changes (the *primary* effects, detected by EEG and MEG) as well as hemodynamic and metabolic changes (the *secondary* effects, detected by PET and fMRI). The present trends in multimodal neural imaging, and future prospects in the use of forward models and inverse techniques in electrophysiological brain mapping, are discussed briefly.

Keywords: PET, fMRI, EEG, MEG, forward models, inverse techniques

1. INTRODUCTION

This article presents a critical appraisal, rather than an exhaustive survey, of various important techniques and methodologies being currently adopted for mapping the functioning brain, and particularly the "mind" which may be defined as its cognitive manifestation. The impact of functional neuroimaging studies has, in recent years, spawned the successful confluence of neuroscience with physics, mathematics and computer science. While physics has contributed to forward models of image signal production, mathematics has contributed to image processing and computer science to analytical and numerical inversion of the forward models for answering several fundamental questions regarding human cognition.

In view of the above, the important methodologies in current practice are individually discussed below, with an emphasis on the underlying principles of physics that relate to neurophysiology

2. EARLY DEVELOPMENTS

Historically, the impact of imaging in the field of neuroscience started with the advent of x-ray computed tomography (CT) in the 1970s, when clinicians started to visualize pathologies in the brains of patients without surgery. Through the small extra step of placing the source of radiation inside the patient, CT evolved into autoradiography, whereby not only structures but also blood flow and metabolism could be monitored in a relatively non-invasive way. The use of radionuclides in studies of regional cerebral blood flow using the techniques of SPECT (Single Photon Emission Computed Tomography) was the next stride forward. SPECT imaging of the brain became established as a radionuclide imaging technique that evaluates regional cerebral blood flow (rCBF) and metabolic activity by using neurospecific pharmaceuticals (Holman and Devous, 1992). An extremely fundamental observation, made over a century ago, that neural activity is accompanied by changes in rCBF (Roy and Sherrington, 1890) has not only been repeatedly underscored as being the single most important gift to the cognitive and brain sciences (van Horn, Grafton, Rockmore, and Gazzaniga, 2004), but has brought into existence the two very important neuroimaging technologies of PET (Positron Emission Tomography) and fMRI (functional Magnetic Resonance Imaging). PET and fMRI measure the metabolic and hemodynamic sequelae, respectively, and shall be examined further in the next section.

In parallel with the aforementioned developments, there has also been a rapid evolution in techniques capable of resolving more direct correlates of neural activity based on the electrophysiology of the brain (Nunez and Srinivasan 2006). Clusters of thousands of synchronously activated pyramidal cortical neurons are the generators of electrical currents in the brain, and the associated electrical scalp potentials are measured as EEG (Electroencephalograms). These currents produce an orthogonally oriented biomagnetic field according to the laws of electromagnetism, and these fields are measured as MEG (Magnetoencephalograms). All neuroimaging strategies of the present day, then, are broadly based on hemodynamic or electrophysiological measures. These will be described further in the following sections.

2.1. Neuroimaging Methodologies Based on Hemodynamic Measures

2.1.1. PET

PET yields images of physiological function through blood flow measurements, and depicts a spatiotemporal distribution of neuronal activity during specific cognitive functions. Isotopes such as ^{11}C, ^{13}N, ^{15}O and ^{18}F decay with the emission of a positron (a particle with the same rest mass as an electron but with a charge of +1). These nuclides can be incorporated into radiopharmaceuticals (molecules with known biological properties which can be injected into the human subject). Once the positron is emitted it travels a short distance

(~ 1 mm) before being annihilated on collision with an electron. This annihilation process creates two γ- photons, each with an energy of 511 keV. In order to conserve momentum, these photons are emitted at virtually 180 degrees to each other, and it is these photons that are detected in a ring of detectors that surround the head of the subject in the scanner gantry. The detector electronics are so arranged that two detection events unambiguously occurring within a 10-nanosec.time window may be called coincident and thus be deemed to have resulted from the same annihilation. These coincidence events are stored in linear arrays (called lines of response, LOR) corresponding to projections through the subject's head. The summation over many such LORs results in line integrals through the radionuclide distribution. For 2- dimensional imaging, these line integrals constitute a discrete approximation of the Radon transform of a cross section of the radionuclide concentration, which is a clear depiction of the region-of-interest within the subject. Typically, PET resolves blood flow on a spatiotemporal scale of ~ 6 mm and 30 sec. and can map either regional cerebral metabolism by using labeled glucose or rCBF using labeled oxygen (^{15}O) (Fox, Mintun, Raichle, and Herscovitch, 1984).

The positron emitters used in PET have half- lives of 2-100 minutes, and this fact necessitates that the isotopes should be made at the site of the scanner, using a cyclotron which is cumbersome and expensive. This modality has been widely used in the last decade and a half as a tool for mapping cognitive function. In the simplest form of its implementation, two cognitive states are imaged, one "active" and the other "resting", and by subtraction of the latter from the former a cortical map representing the task may be constructed. Positron emitting neurotransmitters such as ^{18}F-labeled DOPA enable studies of dopamine function, and this is of much value for neuroimaging subjects with Parkinson's disease.

Although the main drawbacks of PET are its use of radioisotopes and its steep cost, its neurotransmitter mapping ability helps it retain its role in the face of the completely non-invasive, and far more available fMRI, which we shall describe next.

2.1.2. fMRI

Functional MRI (fMRI) is a methodology (Buxton, 2001; Jezzard, Matthews, and Smith, 2003; Raghunathan, 2007) that permits a non-invasive study of brain function in real physiological times. Interestingly, fMRI does not directly detect the functioning of brain cells. Rather, it is a *surrogate* effect in that it depicts the supply of oxygenated blood to these cells. Molecules of oxygenated and deoxygenated hemoglobin (the iron-containing oxygen transport protein) differ in their magnetic susceptibilities. While oxygenated hemoglobin (Hb) has nearly the same magnetic susceptibility as the rest of the tissues or water, deoxyhemoglobin (dHb) of de-oxygenated blood is paramagnetic and can generate magnetic field distortion lines that extend by as much as twice the radius of the vein that is draining the blood flow (for an illustration, see Raghunathan, 1999). This distortion causes MRI signal loss by shortening the relaxation time (T_2^*) of the coherent proton spins that produce the image. As will be seen below, it is this difference in magnetic susceptibility that is detected by the MRI scanner. The operative mechanism of fMRI *vis-à-vis* neurophysiology may be understood as follows (Raghunathan, 2007).

Activation, or increased neural activity in a region of cortical tissue, is accompanied by a depolarization of the neuronal membrane potential. For restoring and re-establishing this potential, there is a metabolic demand by the neurons requiring additional energy. They consume oxygen by inducing a dilation of the capillaries of activated neural tissue and thereby bringing into play all the three hemodynamic parameters (cerebral blood flow, oxygen delivery and cerebral blood volume). First, there is a focal increase in the arterially delivered cerebral blood flow (rCBF) and, to a lesser extent, the oxygen extraction in the brain areas mediating the task performance. Due to the increased rCBF a large pool of proton spins enters the activation area, enhancing the MRI signal intensity (this is the "inflow" or perfusion phenomenon). Additionally, during activation, the increased rCBF *overcompensates* for the cerebral metabolic rate of oxygen consumption ($cMRO_2$) to the extent that these two hemodynamic parameters become uncoupled (Fox and Raichle, 1986).

The overall result of the above is that the capillaries of the activated tissue experience a relative blood oxygen level dependent (BOLD) *decrease* in dHb, leading to less susceptibility distortions near venules, veins and the red blood cells within the veins. Reduced susceptibility in turn causes a prolonged coherence amongst the proton spins (meaning a longer T_2^*). One thus speaks of MR signal intensity enhancement by BOLD contrast (Ogawa and Lee, 1990; Ogawa, Lee, Kay, and Tank, 1990; Ogawa et al., 1993; Thulborn, 1998). Dynamic tracking of the BOLD phenomenon with a high-speed, T_2^*-sensitive MR imaging sequence (such as fast low angle shot, FLASH, or echo planar imaging, EPI) is the basis of fMRI. Although early workers have successfully used FLASH-based gradient echo sequences for mapping functioning neurons (Jagannathan and Raghunathan, 1995; Kleinschmidt, Nitschke, and Frahm, 1997), EPI is almost exclusively used in fMRI today for the following reasons: (i) it can sustain image acquisition periods as short as 20-100 ms, which translates into 10-50 images per second, and (ii) since it is fast, it largely overcomes gross subject movements.

Neural activity associated with a specific cognitive task or state is detected by comparison with a baseline or control condition. Typically, the cognitive activation condition alternates with the control condition in a blocked paradigm design. However, there is also another class of task paradigms designed to measure regional cerebral hemodynamic responses to single sensory or cognitive events. Called "event related" task paradigms, they associate brain processes with discrete (as opposed to blocked) events, which may be randomized to occur at any point in the scanning session (Josephs, Turner, and Friston, 1997). This type of design mimics the format of the more traditional electrophysiological modalities more closely than the blocked design. We shall revert to event-related paradigms shortly.

In effect, then, brain functional maps of cognition can be generated based on the inflow effects that result from a task-related increase in rCBF and a BOLD contrast originating from decreasing dHb content during washout. The respective signatures of these effects often overlap, leading to an overall MR image intensity enhancement of about 4% to 5% at a magnetic field (B_0) of 1.5 T. At higher B_0 fields, one expects better fMRI intensity enhancements because magnetic susceptibility effects are B_0-field dependent and, in fact, bulk T_2^* changes between Hb and dHb specimens are well known to be dependent on B_0^2 (Brooks and Dichiro, 1987). For example, fMRI results from visual stimulation of the brain have shown signal intensity changes of 4.7% at 1.5 T and 15% at 4 T (Turner, et al., 1993).

While implementing event related paradigms in present day fMRI studies, one looks for the occurrence of signal changes that correlate with certain "events" to detect brain activations. In the frequently used general linear model (GLM) approach (Friston et al., 1995) a regressor representing the canonical hemodynamic response function (HRF) is used to detect such correlations. This is a "parametric" model insofar as the *shape* of HRF is assumed and parametrized. Other approaches, which make no *a priori* assumptions regarding the shape of HRF, but are based rather on physiological information, include a non-parametric Bayesian estimation of the HRF (Marrelec et al., 2003). A rather novel semi-parametric model has also been demonstrated more recently where the use of a feed-forward layered artificial neural network has been employed for describing hemodynamic response (Masaki and Miyauchi, 2006).

Neuroscientists rely increasingly on fMRI for exploring almost every conceivable aspect of human cognition. To cite but one example from exhaustive documentations of recent fMRI literature (Cabeza and Kingstone, 2006), imaging studies have provided detailed dynamic measures to suggest that different brain regions are associated with different forms of memory: working memory has been associated with the bilateral prefrontal and parietal regions, semantic memory with the left prefrontal and temporal regions, episodic memory *encoding* with the left prefrontal and medial temporal regions, episodic memory *retrieval* with the right prefrontal, posterior midline and medial temporal regions, and skill learning with the motor, parietal and sub-cortical regions.

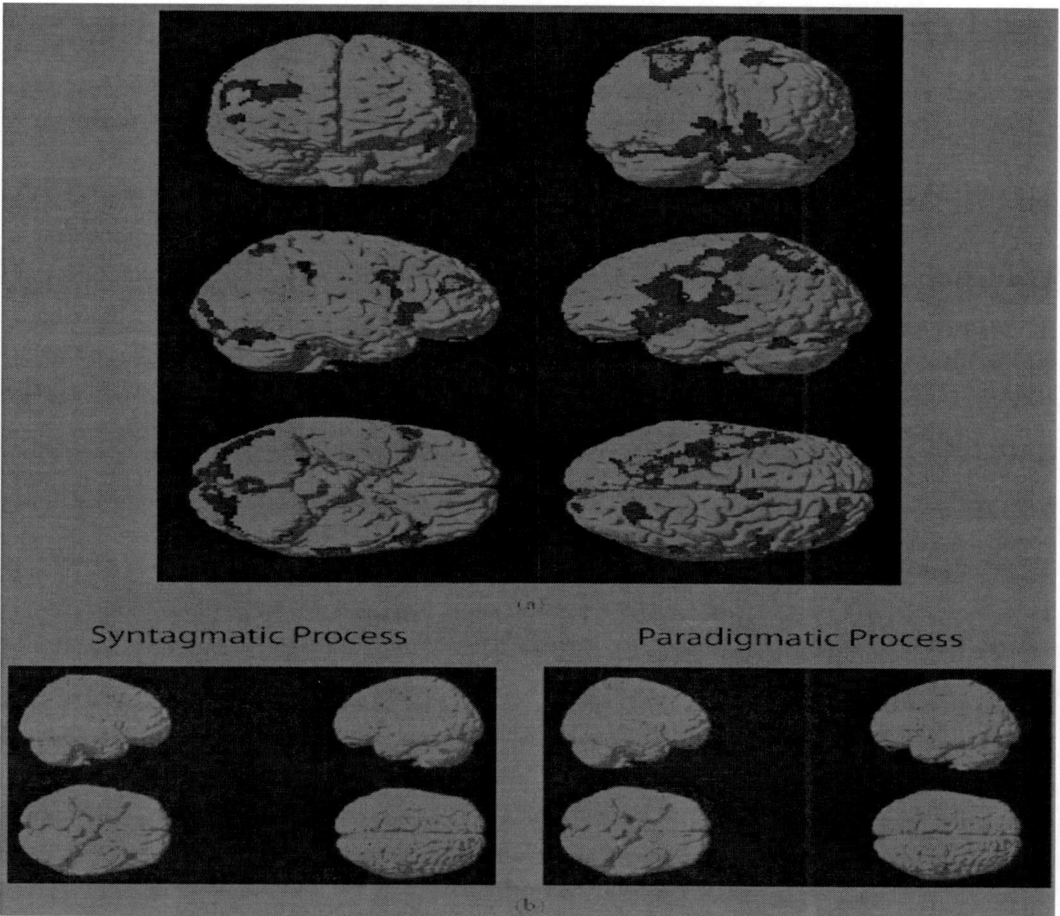

Figure 1. (a) First level fMRI data analysis (single normal subject): the left half of the figure represents, from top to bottom, three tomographic orientations of brain activation due to a syntagmatic word association; the right half of the figure represents activation due to paradigmatic word association. (b) Second level fMRI data analysis on a normal population of 16 subjects (SPM7; with $p < .001$).

Word association is another interesting psycholinguistic aspect of human cognition. It has been long known that responses to stimulus words in a free word-association task tend to be either *paradigmatic*, that is, from the same grammatical form-class as the stimulus word (for example, blue/black or hot/cold), or *syntagmatic*, that is, from a form-class that is frequently found in contiguity with the stimulus words in syntactic sequence (for example, coffee/drink). Functional MRI studies from our laboratory have probed the responses of the normal brain to syntagmatic and paradigmatic cognitive stimuli. Activation due to syntagmatic word association has been found in the bilateral temporal region, while the paradigmatic word association stimulus is seen to activate the pretemporal region (Figure 1).

Thanks to fMRI, it is possible to infer the spatiotemporal distribution of neuronal activity during specific cognitive states (the so-called *forward* inference (Henson, 2006)). We now have sufficient data to address several long-standing issues, which are the *inverse* complements of forward inferences (Poldrack, 2006). What, for example, are the primary functions subserved by particular groups of neurons? How are these different groups recruited and neurally mediated while performing complex brain functions? Are the BOLD responses evoked by sensorimotor and cognitive challenges best characterized by the activation of functionally segregated areas, or are they better understood in terms of anatomically distributed, but functionally integrated systems? And how, in the clinical context, is the fMRI neuroimaging modality helpful in assessing neurological status and neurosurgical risk?

While answers to such questions have started to become available for some time (Stern, Chin, and Travis, 2004), one ought to keep in mind the important limitation of this modality. The BOLD technique is governed by the oxygenation of blood. Following neuronal activation, there is a time delay before the necessary vasodilation can occur to increase rCBF, and for the washout of dHb from the region. The lower limits on the effective resolution of fMRI are therefore physiological, and are imposed by the spatiotemporal organization of *evoked hemodynamic responses* (2-5 mm and 5-8 sec; recall that PET measures blood flow on a spatiotemporal scale of about 6mm and 30 sec). Indeed, the question of what neuroscientists can do and cannot do with fMRI has been recently addressed (Logothetis, 2008).

3. NEUROIMAGING METHODOLOGIES BASED ON ELECTROPHYSIOLOGICAL MEASURES

The electrical activity of active nerve cells in the brain derives from the net effects of ionic currents flowing in the dendrites of neurons during synaptic transmission. This electrical activity produces endogenous *currents* spreading through the head. These currents also reach the scalp surface, and the resulting voltage differences on the scalp can be recorded as the EEG. The currents inside the head produce magnetic fields that can be measured *above* the scalp surface as the MEG. Measurements of EEG and MEG reflect brain electrical activity with *millisecond* temporal resolution, and are the most direct correlates of on-line processing of human cognition obtainable noninvasively.

Unfortunately, the spatial resolution of these methods is limited by physical and geometric constraints. Even with an infinite number of EEG and MEG recordings around the head, an unambiguous localization of the activity inside the brain would not be feasible. This *inverse* problem, namely, the problem of estimating the "source" neural generators of a given potential (in the case of EEG) or magnetic flux distribution (in the case of MEG) is comparable to reconstructing an object from its shadow. Good modelling constraints will therefore have to be imposed, and these will be considered a little later.

3.1. EEG

Electrical signals on the scalp originate from the synchronous firing of neurons in response to a stimulus. The measurement of the potential differences between various locations at the scalp surface is known as EEG. The state-of-the art EEG caps provide 256 electrodes for measurements. The modality provides a direct means of studying the functioning brain. Estimation of the scalp potentials with a known source configuration is customarily termed as the *forward problem*. The forward problem is a part of primary current source localization, which is the *inverse problem*. It is the solution of the inverse problem that provides insights into cognitive and behavioural brain functions.

To solve the forward problem, the human head is modelled as a volume conductor. Two assumptions are implicit in the forward model. The first assumption is that the current distribution in the human head model obeys the quasi-static (namely, frequencies < 1 kHz) Maxwell's equations, and it is justified as follows. In the quasi-static model, volume conduction is defined as the transmission of electric or magnetic fields from a primary electric current source through the biological tissue towards measurement sensors. In the quasi-static low frequency band we are considering, the capacitive component of tissue impedance, the inductive effect and the electromagnetic propagation effect can *all be neglected*, so that the transmission of current to the scalp is purely resistive and is measured in μV.

The second assumption is that the head volume conductor is "spherical" and has a homogeneous isotropic conductivity. This, again, is largely justified because the shape of the neocortex is topologically very close to a spherical shell. If the spherical model were not adopted to get an analytical forward solution, the inverse problem (of defining precise neuronal activity locations) would remain "ill posed" until a more realistic head shape cartography is used. Here is where MR images begin to play a role in the numerical inversion of the forward model. T_1- weighted MR images are typically used to build the correct geometric model of a subject's head. Better still, the advent of MR-based diffusion tensor imaging (DTI) (Raghunathan, 2010) could lead to a map of the conductivity *tensor* of the human brain considering the high correlation between the electrical conductivity tensor and the water diffusion tensor (Liu, Ding, and He, 2006). This model would also remove the assumption of isotropic electrical conductivity implied in the earlier models.

Since the volume conductor model represents the conductivity distribution in the head, we need an accurate conductivity value for each head element. An inaccurate assignment of conductivity contributes a significant error in assessing the source localization. Another limitation of EEG is that the signals are weak (about 50 μv), and therefore care must be taken to reduce electrical interference from external sources.

Several characteristic frequencies are detected in the human EEG. For example, when the subject is relaxed, the EEG consists mainly of evoked potentials (EP) in the frequency range 8-13 Hz, called alpha waves, but when the subject is more alert, event-related potentials (ERP) are detected and their frequencies rise above 13 Hz, called beta waves. A classical example is the measurement of EEG during sleep, which reveals periods of high frequency waves, attributed to rapid eye movement (REM). The measurement has been associated with the "dream" state.

EEG signals are measurable as responses to biorhythms, or the natural frequencies of the brain (Basar, Basar-Eroglu, Karakas, and Schurmann, 2001), as well as to some regularly repeated cognitive stimulus such as number memorizing and recall. The measured response, that is the ERP, is characterized by the delay, or *latency*, of the peak of the signal from the presentation of the stimulus. This is of the order of milliseconds for brain stem responses and several hundred milliseconds for cortical responses. For example, a commonly detected response to an oddball stimulus, the P300, has a latency of 300 milliseconds. Having found an event-related electrical signal of interest, its peak amplitude can be mapped across the scalp, giving an idea of the source.

Relatively high spatial resolution EEGs are now obtained with scalp recordings using a combination of dense electrode arrays (e.g. 131-channel scalp recordings, with ~ 2.3 cm centre-to-centre electrode separation) and computer algorithms of forward models used to project scalp potentials on to the cortical surface. A limitation of studying mental function using EEG is that the signals measured are recovered at the scalp, which may not be representative of the activity in the underlying cortex.

3.2. MEG

Since the magnetic field signals from the firing of the neurons do not need to be conducted to the scalp, their measurement leads to much sharper signal localization than measuring the currents themselves. This is the basis of the much newer neuroimaging technique of MEG. Cohen (Cohen, 1972) made the first successful measurements of these magnetic fields. The flux density of the magnetic signals resulting from the electrical firing of the neurons is 10^{-13} T, for one million synchronously active synapses. Such small signals are detected using a superconducting quantum interference device (SQUID)-based magnetometer first introduced by Zimmerman, Thiene, and Hardings (1970). Neural currents flowing in the brain are measured by an array of multiple channels (about 122) of SQUID sensors that cover the subject's head in a helmet-like arrangement.

A major problem with MEG is interference from stray magnetic fields generated by external sources such as electric power lines, lights, electrical motors, elevators, trains, computers, etc. These fields will swamp the tiny biomagnetic signals. To avoid such stray fields, MEG measurements are carried out in shielded rooms whose floor, ceiling and walls are completely clad in several layers of mu-metal (or a similar high permeability material). Also, the SQUID sensors use low-temperature electronics cooled by liquid helium. Scalp magnetic fields are then typically recorded every millisecond. The resulting data can be visualized as time-evolving magnetic field topographies.

The experimental scans are carried out in much the same way as their EEG counterparts. Having identified the peak of interest, the signals from all the detectors are analyzed to obtain a field map according to a suitable forward model (Henson, Mattout, Phillips, and Friston, 2009). Using this map, numerical methods are then applied to ascertain the source of neuronal activation by solving the inverse problem. Once again, since the inverse problem has no unique solution, assumptions need to be made regarding geometry. The technique offers a time resolution of 1 ms, and a spatial resolution of a few mm.

The theory, instrumentation, and applications of MEG to the working human brain are well discussed by Hamalainen (Hamalainen et al., 1993). Another illustrative example of the application of MEG to cognition is provided by the recent work of Nishimura, Tobinaga, and Tonoike, 2008), who demonstrate the detection of neural activity associated with thinking. In a noteworthy development, MEG equipment has been constructed in this country as a green field project (see Figure 2), and has been demonstrated to give accurate measurements of the cognitive state of 'eyes-closed' wakefulness (alpha rhythm), auditory evoked responses (see Figure 3), as well as binaural response of cognitive processing in the temporal lobes of the brain (Janawadkar et al., 2010). Since MEG is a very satisfactory method for producing high resolution spatiotemporal maps, the inverse solution of magnetometric fields turns out to be a procedure that can give accurate localization of epileptic foci. This mapping approach is of much significance for clinical cognitive neuroscience, since temporal psychomotor epilepsy is often a refractory neurophysiological condition.

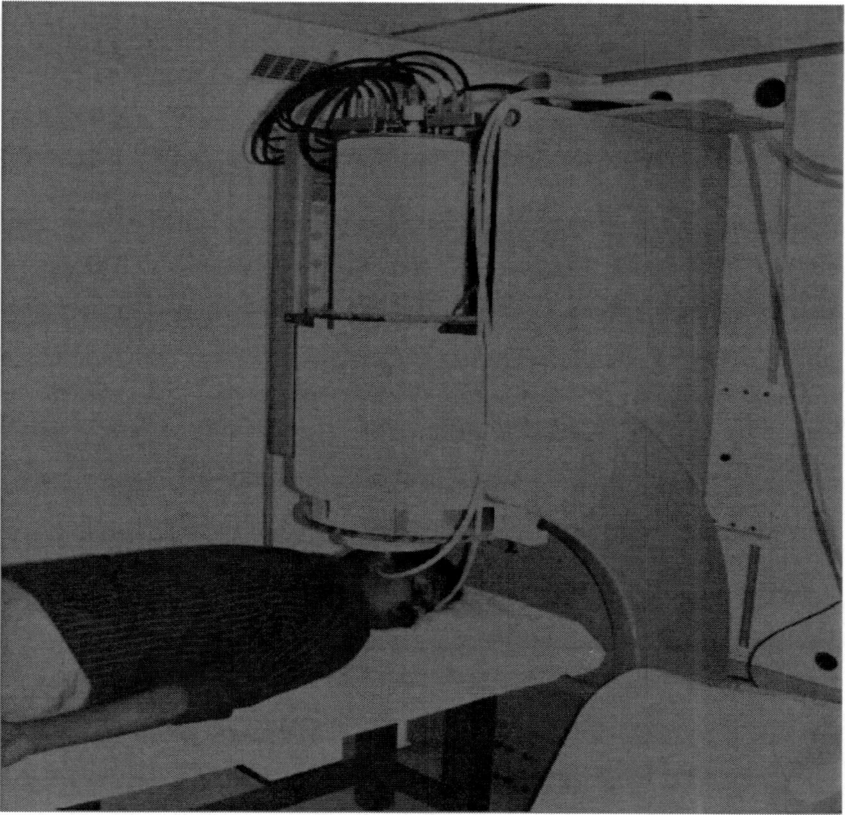

Figure 2. MEG equipment (detector and cryostat), mounted over a gantry inside a magnetically shielded room. Courtesy: IGCAR, Kalpakkam, India.

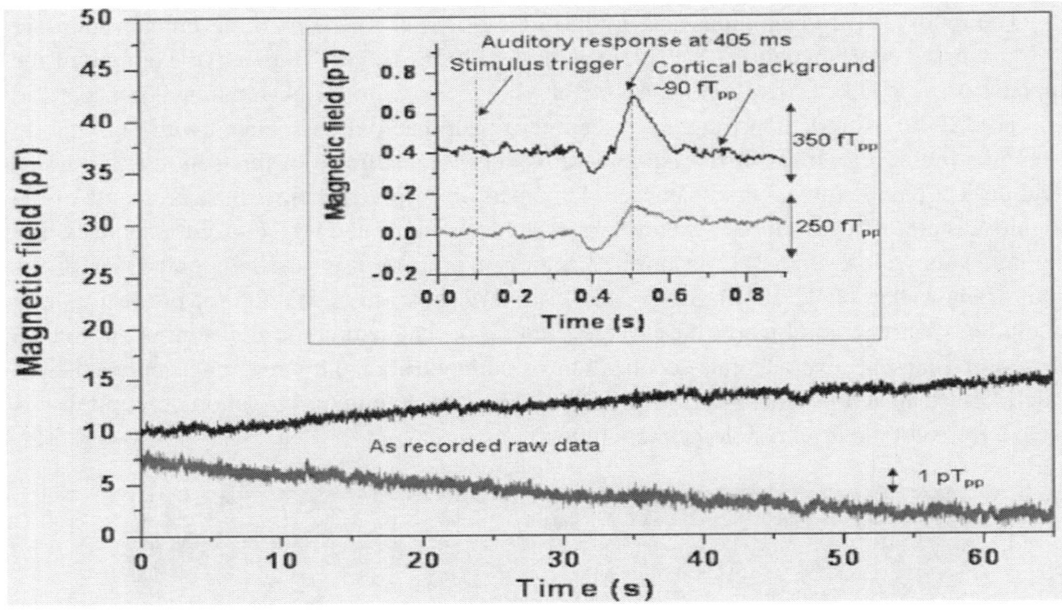

Figure 3. Illustration of how the auditory field is evoked on presentation of a stimulus. Inset shows evoked and averaged response for 500 trials. Courtesy: IGCAR, Kalpakkam, India.

4. FUNCTIONAL NEUROIMAGING AND INSIGHTS INTO COGNITION

Applications of functional neuroimaging to several specific cognitive domains such as attention, skill learning, semantic memory, language, episodic memory, working memory, and executive functions have been documented extensively (Cabeza and Kingstone, 2006). Here we cite a few additional recent examples of how functional neuroimaging has led to further novel insights into cognition.

The excellent neurotemporal resolution offered by ERP measures has been exploited in an investigation on spatial attention (Carlson and Reinke, 2010). Since fearful facial expressions are important non-verbal means for biological communication, this study has used backward masked fearful faces to study spatial attention. In particular, the N170 is a face-sensitive, visually evoked, negative EEG peak amplitude centered at 170 millisec post stimulus; and contralateral N170 amplitudes have been shown to predict the facilitation in reaction times associated with directed spatial attention.

In another interesting study, the functional anatomy of humor has been studied (Goel and Dolan, 2001). Using event-related fMRI, the function of humor has been dichotomized into a cognitive processing component and an affective appreciation component. The former is associated with the juxtaposition of phonological and semantic mental sets; the semantic juxtaposition uses a bilateral temporal network, while the phonological juxtaposition uses a left hemisphere network centered on the regions of speech production. The latter component, namely, affective appreciation of humor, accesses a region in the medial ventral prefrontal cortex that controls reward-related behavior.

The neurocognitive mechanisms of a painful emotion, envy, and a rewarding reaction, *schadenfreude* (the sense of gratification one derives from another's misfortune) have been mapped using fMRI (Takahashi et al., 2009). Envy activates the anterior cingulate cortex, and schadenfreude the ventral striatum.

Considerable effort has also been directed at understanding the resting state network of the brain. Spontaneous fluctuations observed in fMRI BOLD signals during "do nothing" or resting states (e.g. lying quietly with eyes closed or fixated on a spot) have been traced to large-scale intrinsic brain networks (Beckmann, DeLuca, Devlin, and Smith, 2005; Damoiseaux, 2006) and, in particular, to the default mode network (DMN). The DMN, whose functioning has been likened to the working of Freudian ego, comprises regions that show blood flow and metabolic activity at rest but becomes deactivated during goal-directed cognition (Raichle and Snyder, 2007). For example, it has an inverse relationship with the "attention" network. Using extensive results from such fMRI and PET investigations, a very recent review on the neurophysiology of consciousness (including the states of REM sleep, the early and acute psychotic states, the aura of temporal lobe epilepsy, and hallucinogenic drug states) has developed the viewpoint that the brain is a hierarchical inference or "Helmholtz" machine (Carhart-Harris and Friston, 2010).

5. PRESENT TRENDS AND PROSPECTS

From the forward models of hemodynamic response-based imaging modalities (PET, fMRI) described in earlier sections, we may often accurately infer *where* a stimulus-induced activation occurred in the cortex, but not exactly *when*. With electrophysiology-based imaging (EEG, MEG), we know *when* but not *where*. This aspect of mutual exclusion calls for using these techniques as complementary measures. One may effectively combine event-related fMRI with event-related potential (ERP) to map the precise spatiotemporal orchestration of the neural substrates of cognition. For example, it is possible to reconstruct the cortical surface from MRI data (Henson et al. (2009) call this the 'cortical mesh') and use it as a constraint on the EEG/MEG inverse solution. Recent trends in computational neuroimaging strongly suggest the incorporation of hemodynamically identified activation areas into source-localization modeling analyses based on electrophysical (EEG) and magnetic field (MEG) measures. Also, rapid advances in MR-based diffusion tensor imaging (DTI) methodologies (Raghunathan, 2010) offer much promise of mapping the electrical conductivity tensor of the human brain, considering the one-to-one correlation between the electrical conductivity tensor and the water diffusion tensor (Liu et al., 2006).

Recently, the mathematical theory of signal communication using the concept of mutual information, or trans-information flux, has been employed for the analysis and estimation of neural information flow (Roy, Miller, and Majumder, 2002). Using this approach, Roy and Budhachandra (2005) have used the MRI-based modality of electrical conductivity tensor imaging to delineate the direction and path analysis of information flow in the brain. Other biophysical properties of the brain, such as the thermal or electrical conductivity tensor, fluid permeability tensor or blood perfusion tensor, can be computed, and the corresponding tensor imaging modalities can then be constructed using specially devised MRI pulse sequences. For instance, using MRI, one can construct the thermal conductivity tensor image, the perfusion tensor image and the heat convection tensor image of the brain. Based on these, one can get an accurate representation of energy flow maps across the brain using metabolic and electrode stimulation (Budhachandra, Shukla, and Roy, 2010). This energy mapping modality should be of much utility for investigating neural and cognitive disturbances in patients who are subjected to deep brain stimulation therapy (Elwassiff, Vazques, Kong, and Bikson, 2006).

Linear statistical methods of fMRI data analysis make an underlying assumption that the signals are produced by a linear stochastic system (Friston et al., 1995). However, considerations based on brain physiology have led to an increasing realization that the brain acts as a *nonlinear* system that is not completely stochastic (Elbert et al., 1994; Freeman, 1994; Schiff et al., 1994) and therefore the processes generating fMRI data would be expected to be nonlinear. For example, Sheth et al. (2004) report on the nonlinear interactions between the neuronal, metabolic and hemodynamic factors causing the BOLD response. Multivariate nonlinear autoregressive models (Harrison, Penny, and Friston, 2003) and nonlinear principal component analysis (Friston, 2000) have also been published. The viewpoint that the brain is a nonlinear system is likely to gain centre-stage in the days to come.

Finally, the following ontological question is also being debated (Poldrack,2006): given neurological data, can cognitive processes be inferred? Using extensive data sets on the brain areas (in excess of 200), and the fiber tracks (in excess of 30,000) that identify their locations, boundaries, volumes of tissue, connective topology, causal links and core functions, can we *reverse engineer* the brain system? Research efforts in several neuroscience laboratories of the world, including ours, are looking for an answer.

REFERENCES

Basar, E., Basar-Eroglu, C., Karakas, S., and Schurmann, M. (2001). Gamma, alpha, delta, and theta oscillations govern cognitive processes. *International Journal of Psychophysiology, 26*, 5-29.

Beckmann, C.F., DeLuca, M., Devlin, J.T., and Smith, S.M. (2005). Investigation into resting-state connectivity using independent component analysis. *Philosophical Transactions of the Royal Society B: Biological Sciences, 360*, 1001-1013.

Brooks, R.A., and Dichiro, G. (1987). Magnetic resonance imaging of stationary blood. A review. *Medical Physics, 14*, 903-913.

Budhachandra, K., Shukla, V., and Roy, P. (2010). Thermal conduction tensor imaging and energy flow mapping of brain using MRI. *Annals of Biomedical Engineering, 38*, 3070-3083.

Buxton, R. (2001). *An introduction to functional magnetic resonance imaging: principles and techniques.* Cambridge, UK: Cambridge University Press.

Cabeza, R., and Kingstone, K. (2006). *Handbook of Functional Neuroimaging of Cognition.* Massachusetts: The MIT Press.

Carhart-Harris, R.L., and Friston, K.J. (2010). The default-mode, ego-functions and free-energy: a neurobiological account of Freudian ideas. *Brain, 133*, 1265-1283.

Carlson, J.M., and Reinke, K.S. (2010). Spatial attention-related modulation of the N170 by backward masked fearful faces. *Brain and Cognition, 73*, 20-27.

Cohen, D. (1972). Magnetoencephalography: detection of the brain's electrical activity with a superconducting magnetometer. *Science, 175*, 664-666.

Damoiseaux, J.S., Rombouts, S.A., Barkhof, F., Scheltens, P., Stam, C.J., Smith, S.M., and Beckmann, C.F. (2006). Constituent resting-state networks across healthy subjects. *Proceedings of the National Academy of Sciences USA, 103*, 13848-13853.

Elbert, T., Ray, W.J., Kowalik, Z.J., Skinner, J.E., Graf, K.E., and Birbaumer, N. (1994). Chaos and physiology: deterministic chaos in excitable cell assemblies. *Physiological Review, 74*, 1-47.

Elwassiff, M., Vazques, M., Kong, Q., and Bikson, M. (2006). Bioheat transfer model of deep brain stimulation induced temperature changes. *Journal of Neural Engineering, 3*, 306-315.

Fox, P.T., and Raichle, M.E. (1986). Focal physiological uncoupling of cerebral blood flow and oxidative metabolism during somatosensory stimulation in human subjects. *Proceedings of the National Academy of Sciences USA, 83*, 1140-1144.

Fox, P.T., Mintun, M.A., Raichle, M.E., and Herscovitch, P. (1984). A noninvasive approach to quantitative functional brain mapping with $H_2{}^{15}O$ and positron emission tomography. *J. Cereb. Blood Flow Metab.* 4: 329-33.

Freeman, W.J. (1994). Role of chaotic dynamics in neural plasticity. *Progress in Brain Research, 102*, 319-333.

Friston, K.J. (2000). Bayesian estimation of dynamical systems: an application to fMRI. *NeuroImage, 16*, 513-530.

Friston, K.J., Holmes, A.P., Poline, J.P., Grasby, P.J., Williams, S.C., Frackowiak, R.S., and Turner, R. (1995). Analysis of fMRI time series revisited. *NeuroImage, 2*, 45-53.

Goel, V., and Dolan, R.J. (2001). The functional anatomy of humor: segregating cognitive and affective components. *Nature Neuroscience, 4*, 237-238.

Hamalainen, M.S., Hari, R., Ilmoniemi, R.J. Knuutila, J., and Lounasmaa, O. V. (1993). Magnetoencephalography-theory, instrumentation, and applications to noninvasive studies of the working human brain, *Reviews of Modern Physics, 65*, 413-497.

Harrison, L., Penny, W.D., and Friston, K. (2003). Multivariate autoregressive modeling of fMRI time series. *NeuroImage, 19*, 1477-1491,

Henson, R. (2006). Forward inference using functional neuroimaging: dissociation versus association. *Trends in Cognitive Sciences, 10*, 64-69.

Henson, R.N., Mattout, J., Phillips, C., and Friston, K.J. (2009). Selecting forward models for MEG source-reconstruction using model-evidence. *NeuroImage, 46*, 168-176.

Holman, B.L., and Devous, M.D. Sr (1992). Functional brain SPECT: the emergence of a powerful clinical method. *Journal of Nuclear Medicine, 33*, 1888-1904.

Jagannathan, N.R., and Raghunathan, P. (1995). Task activation of the human brain studied in real time by functional MR imaging. *Current Science, 69*, 448-451.

Janawadkar, M., Radhakrishnan, T., Gireesan, K., Parasakthi, C., Sengottuvel, S., Patel, R., Sundar, C. S., and Raj, B. (2010). SQUID based measurement of biomagnetic fields. *Current Science, 99*, 36-45.

Jezzard, P., Matthews, P.M., and Smith, S.M. (2003). *Functional magnetic resonance imaging*. Oxford University Press, Oxford, UK.

Josephs, O., Turner, R., and Friston, K. (1997). Event-related fMRI. *Human Brain Mapping, 5*, 243-248.

Kleinschmidt, A., Nitschke, M.F., and Frahm, J. (1997). Somatotopy in the human motor cortex hand area. A high resolution MRI study. *European Journal of Neuroscience, 9*, 2178-2186.

Liu, Z., Ding, L., and He, B. (2006). Integration of EEG/MEG with MRI and fMRI in functional neuroimaging. *IEEE Engineering Medicine Biology Magazine, 25*, 46-53.

Logothetis, N.K. (2008). What we can do and what we cannot do with fMRI. *Nature, 253*, 869-878.

Marrelec, G., Benali, H., Ciuciu, P., Pélégrini-Issac, M., and Poline, J.B. (2003). Robust Bayesian estimation of the hemodynamic response function in event-related BOLD fMRI using physiological information. *Human Brain Mapping, 19*, 1-17.

Masaki, M. and Miyauchi, S. (2006). Application of artificial neural network to fMRI regression analysis. *NeuroImage, 29*, 396-408.

Nishimura, K., Tobinaga, Y. and Tonoike, M. (2008). Detection of neural activity associated with thinking in frontal lobe by MEG. *Progress in Theoretical Physics, 173*, 332-341.

Nunez, P,L. and Srinivasan, R. (2006). *Electrical fields of the brain: The neurophysics of EEG*, 2nd edn. Oxford University Press, Oxford, UK.

Ogawa, S., and Lee, T. (1990). Magnetic resonance imaging of blood vessels at high fields: in vivo and in vitro measurements of image simulation. *Magnetic Resonance in Medicine, 16*, 9-18.

Ogawa, S., Lee, T.M., Kay, A.R., and Tank, D.W. (1990). Brain magnetic resonance imaging with contrast dependent on blood oxygenation. *Proceedings of the National Academy of Sciences USA, 87*: 9868-9872.

Ogawa, S., Menon, R.S., Tank, D.W., Kim, S.G., Merkle, H., Ellermann, J.M., and Ugurbil, K. (1993). Functional brain mapping by blood oxygenation level-dependent contrast magnetic resonance imaging. *Biophysical Journal, 64*, 803-812.

Poldrack, R.A. (2006). Can cognitive processes be inferred from neuroimaging data? *Trends in Cognitive Sciences, 10*, 59-63.

Raghunathan, P. (1999). Magnetic resonance imaging and spectroscopy in biomedicine. *Proceedings of the National Academy of Sciences USA, 65A*: 699-729.

Raghunathan, P. (2007). *Magnetic resonance imaging and spectroscopy in medicine: concepts and techniques*. Orient Longman, Chennai, India.

Raghunathan, P. (2010). Diffusion magnetc resonance imaging concepts in neuroscience: the propagator, tensor, q-space and q-ball. In: Khetrapal CL, Anil Kumar, Ramanathan KR (ed.) *Future directions of NMR*. Springer

Raichle, M.E. and Snyder, A.Z. (2007). A default mode of brain function: a brief history of an evolving idea. *NeuroImage, 37*, 1083-1090.

Roy, C.S. and Sherrington, C.S (1890). On the regulation of blood supply of the brain. *Journal of Physiology, 11*, 85-108.

Roy, P., and Budhachandra, K. (2005). Informational flux mapping and connectivity in neuroimaging: a biothermodynamic tensorial approach, *IEE Proceedings of Biomedical Engineering, 12*, 2406-2409.

Roy, P., Miller, J., and Majumder, D.D. (2002). A control analysis of neuronal information processing: a study of electrophysiological experimentation. *Springer Lecture Notes on Computer Science, 2275*, 191-203.

Schiff, S.J., Jerger, K., Duong, D.H. et al. (1994). Controlling chaos in the brain. *Nature, 370*, 615-620.

Sheth, S.A., Nemoto, M., Guiou, M., Walker, M., Pouratian, N., and Toga, A.W. (2004). Linear and nonlinear relationships between neuronal activity, oxygen metabolism, and hemodynamic responses. *Neuron, 42*, 347-355.

Stern, P., Chin, G., and Travis, J. (2004). Neuroscience: higher brain functions. *Science, 306*, 431.

Takahashi, H., Kato, M., Matsuura, M., Mobbs, D., Suhara, T., and Okubo, Y. (2009). When your gain is my pain and your pain is my gain: neural correlates of envy and schadenfreude. *Science, 323*, 937-939.

Thulborn, K.R. (1998). A 'BOLD' move for fMRI. *Science (Medicine), 4*, 155-158.

Turner, R., Jezzard, P., Wen, H., Kwong, K.K., Le Bihan, D., Zeffiro, T., and Balaban, R.S. (1993). Functional mapping of the human visual cortex at 4 and 1.5 T using deoxygenation contrast EPI. *Magnetic Resonance in Medicine, 29*, 277-279.

Van Horn, J.D., Grafton, S.T., Rockmore, G., and Gazzaniga, M.S. (2004). Sharing neuroimaging studies of neural cognition. *Nature Neuroscience, 7*, 473-481.

Zimmerman, J.E., Thiene, P., and Hardings, J. (1970). Design and operation of stable rf-biased superconducting point-contact quantum devices. *Journal of Applied Physics, 41*, 1572-1580.

Chapter 4

ELECTROPHYSIOLOGICAL MEASUREMENT: INSIGHTS ABOUT THE TEMPORAL DYNAMICS OF BRAIN AND COGNITION

Bhoomika Rastogi Kar
Centre of Behavioural and Cognitive Sciences, University of Allahabad,
Allahabad, India

ABSTRACT

Electroencephalography (EEG) and recording of the event related potentials (ERP) is a method in cognitive neuroscience research to investigate the temporal dynamics of neural and cognitive process. EEG/ERP measures the chronometry of neural events underlying mental events. This chapter discusses about how an EEG signal is generated and how ERP works. Different ERP components and their relevance to the study of brain and cognition as well as the issues related to the design and interpretation of EEG/ERP and analysis are also discussed. EEG/ERP has been applied to the study of various cognitive processes such as perception, attention, language, memory, and certain control processes as well as to investigate the neurocognitive mechanisms that underlie abnormal cognition, meditation and creativity. As an example, results of one of the studies on electrophysiological correlates of remediation in dyslexia and one on meditation are discussed to demonstrate plasticity effects related to changes in amplitude and latency investigated through EEG/ERP.

1. INTRODUCTION

Information processing depends on mental representations. Cognitive psychology explains how we manipulate representations. Each cognitive task may involve a set of mental operations. Cognitive neuroscience focuses on functional neuroanatomy that mediates mental operations (Gazzaniga, Ivry, and Mangun, 2002). Functional neuroimaging techniques like Positron Emission Tomography (PET) and functional magnetic resonance imaging (fMRI) exploit the changes in local blood flow while a cognitive task is being performed. On the other hand, investigating the time course of cognitive processes is an important method in cognitive psychology and cognitive neuroscience. Electroencephalography (EEG) and event related potential (ERP) informs about the time course of information processing.

EEG/ERP is the measurement of electrical activity produced by the brain as recorded from electrodes placed on the scalp. The electrical activity of active nerve cells in the brain produces currents spreading through the head. These currents also reach the scalp surface, and resulting voltage differences on the scalp can be recorded as the electroencephalogram (EEG). EEG reflects brain's electrical activity with millisecond temporal resolution, and is the most direct correlate of on-line brain processing obtainable non-invasively. Unfortunately, the spatial resolution of EEG is limited for principal physical reasons. Even with an infinite amount of EEG recordings around the head, a non-ambiguous localization of the activity inside the brain would not be possible. This "inverse problem" is comparable to reconstructing an object from its shadow: only some features (the shape) are uniquely determined, others have to be deduced on the ground of additional information. However, by imposing reasonable modelling constraints or by focusing on rough features of the activity distribution, useful inferences can be made.

The continuous or spontaneous EEG (i.e. the signal which can be viewed directly during the recording) can be helpful in clinical environments, e.g. for diagnosing epilepsy or tumors, predicting epileptic seizures, detecting abnormal brain states or classifying sleep stages. The investigation of more specific perceptual or cognitive processes requires more sophisticated data processing, like averaging the signal over many trials. The resulting "event-related potentials" (ERPs) or "event-related magnetic fields" are characterized by a series of deflections or "components" in their time course, which are differentially pronounced at different recording sites on the head. They are classically labelled according to their polarity (positive/negative) at specific recording sites and the typical latency of their occurrence (e.g. N100 refers to a negative potential around 100ms, similarly for P300, N400 etc.). It is to be noted that polarity of peaks is of no real significance cognitively or neurophysiologically (Otten and Rugg, 2005).

The amplitude and pattern of these components can be interpreted as dependent variables, which are distinguished by their dependency on the task, stimulus parameters, degree of attention etc. Components related to basic perceptual processing (somato-sensory, visual, auditory, which are also termed "evoked potentials" or "evoked magnetic fields") can also be used by clinicians to examine the connection of the periphery to the cortex. Event-related potentials and magnetic fields are widely used to characterize the spatio-temporal pattern of brain activity in basically all areas of neuroscience. Latest developments include the combination of EEG with functional magnetic resonance imaging, for combining the superior spatial resolution of the latter with the better temporal resolution of the former.

2. GENERATION OF THE SIGNALS

If an electrical signal is transmitted along an axon or dendrite, electrical charges are separated along very short distances on the corresponding cell membranes. Separated charges act as a small temporary "battery", with positive polarity at the side where the current is leaving and with negative polarity where the current is returning. These "batteries" are called "primary currents" since they are the sources of interest that reflect directly the activity of the corresponding neurons. The primary currents are embedded in a conductive medium, i.e. the brain tissue and brain liquor. Therefore a current is induced which flows through these media, as well as the skull and scalp, called "secondary" or "volume" currents. EEG would not be possible without volume currents, which also reach the scalp surface and cause voltage differences at the scalp that can be picked up by EEG electrodes. Possible contributors to the measurable signals are: 1) Action potentials along the axons connecting neurons, 2) currents through the synaptic clefts connecting axons with neurons/dendrites, and 3) currents along dendrites from synapses to the soma of neurons (Nunez, 1981).

Action potentials are "quadrupolar", i.e. two opposing currents occur in vicinity of each other, the effects of which cancel each other out. In addition, action potentials last for only a few milliseconds, making it more difficult to achieve a larger number of them being active simultaneously and to sum up to a measurable signal at a larger distance. Synapses are very small and the current flow is relatively slow. Furthermore, they can be located nearly randomly around a neuron and its dendrites, so that the contributions of different synapses are likely to cancel each other out. The current flow along the post-synaptic dendrites is dipolar, and at least the apical dendrites in cortical pyramidal cells are generally organized parallel to each other. They are typically active for about 10ms after the synaptic input. For these reasons, the apical dendrites of the cortex are assumed to contribute strongest to the measurable EEG signals. However, one dendrite would be far too weak to produce a measurable signal; instead tens of thousands are required to be active synchronously. Consequently, EEG is sensitive to coherent simultaneous activity of a large number of neurons. In nut shell, excitatory neurotransmitter is released by pre-synaptic axons. Positive ions flow into post-synaptic apical dendrites leaving negative voltage outside the cell. To complete a circuit, positive ions flow out of the cell body and basal dendrites. This creates a tiny dipole (and voltage) that is too small to measure from the scalp. But if 1000s of cells are aligned, and activate at the same time, the voltage sum may reach to a size that can be detected by electrodes at the scalp (Hauk, 2003-2008).

3. EEG RECORDING

3.1. Referencing

Electrical potentials are only defined with respect to a reference, i.e. an arbitrarily chosen "zero level". The choice of the reference may differ depending on the purpose of the recording. This is similar to measures of height, where the zero level can be at sea level for the height of mountains, or at ground level for the height of a building, for example.

For each EEG recording, a "reference electrode" has to be selected in advance. Ideally, this electrode would be affected by global voltage changes in the same manner as all the other electrodes, such that brain unspecific activity is subtracted out by the referencing (e.g. slow voltage shifts due to sweating). Also, the reference should not pick up signals which are not intended to be recorded, like heart activity, which would be "subtracted in" by the referencing. In most studies, a reference on the head but at some distance from the other recording electrodes is chosen. Such a reference can be the ear-lobes, the nose, or the mastoids (i.e. the bone behind the ears). With multi-channel recordings (e.g. >32 channels), it is common to compute the "average reference", i.e. to subtract the average over all electrodes from each electrode for each time point. This distributes the "responsibility" over all the electrodes, rather than assigning it to only one of them. If a single reference electrode was used during the recording, it is always possible to re-reference the data to any of the recording electrodes (or combinations of them, like their average) at a later stage of processing. In some cases "bipolar" recordings are carried out, where electrode pairs are applied and referenced against each other for each pair (e.g. left-right symmetrical electrodes) (Hauk, 2003-2008).

3.2. Electrode Locations

To compare the results over different studies, electrode locations must be standardized. These locations should be easily determinable for individual subjects. Common is the extended 10/20 system, where electrode locations are defined with respect to fractions of the distance between nasion-inion (front-back) and the pre-auricular points (left-right). Obviously, the relative locations of the electrodes with respect to specific brain structures can only be estimated very roughly. If this is a crucial point, electrode locations can be digitized together with some anatomical landmarks (like the pre-auricular points, the ear-holes, nasion or inions). These landmarks can also be detected in individual MRI or CT scans if available. The landmarks of the different co-ordinate systems (EEG digitization, MRI/CT scan) can then be matched, such that both data sets are in the same co-ordinate system. Another possibility is to digitize a large number of points (at least several hundreds) evenly distributed over the scalp surface, and match those to the scalp surface segmented from the individual MRI/CT image. This makes possible more realistic modeling of the head geometry for sophisticated source estimation, and furthermore the visualization of the results with respect to the individual brain geometry.

3.3. The Inverse Problem and the Inverse Solution

Even if the EEG is measured simultaneously at many points around the head, the information would still be insufficient to uniquely compute the distribution of currents within the brain that generated these signals. The so-called "inverse problem", i.e. the estimation of the sources given the measured signals, is "under-determined" or "ill-posed" (Ward, 2006). Inverse problem can be briefly explained as follows: If we knew the locations and orientations of a set of dipoles, we could use a set of equations to predict the distribution of voltages at the scalp. This is called the forward problem that is relatively easy to solve. If we knew the distribution of voltages at the scalp, we could not use a set of equations to predict (with any certainty) the locations of orientations of a set of dipoles because an infinite number of these could produce any one distribution of voltages at the scalp. This is called the inverse problem that is very hard to solve.

By imposing appropriate modeling constraints, one can often derive valuable information about spatial features of the source distribution from the data. The main questions in this context are: Do we have enough information from other sources (neuropsychology, other neuro-imaging results) to narrow down the possible generators? Or could we ask our questions such that it is not necessary to know the exact source distribution, but only some rough features that can be determined without further modelling constraints?

One can distinguish between two strategies to tackle the inverse problem: 1) Approaches that make specific modelling assumptions, like the number of focal sources needed to produce the recorded data and their approximate location, and 2) approaches that make as little modelling assumptions as possible, and focus on rough features of the current distribution that are determined by the data rather than specific modelling assumptions (like laterality). Under 1) fall the 'dipole models': The number of active dipoles, i.e. small pieces of activated cortex, is assumed to be known (usually only a few). An initial guess about their location and orientation is made, and these parameters are then adjusted step by step until the predicted electric potential or magnetic field resembles the measured one within certain limits. These methods have been successfully applied in studies of perceptual processes and early evoked components. The naturally continuous real source distribution is approximated by a large number (several hundreds or thousands) of sources equally distributed over the brain. The strengths of all these sources cannot be estimated independently of each other, but at least a "blurred" version of the real source distribution can often be obtained. Peaks in this estimated current distribution might correspond to centers of activity in the real current distribution. These methods are preferred if modeling assumptions for dipole models cannot be justified, like in cognitive tasks or in noisy data. One can use constraints to try and reduce the number of likely of dipoles, but there is currently no widely-accepted mathematical technique that identifies the source of ERPs with a small and well-justified margin of error (Luck, 2005a). Ideally, one would like to identify the precise neural sources that generate the ERPs (known as the "inverse solution"). Unfortunately, the inverse solution is impossible to compute with certainty, because any given scalp distribution could, in principle, be generated by any number of source configurations within the brain. However, researchers have developed powerful tools that provide good estimates of these neural sources, given some reasonable assumptions.

4. EVENT RELATED POTENTIALS

Event-related brain potentials (ERPs) are positive and negative voltage fluctuations (or components) in the ongoing EEG that is time-locked to the onset of a sensory, motor, or cognitive event. ERPs reflect brain activity that is specifically related to some stimulus or other event. This activity cannot be directly observed in the EEG. EEG is a composite of simultaneously occurring brain activity. It doesn't reflect just the activity associated with the event of interest. In other words, the "signal" (the brain response to some event) is swamped by the "noise" (the brain activity that is unrelated to that event). The solution to this problem is to present not just one instance of the event of interest, but many instances. The first sensory ERP record on humans was obtained by Pauline and Davis in 1935-36.

Event related potentials are signals embedded within the EEG. ERPs are neural responses associated with specific sensory, cognitive and motor events and it is possible to extract these responses from overall EEG by averaging. Epochs of brain activity, each one time-locked to the onset of an event, are then averaged together. The "random" activity washes out during averaging, whereas the brain activity of interest - namely, what is constant over presentations of the event of interest - stays in the signal. Through this signal-averaging procedure, it is possible to isolate the brain response that is specifically elicited in response to some event of interest. The results are represented graphically depicting the change in voltage as a function of time. The positive and negative peaks are labelled as P or N and their corresponding number. P1, P2, P3 refer to the first, second, third largest positive peaks. They could also be labeled as P300, N400 which refers to a positive peak at 300 ms or negative peak at 400 ms respectively. Polarity of peaks depends on the baseline electrical activity and position of the reference electrodes. Our interest in ERP data lies in the change in timing and amplitude of the peaks.

Behavioural experiments also provide a measure of mental chronometry of cognitive events through reaction time measurements. ERPs provide a continuous measure of processing between the stimulus and response and help to determine which stage of processing is affected by experimental manipulation. On the other hand, overt responses on behavioural measures reflect output of a large number of cognitive processes. ERPs are online measures of information processing. However, functional significance of ERPs is not as clear as that of behavioural measures as we do not know the biophysical events which underlie the production of ERPs. ERP as a method in cognitive science and cognitive neuroscience has poor spatial resolution and better temporal resolution as compared to brain imaging techniques like fMRI that are based on hemodynamic measures.

5. ERP COMPONENT

An ERP component has been defined as "Scalp-recorded neural activity that is generated in a given neuroanatomical module when a specific computational operation is performed" (Luck 2005a). Peaks are not necessarily the same as components; "peaks are not special". Peaks are comprised of summation of latent components that are not observable, how we analyze our ERP data will relate to the validity and accuracy of our observations. There may not be a simple mapping between an ERP component and a cognitive component. For example one cognitive component may reflect the activity of many neural populations or several cognitive components may affect one ERP component.

5.1. Major ERP Components

5.1.1. Visual Sensory Responses

C1component shows higher amplitudes at posterior sites generated in V1 that is folded in the calcarine fissure and it is not labelled as P or N as its polarity varies (Clark, Fan, and Hillyard, 1995). Onset of C1 is between 40-60 ms and it peaks at 80-100 ms. C1 is followed by P1 and P1 is largest at lateral occipital sites. Its onset is between 60-90ms and it reaches its peak between 100-130 ms. P1 latency varies depending on stimulus contrast. P1 is followed by N1 wave. There are several visual N1 components one arising from lateral occipital sites during discrimination tasks whereas the other N1 components arises from the parietal cortex and is involved in spatial attention (Hillyard et al., 1998). P2 waveform follows N1 component at anterior and central sites. P2 is larger for stimuli with target features. Anterior

P2 effects occur when target is defined by simple stimulus features. Faces elicit a negative potential than non-faces at lateral occipital sites especially right hemisphere with a peak at 170 ms. N170 component is later and larger for inverted faces (Rossion, et al., 2002).

5.1.2. Auditory Sensory Responses

It is possible to observe a sequence of ERPs within the first 10 ms of the onset of an auditory stimulus. These peaks are called brain stem evoked response (BER) or auditory brain stem response (ABR). BER is followed by midlatency component between 10-50 ms and this is followed by auditory P1 wave at 50 ms being largest in the frontal central sites. Auditory N1 is a frontocentral component which peaks around 75ms. Mismatch Negativity (MMN) component is a measure of pre-attentive processing. MMN is observed when there is a mismatch between the target and the perceptual context by presenting an occasional target amidst a repetitive train of identical stimuli. MMN occurs when the participants is not using the stimulus stream for the task.

5.1.3. P3 Component

Squire and Hillyard (1975) first identified the major P3a component occurring at frontal sites and P3b component at parietal sites. Unexpected task irrelevant stimuli within an attended stimulus train elicited the frontal P3 component. Frontal P3 is elicited by a truly unexpected stimulus. Three factors influence the amplitude and latency of P3 component: a) target probability: larger amplitudes as the probability of the target gets smaller; b) uncertainty; c) resource allocation. P3 amplitude represents an interactive influence of probability, uncertainty and resource allocation. Since P3 depends on the probability of task defined category of stimuli and not the probability of physical stimuli, its amplitude is larger if the target has been preceded by more non-targets. P3 amplitude is larger when more effort is devoted to the task which is a measure of resource allocation. P3 is generated only once the stimulus has been categorized and is not sensitive to the time required to select and execute response once the stimulus has been categorized. Thus, P3 can be used to determine if an experimental manipulation influences the processes leading to stimulus categorization or processes related to response selection.

P3a and P3b are the two subcomponents of P300. The P3a component is a positive-going scalp-recorded brain potential that has a maximum amplitude over frontal/central electrode sites with a peak latency falling in the range of 250-280 ms. The P3a has been associated with *brain* activity related to the engagement of *attention* (especially *orienting* and involuntary shifts to changes in the environment) and the processing of novelty. P3a latencies often occur 75-100 ms earlier than P3b peak latencies, and around 250-280 ms. P3b is a positive-going ERP whose *latency* at peak *amplitude* is usually about 300ms. Truly improbable events will elicit a P3b, and the less probable the event, the larger the P3b. However, in order to elicit a P3b, the improbable event must be related to the task at hand in some way (for example, the improbable event could be an infrequent target letter in a stream of letters, to which a subject might respond with a button press). The P3b can also be used to measure how demanding a task is on *cognitive workload.* Odd ball paradigm and dual task paradigms have been used to elicit P3b component. The scalp distribution of P3b is generally larger over parietal areas (Polich, 2007).

5.1.4. Language Related ERP Components

N400 is the best studied language related component first reported by Kutas and Hillyard (1980). N400 is a negative going waveform found to have larger amplitudes over central and parietal sites. It occurs in response to semantic violations for example: "While visiting my old town I had lunch with my old *shirts*". Nonlinguistic stimuli can also elicit N400 if they are meaningful. N400 component has also been reported with studies on language comprehension looking at the effect of word order variations on comprehension (Bornkessel et al., 2004). P600 is a positive going waveform associated with syntactic violations such as the broker persuaded to sell the stock as compared to the broker hoped to sell the stock (former sentence having a syntactic violation.

Left Anterior Negativity (LAN): syntactic processing is reflected in other brain waves as well. Munte and colleagues (1993) described a negative wave over the left frontal areas of the brain. This brain wave has been labeled as LAN. It has been observed when words violate the required word category in a sentence for example in the sentence "the red eats" a noun instead of a verb lemma is required. It is also observed when morphosyntactic features are violated. The LAN has the same latency as N400 but a different voltage distribution over the scalp. Early Left Anterior Negativity (ELAN) component is another language related ERP component associated with sentence processing. It is characterized by a negative-going wave that peaks around 200 milliseconds after the onset of a stimulus, and most often occurs in response to linguistic stimuli that violate word-category or *phrase structure*. ELAN might also occur in response to nonlinguistic stimuli.

One issue which has been of concern for psycholinguists is whether semantic and syntactic processes are modular or interactive. The distinct electrophysiological signatures help to address such concerns and to understand the language structure of the human language system.

5.1.5 Error Detection Related ERP Components

Error related negativity (ERN): The process of error monitoring involves error detection followed by subsequent adjustment to performance. It is one of the executive control processes that provide top-down adjustment of elementary mental operations. These waveforms are elicited by comparing error trials with waveforms on correct trials. Through this component we could learn about the brain's response following the detection of error. ERN is a negative going waveform (Falkenstein et al., 1990) followed by a positive deflection called *Pe* which reflects the activity of a system which monitors one's response and is sensitive to conflict between intended and actual response.

5.1.6. Design and Interpretation of ERPs

Earlier ERP was defined on the basis of polarity, latency, and general scalp distribution. These are superficial features as they represent the underlying latent cognitive component processes. Thus, any physical change in the amplitude of the waveform when it reaches its maximum may not reflect the latent component process. Luck defined ERP as "scalp recorded activity that is generated in a given neuroanatomical module when a specific computational operation is performed" (Luck, 2005a). A component may occur at different times under different conditions as long as it arises from the same module and represents some cognitive process. Scalp distribution and polarity may also be different for the same cognitive process depending on the different cortical modules involved. ERP waveforms include a series of peaks representing a sum of several independent latent components. However, voltage peaks are not special as our purpose is to know how an experimental manipulation influences a latent component. Since we do not have direct access we make inferences about them from observed ERP waveforms. Luck states five rules while we draw a relationship between an ERP waveform and latent cognitive processes. 1) Peaks and components are not the same. 2) It is impossible to estimate the time course latency of a latent component by looking at a single ERP waveform as there is a difference in the amplitude of summed up components and a single waveform. 3) We should not compare an experimental effect with raw ERP. 4) Differences in peak amplitude do not necessarily correspond with differences in component size. 5) We should never assume that an averaged ERP accurately represents individual waveforms which were averaged together (Luck, 2005a).

5.1.7. Issues of Concern while Interpreting the ERP Components

It is important that the initial hypothesis places constraints on the ERP component of interest rather than predicting non-specific ERP changes. This is because large dataset gets generated in an ERP experiment and one may have significant results which may be difficult to justify based on a theory. Various factors affect the relationship between the ERPs and latent cognitive components. We could use certain strategies to avoid these ambiguities: 1) one should focus on a specific ERP component. 2) Well studied experimental manipulations should be used to derive well characterized ERP components. 3) It is advisable to focus on large ERP components like P300 as they dominate the waveform. 4) Difference waveforms should be used to isolate components. Difference waveforms also help to isolate the experimental effects as they enhance the difference in amplitude and timing very clearly as compared to showing two separate waveforms. For example, as depicted in figure 1 the difference waveform reflects the post remediation effect very clearly once the pre remediation ERP was subtracted from the post remediation ERP on the temporal order judgement task. 5) Another strategy is to focus on components that are easily isolated, for example, the lateralized readiness potential which is associated with contra lateral movement and is distinguished by its contralateral scalp distribution. 6) Best thing to do is to use component independent experimental design where it does not matter which latent cognitive component is responsible for the observed changes in the ERP waveform (Luck, 2005b). Most difficult aspect of ERP experiments is related to the problem of assessing a latent component on the basis of observed ERP waveforms. Most fundamental principle of experimentation is to ensure that a given experimental effect has only one possible cause and that the experimental manipulation does not have any secondary effects. The Hillyard principle says that we should always compare ERPs elicited by the same physical stimuli while varying only the psychological conditions (Luck, 2005a).

6. EEG/ERP ANALYSIS

6.1. Artifact Rejection

Various artifacts including power line interference, muscle activity, blinks, eye-movements, skin potentials and so on affect the EEG signals. One could either reject the trials which contain the artefact or employ techniques to estimate and remove the artefacts. Rejection of trials with artefacts may result in better estimation but the number of artifact-free trials needs to be considered when choosing this alternative. If we are not having a sufficient number of such trials, we will have a poor estimate of ERPs. There are three primary sources of artefacts: eye movements, instrumentation artefacts, and overlap from adjacent trials (Handy, 2005). Overlap from adjacent trials typically occurs at fairly high stimulus presentation rates (more than one stimulus per second). The effect of overlap can be reduced by introducing a temporal jitter (varying the duration) between stimuli and then averaging the waveforms obtained with different inter-stimulus durations. Other methods include estimating the overlap or including *no-stimulus* trials (Woldorff, 1988; 1993; Talsma and Woldorff, 2005; Vogel, Luck, and Shapiro, 1998). High stimulus presentation rates are fairly unusual in creativity research and may not be a significant problem in creativity research.

We could use filters to remove the electrical noise present during the recording especially if the noise contains higher frequencies, as we are interested in lower frequencies. One could also use a Faraday cage to reduce noise (Luck, 2005a). Artefacts due to instrumentation consist of both high-frequency artefacts as well as low frequency artefacts. Loose electrode connections can also cause artefacts. Sometimes a pulse-wave artefact can be seen due to blood pulse waves affecting the electrode connections. A major source of artefact is power line interference which is picked up by loosely connected electrodes. The low frequency waveforms could alter the baseline of EEG waveforms and may result in amplifier saturation.

Another major source of noise in EEG data is eye movements. Eye movement produces electrical interference that causes a large change in the EEG waveform. Blinks (eye-lid movements) and eye movements produce substantially larger amplitude in the frontal electrodes than in the background EEG. One possible way to eliminate blinks is to use the Electrooculogram (EOG) signal obtained by placing electrodes near the eyes. Participants in experiments are typically instructed to fixate and minimize blinks. A threshold can be used to eliminate epochs which contain larger than threshold differences between the minimum and maximum peaks in the EOG signal. A typical eye blink response consists of a peak in the order of 50-100 μV and lasts somewhere between 200 and 400 milliseconds. A number of correction techniques have been proposed for blink artefacts including regression, dipole source modelling and independent component analysis (Talsma and Woldorff, 2005).

6.2. Epochs

We should weigh the advantages and disadvantages of the role epochs play in the recordings of event related potentials. So far there has been no standard on the number of trials, jitter, averaging amplitudes, or the appropriateness of single trial analysis. For instance, the signal-to-noise ratio (SNR) increases with the number of trials, that is, the number of epochs; however, habituation can affect the results of some studies. We propose that a document outlining these categories would benefit future studies in terms of comparison and regularization. The neuroelectric signals are buried in spontaneous EEGs with signal-to-noise ratios as low as 5 dB. In order to decrease the noise level and find a template Evoked Potential (EP) signal, an ensemble-average (EA) is obtained using a large number of repetitive measurements (Bunge et al., 2001). This approach treats the background EEG as additive noise and the EP as an uncorrelated signal (Donchin, and Coles, 1998). The magnitudes and latencies of EP waveforms display large inter-individual differences and changes depending on the psychophysiological factors for a given individual (Knight, 1997). Thus, one goal in the methodological EP research is to develop techniques to extract the true EP waveform from a single sweep.

6.3. Averaging

Since ERPs are embedded within the EEG signal we depend on some averaging procedure to minimize the EEG noise. First of all, the EEG epochs following a certain stimulus event are extracted from the ongoing EEG. These epochs are averaged in a point by point manner. The EEG data collected on a single trial may consist of an ERP waveform with random noise. As more and more trials are averaged the noise in the averaged waveform gets smaller. The remaining noise in an average decreases as a function of the square root of the number of trials (Luck, 2005a). Since the signal is assumed to remain unaffected by the averaging process, the signal to noise ratio increases as a function of the square root of the number of trials. The signal averaging approach is based on the assumption that the neural activity related to the event is the same on every trial and it's only the EEG noise that varies from trial to trial. Trial to trial variability in amplitude is not a problem but trial to trial variability in latency is sometimes a problem. One could use certain techniques like response-locked averages or time locked spectral analysis to reduce the effects of latency variability. Another concern while we average across trials is the overlapping ERPs from previous and subsequent stimuli that may distort the averaged ERP waveforms in ways that are less obvious (Woldorff, 1988). One could use techniques like introducing no stimulus trials, using the broadest range of time delay (about 1000 ms) between stimuli.

6.4. Source Localization

The purpose of source localization is to identify neural sources/dipoles that may underlie the EEG signals obtained by recording from the scalp (Luck, 2005a; Slotnick, 2005). Usually the number of dipoles is not known. One way is to assume a certain number of dipoles, obtain the signals for the given set of dipoles, compare the EEG signals predicted by the model with the observed EEG signals and then make changes to the dipole model until we find a match between the predictions of the model and the actual observations. The most commonly used technique for localizing dipoles is the brain electrical source analysis (BESA). Low resolution brain electromagnetic tomography (LORETA) is another approach for source localization which models the electrical activity as a resultant of a number of voxels (Pascual-Marqui, Esslen, Kochi, and Lehmann, 2002).

7. EEG/ERP Studies on Brain and Cognition

EEG/ERP has been employed to investigate the time course of cognitive processes such as perceptual processing (including objects, faces or words), attention, language processing, executive control, and so on. For instance, ERP studies have investigated the way faces are processed. An initial stage consists of perceptual coding of the facial image followed by a stage in which the facial identity is computed. There is evidence for an ERP component that is relatively selective to the processing of faces compared with other visual objects. This has been termed as N170 and is strongest over right posterior temporal sites. This component is not influenced by the fact if face is famous or not (Bentin and Deouell, 2000) and is found for smiley faces (Sagiv and Bentin, 2001). It is reduced if the face is perceptually degraded. ERP also enables electrophysiological changes associated with unattended stimuli to be measured whereas a reaction time measure always requires an overt behavioural response.

Neural mechanisms of visual selective attention have been investigated using EEG/ERP. ERPs directly measure the activity generated by visually responsive neurons in the brain to indicate stages of processing that can be affected by selective attention. Studies have reported that modulations in the amplitudes of visual ERPs can begin at 70 ms after the onset of a visual stimulus. Directing voluntary covert attention towards a stimulus leads to modulations in the amplitudes of ERPs between 70-90 ms from the onset of the stimulus. These early P1 attention effects have been obtained during spatial attention. Attention effects for more complex tasks and objects appear later (>120ms). Luck and Hillyard (1994) reported the N2pc (second large negative posterior contralateral component) component for shift of spatial attention based on their studies on visual search. It is an index of the process of spatial filtering.

Electrophysiological studies have also investigated changes in brain activity in response to auditory stimuli. Electrophysiological data can provide temporal information about the on-line brain function by computing event related potentials during a phonological/ rhyming/orthographic task. An electrophysiological measurement could inform about the temporally mediated changes in brain activity with respect to phonological and temporal processing involved in reading. In this context, the preliminary findings of one of our studies are discussed here, in which we explored the cognitive and electrophysiological mechanisms of remediation in dyslexia particularly looking at the plasticity effects related to auditory processing and phonological processing. The study aimed to determine whether remediation in children with dyslexia could alter the disrupted neural response to phonological demands. The study also aimed to provide evidence for neural plasticity effects generalized from the specific strategies learned from the training program to early and late stage processing shown by EEG and ERP. Amplitude and latency related changes with respect to N100 and P300 were expected while comparing the ERPs before and after remediation among dyslexic children. Participants included eight normally progressing readers and two dyslexics with remediation and two dyslexics with no remediation. All the participants were attending regular instruction in the same school. The remediation program was carried out for six weeks with two dyslexic children. All the participants were assessed on a The cognitive assessment system, a battery consisting of 9 cognitive tasks (2 planning tasks, 2 attention tasks, 2 simultaneous processing tasks, and 3 successive processing tasks taken from the standardized version of the Das-Naglieri Cognitive Assessment System (Naglieri and Das, 1997) and 2 phonological processing tasks. Two of the four dyslexic children were offered the remediation program called PASS reading enhancement program (Das, Mishra, and Pool, 1995). The PREP group received the following eight tasks in the order listed: Joining Shapes, Shapes and Objects, Shape Design, Sentence verification, Connecting letters, Window Sequencing and Related memory.

EEG/ERP recording was conducted before starting the remediation program for the remediation group. EEG recording was also conducted for the no remediation group in the beginning of the study and same was followed for the normally progressing readers. EEG recording was repeated with same experiments as the first session for the remediation group after the completion of the remediation program after a period of 6 weeks from the time of first recording to study the effects on remediation on neural and cognitive mechanisms. EEG recording was also repeated with same experiments for the no remediation group and normal participants after a period of 6 weeks from the time of first recording.

The ERP experiment consisted of an auditory temporal judgment paradigm in which two pure tones (300 Hz and 1200 Hz; duration: 75ms each) appeared in rapid succession separated by three different ISIs in separate blocks (150ms, 250ms and 350ms). Similarly a set of two phonemes (Ka-Ga) was also presented across three ISIs (350ms, 450ms and 550ms). Tones and phonemes were presented at 90 dB level. EEG data was recorded with 64 channel (Neuroscan) EEG system digitized at 256 Hz sampling rate using Scan 4.3 program. An epoch of 600ms (100ms pre-stim to 500ms post stim) was taken to analyze the EEG data recorded during the temporal order judgement (TOJ) task. The FCZ electrode as one of the central midline electrode was used to compare the ERPs across the three groups of participants. Figures 1 and 2 present the pre and post remediation comparisons for the temporal order judgment task with two pure tones with respect to the ERP related changes observed in case of the remediation group as compared to the no remediation group. ERP components observed included N100 and P300 for two kinds of acoustic stimuli tones and phonemes. Figure 1 and 2 present the reduction in the amplitudes of N100 component indicating an improvement with respect to early stage processing (in this case auditory processing). We also observed an increase in the amplitudes of P300 (for both tones and phonemes as stimuli) on post remediation recordings of dyslexics indicating an improvement in higher cognitive processes underlying phonological processing particularly auditory attention.

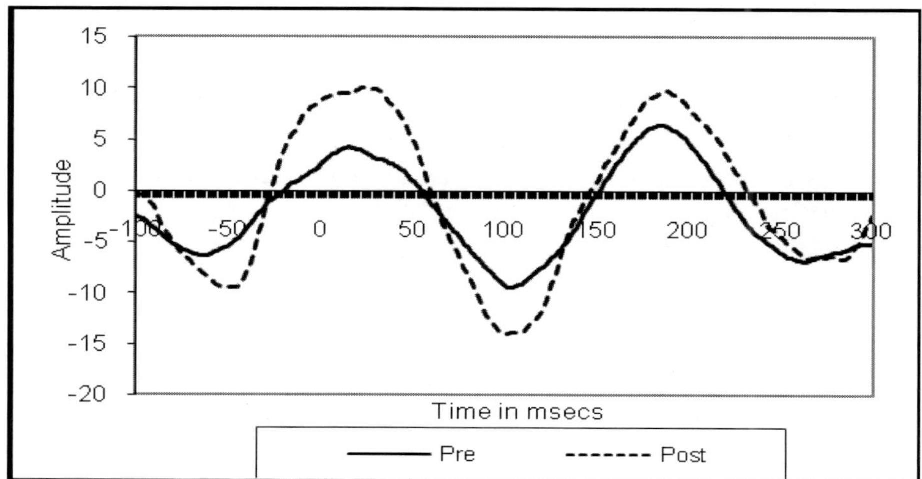

Figure 1. Pre and post ERPs for the temporal order judgment task with tones among dyslexics in the no remediation group.

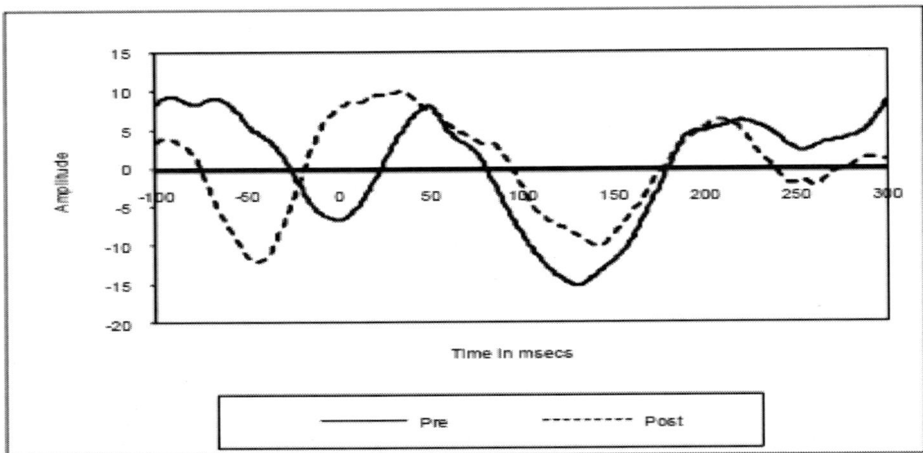

Figure 2. Pre and post remediation ERPs for the temporal order judgment task with tones among dyslexics in the remediation group.

Studies have reported differences between dyslexics and non-dyslexics particularly in early stages of auditory processing. N100 amplitudes have been found to be larger in dyslexics (Taylor et al., 2003). Thus a reduction in N100 amplitude among dyslexics after remediation is suggestive of an electrophysiological evidence of plasticity. Moreover, the early components that reflect early stages of processing have been reported to be more sensitive to differentiate dyslexics from non-dyslexics (Khan et al., 1998). At the level of sound discrimination the deficit is either in terms of processing of auditory stimuli or increased cortical recruitment to process the stimuli correctly. We observed reduction in the N1 amplitude in dyslexics after remediation as compared to dyslexics with no remediation and non-dyslexics which further confirms the early stage processing deficits in dyslexia.

The preliminary results have shown some interesting trends with respect to auditory temporal processing - and also the ERP related changes as an effect of remediation which should be further examined with a larger data set. Since there are indications for positive effects of remediation on auditory temporal processing as well as phonological awareness in children with dyslexia who were taken for remediation, it appears that auditory temporal processing could be an underlying deficit in dyslexia. It is to be noted that the remediation program was based on phonological awareness tasks and not auditory attention enhancement. We further need to look at the orthography related differences in the bi-literate child's response to remediation based on training in phonological awareness and visuospatial ability and also the differences in electrophysiological effects of remediation with a larger data set. These findings suggest that early stages of sensory processing may require lesser recruitment of neuronal resources whereas the late stage processing involving decisions about temporal sequence of sounds or discrimination may require auditory attention and hence shows an increase in amplitudes after remediation. These indications based on this preliminary study need to be tested with a larger data set and a remediation program which directly works on auditory processing and auditory attention.

EEG/ERP studies have investigated plasticity related issues in studies on meditation. Meditators show improvement in automatic change detection as indexed by larger MMN amplitudes compared to age-matched controls. Meditators also showed larger amplitudes immediately after meditation indicating state changes in the sensitivity of the perceptual system due to meditation. In addition, EEG studies have investigated spectral properties of EEG during meditation. Increase in theta activity has been reported especially in the frontal areas and corresponding decrease in activity in the parietal areas especially in the middle period of doing meditation (Baijal and Srinivasan, 2010).

Thus, EEG/ERP has been applied to the study of various cognitive processes such as perception, attention, language, memory, and certain control processes as well as to investigate the neurocognitive mechanisms that underlie abnormal cognition, meditation and creativity. In the recent years the simultaneous use of functional magnetic resonance imaging (fMRI) and EEG/ERP has generated interest among cognitive neuroscientists in order to ask the where, when and how questions about cognitive processes more effectively.

8. SUMMARY

EEG measures electrical changes in groups of neurons. It measures the difference between the electrical potential of two electrode sites. These changes are measured in terms of changes in frequency, amplitude, and specific wave-types. ERPs are electrical potentials associated with specific sensory, perceptual, cognitive, or motor events. ERP is a Time-locked average of EEG from many trials involving same 'event'. EEG could provide a measure of electrical changes ranging between 20-50µv and ERP between 1-10 µv. Studies on face processing, feature extraction, attention, language and memory on one hand and those on meditation, consciousness, awareness, neural plasticity on the other have informed about the time course of cognitive and neural events underlying these processes.

REFERENCES

Bentin, S., and Deouell, S. Y. (2000). Structural encoding and identification in face processing. ERP evidence for separate mechanisms. *Cognitive Neuropsychology, 17*, 35-54.

Bai, O., Nakamura, M., Nagamine, T., and Shibasaki, H. (2001). Parametric modeling of somatosensory evoked potentials using discrete cosine transform. *IEEE Transactions on Biomedical Engineering, 48*, 1347–1351.

Baijal, S., and Srinivasan, N. (2010). Theta activity and meditative states: Spectral changes during Sahaj Samadhi meditation. *Cognitive Processing, 11*, 31-38.

Bornkessel, I., Fiebach, C. J., Friederici, A. D. (2004). On the cost of syntactic ambiguity in human language comprehension: an individual differences approach. *Cognitive Brain Research, 21*, 11 – 21.

Bunge, S.A., Ochsner, K.N.. Desmond, J.E Glover, G.H. Gabrieli, J.D. (2001). *Brain, 124*, 2074-2086.

Clark, V. P., Fan, S., and Hillyard, S. A. (1995). Identification of early visually evoked potential generators by retinotopic and topographic analysis. *Human Brain Mapping, 2*, 170-187.

Das, J. P., Mishra, R. K., and Pool, J. E. (1995). An experiment on cognitive remediation or word-reading difficulty. *Journal of Learning Disabilities, 28*, 66-79.

Davila, C. E., and Srebro, R. (2000). Subspace averaging of steady-state visual evoked potentials. *IEEE Transactions on Biomedical Engineering, 47*, 720–728.

Donchin, E., and Coles, M.G.H. (1998). Is the P300 component a manifestation of context updating? *Behavioural and Brain Sciences, 21*, 152–154.

Falkenstein, M., Hohnsbein, J., Joormann, J., and Blanke, L. (1990). Effects of errors in choice reaction tasks on the ERP under focused and divided attention. In C. H. M. Brunia, A. W. K. Gaillard, and A. Kok. (Eds.), *Psychophysiological Brain Research* (192-195). Amsterdam: Elsevier.

Hagoort, P. (2003). "How the brain solves the binding problem for language: a neurocomputational model of syntactic processing". *NeuroImage, 20*, S21.

Handy, T. C. (2005). *Event-related potentials: A Methods Handbook*. MIT Press.

Hauk, O. (2003-2008). Introduction to EEG and MEG. Basics of EEG and MEG: Physiology and data analysis, http://www.mrc-cbu.cam.ac.uk/research/eeg/eeg_intro.html.

Hauk, O. (2004). Keep it simple: A case for using classical minimum norm estimates in the analysis of EEG and MEG data. *Neuroimage, 21*, 1612-1621.

Hillyard, S.A., Vogel, E. K., and Luck, S.J. (1998). Sensory gain control (amplification) as a mechanism of selective attention: electrophysiological and neuroimaging evidence. *Philosiphical transactions of the Royal Society: Biological sciences, 353*, 1257-1270.

Khan, S.C., Frisk, V. and Taylor, M.J. (1999). Neurophysiological measures of reading difficulty in very low birth weight children. *Psychophysiology, 36*, 1-10.

Knight, R.T. (1997). Distributed cortical network for visual attention. *Journal of Cognitive Neuroscience, 9*, 75–91.

Kutas, M. and Hillyard, S.A. (1980). Reading senseless sentences: Brain potentials reflect semantic incongruity. *Science, 207*, 203-205.

Luck, S. J. and Hillyard, S. J. (1994). Spatial filtering during visual search: Evidence from human electrophysiology. *Journal of Experimental Psychology: Human Perception and Performance, 20*, 1000-1014.

Luck, S.J. (2005a). An introduction to the event-related potential technique. MA: MIT Press.

Luck, S. J. (2005b). Ten simple rules for designing ERP experiments. In T. C. Handy (Ed.), Event related potentials: A methods handbook, Cambridge, MA: MIT Press.

Nagleri, J. A. and Das, J. P. (1997). Cognitive Assessment System. Illinois: Riverside publishing.

Niedermeyer E and Lopes da Silva FH (Eds) (2005) Electroencephalography. Basic Principals, Clinical Applications, and Related Fields. Fifth Edition. London: Williams and Wilkins.

Nunez P. L. (1981). *Electric fields of the brain: The Neurophysics of EEG*. London: Oxford University Press.

Nunez P. L. (1989) Generation of human EEG by a combination of long and short range neocortical interactions. *Brain Topography, 1*, 199-215.

Otten, L. J. and Rugg, M. D. (2005). Interpreting event related brain potentials. In T. C. Handy (Ed), *Event- related potentials: A methods handbook*. Cambridge, MA: MIT Press.

Pascual-Marqui, R.D. Esslen, M. Kochi, K. Lehmann, D. (2002). Methodological Findings in Experimental and Clinical. *Pharmacology. 24*, 91-95.

Picton, T. W. (1988). *Handbook of Electroencephalography and Clinical Neurophysiology: Human Event-Related Potentials.* Amsterdam: Elsevier.

Polich, J. (2007). Updating P300: An integrative theory of P3a and P3b. *Clinical Neurophysiology, 118,* 2128-2148.

Pulvermüller, F., Yury S. Anna S. H. and Robert P. C. (2008). "Syntax as a reflex: Neurophysiological evidence for the early automaticity of syntactic processing". *Brain and Language 104,* 245.

Rossion, B., Curran, T., and Gauthier, I. (2002). A defense of the subordinate level expertise account for the N170 component. *Cognition, 85,* 189-196.

Sagiv, N. and Bentin, S. (2001). Structural encoding of human and schematic faces: Holistic and part based processes. *Journal of Cognitive Neuroscience, 13,* 1-15.

Slotnick, S.D. (2005). In, Handy, T.C. (Ed.) *Event-related potentials: A methods handbook,* Cambridge, MA: MIT Press, pp. 149-166.

Talsma, D., and Woldorff, M. (2005). In Handy, T. (Ed.). *Event related potentials: A methods handbook,* MIT press, pp. 115-148.

Taylor, M. J., Batty, M., Chaix, Y., and Demonet, J. (2003). Neurophysiological measures and developmental dyslexia: Auditory segregation analysis. *Current Psychology Letters, 10.*

Vogel, E.K., Luck, S.J., and Shapiro, S.L. (1998). *Journal of Experimental Psychology Human Perception and Performance, 24,* 1656-1674.

Ward, J. (2006). *The Student's Guide to Cognitive Neuroscience.* New York: Psychology Press.

Woldorff, M. (1988). Adjacent response overlap during the ERP averaging process and a technique for its estimation and removal. *Psychophysiology. 25,* 490.

Woldorff, M. (1993). Distortion of ERP averages due to overlap from temporally adjacent ERPs: Analysis and correction. *Psychophysiology. 30,* 98-119.

Section 2. Linking Cognitive, Neural, and Cultural Processes

In: Expanding Horizions of the Mind Science(s)   ISBN: 978-1-62808-705-5
Editors: P.N. Tandon, R.C. Tripathi and N. Srinivasan   ©2013 Nova Science Publishers, Inc.

Chapter 5

BIOLOGICAL PROPERTIES OF TRANSDUCTION MECHANISMS IN SENSORY SYSTEMS

Avi Chaudhuri

Department of Psychology, McGill University, Montral, Canada

ABSTRACT

The fundamental basis of all behaviour lies with the acquisition of sensory signals by an organism. Those signals are then processed by the nervous system through a cascade of neurological hierarchies to produce an internal representation of the external world. This representation is referred to as *sensory perception*, a phenomenon that in turn is responsible for mediating higher cognitive aspects of our behaviour, such as the storage and recall of information (memory), goal-directed actions (motivation and executive functions), higher aspects of our internal state (emotion), and ultimately the highest aspect of our sentient being (consciousness). Here, I focus on the very first stage of this process — the acquisition and conversion of external physical stimuli into biological signals. In each of the sections of this chapter, I review the fundamental mechanisms of transduction — the process of converting physical into biological signals — for each of the five primary sensory systems.

1. INTRODUCTION

The fundamental basis of all behaviour lies with the acquisition of sensory signals by an organism. Those signals are then processed by the nervous system through a cascade of neurological hierarchies to produce an internal representation of the external world. This representation is referred to as sensory perception, a phenomenon that in turn is responsible for mediating higher cognitive aspects of our behaviour, such as the storage and recall of information (memory), goal-directed actions (motivation and executive functions), higher aspects of our internal state (emotion), and ultimately the highest aspect of our sentient being (consciousness).

In this chapter, I focus only on the very first stage of this process — the acquisition and conversion of external physical stimuli into biological signals. Although the content will lie largely within the domain of sensory neuroscience, it is important for psychologists to understand these processes because of the impact they have on perceptual function and behaviour. The ensuing chapters of this book will be devoted to many of the higher aspects of behaviour, which as noted above, must arise from the very first sequence of events at the interface between physics and biology. In each of the sections below, I review the fundamental mechanisms of transduction — the process of converting physical into biological signals — for each of the five primary sensory systems. The contents of this chapter have been excerpted from a book titled Fundamentals of Sensory Perception, written by the author and to be published by Oxford University Press in 2010.

2. SOMATOSENSORY TRANSDUCTION

The perception of touch occurs when we physically interact with an object in the environment. That object may emit energy, which we then capture through sensory receptors. Alternatively, we may have a situation where energy is reflected off the object and again captured by our receptors. Either way, we must possess a mechanism to capture that energy, whether it is in the form of mechanical pressure, vibration, heat, etc., and transform it into biological signals that can be processed by the central nervous system. Historically, two main ideas have been proposed to account for the way in which touch signals are generated and transmitted to the brain (Kruger, 1996; Morgan, 2009).

The theory of receptor specificity is borrowed directly from Müller's doctrine of specific nerve energies whereby individual receptors and nerve endings are selectively sensitive to a particular form of energy impinging on the skin. Each type of receptor not only produces a particular touch sensation but also indicates where on the skin surface the stimulation occurred. The influential German physiologist Max von Frey, who contributed much to the field of touch perception, took this concept one step further and proposed that there were specific receptors for heat, cold, pressure, and pain.

A somewhat different view was offered by the pattern theory in which the specificity of touch sensations was believed to arise not from individual receptors but rather through the overall pattern of activity across a broad spectrum of receptors. Different types of touch, according to this theory, generated different patterns of activity and somehow the brain is able to distinguish those patterns to produce the different touch perceptions.

We now know that touch signals are indeed highly specific and selectively generated by a particular type of receptor. These receptors, broadly classified as mechanoreceptors, are present in the skin throughout the body. Although receptor specificity is now accepted as the basis for touch perception, it should be clear that most forms of touch usually involve stimulation of multiple types of receptors. Therefore, although the individual receptors convey a unitary tactile sensation, the total perception of touch is quite complex, often involving multiple attributes of the stimulus and accordingly, a multitude of receptor types.

2.1. Types of Mechanoreceptors

The human skin contains a variety of mechanoreceptors that can be classified into three types, as shown in Figure 1. The so-called encapsulated receptors have a specialized capsule that surrounds a nerve ending (Kandel, Schwartz, and Jessell, 2000; Purves, 2007; Squire, Roberts, Spitzer, Zigmond, 2002). The Pacinian, Meissner, and Ruffini corpuscles are examples of mechanoreceptors that fall into this category. The capsule itself serves specific functions that affect the kinds of mechanical stimulation that are optimal for that particular receptor. For example, the onion-like structure of the Pacinian corpuscle is believed to act as a mechanical filter, which aids in the transmission of vibrational stimuli, whereas the Meissner and Ruffini capsules are designed for transmitting light touch and steady pressure sensations, respectively.

A second type of mechanoreceptor is composed of a sensory nerve fiber in conjunction with a separate accessory structure. The Merkel disc, for example, is a type of mechanoreceptor where the sensory nerve ending is associated with a type of epithelial cell (the so-called Merkel cell). This complex is responsible for the detection of light touch to the skin. Another example is the Lanceolate endings, which are found in hairy skin where they run along the hair shaft and are therefore triggered only by movement of the hair follicles.

The third type of mechanoreceptor is made up of various types of free nerve endings that do not have any specialized terminal structures or other associations. These nerve ending are responsible for detecting thermal changes — warmth and cold — as well as pain. The pain receptors, also called nociceptors, can be further subdivided on the types of pain that are detected, e.g. mechanical, thermal, or mixed forms of pain.

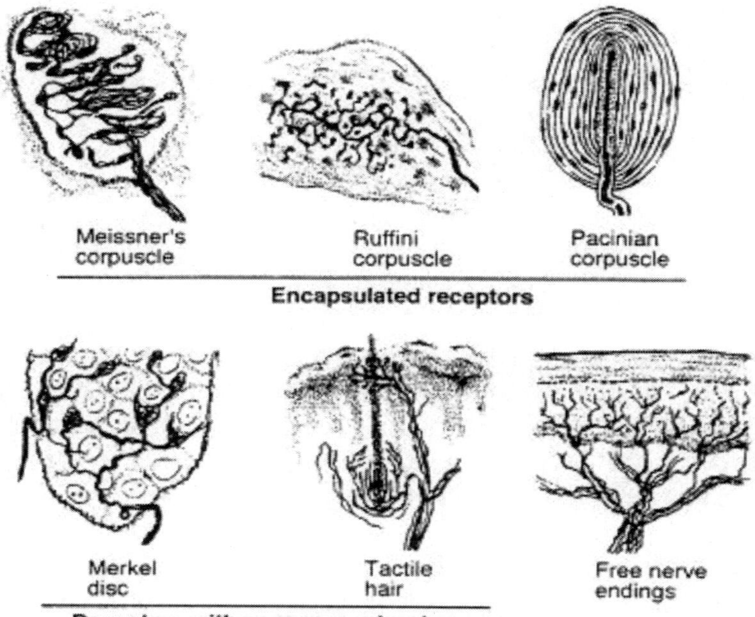

(Modified from Cajal's Histology, William Wood and Co).

Figure 1. Drawings of several types of mechanoreceptors found in the skin. The receptors can be broadly classified as being encapsulated, having an accessory structure, or as free nerve endings.

The distribution of these different types of mechanoreceptors can vary with skin type. In general, the receptors are found either in superficial skin, near the interface of the dermis and epidermis, or more deeply in the dermis and even below it. Superficial receptors found in hairy skin include Merkel discs, hair receptors, and free nerve endings. Both Pacinian and Ruffini corpuscles are found in the deep zone of hairy and hairless skin (Marieb and Hoehn, 2006).

2.2. Signal Transduction in Mechanoreceptors

Our knowledge of transduction mechanisms in mechanoreceptors is somewhat limited because of the way in which they are embedded in skin tissue. It is difficult to isolate an individual receptor and study the way in which it generates electrical signals in response to mechanical stimulation. What we do know is largely derived from studies of the Pacinian corpuscle. Its relatively large size and location deep within the skin has helped it to serve as a model mechanoreceptor for tactile transduction studies.

When a mechanical stimulus is applied to the skin, the pressure is ultimately transmitted to the receptor itself where the strain produces opening of Na^+ channels (Hudspeth, 1999; Mombaerts, 1999). The resulting movement of Na^+ into the mechanoreceptor will result in a depolarizing receptor potential. If this potential is large enough, it will cause action potentials to be generated. But exactly where are these action potentials produced, given that all we have within the skin is a nerve ending, possibly with an accessory capsule? To answer that, we have to examine a rather unusual characteristic of the neurons that give rise to mechanoreceptors.

2.3. Mechanoreceptors are Terminals of Modified Bipolar Neurons

The vast majority of neurons in the nervous system have a multipolar structure whereby a single axon and various dendrites emerge from the cell body. The number and pattern of arborization of the dendrites together are responsible for producing the diverse structural types of neurons in the brain. Mechanoreceptors, however, emerge from a separate class known as bipolar neurons. These have a single axon and only a single dendrite that generally protrude from opposite ends of the cell body. At some point during development, the two processes fuse and produce a single emergent fiber that quickly splits into two just outside the cell body (Figure 2). One of the fibers is called the peripheral branch and proceeds to the skin where it ends up either as free nerve endings or in some kind of encapsulated form, e.g., a Pacinian corpuscle, as shown in Figure 2. The other fiber is called the central branch and carries touch signals to the spinal cord (Catania and Henry, 2006; Squire, Roberts, Spitzer, and Zigmond, 2002)[5,9].

Where then is the cell body of the mechanoreceptor neuron located? It turns out that the cell body is actually situated quite far away from the skin in a structure called the dorsal root ganglion (DRG). The DRG itself is located right next to the spinal cord. As we have already noted, touch stimulation will produce a depolarizing receptor potential in the terminal end of the peripheral branch (i.e., the mechanoreceptor), the only site in the DRG neuron that is sensitive to stimulus energy.

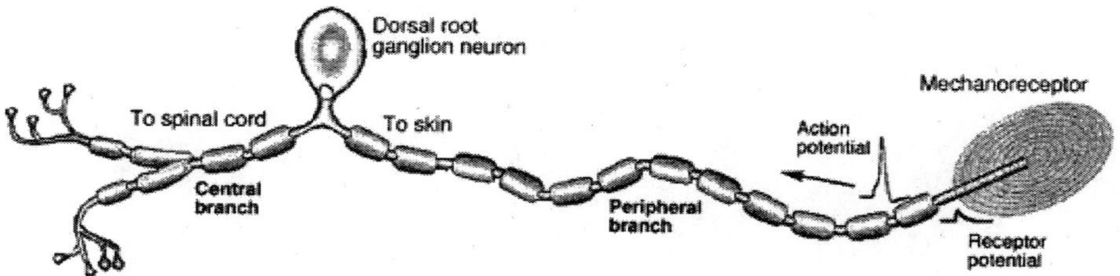

Figure 2. Mechanoreceptors belong to a specialized class known as bipolar neurons. The cell body, which is located in the dorsal root ganglion, gives off two processes.

What makes the DRG neuron unusual is that once the receptor potential reaches threshold, an action potential is generated in the immediate vicinity of the end organ itself[10]. That is, unlike conventional multipolar neurons where the depolarizing signal must first reach the cell body and can thereafter produce an action potential only at the axon hillock, DRG neurons are capable of producing action potentials right near the terminal end of the peripheral branch. These action potentials then flow along the peripheral and central branch to reach the spinal cord, from where they are transmitted to the brain through dedicated fibre pathways.

3. GUSTATORY TRANSDUCTION

The taste system in humans has traditionally been thought of as having four basic qualities — sweet, sour, salt, and bitter. More recently, a fifth category of taste stimuli, called *umami*, has been added to the list of primary taste sensations. The signals that produce these primary sensations arise from specialized sensory organs that reside in the oral cavity. It has been known since the middle of the 19th century that tiny structures in the tongue that resemble flower buds are responsible for producing taste sensations. In addition to the tongue, these so-called *taste buds* are also found in the roof and back of the mouth, though in much smaller numbers. We have only recently come to learn the remarkable way in which these taste organs function and how they produce the signals that ultimately generate the different types of taste qualities that we experience.

3.1. Taste Receptor Cells are Located Inside the Taste Bud

Although taste buds are considered to be the primary sensory organ of taste, it is the specialized receptor cells contained within it that are responsible for transducing the chemical stimuli into electrical signals. The dissolved chemicals, also known as *tastants*, infiltrate the taste bud through a narrow pore that opens into the grooves (see Figure 3). The receptor cells inside the taste bud are similar to the cells found in skin tissue, though with one important difference. The receptor cells actually behave as neurons in that sensory stimulation — in this case the binding of a tastant molecule — produces a change in the membrane potential that leads to neurotransmitter release (Genders, 1972; Gibbons, 1986).

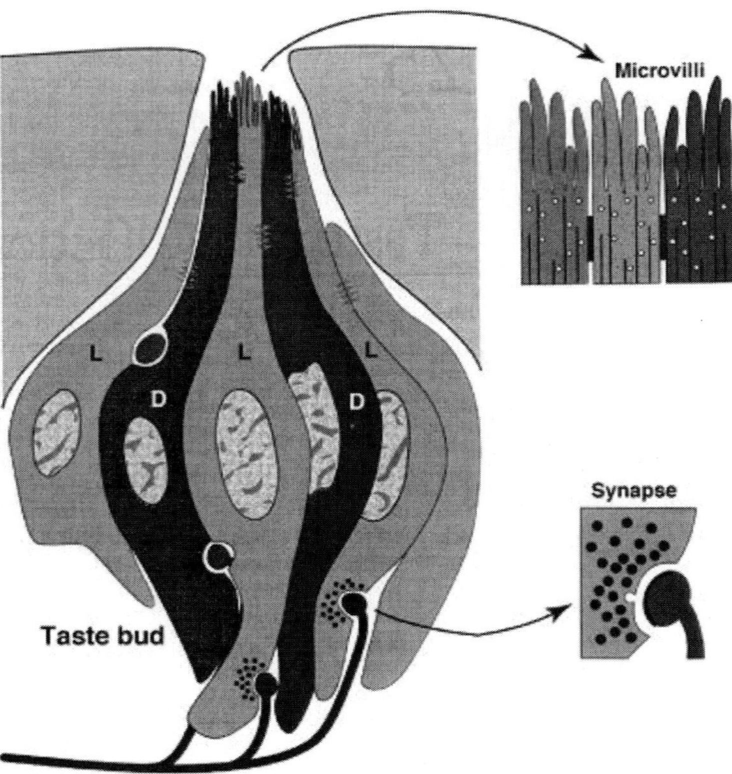

Figure 3. Among the types of receptor cells found in the taste bud complex are the light (L) and dark (D) cells. The receptor cells become tapered toward the pore through which tastants enter and then interact with the cells' microvilli, leading to neurotransmitter release at the other end of the receptor cell.

Each taste bud contains about 50 - 150 receptor cells. There are two general types of receptor cells that can be distinguished by their appearance under a microscope — the so-called *dark* and *light* cells (Squire, Roberts, Spitzer, and Zigmond, 2002). As Figure 3 shows, both types become tapered at the pore where the cells' membrane becomes invaginated into a series of folds or *microvilli*. Once tastants enter the taste bud, they interact with the receptor cell membrane at the microvilli. The ensuing transduction process unfolds through a series of molecular events that results in neurotransmitter release at the opposite end of the receptor cell.

3.2. Taste Receptor Cells are Innervated by Sensory Neurons

A synapse is present at the interface between the receptor cell and gustatory nerve fiber where the release of neurotransmitter substance leads to action potentials in the nerve fibers. These fibers then carry the action potentials along one of two cranial nerves, depending on the part of the tongue in question. Gustatory signals from taste buds in the front two-thirds of the tongue are transmitted through the facial nerve (also known as cranial nerve VII) whereas signals from the back one-third of the tongue are carried by fibers that belong to the glossopharyngeal nerve (also known as cranial nerve IX). Another cranial nerve, the vagus nerve (cranial nerve X), carries gustatory signals that arise from the few taste buds located at the back of the throat (Breslin, 2000; Miller, 1995).

The organizational layout of these three cranial nerves is similar to that of the spinal nerves. The major difference, however, is that these cranial nerves emerge from the brainstem rather than the spinal cord. Associated with each nerve is a ganglion, or collection of neurons. It is these sensory neurons that give rise to the gustatory fibers that innervate the tongue and throat. It should be noted that in addition to transmitting taste messages, the fibers in these nerves also carry non-taste signals such as touch and pain. Although these facets of sensation are often not discussed within the context of gustation, they are nevertheless very important in delivering the total experience of food ingestion. One example of this occurs when we consume spicy hot foods. The resulting burning sensation is caused by the active ingredient, *capsaicin*, which directly stimulates pain-sensitive neurons in the tongue and mouth.

3.3. Signal Transduction Mechanisms

We return to the picture at the receptor cell level to explore the events by which tastants produce a neural signal in these cells. This process had remained a mystery until recent advances in electrophysiological recording techniques coupled with the explosive developments in molecular technology. The picture that has emerged is that the primary taste qualities each have their own transduction mechanism. These mechanisms fall into two general categories — ionic channel and receptor-linked (Chandrashekar, Hoon, Ryba, and Zuker, 2006). Regardless of the specific mechanism involved, the end result of chemical stimulation upon a receptor cell is the same — membrane depolarization that results in neurotransmitter release upon the gustatory fiber.

3.4. Ionic Channel Mechanisms Code for Salt and Sour Tastes

The taste of salt is mediated by certain free ions, among the most potent of which is sodium (Na+). The transduction of sodium ions by taste receptor cells is fairly straightforward. All cells in the body have a lower concentration of Na+ inside the cell than outside, creating an electrochemical gradient that favors the entry of Na+ into the cell. This is precisely what happens in receptor cells when we ingest salt. The upper panel in Figure 4 shows the entry of Na+ into the taste receptor cell through a sodium channel in the membrane, an event that results in a brief membrane depolarization that travels to the opposite end of the cell where it causes neurotransmitter substance to be released into the synaptic space (Beauchamp and Bartoshuk, 1997; Finger, Silver, and Restrepo, 2000).

Sour taste is produced by acidic substances, which release free hydrogen ions or protons (H+) in solution. The H+ ions enter the receptor cells simply by proceeding through the sodium channels in the microvilli (Figure 4). As with Na+ ions, H+ also depolarizes the receptor cell once inside and therefore leads to neurotransmitter release. However, the sodium channel can only be used for this purpose if the Na+ concentration in saliva is quite low. Otherwise, the two ions compete for the channel and one ion can block the entry of the other. It is well known that acids from sour foods reduce the perceived taste of salt in humans.

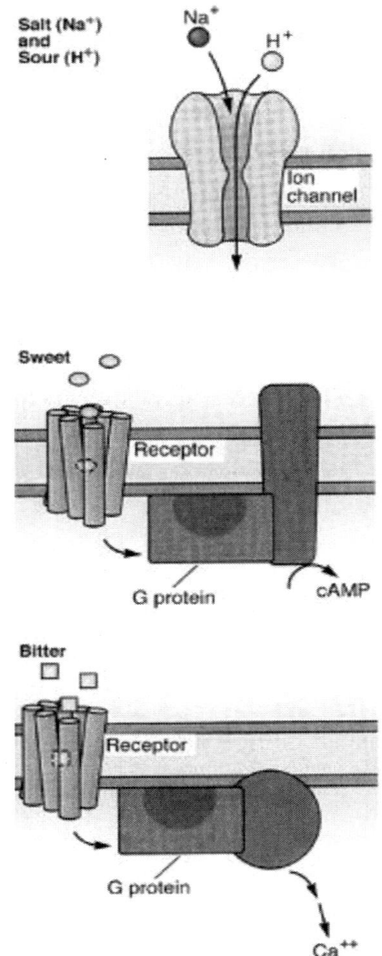

Figure 4. The primary taste sensations are transduced by either ionic channel (salt and sour) or receptor-mediated mechanisms (sweet and bitter). The movement of Na+ or H+ through a pore in the receptor cell membrane produces depolarization and neurotransmitter release. The same occurs when sweet or bitter tastants bind to specific receptors. A cascade of biochemical events is then triggered by way of a G protein within the receptor cell.

3.5. Receptor-Linked Mechanisms Code For Sweet and Bitter Tastes

The chemicals that produce sweet and bitter tastes are more complex than the simple ionic tastants. In fact, there are many different compounds that can give rise to both taste sensations. The sweet taste, for example, can be generated by simple sugars, such as glucose, fructose, and sucrose (ordinary table sugar), as well as more complex compounds, such as various amino acids, peptides, and artificial sweeteners (e.g., aspartame, saccharine). However, it is the way in which sweet and bitter chemicals generate taste signals that provides the major framework for our discussion here (Finger, Silver, and Restrepo, 2000; Squire et al., 2002).

Sweet tastants in general bind to specific receptors that are found in the membranes of taste receptor cells at the microvilli. These receptors, which are large protein molecules that actually span the cell membrane, are coupled to a so-called G protein (see bottom tow panels of Figure 4). The binding of a sweet tastant to the receptor triggers the G protein and this ultimately results in the production of cAMP within the receptor cell. The cAMP serves as a so-called second messenger, that is, it triggers a further set of biochemical events within the cell. In this case, it is believed that cAMP works directly upon the ionic channels to produce membrane depolarization, an event that produces neurotransmitter release and hence signal generation in gustatory fibers.

Bitter tastes can be produced by a variety of chemicals, especially a family of nitrogen containing compounds called alkaloids. These complex chemicals are usually found in plants. The most commonly cited example is quinine, an extremely bitter compound that is obtained from the bark of the South American cinchona tree and first used by Amerindians for the treatment of malaria. Some other plants that contain high levels of alkaloids are the periwinkle, California poppy, wolf berry, yellow pond lily, and tobacco.

As with sweet tastants, bitter tasting chemicals first bind to specific receptors, which then produces a series of G protein coupled intracellular events. In this case, the events appear to culminate in the release of Ca++ ions, which in turn triggers the release of neurotransmitters into the synaptic space. It is likely that different transduction mechanisms may be involved for different types of bitter substances. One compound that is particularly effective in unleashing this cascade is denatonium benzoate, the bitterest known substance.

3.6. Umami

The umami taste, which is produced by the flavor enhancer monosodium glutamate, is considered by many to represent a fifth primary taste (Doty, 2003; Finger, Silver, and Restrepo, 2000). Thus, there should be a specific receptor that mediates the transduction process in taste cells. A specific receptor for umami has indeed been identified — one that specifically binds glutamate molecules and, as with the receptors for sweet and bitter tastants, triggers a series of biochemical events in the taste cell. The identification of the umami receptor has given further credibility to establishing it as a primary taste similar to the other four. It would make sense that umami would have evolved to be a primary taste because glutamate is among the most abundant of amino acids and found at high levels in most food proteins.

4. OLFACTORY TRANSDUCTION

We next turn our attention to the second chemosensory system — smell perception or olfaction. It was generally believed throughout much of history that our sense of smell occurred via direct access of odors to the brain. This notion was not questioned until the 17th century, although the finer details only began to appear around the middle of the 1800s. It was shown then that human olfaction occurred through a structure — the so-called olfactory epithelium — located at the upper margins of the nasal cavity. It is here that airborne or volatile chemicals, known as *odorants*, trigger the neural signals that ultimately reach the brain to produce olfactory perception (Beauchamp and Bartoshuk, 1997; Squire et al., 2002; Tortora and Derrickson, 2008). However, to get to the epithelium, odorants must pass through the complex conduits of the nasal channels. There are two possible routes that can be taken. The orthonasal route proceeds from the nostrils, up the two nasal passages, around a set of bony protuberances or turbinates to reach the olfactory epithelium. The retronasal route allows food-based odorants to reach the epithelium through a separate channel at the very back of the throat. In the remainder of this section, we follow the events from this point onwards by discussing the neural processes that underlie olfactory perception.

4.1. The Olfactory Epithelium

The very first events in olfactory signal processing occur in the olfactory epithelium. This sense organ is composed of several layers of cells and located at the uppermost recesses of the nasal cavity, as shown in Figure 5. Odorants carried in air must first pass through a thin mucus barrier that covers the epithelium before reaching the sensory neurons. The mucus, which serves a protective function, is composed of water, electrolytes, and protein. Because of its high water content, the mucus layer provides a barrier for hydrophobic or fat-soluble odorants. It turns out, however, that a chemical must have a certain amount of fat solubility in order for it to be odorous. Thus, without some assistance, most odorants will not easily reach the epithelium. A special protein known as *odorant binding protein (OBP)*, which is found in fairly high concentration within the mucus, is believed to attach to many different kinds of hydrophobic odorants and help them across the mucous layer.

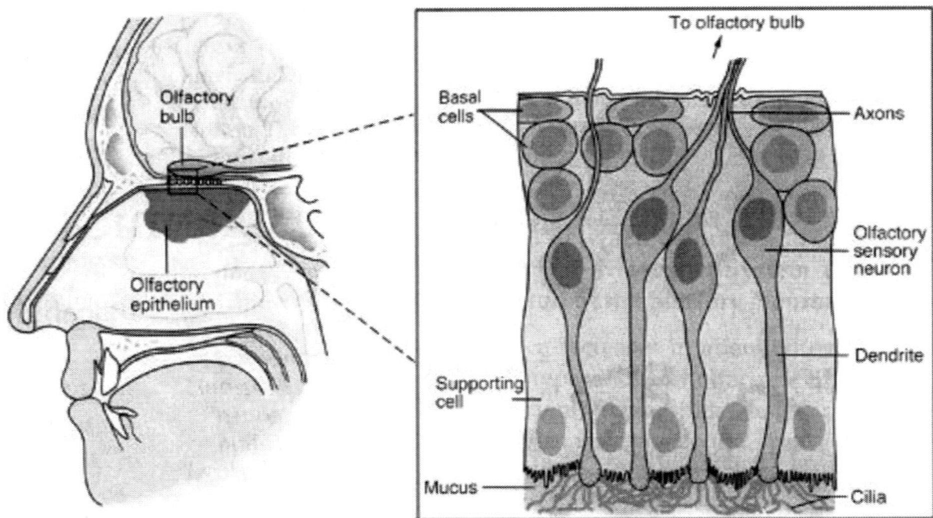

Figure 5. The olfactory epithelium is located at the uppermost margins of the nasal chamber and is composed of three main cell types — olfactory sensory neurons, supporting cells, and basal cells. The cilia belonging to the sensory neurons protrude into a thin layer of mucus substance that coats the lower margins of the epithelium. Axons belonging to the sensory neurons project to the olfactory bulb.

4.2. Cellular Composition of the Epithelium

Once the odorants pass through the mucus layer, they encounter the *olfactory sensory neuron*, one of three kinds of cells found in the epithelium. There are approximately 12 million sensory neurons within the roughly 5 sq cm area of the olfactory epithelium (Doty, 2003). These sensory cells are bipolar neurons, that is, they have one dendrite and one axon that emerge from opposite sides of the cell body. The axons from several neurons join together to form a small nerve bundle that projects to the *olfactory bulb*, which is the next structure in the neural hierarchy of olfactory processing. The dendrites of the sensory neurons proceed toward the lower margin of the epithelium. Projecting from the terminal ends of the dendrites are minute hair-like filaments or *cilia*. The cilia from all the dendrites together form an intricate matted arbor that invades the mucus layer. The interaction between odorants and olfactory sensory neurons occurs at the cilia.

One particularly interesting property of the olfactory sensory neuron is its rapid turnover, approximately every 30 - 60 days. That is, each sensory neuron has only a limited lifetime after which it is replaced (Doty, 2003; Giboons, 1986). Although cell death is not unusual in the nervous system, a somewhat unique feature of this process in the olfactory epithelium is that the dying neurons are rapidly replaced by new sensory neurons, unlike elsewhere in the nervous system. The other notable exception is that of the gustatory receptor cells where a similar, though more rapid, turnover occurs. However, there is one striking difference between the two chemosensory neurons in this respect. Gustatory receptor cells merely release neurotransmitter substance onto a nerve fiber, and therefore its replacement can be easily envisioned, whereas olfactory sensory cells have an axon that projects to another brain area. One of the major unresolved questions in this field concerns the mechanisms by which newly generated sensory neurons establish functional connections that maintain the integrity of prior olfactory circuits.

The second type of cell found in the olfactory epithelium is the *basal cell*. It is these cells that provide the source for the continuous replacement of the dying sensory neurons. The basal cells make up less than 5% of the total cellular population in the epithelium. The third type of cell is the *supporting cell*. It is believed that these cells are partly responsible for producing the mucus and regulating its ionic content. Additional functions of the supporting cell include degradation of odorant molecules and other substances that may penetrate the mucus barrier as well as removing the dead or dying neurons in the epithelium.

4.3. Sensory Transduction in the Olfactory Epithelium

The interaction between odorants and sensory neurons occurs at the cilia. It has been shown that the entire transductional machinery resides within these thin filaments. It is here that odor molecules become bound to specific receptors. There are approximately 1000 different types of receptors, each being sensitive to a broad set of odorants, though there appears to be some affinity for a specific chemical structure. Each olfactory sensory neuron expresses only one type of receptor and it therefore produces a signal only when that receptor is stimulated. The painstaking work of uncovering the nature of olfactory receptors and their associated genes was carried out by the American neuroscientists Richard Axel and Linda Buck, for which they received the Nobel Prize in 2004 (Buck, 2000; Buck and Axel, 1991).

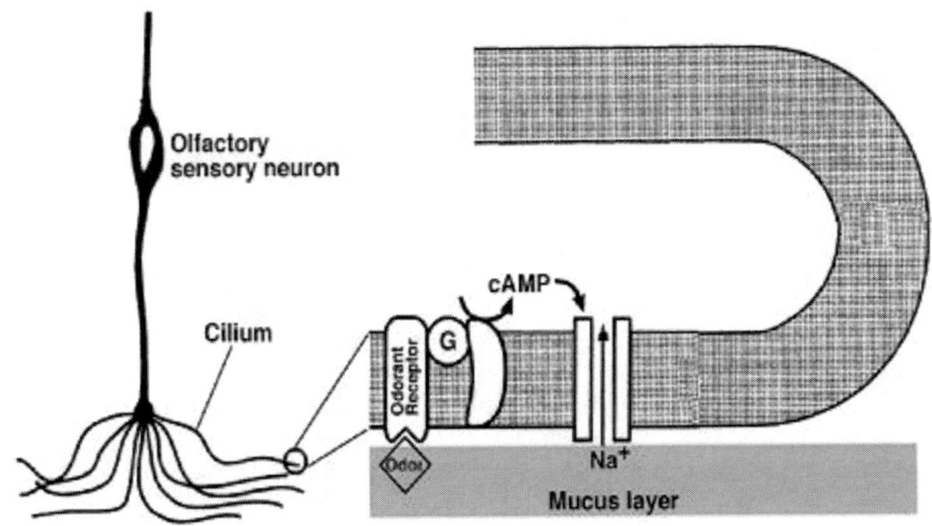

Figure 6. Odorant binding to a membrane receptor in a cilium triggers a G protein and leads to the production of cAMP. The cAMP then increases Na+ entry through an ionic channel that produces membrane depolarization and, if threshold is reached, action potentials in the olfactory sensory neuron.

The manner in which olfactory receptors function is similar to the process we saw earlier for sweet taste reception. The receptors in the olfactory sensory neuron are large protein molecules that span the cell membrane and are coupled to a so-called G protein (Laurent, 1999; Mombaerts, 1999). As Figure 6 shows, the binding of an odorant molecule to the receptor triggers the G protein and this results in the production of a second messenger. As with the sweet receptor system in gustation, the second messenger in olfactory sensory neurons is cAMP. It is believed that cAMP then works directly upon the ionic channels to increase the entry of Na+ ions into the neuron and produce membrane depolarization. An action potential can be produced if odorant binding should cause sufficient membrane depolarization to reach threshold. Once produced, action potentials are then transmitted via the *olfactory nerve* to the olfactory bulb and then on to higher areas of the brain that together are known as the olfactory cortex.

5. AUDITORY TRANSDUCTION

Any effort to understand auditory function must begin with a study of the nature of sound — how it is produced, what are its physical characteristics, how it travels, and how it interacts with other objects. For us to hear a sound, there are three essential requirements. First, there must be something that creates sound. Second, the sound must propagate through a medium from its source. And third, there must be a mechanism to translate sound energy into a biological signal that ultimately generates the perceptual experience we refer to as *hearing*. To understand the latter more thoroughly, we need to first come to grips with exactly what sound is.

It was not until the 16th century that humanity first began to understand sound. The first major advance originated with Galileo Galilei whose interest in sound was inspired by his father, who was a mathematician, musician, and composer. Galileo believed that sound can be considered as a *vibrational* event and that alterations in the frequency of those vibrations are correlated with differences in pitch perception. His idea that sound occurred due to waves in air was quite contentious because of the view held by many philosophers and scientists of that time that sound was created by invisible particles originating at its source and which interacted with the ear to produce hearing. The dispute was resolved by the famous *bell jar* experiment conducted by the British chemist Robert Boyle in 1660. The experiment, which even today remains a common demonstration in high school physics classes, showed that the sound produced by a bell becomes less audible if enclosed in a jar from which the air is slowly pumped out, eventually becoming silent when the jar is free of air. Although the actual reasons for the reduction in sound intensity are somewhat complicated, this experiment nevertheless convincingly demonstrated that air was required to serve as a medium for the transmission of sound.

Sound waves are systematically transmitted through a series of structures within the ear. Each of these structures must retain the frequency characteristics of the original sound stimulus and transmit it in a way that causes the least disruption in amplitude. Although we usually refer to the ear by the flap of cartilaginous tissue on the side of the head, the human ear actually encompasses many more structures that are contained in one of three well-defined structural segments — the outer, middle, and inner ear. It is within this last element where a special structure, known as the *cochlea*, resides and which is responsible for all of the functions related to auditory transduction.

5.1. Structural and Functional Elements of the Cochlea

The cochlea is a coiled fluid-filled bony structure that contains the sensory transduction apparatus of hearing. It is here that the physical energy contained in sound waves imparts an identical set of waves in the cochlear fluids. This in turn gives rise to the electrochemical signals that are transmitted to the brain. The spiral form of the cochlea begins at its basal end and makes almost two and one-half turns before ending at the apex. If the cochlea were to be uncoiled and stretched out, it would be about 35 mm in length. The diameter at the base would be about 9 mm and somewhat less at the apex due to its tapering nature.

The interior of the cochlea is divided into three separate fluid-filled channels, as shown in Figure 7. The largest of the three is called the *scala vestibuli*. In a cross-sectional view of the cochlea, it would appear as the uppermost of the three channels extending from the base to the apex. The lowermost of the three channels is the *scala tympani*. The basal end of this channel contains a very thin membrane that forms the *round window*.

The scalae vestibuli and tympani are actually connected to each other through a small aperture at the apex known as the *helicotrema*. Thus, both of these channels contain the same watery fluid. Sandwiched between these two channels is the *scala media*, often referred to as the *cochlear duct*. This is the smallest of the three channels and occupies only about 7% of the total volume of the cochlea. The *basilar membrane* separates the cochlear duct from the scala tympani, which lies below it (see Figure 7). Unlike the other two channels, the cochlear duct is a self-contained chamber that is filled with a different fluid. The two fluids do not come into contact with each other.

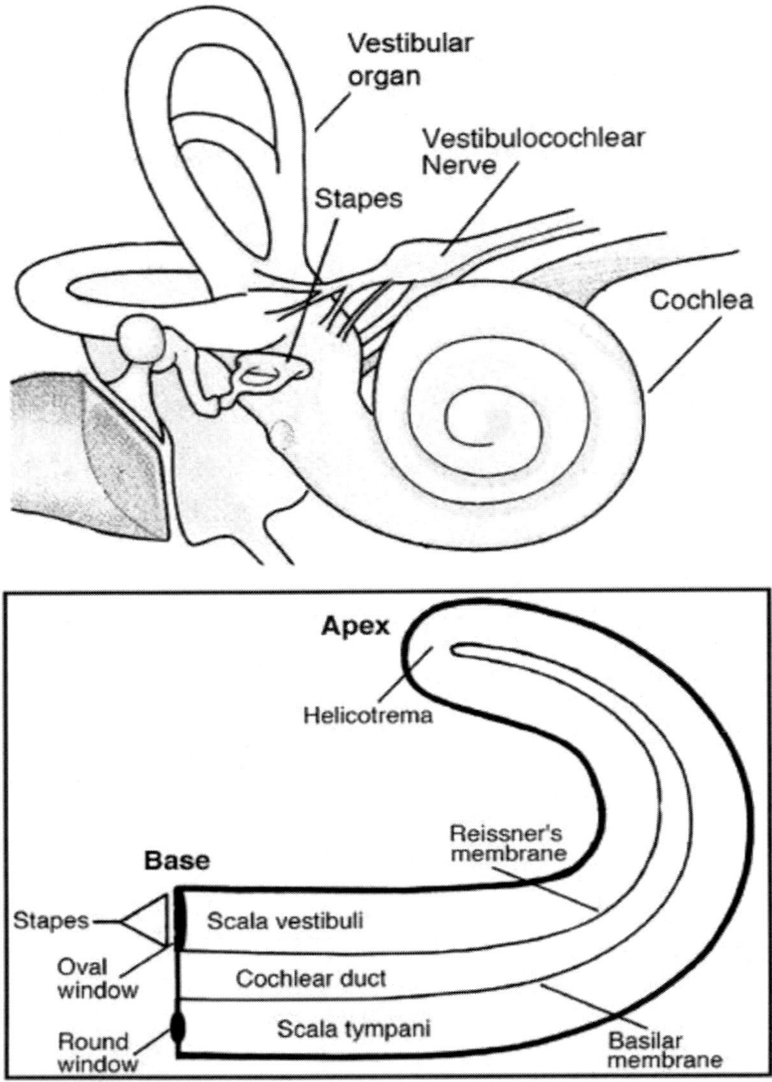

Figure 7. The cochlea is a coiled fluid-filled bony structure embedded within the temporal bone. The internal structure of the cochlea is shown in uncoiled form in the inset. The fluid in the scala vestibuli can mix with that in the scala tympani through an opening at the apex known as the helicotrema. A separate fluid is contained within the cochlear duct, which is a self-contained chamber separated from the scalae vestibuli and tympani.

The conversion of vibrational energy (sound) into neural signals occurs within a structure that lies on top of the basilar membrane. The *organ of Corti*, named after the Italian anatomist Alfonso Corti who first described it in 1851, is group of cells and associated structures that spans the entire length of the cochlea (Bear, Connors, and Paradiso, 2006; Luce, 1993; Spoendlin, 1974). Because of its close proximity to the basilar membrane, any vibrational disturbance there is imparted onto the organ of Corti. It is this mechanical stimulation that produces the initial bioelectric response through a remarkable process, thereby generating the first event in the neural processing of sound.

5.2. The Organ of Corti

The organ of Corti is located entirely within the cochlear duct and is therefore immersed in the fluid that is contained in this chamber. Figure 8 shows a cross-section of the cochlea, revealing the relationship of the organ of Corti with respect to other cochlear structures. The finer structural elements within the organ of Corti are shown in the bottom panel of this figure. One of the most prominent features is the *arch of Corti*, so named because of its shape. This is a rigid structure that appears as an inverted 'V'. The pillars rest upon the basilar membrane in a longitudinal series that runs the length of the cochlea.

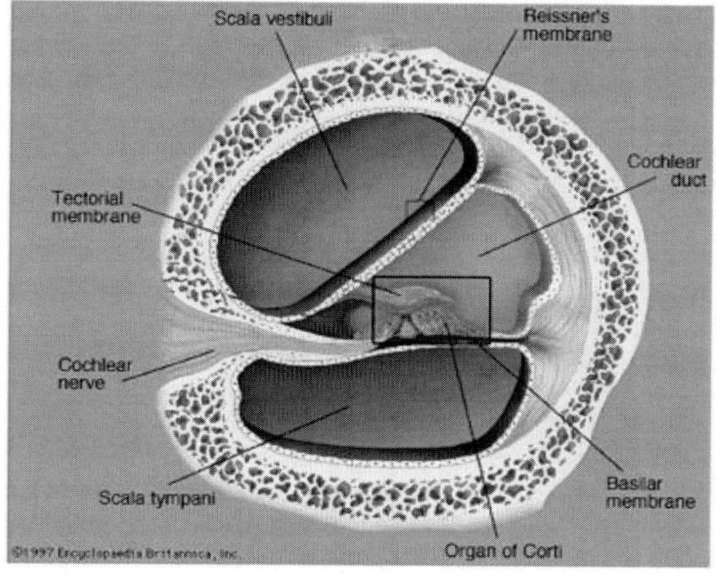

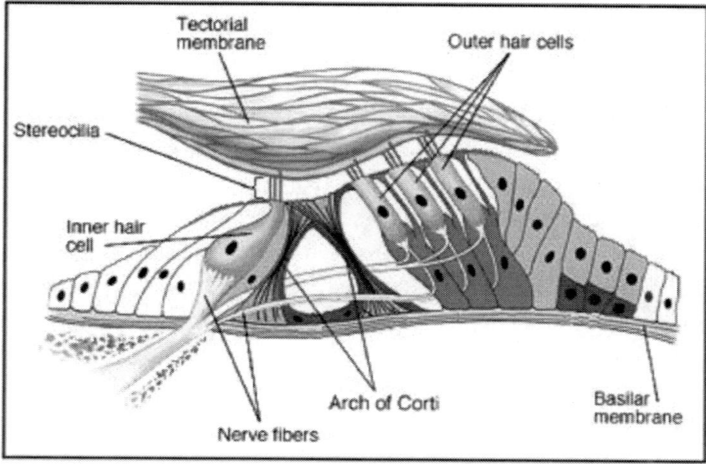

Figure 8. A cross-sectional view of the cochlea showing all three fluid chambers (upper panel). The organ of Corti can be seen to be resting on the basilar membrane within the cochlear duct. The cochlear nerve emerges from the organ of Corti and follows the coiled excursion of the cochlea. The area within the rectangular outline is shown in higher magnification in the lower panel, revealing the cellular and structural elements of the organ of Corti.

There are several different cell types within the organ of Corti, most of them serving a supporting function. The most important cells with regard to auditory signal processing are *the hair cells* (Hudspeth, 1997). As their name implies, hair cells contain fine filaments, called *stereocilia,* that protrude from their upper surface. A soft gelatinous structure called the *tectorial membrane* lies above the cellular components of the organ of Corti (Hudspeth, 1997; Luce, 1993; Richardson, Lukashkin, and Russell, 2008; Spoendlin, 1974). It too runs the entire length of the cochlea, being attached to its inner margin. The stereocilia of the outer hair cells are connected with the tectorial membrane. Both the tectorial membrane and the arch of Corti serve a critical function in transferring the vibrational energy in the basilar membrane throughout the organ of Corti, with important consequences for the hair cells where that energy is ultimately transduced into a neural signal.

The propagation of a traveling wave in the basilar membrane produces a complex set of mechanical events within the organ of Corti (Hudspeth, 1999). The result of the up and down movement of the basilar membrane is to produce lateral shearing forces by the tectorial membrane and arches of Corti. The net effect upon the stereocilia of the hair cells is that they will bend in the direction of the force — outward for an upward deflection and inward for a downward deflection. These different movements of the stereocilia have a dramatic effect upon signal generation by the hair cells.

5.3. Transductional Mechanism in Auditory Hair Cells

The shearing forces serve as the ultimate event that triggers a neural signal by the hair cells. As noted before for the various other systems, sensory transduction in general involves the production of a bioelectric current in response to some sort of external stimulation (Bear, Connors, and Paradiso, 2006; Gulick, Gescheider, and Frisina, 1989). In the case of hair cells, it is the bending of the stereocilia on hair cells that leads to the opening of ionic pores and subsequent depolarization. Our knowledge of the remarkable way that this happens has been advanced largely through the work of James Hudspeth at Rockefeller University (Hudspeth, 1999). The individual filaments of the stereocilia are themselves connected to each other by way of a very thin fiber known as a *tip link*. Rightward bending of the stereocilia due to upward movement of the basilar membrane increases the tension on the tip links. As Figure 9 illustrates, these minute fibers are actually connected to gates on ionic pores that are normally in the closed position. Hudspeth showed that these gates serve as a trapdoor that controls the flow of ions into the hair cell. Increased tension on the tip links opens the gates, thereby allowing ions such as potassium (K+) and calcium (Ca2+) to flow into the hair cell through the stereocilia. Increased intracellular levels of K+ in the hair cell have the effect of depolarizing its membrane potential. If the stereocilia are bent in the opposite direction due to downward movement of the basilar membrane, then the tension on the tip links is reduced and the ionic gates remain closed. This has the effect of hyperpolarizing the hair cell since even spontaneous openings are less likely to occur.

This unusual mechanism has one key advantage that is of great importance to hearing — rapidity (Vollrath, Kwan, and Corey, 2007). The auditory system must contend with rapid changes in the stimulus that typically occurs with high sound frequencies. Such rapid events imposed by the very nature of the sound stimulus also require that the transductional mechanisms show an extremely rapid response.

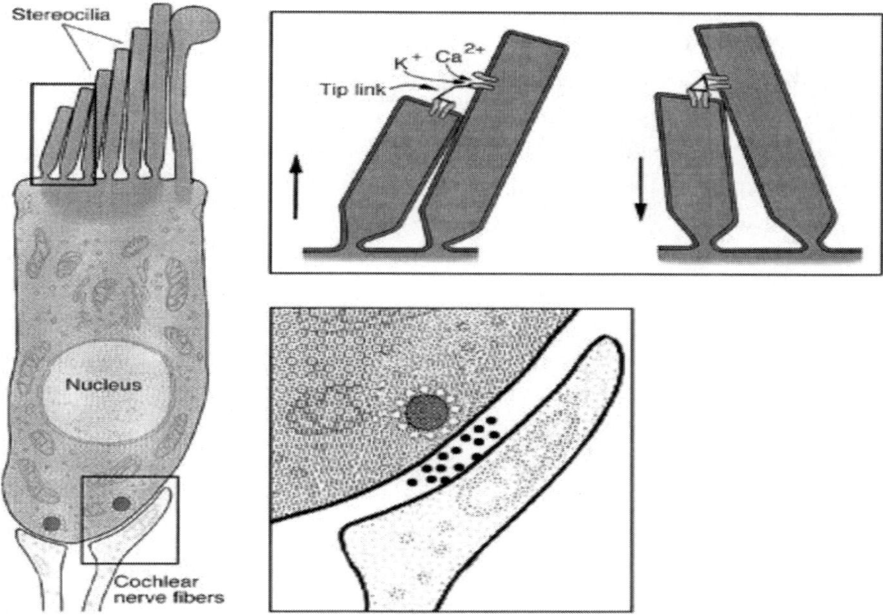

(Adapted from Hudspeth (1999)).

Figure 9. Auditory hair cells are cylindrical in shape and have stereocilia that appear as a beveled set of filaments. Each hair cell is innervated by cochlear nerve fibers at its base. The upper inset shows movement of the stereocilia in response to deflection of the basilar membrane. Upward deflection produces bending of the stereocilia toward the taller edge, which opens ionic pores in the stereocilia to allow potassium (K+) and calcium (Ca2+) ions to enter the hair cell. This produces depolarization leading to neurotransmitter release upon cochlear nerve fibers (lower inset). A downward movement of the basilar membrane bends the stereocilia in the opposite direction, thereby closing the ionic pores.

Recall that signal transduction in the chemosensory systems involve membrane bound G proteins and second messengers. Such a mechanism would be too slow for the auditory system and simply not allow the hair cells to provide the rapid response that is necessary. Instead, the direct gating action on ionic pores through the tip links provides the fastest possible response to mechanical deflection of the stereocilia.

The initial depolarization caused by bending of the stereocilia causes two things to happen. First, the depolarization leads to further opening of ionic gates within the hair cells, thereby producing greater depolarization. And second, the depolarization along with the increased Ca2+ levels lead to neurotransmitter release at the base of inner hair cell. The neurotransmitter, which is believed to be glutamate, crosses the synaptic space to depolarize cochlear nerve fibers that innervate the hair cells (see Figure 9). The resulting action potentials produced in these nerve fibers carry auditory information out of the cochlea to higher centers in the auditory nervous system.

6. VISUAL TRANSDUCTION

In previous sections we reviewed the various different ways by which external stimulation can trigger neural activity at the very first stages of sensory processing. The transductional mechanism is unique for each sensory system. The way that mechanical stimulation induces neural activity in mechanoreceptor neurons, for example, is quite distinct from the way that chemicals trigger activity in gustatory and olfactory receptor neurons, which in turn is very different from the vibrational triggering of electrical activity in cochlear hair cells. The neural processing of visual information must also begin with the very first act of transforming energy in the stimulus (light) into neural activity within those neurons (photoreceptors) located at the very first stage of the visual pathway. This process, which is referred to as *phototransduction*, is far different from any of the others we have covered thus far in that it produces a very unconventional outcome in terms of electrical activation.

All of the process involved in phototransduction occur within the *retina*, a thin film of cells that lines the inside back wall of the eyeball. The retina can be thought of in three dimensions as one half of a sphere that extends from the very back of the eyeball all the way forward to just a little past the midline along both the vertical and horizontal axes. An optical image is cast upon this entire area by the refractive elements of the eye. All vertebrates have a retina that is made up of similar types of cells, though there is variability in the composition and layout of these cells. The dominant cells in the retina are actually neurons. In fact, the retina forms as an outgrowth of the central nervous system that invades the eyeball during embryonic development.

6.1. Photoreceptors Contain Photopigment Material

The first event in phototransduction is the capture of light photons by photoreceptors to generate an electrical signal. This is accomplished by a special material, known generically as photopigment, which is found in high concentration within the two types of photoreceptors — rods and cones. The photopigment in all mammalian photoreceptors is called *rhodopsin*. Figure 10 shows that rod and cone photoreceptors can be divided into two segments, an inner and an outer segment (Batschauer, 2004; Tessier-Lavigne, 2000). It is the outer segment that is embedded within the fibrous matrix of the eyeball and which contains a large concentration of rhodopsin molecules.

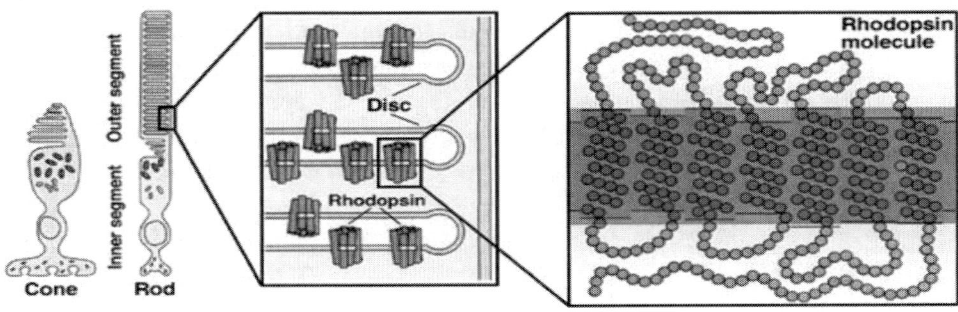

(Adapted from Kandel et al. (2000)).

Figure 10. Rod and cone photoreceptors can be broadly divided into two compartments — the inner and outer segment. The outer segment of rods contains a set of membranous discs that are stacked like pancakes. Each disc in turn contains millions of molecules of the photopigment known as rhodopsin. Light absorption by rhodopsin triggers a cascade of events that leads to the production of a neural signal in both rods and cones.

The outer segment of rods contains a stack of discs whereas in cones, the outer segment is invaginated by a series of infoldings. The rhodopsin molecules are located within the membrane of each of the discs or infoldings, as shown in the right panel of Figure 10. The American biochemist George Wald was responsible for much of our understanding on the chemical makeup and function of rhodopsin in absorbing light, for which he received the Nobel prize in 1967 (Wald, 1968).

6.2. The Dark Current

In total darkness, rhodopsin molecules are maintained in an inactive form. However, there are still a number of very interesting things going on inside the photoreceptor cells, as outlined in Figure 11.

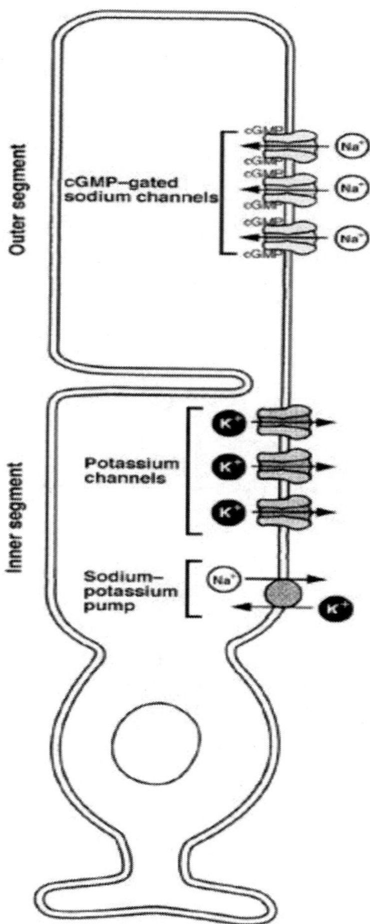

Figure 11. Photoreceptor outer segments contain cGMP-gated sodium channels that allow sodium ions to flow into the cell in the absence of light. This constitutes the so-called dark current. To maintain charge balance, potassium ions are ejected from the inner segment. A sodium-potassium pump exchanges these ions in the opposite direction to maintain their concentrations both inside and outside the cell.

In the absence of any light whatsoever, there is a steady flow of sodium ions (Na+) into the outer segment through what are known as cGMP-*gated sodium channels* (MacLeish, Shepherd, Kinnamon, and Santos-Sacchi, 1999; Wensel, 2008; Yau, 1994). cGMP is an organic molecule found in high concentration within the outer segment. The sodium channels located here are somewhat special in that they need to bind cGMP in order to allow sodium to flow in. In other words, the sodium gates are only open when there is enough cGMP around. The entry of sodium ions into the outer segment by this mechanism in the absence of light is known as the *dark current.*

The entry of positively charged sodium ions into the photoreceptor is accompanied by an outward flow of potassium ions (K+) in the inner segment. This maintains a net balance of ionic charges. However, continued inflow and outflow of these two ions will result in an excess accumulation of Na+ inside and K+ outside the cell. To prevent this from happening, a sodium-potassium pump exchanges these ions in an opposite manner — i.e., sodium ions are ejected from the photoreceptor and potassium ions are brought back inside. The net result of this entire series of ongoing events is that the photoreceptor membrane potential is maintained at a steady level of around −40 mV.

6.3. Phototransduction Involves a Cascade of Biochemical Events

The ionic exchanges described above take place as long as rhodopsin remains in an inactive state due to darkness. We now turn to the events that take place when photoreceptors are exposed to light (Batschauer, 2004; Fiesler and Kisselev, 2007). Figure 12 shows the activation of a single rhodopsin molecule after light absorption and the subsequent sequence of events that take place in its immediate vicinity. Light photons have to pass through the inner segments of the photoreceptors before reaching the stack of discs in the outer segment containing the rhodopsin molecules.

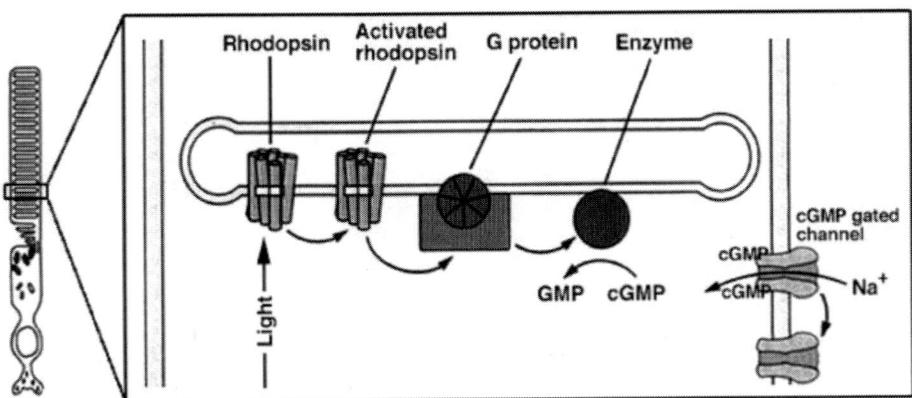

Figuer 12. Rhodopsin molecules in the outer segment are activated after light absorption. This in turn causes a membrane-bound G protein to become activated. This is followed by activation of an enzyme that converts cGMP into GMP. The reduction in cGMP means that less of it is available to bind to the sodium channels that need it in order to remain open. Thus, light activation causes the sodium channels to shut down thereby reducing the dark current (inward sodium flow).

The large number of discs ensures that a light particle will likely be absorbed at some point. When this happens, the energy in the photon is imparted to a rhodopsin molecule causing a slight change in its physical structure or conformation. Rhodopsin is now in its activated state. The activated rhodopsin then interacts with a so-called *G protein,* which also resides in the disc membrane. The G protein that is activated by this interaction in turn activates a specific enzyme. The function of this enzyme is to convert the cGMP within the outer segment into ordinary GMP, which are unable to bind to the sodium channels. Thus, the cascade of biochemical events that takes place in quick order after light absorption by rhodopsin results in the removal of cGMP within the outer segment.

6.4. Light Absorption Leads to Photoreceptor Hyperpolarization

We saw above that cGMP is essential for the sodium channels to remain open, thereby allowing a steady stream of sodium ions into the photoreceptor outer segment (the dark current). The removal of cGMP by the activated enzyme means that there is less of it available to bind to the sodium channel. This in turn causes the sodium channels to shut down, preventing the entry of sodium ions and thereby abolishing the dark current. However, the potassium ions continue to get ejected from the inner segment. Recall that the potassium outflow was necessary to offset the sodium influx and therefore maintain the charge balance. With the sodium channels now closed, the continued outflow of the positively charged potassium ions means that the membrane potential will become more negative. In other words, the photoreceptor actually becomes hyperpolarized after absorbing light (Yau, 1994; Baylor, Lamb, and Yau, 1979).

This is an astonishing result. The standard view of neuronal activation is for it become depolarized, i.e., for its membrane potential to shift toward the positive end. Indeed, the transducing neurons in all of the other sensory systems behave in this manner. And yet, photoreceptor activation following light absorption actually produces membrane hyperpolarization, an effect we normally associate with inhibition. The greater the amount of light absorbed, the greater is the impact of the above series of biochemical events, leading to a greater hyperpolarization. The normal resting membrane potential of −40 mV in photoreceptors can become as low as −70 mV under bright light conditions.

6.5. Light Activation of Photoreceptors Causes Less Neurotransmitter Release

Perhaps just as surprising as the electrical behavior of photoreceptors after light absorption is the subsequent effects on neurotransmitter release. We normally associate neuronal activation with greater neurotransmitter release. However, the hyperpolarization of photoreceptors causes just the opposite effect at its synaptic junction with the next level of neurons in the retina. Both rod and cone photoreceptors produce and release the excitatory neurotransmitter *glutamate* (Fiesler and Kisselev, 2007; Tessier-Lavigne, 2000). Under dark conditions, photoreceptors display a steady release of glutamate. Light absorption and subsequent hyperpolarization of photoreceptors actually causes them to release less glutamate. Thus, the next level of neurons in the retina must somehow interpret the reduction in glutamate release as a signal that photoreceptor activation has occurred. A description of exactly how that happens is beyond the scope of this chapter. Nevertheless, the subsequent steps are part of the fascinating sequence of events by which neural signals are generated in response to light stimulation, to ultimately produce the perceptual phenomenon that we experience as *vision*.

ACKNOWLEDGMENT

I am grateful to Oxford University Press for granting permission to excerpt text and figures from my book titled *Fundamental of Sensory Perception* published in 2010.

REFERENCES

Batschauer, A. (2004). *Photoreceptors and Light Signalling*. Royal Society of Chemistry, London.

Baylor, D.A., Lamb, T.D., and Yau, K.W. (1979). Response of retinal rods to single photons. *Journal of Physiology, London, 288*, 613-634.

Bear, M.F., Connors, B., Paradiso, M. (2006). *Neuroscience: Exploring the Brain*, 3rd Ed. Lippincott Williams and Wilkins, New York.

Beauchamp, G.K., and Bartoshuk, L . (1997). *Tasting and smelling*. Academic Press, San Diego.

Breslin, P.A.S. (2000). Human gustation. In. *Neurobiology of Taste and Smell* (Eds.) Finger TE, Silver WL, Restrepo D. Wiley-Liss, New York.

Buck, L.B. (2000). The molecular architecture of odor and pheromone sensing in mammals. *Cell, 100*, 611-618.

Buck, L., and Axel, R. (1991). A novel multigene family may encode odorant receptors: a molecular basis for odor recognition. *Cell, 65*, 175-187.

Catania, K.C., and Henry, E.C. (2006). Touching on somatosensory specialization in mammals. *Current Opinion in Neurobiology, 16*, 467-473.

Chandrashekar, J., Hoon, M.A., Ryba, N.J., and Zuker, C.S. (2006). The receptors and cells for mammalian taste. *Nature, 444*, 288-294.

Doty, R.L. (2003). *Handbook of Olfaction and Gustation*, 2nd Ed. Informa HealthCare, New York.

Fiesler, S.J., and Kisselev, O.G. (2007). *Signal Transduction in the Retina*. CRC Press, Boca Raton, FL.

Finger, T.E., Silver, W.L., and Restrepo, D. (2000). *Neurobiology of Taste and Smell*. Wiley-Liss, New York.

Genders, R. (1972). *A History of Scent*. Hamish Hamilton, London.

Gibbons, B. (1986). *The intimate sense of smell*. National Geographic 170, 324-361.

Gulick, W.L., Gescheider, G.A., and Frisina, R.D. (1989). *Hearing: Physiological Acoustics, Neural Coding, and Psychoacoustics*. Oxford University Press, New York.

Hudspeth, A. J. (1997). How hearing happens. *Neuron, 19*, 947-950.

Hudspeth, A.J. (1999). Sensory transduction in the ear. In *Principles of Neural Science*, 4th edition, (Eds.) E.R. Kandel, J.H. Schwartz, and T.M. Jessell, Appleton and Lange, Norwalk, CT.

Kandel, E.R., Schwartz, J.H., and Jessell, T.M. (2000). *Principles of Neural Science*, 4th Ed. McGraw-Hill, New York.

Kruger, E. (1996). *Pain and Touch. Handbook of Perception and Cognition*, 2nd Ed. Academic Press, New York.

Laurent, G. (1999). A systems perspective on early olfactory coding. *Science, 286*, 723-728.

Luce, R.D. (1993). *Sound and Hearing.* Lawrence Erlbaum, Hillsdale, NJ.

Lumpkin, E.A., and Caterina, M.J. (2007). Mechanisms of sensory transduction in the skin. *Nature, 445*, 858-865.

MacLeish, P.R., Shepherd, G.M., Kinnamon, S.C., and Santos-Sacchi, J. (1999). Sensory transduction. In: *Fundamental Neuroscience*, Eds. MJ Zigmond, FE Bloom, SC Landis, JL Roberts, LR Squire. Academic Press, San Diego.

Marieb, E., and Hoehn, K. (2006). *Human Anatomy and Physiology*, 7th Ed. Benjamin Cummings, New York.

Miller, I.J. Jr. (1995). *Anatomy of the peripheral taste system.* In. Handbook of Olfaction and Gustation. Ed. RL Doty. Marcel Dekker, New York.

Mombaerts, P. (1999). Molecular biology of odorant receptors in vertebrates. *Annual Review of Neuroscience, 22*, 487-509.

Morgan, M.J. (2009). *Molyneux's Question: Vision, Touch and the Philosophy of Perception.* Cambridge University Press, Cambridge, UK.

Morley, J.W. (1998). *Neural aspects in tactile sensation.* Elsevier, New York.

Purves, D. (2007). *Neuroscience*, 4th Ed. Sinauer Associates, Sunderland, MA.

Richardson, G.P., Lukashkin, A.N., and Russell, I.J. (2008). The tectorial membrane: one slice of a complex cochlear sandwich. *Current Opinion in Otolaryngology, Head, and Neck Surgery, 16*, 458-464.

Squire, L.R., Roberts, J.L., Spitzer, N.C., and Zigmond, M.J. (2002). *Fundamental Neuroscience*, 2nd Ed. Academic Press, San Diego.

Smith, C.U.M. (2000). *The biology of sensory systems.* John Wiley, New York.

Spoendlin, H. (1974). Neuroanatomy of the cochlea. In: *Facts and Models in Hearing*, Eds. E. Zwicker, E. Terhardt. Springer-Verlag, New York.

Tessier-Lavigne, M. (2000). Visual processing by the retina. In: *Principles of Neural Science*, 4th ed.Eds: ER Kandel, JH Schwartz, TM Jessell. McGraw-Hill, New York.

Tortora, G.J., and Derrickson, B.H. (2008). *Principles of Anatomy and Physiology.* Wiley, New York.

Vollrath, M.A., Kwan, K.Y., and Corey, D.P. (2007). The micromachinery of mechanotransduction in hair cells. *Annual Review of Neuroscience, 30*, 339-365.

Wald, G. (1968). The molecular basis of visual excitation. *Nature, 219*, 800-807.

Wensel, T.G. (2008). Signal transducing membrane complexes of photoreceptor outer segments. *Vision Research, 48*, 2052-2061.

Yau, K-W. (1994). Phototransduction mechanisms in retinal rods and cones. *Investigative Ophthalmology and Visual Science, 35*, 9-32.

In: Expanding Horizions of the Mind Science(s)
Editors: P.N. Tandon, R.C. Tripathi and N. Srinivasan

ISBN: 978-1-62808-705-5
©2013 Nova Science Publishers, Inc.

Chapter 6

PLASTICITY OF THE SOMATOSENSORY SYSTEM FOLLOWING SPINAL AND PERIPHERAL INJURIES

Neeraj Jain and Shashank Tandon
National Brain Research Centre, Manesar, Haryana

ABSTRACT

Brains of adult mammals retain remarkable ability to undergo plastic reorganization. Deafferentations due to peripheral nerve or spinal cord injuries, result in parts of the somatosensory cortex which lose their peripheral inputs, acquiring novel inputs originating in the intact parts of the system. For example, after injuries to the dorsal columns of the spinal cord at cervical levels, intact inputs from the face expand to reactivate neurons in the deafferented hand cortex. The reorganization results in changes in the topographic organization of the brain, which takes place at multiple levels of the system - the cortex, the thalamic nuclei and the brain stem nuclei. The ability of the brain to reorganize has the potential to be utilized for facilitating functional recoveries by potentiation of remaining neuronal circuits after injuries. Brain reorganization can also have undesirable perceptual consequences such as phantom sensations. Experiments from many laboratories have led to our current understanding of the nature of reorganization, and have given tantalizing clues about the possible mechanisms of adult brain plasticity.

1. INTRODUCTION

Most neurons in adult brains do not divide and recoveries from brain injuries are partial at best. These observations imply that adult brains have only limited abilities to undergo reorganization. Since the normal brain function depends upon the integrity of the complex wiring between neurons and the fidelity of information transmission across this wiring, the limited plasticity of the adult brains can be regarded as an evolutionary 'trade-off' to maintain this fidelity.

However, contrary to the dogma prevailing at that time, it was established in early 1980's that large parts of the adult mammalian brains do remain plastic. Kaas, Merzenich and their colleagues for the first time provided an irrefutable evidence of plasticity of the adult primate brains. They showed reorganization in the primary somatosensory cortex of owl and squirrel monkeys following injury to the median nerve to the hand (Merzenich et al., 1983b; Merzenich et al., 1984). Following up on this discovery, a body of research work has accumulated on plasticity of different sensory and motor systems in a variety of mammalian species. In this review we will focus on the plasticity in the somatosensory system in mammals following peripheral nerve or spinal cord injuries. We will examine the evidence of reorganization of the somatosensory system in different primates, including humans, and its perceptual consequences. Our goal here is to give a comprehensive view of the current knowledge using selected examples from the published literature, rather than provide an exhaustive list of citations. The focus of this chapter is the primate somatosensory system, although other mammalian species will be described where relevant. A very brief overview of the mammalian somatosensory system precedes the discussion of the plasticity of the system.

2. The Mammalian Somatosensory System

The primate skin has many different kinds of tactile receptors such as the Pacinian corpuscles, Meissner corpuscles, Merkel receptors and Ruffini endings, which detect and transmit different kinds of tactile sensations such as touch, pressure, flutter, or vibration. Neurons, whose cell bodies lie in the dorsal root ganglia, carry information from these receptors to the spinal cord via the dorsal roots. The major ascending spinal pathway that is responsible for transmission of information for fine touch, discrimination and proprioception is the dorsal column-medial lemniscal pathway (Figure 1). This pathway, which lies in the dorsal funiculus of the spinal cord has been found in all mammals (Lund and Webster, 1967; Bowsher, 1958).

Dorsal column pathway terminates on neurons in the brainstem nuclei. Neurons in the cuneate nucleus and the gracile nucleus of the brain stem receive inputs from the upper limb and the lower limb respectively. The third major brainstem nucleus, the trigeminal nucleus receives sensory inputs from the face through the trigeminal nerve. Axons of the second order neurons whose cell body lies in the brain stem nuclei cross the midline, ascend in the medial lemniscus, and project to neurons in the contralateral ventroposterior (VP) nucleus of the thalamus. Subdivisions of the VP nucleus receive inputs from different body parts, the ventroposterior lateral (VPL) subnucleus from the limbs and trunk, and the ventroposterior medial (VPM) subnucleus from the face. Third order neurons in the VP nucleus project to the ipsilateral primary somatosensory cortex.

In monkeys, the anterior parietal somatosensory cortex consists of four distinct Brodmann's areas namely 3a, 3b, 1 and 2. These areas were initially described as consisting of a single somatosensory area, S1. Later mapping studies using microelectrodes showed that each of these Brodmann's areas has a separate complete representation of the contralateral half of the body (Merzenich et al., 1978; Nelson et al., 1980). Neuroanatomical studies showed that majority of the thalamic inputs from the VPL and VPM nuclei terminate in area 3b and 1. Area 3a receives cutaneous inputs from area 3b, apart from inputs from muscle spindles directly through the ventroposterior superior nucleus (VPS) of the thalamus.

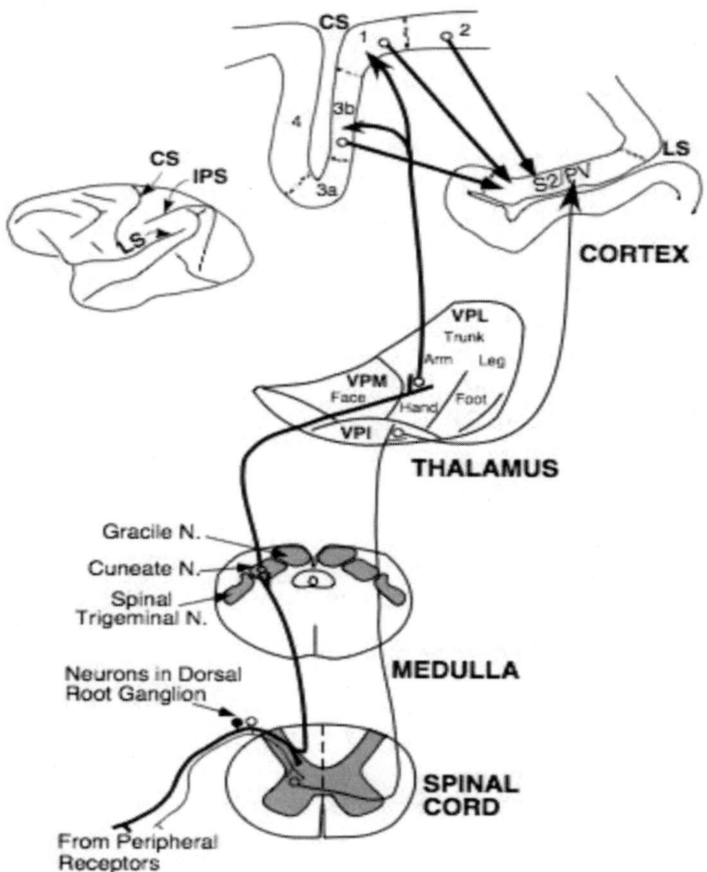

Figure 1. The somatosensory system of a macaque monkey. The pathway shown by thick lines is the dorsal column medial lemniscus pathway, and the pathway shown by thin lines is the spinothalamic pathway. The ventroposterior superior nucleus and its connections are not shown. Inset on the top left shows a drawing of the lateral view of the macaque brain showing locations of various sulci. N., nucleus; VPM, ventroposterior medial nucleus; VPL, ventroposterior lateral nucleus; VPI, ventroposterior inferior nucleus; CS, central sulcus; LS, lateral sulcus; IPS, intraparietal sulcus.

Area 2 receives cutaneous input primarily from area 1, fewer inputs from area 3b, and other inputs through VPS. Based upon these neuroanatomical and mapping studies it was proposed that area 3b of primates is the true homologue of the S1 cortex of non-primates (Kaas, 1983).

There is an orderly somatotopic representation of the body surface in area 3b. The oral cavity representation is lateral-most followed medially by the face representation (Jain et al., 2001). Medial to the face representation, the hand, the arm and the trunk are represented in a lateral to medial order. The leg and the foot representations are medial-most. The detailed representation of each body part is also somatotopic. For example, within the hand representation, digit 1 is represented most laterally and the digits 2, 3, 4 and 5 progressively more medially. Likewise a precise somatotopy is maintained within the representation of the face, foot and other body parts (Kaas et al., 2001).

Besides the somatosensory areas of the anterior parietal cortex, other somatosensory areas are located in the lateral parietal cortex in the upper bank of the lateral sulcus. These somatosensory areas, the secondary somatosensory area (S2) and the parietal ventral area (PV) also have a topographic representation of the body surface, although the representations are less detailed and the receptive fields of neurons are much larger than in area 3b (Krubitzer et al., 1995; Tandon et al., 2009).

Besides the dorsal column-medial lemniscus pathway, the other major ascending pathway of the spinal cord is the spinothalamic tract. The fibers of the spinothalamic tract are located in the ventro-lateral quadrant of the spinal cord. The afferents of the dorsal root ganglion cells, which project via the spinothalamic pathway synapse in the ipsilateral dorsal horn. Axons of the second order neurons cross the midline in the spinal cord and ascend in the spinothalamic tract to the ventroposterior nucleus of the thalamus, terminating in the ventroposterior inferior and other nuclei. From the thalamus, the neurons project to the somatosensory regions in the anterior parietal cortex and the lateral parietal cortex. Spinothalamic pathway carries information about crude touch, besides pain and temperature.

3. REORGANIZATION IN THE SOMATOSENSORY SYSTEM

Reorganization of the primary somatosensory cortex: Experiments in adult owl and squirrel monkeys showed that the topographic maps in the primary somatosensory cortex can reorganize following peripheral nerve injury. Merzenich et al. (1983a) transected the median nerve in the forearm, which supplies the radial (thumbward) half of the glabrous hand. They ligated the ends of the cut nerve to prevent regeneration. The lateral half of the hand representation in the primary somatosensory cortex was thus deprived of its normal peripheral inputs. Two to nine months after the nerve was cut, neurons in the region of area 3b which normally receive median nerve inputs were reactivated by inputs from the radial nerve, which innervates hairy skin on the back of the hand (Figure 2E and 2F). This lead to reactivation of neurons in the deprived cortex by stimulation of the hairy skin on the back of the hand (Merzenich et al., 1983b).

A number of other experiments were done in different laboratories to explore the limits of plasticity after lesions that deprive different extents of the somatosensory cortex. Few examples are given below.

Garraghty and Kaas (1991) cut both the median and the ulnar nerves to the hand in squirrel monkeys to deprive entire representation of the glabrous hand. They found that neurons in the entire deprived cortex becomes reactivated by inputs from the skin of the dorsal hand supplied by the radial nerve as in case of deafferentation by cutting the median nerve. Thus the entire deafferented glabrous hand representation can be reactivated by the adjacent radial nerve inputs. Merzenich and colleagues showed that area 3b reorganizes after amputation of digits in adult owl monkeys (Figure 2B and 2C). Two months following the amputation of one or two digits of the hand, inputs from the adjacent digits, stump and the palm expanded to reactivate neurons in the deafferented region (Merzenich et al., 1984). Here, unlike in case of deprivations following injuries to the nerves to the hand, representations in the margins of the deprived cortex expanded to reactivate the neurons.

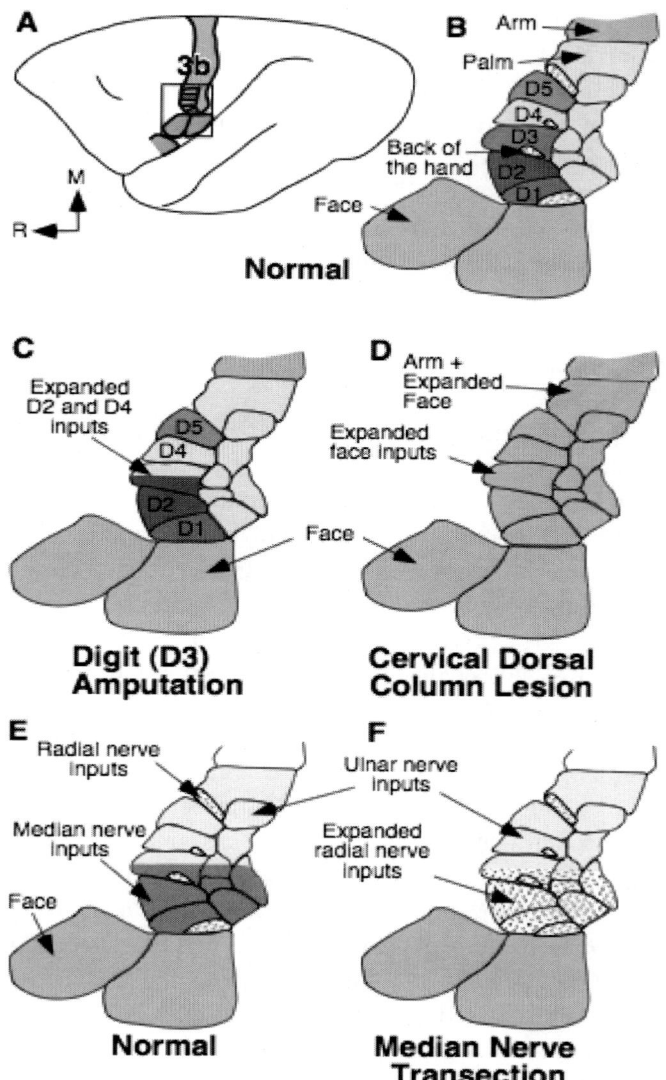

Figure 2. Reorganization in the primary somatosensory cortex or area 3b after different kinds of deafferentations. (A) A lateral view of the owl monkey brain showing location of area 3b (shaded in grey). The brain of owl monkey, which is a New World monkey, is relative free of sulci as compared to the macaque monkey brain (c.f. Fig. 1). The box marks the region of the hand and the face representations, which is shown enlarged in 'B'. R, rostral; M, medial. The orientation arrows apply to all the panels of the figure. (B) Details of the face, digits, palm and part of the arm representations in a normal adult owl monkey. Regions marked by D1, D2 etc., are representations of digit 1 (thumb), digit 2 (first finger) etc. Regions with representation of the inputs from the hairy skin on the back of the hand are stippled. (C) Organization of area 3b following amputation of digit 3 (D3). Representations of the adjacent digits, D2 and D4 expand into the deafferented D3 representation. (D) Organization of area 3b following transection of the dorsal columns at cervical levels. Many months after the lesion, the face representation expands into the deafferented hand and arm representations. (E and F) Reorganization of area 3b following median nerve transection. 'E' shows normal location of the median, ulnar and radial nerve inputs. 'B' shows representation of these inputs in a monkey many weeks after transection of the median nerve. Note expansion of the radial nerve inputs into the median nerve territory.

Reorganization of the somatosensory cortex following peripheral injuries was shown in a number of other mammalian species including flying foxes (Calford and Tweedale, 1991), rats (Wall and Cusick, 1984) and raccoons (Turnbull and Rasmusson, 1991) showing that brain plasticity is not limited to a particular group of mammals.

The extent of reorganization that was reported so far was limited to the same body part representation. Such reorganization could be mediated by expression of pre-existing inputs, which are normally suppressed by the dominant input. Nerve injuries or amputations remove the dominant inputs releasing the inhibition and leading to the expression of the non-dominant but pre-existing inputs.

However, Pons and colleagues (Pons et al., 1991) found a large shift in the representational boundaries in the somatosensory cortex. They mapped area 3b of macaque monkeys in which dorsal roots at spinal segments C2-T4 were cut more than twelve years earlier. Such injuries remove sensory inputs from the hand and the arm. Pons and colleagues observed a massive reorganization in area 3b, such that the face inputs expanded into the deafferented hand region, i.e. a shift in the boundary of the face representation by as much as 11 mm, which was almost an order of magnitude more than previously known.

Subsequently, large-scale expansion of the face representation into the hand region of the area 3b was reported following unilateral lesions of the dorsal columns of the spinal cord at C3-C5 levels (Figure 2B and 2D) (Jain et al., 1997; Jain et al., 2008). Unlike cutting of the dorsal roots, these injuries deprive the brain of only part of the tactile inputs, since the spinothalamic pathways remain intact. Such reorganization was seen in both New World owl monkeys and Old World macaque monkeys many months after lesions of the dorsal columns. In one monkey Jain et al., reported appearance of face responses as far as 20 mm from the original hand-face border, effectively spanning the entire length of area 3b as far medial as the representation of the foot (Jain et al., 2008). No anatomical studies have demonstrated pre-existing connections that span these extents of representations. Therefore, it is likely that these kinds of large-scale reorganizations involve neuronal growth (Jain et al., 2000).

Reorganization of other brain areas. Since area 3b and the surrounding cortex undergo reorganization in adults, the obvious question was if subcortical somatosensory regions are also plastic, and if the nature of reorganization subcortically is similar to that seen in the cortex. Using the paradigm of transection of the cutaneous nerves to the hand, Garraghty and Kaas (1991) showed that after cutting of the median and ulnar nerves, neurons in the deafferented portions of the VPL nucleus of the thalamus get reactivated by inputs from the normally innervated skin of the back of the hand, just as in area 3b of the cortex. Similarly, in adult macaque monkeys with amputation of a limb, the representation of the skin of the stump expanded into the deafferented portions of the VPL (Florence et al., 2000). Jain et al., (2008) reported that the VP nucleus is also capable of large-scale reorganization after transection of the dorsal columns of the spinal cord in macaque monkeys. They reported expansion of the face representation into the deafferented hand region of VPL of thalamus. Thus the nature of reorganization in the thalamus is largely similar to that seen in the cortex.

Recent experiments have shown that reorganization in the brain following dorsal column lesions is widespread and it takes place at multiple sites. In addition to area 3b, upstream somatosensory areas - the second somatosensory area (S2) and the parietal ventral area (PV) also show an expansion of the face inputs into the hand region (Tandon et al., 2009). Interestingly, area S2 and PV get reorganized although these areas continue to receive intact peripheral inputs via the spinothalamic pathway after dorsal column lesions indicating that neurons in area 3b drive the reorganization of the upstream areas. Other experiments show that similar reorganization also takes place in the ventroposterior nucleus of the thalamus (Jain et al., 2008)

Reorganization in the human brain. Experiments have been done to determine if the human brain also undergoes reorganization following chronic peripheral and spinal cord injuries. In the somatosensory cortex of patients with median or ulnar nerve transections somatosensory evoked responses showed enlargement of representation of the neighboring digits into the deafferented digit regions (Diesch et al., 2001). In patients with carpal tunnel syndrome, who have chronic median nerve damage, magnetic source imaging demonstrated an invasion of the deprived area by cortical regions receiving inputs from the little finger (supplied by the ulnar nerve) and from the dorsum of the thumb (innervated by the radial nerve) thus indicating topographic rearrangement in the human primary somatosensory cortex (Druschky et al., 2000). Functional magnetic resonance imaging studies of the sensorimotor cortex 5–7 years after complete thoracic spinal cord injuries have shown invasion of the intact forelimb inputs into the deafferented lower limb region (Perani et al., 2001). In humans with amputations the affected somatosensory cortices show marked intrusion of the face representations into the digit and hand area (Elbert et al., 1997). Thus the human somatosensory cortex is capable of reorganization after spinal cord injury and peripheral deafferentation just as in non-human primates.

As for non-human primates, studies in humans have also shown reorganization at subcortical levels of the somatosensory system. In patients with pain after spinal cord transection, anesthetic areas of principal sensory nucleus of the thalamus often exhibited increased representations of the border of the anesthetic part of the body in comparison with the representation of the same parts of the body in normals (Lenz et al., 1994). Chronic spinal transections (C3–T12) also resulted in enlargements in representation of uninjured head, arm, neck, or trunk inputs in parts of the thalamic body map that had lost normal inputs (Tasker et al., 1987; Hua et al., 2000).

Thus following loss of sensory inputs due to deafferentations, adult somatosensory system undergoes reorganization at multiple levels including brain stem nuclei, the thalamus, primary somatosensory cortex and higher somatosensory areas. If the extent of deafferentation is small, such as after cutting of a nerve to the hand, or amputation of one or more digits, the adjacent intact hand inputs reactivate neurons in the deafferented hand representations. However, if the injury deprives large expanse of the cortex, such as after transection of the dorsal roots or the dorsal columns, even distant inputs from the face area are capable of reactivating the deprived regions of the cortex and the thalamus. Such changes can have important consequences such as perceptual mislocalizations and might have a bearing on the extent of recovery after therapeutic rehabilitative interventions.

4. CORTICAL REORGANIZATION AND ITS RELATION TO PHANTOM SENSATIONS

Following the observations of Pons and others (Pons et al., 1991) that in monkeys with transection of dorsal roots there is an extensive reorganization in the primary somatosensory system it was proposed that such reorganization could lead to phantom sensations perceived by patients with limb amputations and spinal cord injuries (see Jain, 2002).

Ramachandran et al. (1992a; 1992b) reported that in patients with unilateral arm or hand amputation, tactile non-painful stimulation of the face and the skin area above the stump could elicit sensations in the missing phantom limb. They observed that the touch locations that evoke such sensations on the face are topographically organized. Specific sites on the face evoked sensations of touch on specific parts of the amputated hand such as a particular finger. The phantom sensations were also submodality specific. For example, movement of a stimulus across the skin of the face evoked sensations of tactile movement on the missing hand. From these clinical observations Ramachandran proposed the topographical remapping hypothesis for phantom sensations according to which the phantom sensations in these patients are due to the cortical reorganization observed following peripheral or spinal cord injuries (Ramachandran and Rogers-Ramachandran, 2000).

However, later studies trying to understand the link between referred phantom sensation and the reorganization in S1 in upper limb amputees did not find a simple correlation between the two. These studies instead found a strong correlation between the intensity of phantom pain and the extent of reorganization in SI (Elbert et al., 1994; Flor et al., 1995; Grusser et al., 2004). Moreover, there was a relationship between the percentage of sites from which painful referred sensations could be elicited and the degree of cortical reorganization (Grusser et al., 2004).

MacIver et al., (2008) using fMRI showed that training with mental imagery in patients with phantom limb pain decreased phantom pain intensity, which correlated with reduction in cortical reorganization. Lotze et al. (2001) observed that in patients with phantom limb pain when the lip is moved the activation is in both the lip and the hand area. In these patients imagined movement of the phantom hand activated the face area (Lotze et al., 1999). These studies further confirm the correlation between phantom limb pain and cortical reorganization.

Likewise in spinal cord injury patients a positive correlation between cortical reorganization and phantom sensations is seen. A study by Wrigley et al. (2009) showed that subjects with complete spinal cord injury and neuropathic pain below the injury level had reorganization of the primary somatosensory cortex that correlates with pain intensity. In patients with complete thoracic spinal cord injury, there was a medial shift of the thumb and the little finger representation into the leg representation compared with controls. This shift was much more extensive in patients with painful phantom sensations as compared to the patients with no neuropathic pain.

Thus a strong correlation exists between cortical reorganization and phantom sensations. An understanding of the mechanisms of plasticity will help devise effective therapeutic interventions for such disabilities, and also help in developing strategies for better recoveries by effectively using the potential of the brain to reorganize.

ACKNOWLEDGMENTS

NJ's research has been supported by International Senior Research Fellowship from the Wellcome Trust, UK (No. 063259/Z/00/Z) and a grant from DRDO (No. ERIP /ER/0501094/M/01/1248/D(R&D)).

REFERENCES

Bowsher, D. (1958). Projection of the gracile and cuneate nuclei in Macaca mulatta: an experimental degeneration study. *The Journal of comparative neurology, 110*, 135-155.

Calford, M.B., and Tweedale, R. (1991). Acute changes in cutaneous receptive fields in primary somatosensory cortex after digit denervation in adult flying fox. *Journal of Neurophysiology, 65*, 178-187.

Diesch, E., Preissl, H., Haerle, M., Schaller, H.E., and Birbaumer, N. (2001). Multiple frequency steady-state evoked magnetic field mapping of digit representation in primary somatosensory cortex. *Somatosensory and Motor Research, 18*, 10-18.

Druschky, K., Kaltenhauser, M., Hummel, C., Druschky, A., Huk, W.J., Stefan, H., and Neundorfer, B. (2000). Alteration of the somatosensory cortical map in peripheral mononeuropathy due to carpal tunnel syndrome. *Neuroreport, 11*, 3925-3930.

Elbert, T., Flor, H., Birbaumer, N., Knecht, S., Hampson, S., Larbig, W., and Taub, E. (1994). Extensive reorganization of the somatosensory cortex in adult humans after nervous system injury. *NeuroReport, 5*, 2593-2597.

Elbert, T., Sterr, A., Flor, H., Rockstroh, B., Knecht, S., Pantev, C., Wienbruch, C., and Taub, E. (1997). Input-increase and input-decrease types of cortical reorganization after upper extremity amputation in humans. *Experimental Brain Research, 117*, 161-164.

Flor, H., Elbert, T., Knecht, S., Wienbruch, C., Pantev, C., Birbaumer, N., Larbig, W., and Taub, E. (1995) Phantom-limb pain as a perceptual correlate of cortical reorganization following arm amputation. *Nature, 375*, 482-484.

Florence, S.L., Hackett, T.A., and Strata, F. (2000). Thalamic and cortical contributions to neural plasticity after limb amputation. *Journal of Neurophysiology, 83*, 3154-3159.

Garraghty, P.E., and Kaas, J.H. (1991). Large-scale functional reorganization in adult monkey cortex after peripheral nerve injury. *Proceedings of National Academy of Sciences USA, 88*, 6976-6980.

Grusser, S.M., Muhlnickel, W., Schaefer, M., Villringer, K., Christmann, C., Koeppe, C., and Flor, H. (2004). Remote activation of referred phantom sensation and cortical reorganization in human upper extremity amputees. *Experimental Brain Research, 154*, 97-102.

Hua, S.E., Garonzik, I.M., Lee, J.I., and Lenz, F.A. (2000). Microelectrode studies of normal organization and plasticity of human somatosensory thalamus. *Journal of Clinical Neurophysiology, 17*, 559-574.

Jain, N. (2002). Adult brain plasticity - what is revealed is exciting, what is hidden is critical. *Journal of Biosciences, 27*, 439-442.

Jain, N., Catania, K.C., and Kaas, J.H. (1997). Deactivation and reactivation of somatosensory cortex after dorsal spinal cord injury. *Nature, 386*, 495-498.

Jain, N., Florence, S.L., Qi H-X., and Kaas, J.H. (2000). Growth of new brain stem connections in adult monkeys with massive sensory loss. *Proceedings of National Academy of Sciences USA, 97*, 5546-5550.

Jain, N., Qi H-X, Catania, K.C., and Kaas, J.H. (2001). Anatomical correlates of the face and oral cavity representation in somatosensory area 3b of monkeys. *Journal of Comparative Neurology, 429*, 455-468.

Jain, N., Qi HX, Collins, C.E., and Kaas, J.H. (2008). Large-Scale Reorganization in the Somatosensory Cortex and Thalamus after Sensory Loss in Macaque Monkeys. *Journal of Neuroscience, 28*, 11042-11060.

Kaas, J.H. (1983). What if anything is S1? The organization of the "first somatosensory area" of cortex. *Physiological Review, 63*, 206-231.

Kaas, J.H., Jain, N., and Qi HX. (2001). The organization of the somatosensory system in Primates. In: *The Somatosensory System* (Ed.) R.J. Nelson, New York: CRC Press.

Krubitzer, L., Clarey, J., Tweedale, R., Elston, G. and Calford, M. (1995). A redefinition of somatosensory areas in the lateral sulcus of macaque monkeys. *Journal of Neuroscience, 15*, 3821-3839.

Lenz, F.A., Kwan, H.C., Martin, R., Tasker, R., Richardson, R.T., and Dostrovsky, J.O. (1994). Characteristics of somatotopic organization and spontaneous neuronal activity in the region of the thalamic principal sensory nucleus in patients with spinal cord transection. *Journal of Neurophysiology, 72*, 1570-1587.

Lotze, M., Laubis-Herrmann, U., Topka, H., Erb, M., and Grodd, W. (1999). Reorganization in the primary motor cortex after spinal cord injury-a functional magnetic resonance (fMRI) study. *Restorative Neurology and Neuroscience, 14*, 183-187.

Lotze, M., Flor, H., Grodd, W., Larbig, W., and Birbaumer, N. (2001). Phantom movements and pain. An fMRI study in upper limb amputees. *Brain, 124*, 2268-2277.

Lund, R.D., and Webster, K.E. (1967). Thalamic afferents from the dorsal column nuclei. An experimental anatomical study in the rat. *The Journal of Comparative Neurology, 130*, 301-312.

MacIver, K., Lloyd, D.M., Kelly, S., Roberts, N., and Nurmikko, T. (2008). Phantom limb pain, cortical reorganization and the therapeutic effect of mental imagery. *Brain, 131*, 2181-2191.

Merzenich, M.M., Kaas, J.H., Sur, M., and Lin, C.S. (1978). Double representation of the body surface within cytoarchitectonic areas 3b and 1 in "SI" in the owl monkey *(Aotus trivirgatus). Journal of Comparative Neurology, 181*, 41-73.

Merzenich, M.M., Kaas, J.H., Wall, J.T., Sur, M., Nelson, R.J., and Felleman, D.J. (1983a). Progression of change following median nerve section in the cortical representation of the hand in areas 3b and 1 in adult owl and squirrel monkeys. *Neuroscience, 10*, 639-665.

Merzenich, M.M., Kaas, J.H., Wall, J., Nelson, R.J., Sur, M., and Felleman, D. (1983b). Topographic reorganization of somatosensory cortical areas 3B and 1 in adult monkeys following restricted deafferentation. *Neuroscience, 8*, 33-55.

Merzenich, M.M., Nelson, R.J., Stryker, M.P., Cynader, M.S., Schoppmann, A., and Zook, J.M. (1984). Somatosensory cortical map changes following digit amputation in adult monkeys. *Journal of Comparative Neurology, 224*, 591-605.

Nelson, R.J., Sur, M., Felleman, D.J., and Kaas, J.H. (1980). Representation of the body surface in postcentral parietal cortex of *Macaca fascicularis. Journal of Comparative Neurology, 192*, 611-643.

Perani, D., Brunelli, G.A., Tettamanti, M., Scifo, P., Tecchio, F., Rossini, P.M., and Fazio, F. (2001). Remodelling of sensorimotor maps in paraplegia: a functional magnetic resonance imaging study after a surgical nerve transfer. *Neuroscience letters, 303*, 62-66.

Pons, T.P., Garraghty, P.E., Ommaya, A.K., Kaas, J.H., Taub, E., and Mishkin, M. (1991). Massive cortical reorganization after sensory deafferentation in adult macaques. *Science, 252*, 1857-1860.

Ramachandran, V.S., and Rogers-Ramachandran, D. (2000). Phantom limbs and neural plasticity. *Archives of Neurology, 57*, 317-320.

Ramachandran, V.S., Rogers-Ramachandran, D., and Stewart, M. (1992a). Perceptual correlates of massive cortical reorganization. *Science, 258*, 1159-1160.

Ramachandran, V.S., Stewart, M., and Rogers-Ramachandran, D.C. (1992b). Perceptual correlates of massive cortical reorganization. *NeuroReport, 3*, 583-586.

Tandon, S., Kambi, N., Lazar, L., Mohammed, H., and Jain, N. (2009). Large-scale expansion of the face representation in somatosensory areas of the lateral sulcus after spinal cord injuries in monkeys. *Journal of Neuroscience, 29*, 12009-12019.

Tasker, R.R., Gorecki, J., Lenz, F.A., Hirayama, T., and Dostrovsky, J.O. (1987). Thalamic microelectrode recording and microstimulation in central and deafferentation pain. *Applied neurophysiology, 50*, 414-417.

Turnbull, B.G., and Rasmusson, D.D. (1991). Chronic effects of total or partial digit denervation on raccoon somatosensory cortex. *Somatosensory and Motor Research, 8*, 201-213.

Wall, J.T., and Cusick, C.G. (1984). Cutaneous responsiveness in primary somatosensory (S-I) hindpaw cortex before and after partial hindpaw deafferentation in adult rats. *Journal of Neuroscience, 4*, 1499-1515.

Wrigley, P.J., Press, S.R., Gustin, S.M., Macefield, V.G., Gandevia, S.C., Cousins, M.J., Middleton, J.W., Henderson, L.A., and Siddall, P.J. (2009). Neuropathic pain and primary somatosensory cortex reorganization following spinal cord injury. *Pain, 141*, 52-59.

In: Expanding Horizions of the Mind Science(s)   ISBN: 978-1-62808-705-5
Editors: P.N. Tandon, R.C. Tripathi and N. Srinivasan   ©2013 Nova Science Publishers, Inc.

Chapter 7

SELECTIVE ATTENTION: INSIGHTS FROM MULTIPLE METHODOLOGIES

Priyanka Srivastava and Narayanan Srinivasan
Centre of Behavioural and Cognitive Sciences, University of Allahabad, India

ABSTRACT

Attention enables us to selectively focus on the relevant information amidst the distracting stimuli, for efficient performance. Selective attention has been investigated using various behavioural, computational, and neuroscientific approaches. The chapter discusses multiple issues associated with selective attention including the reasons for selection, the locus of selection and types of selection. Selection is based on both spatial and object's information and recent studies have elucidated the mechanisms involved in both types of selection. Also, selection can be focused on one object or divided across multiple locations/objects over space and time. Two major paradigms, visual search and attentional blink and results from those paradigms are discussed in the chapter. Moreover, the chapter discusses the distinction between focused and distributed attention with evidence from multiple studies on emotions and hierarchical processing and finally concludes with a discussion of the computational models for selective attention.

1. INTRODUCTION

Imagine driving in a narrow congested Indian road. You have to drive safely back home and to do that you have to monitor traffic lights and other vehicles that range from bicycles to other cars. You might be listening to music or chatting with your friend sitting in the front seat next to you. Depending on what the vehicles and people around you are doing, you have to change gears and speed up or slow down. When somebody or some vehicle suddenly jumps in front of your vehicle, you have to take immediate action by breaking or avoid by going around the obstacle. The environment is complex and many attentional processes come into play in driving safely back home without injuries to self, others, the car and to other objects in the traffic. In this chapter, we focus on various aspects of selective attention employed for efficient perception and action using multiple methodologies: behavioural, neurophysiological and computational studies on selective attention. The chapter discusses the need for selection and stage at which selection occurs. The following sections also discuss selection based on object and spatial representation and orienting of attention and then attentional processing divided over multiple objects/locations over space and time. Further, we discuss the differences between focused and distributed attention and conclude with a discussion on the computational models for selective attention.

2. THE WHY OF SELECTION

At any given time, our sensory system encounters far more perceptual information than what can be effectively processed simultaneously. To cope with this potential overload, the brain is equipped with a variety of attentional mechanisms which serve two critical roles. First, the selection of relevant information, thoughts, or actions while ignoring irrelevant, distracting or interfering information or thoughts. Second, enhance the further processing of the selected information according to the state and goals of the perceiver. Selection of perceptual information about objects or events might also result in awareness of those objects or events. Even assuming that all available perceptual information can be processed simultaneously, a specific action needs to be selected amongst many possible actions to perform a given goal-directed behaviour.

Given the importance of attentional mechanisms for awareness and efficient behaviour, it is very important to understand the nature of visual selective attention. Selective attention can be discussed around varieties of themes, such as the locus of selection (where it occurs in the information processing stream), the basis of selection (object or location or actions), the result of selection, the shifts of visual attention, and the way selective attention affects other cognitive processes. The next section discusses about the possible stage(s) in which selection occurs in the information processing stream.

3. LOCUS OF SELECTION

Early studies on selection focused on auditory processing and the stage in which selection occurred. Cherry (1953) examined the real life experience of following a conversation in a room full of people talking at the same time in the lab settings. Two auditory messages were presented simultaneously, one to each ear (dichotic presentation) and participants were asked to attend to one stream of messages and repeat it loudly after the message is over (shadowing) while ignoring the message from the other ear/stream. Participants were able to report only the attended messages but could not give any details of the unattended messages. The result suggested that attending to a particular stream of inputs resulted in better encoding in the attended stream compared to the unattended stream in which there was degradation or loss of information.

A pertinent question about selection is to what extent information from an unselected stream was processed. Cherry (1953) observed that participants were able to report the pitch (when there was switch in the gender of the speaker) and tone (when speech was replaced with a 400-hz tone) but could not report other details like change in language, individual words or semantic associations in the unattended stream. These results suggest that certain statistical properties are obtained from the unattended channel or stream but not minute details that need selective attention.

The earliest theory of selection was proposed by Broadbent (1958) to explain the findings from studies on auditory attention. Broadbent conceptualized the brain as information processing system with limited capacity, which allows only a certain amount of information to be processed for semantic identification. Broadbent (1958) suggested that the filtering of irrelevant sensory information is based on physical attributes such as pitch, tone, location, colour etc. An early-selection theory posits that unattended, filtered information is not processed beyond its initial physical attributes, for instance, the pitch and tone discrimination of the unattended stream.

However, the alternative, late-selection view holds that selection occurs only after categorization and semantic analysis of all input has occurred (Duncan, 1980). Evidence for late selection view comes from studies that demonstrate semantic processing in the unattended channel (Moray, 1959). Moray (1959) showed that if the unattended message contained highly familiar information such as participant's own name, then participants noticed it (but only on 33% of the trials). The late selection view indicates that all information is processed in parallel up to semantic analysis. However, the late selection view does not imply that there are no limitations to processing (an example would be loss in sensitivity in the periphery of the retina) but that there are no central capacity limitations for perception. The late selection view is consistent with approaches arguing that the purpose of selection is to select the appropriate action based on the semantic analysis of incoming information.

Interestingly, both views of selection (early and late selection) take extreme positions on processing in the unattended channel, as though it is an all or none process. Sometimes we may identify information in the unattended channel, if it is extremely relevant to the current task or behaviour or salient. The filter theories would not be able to explain such effects and therefore Treisman (1964) proposed an intermediate view, which suggests that the filter (or selection process) does not completely reject the unattended information but rather attenuates it. She further added that the attenuated information can reach to the level of semantic analysis, if it is important enough. It might be possible for processing to switch to the unattended stream based on the results of semantic analysis that may lead to awareness.

To obtain evidence for early or late selection, researchers have turned to cognitive neuroscience techniques and some evidence of early selection also comes from neurophysiological studies on attention. An event related potentials (ERP) study using peripheral cues with short SOAs has shown enhancement of the components associated with the early stimulus processing when the target was presented at the cued location (Hopfinger and Mangun, 1998), supporting the early selection view. Effects on P1 (positive peak around 100 ms from stimulus onset) have been found for both non-predictive (Hopfinger and Mangun, 1998) and predictive (van der Lubbe and Woestenburg, 1997) exogenous cues at short SOAs (< 300 ms).

Heinz et al. (1990) have also reported that covert orienting affects early perceptual processing. In this experiment, participants were instructed to direct their attention to one side of the visual display of a block of trials with a fixation at the centre of the screen. Four letter arrays were used as stimuli and presented either to right or left visual field. Participants were required to respond to a specific combination of letters presented on the attended side (25%).

On these trials, the occipital P1 ERP component, which is associated with early stimulus processing, was enhanced on the contralateral side (indicating the attended side) relative to the ipsilateral side (indicating the processing of unattended side) indicating early selection in the information processing stream. The results suggest that covert orienting of attention effects early stages of visual processing.

An important issue that is different from the nature of processing in the unattended channel is related to the number of attended locations or objects. Can we attend to more than one object or location simultaneously? According to the late selection models, we can process semantic information from any number of locations or objects. According to strict early selection models, we can process semantic information from only one location or object. A more interesting and plausible position based on recent evidence is to assume that there are capacity limitations but we can also attend to certain (possibly small) number of locations or objects (Pashler, 1998). This position has been referred to as controlled parallel view of selective attention (Pashler, 1998).

Given that early or late selection view has been supported separately by many studies (Heinz et al., 1990; Pashler, 1998; van der Lubbe and Woestenburg, 1997), Lavie and colleagues have shown that selection processes are affected by task demands (Lavie, 2001; Lavie and Tsal, 1994). Manipulating attentional load to the central task, Lavie and Tsal (1994) found that only selected items were identified in the high load (more difficult) condition compared to the low load condition in which all the items in a display were identified. They have argued under high load conditions, selection is efficient and takes place early in the information processing stream and only selected items are identified further. In the low load condition, filtering occurs late in the information processing stream and allows the identification of all the elements presented in the display (Lavie and Tsal, 1994). A prediction of the load theory (Lavie, 2001) is that distractors also get processed in the low-load condition leading to distractor interference under this condition but not in the demanding high-load condition.

ERP studies have also supported the view of load theory (Lavie, 2001) that task demands determine the locus of selection (Handy and Mangun, 2000). Handy and Mangun (2000) using a cueing paradigm to investigate the effect of perceptual load on level of selection, showed a display containing an arrow pointing towards either right or left visual field, indicating the location of the target (the letter A or H). The task was to identify the letters. The cue was valid on 73% of the trials and invalid on 27% of the trials. The letters were kept intact in the low load condition. In the high load condition, the arms of the letters were pulled apart slightly and the arms of the H were tilted inward to make the discrimination more difficult. They found larger P1 and N1 amplitudes (to some extent) for validly cued targets in the high load condition compared to the low load condition, suggesting that selection occurs early in the high load condition.

To put the locus of selection issue into perspective, it appears that selection processes might depend on task demands with early selection in difficult tasks and late selection in easy tasks. Based on the above discussion it might be suggested that it is also possible to select more than one location or object at any given time. Now, moving away from the debate on locus of selection, the next section focuses on the important issues in selective attention such as the basis on which selection (location or object) is made, orientation of attention, and shifting attention across location/objects for efficient behaviour.

4. TYPE OF SELECTION

Selective attention can be characterized in many ways. One is to distinguish between exogenous and endogenous attention. Exogenous cueing or reflexive orienting is involuntary in nature and depends on the properties of object in space. Endogenous attention is voluntary and is typically based on semantic analysis of the cue. The cuing effects in both exogenous and endogenous can be seen even without accompanying eye movements and is typically referred to as covert orienting. The selection processes depend on information obtained from a given visual display or defined through task instructions. Selection in visual attention is based on location-based information or object-based information.

4.1. Spatial Attention

Active attentional selection occurs over space and time. In space-based attention, the selection is based on the location of an object in space. Spatial attention has been metaphorically compared with a spotlight, highlighting the selected information while leaving the information outside the focus in the dark (Eriksen and Hoffman, 1973). The focus of attention can be adjusted to smaller or larger region of space (LaBerge, 1983). LaBerge (1983) displayed a string of letters and words. The participants' task was to either report the target letter located at the centre of the string or indicate whether the word was a proper noun. On some trials participants were asked to detect the probe letter/digit displayed along with #s, as soon as possible. The position of probe varied across trials with respect to the centre. LaBerge found fastest probe detection at the centre and slowest probe detection located farthest away from the centre (focus of attention) as the primary task was to report the middle letter. However, no difference was found between the probe positions, when participants were asked to report the word, indicating that spotlight can be adjusted according to the task demands (LaBerge, 1983).

The cueing task has been effectively used to study selective attention (Posner, 1980). In the cueing paradigm, participants are required to respond as quickly as possible to the onset of a visual target. The target stimulus is preceded by a "cue" whose function is to draw attention to the occurrence of a target in space. Cues come in various forms, e.g. the brightening of an outline object (Posner and Cohen 1984), the onset of some simple stimulus (Averbach and Coriell, 1961; Eriksen and Hoffman, 1973), or a symbol, like an arrow, indicating where attention should be deployed (Jonides 1981; Posner and Cohen, 1984). Although the mechanisms are debated, generally, cues facilitate the response to stimuli presented at the cued location (Luck et al., 1996). Cues can be exogenous or endogenous in nature and it can orient attention covertly or overtly to the location or object of interest in space.

The typical facilitatory effect at the validly cued location compared to the invalidly cued location due to exogenous cueing is largest at a SOA of 100ms (Cheal and Lyon, 1991; Muller and Rabbit, 1989; Posner, 1980; Yantis and Jonides, 1984). In contrast to exogenous cuing, spatial attention has been studied with endogenous cuing that depends on volition or more precisely explicit semantic processing of the cue information. Studies show that voluntary cues also facilitate processing at the cued location compared to the uncued location but the optimal effect has been reported with longer cue-to-target intervals.

The differences between exogenous and endogenous effect on target detection have been in the same experiment by presenting a display consisting of central fixation and four boxes at the periphery (Muller and Rabbitt, 1989). The task was to discriminate the orientation of the letter 'T'. Cue was valid 50% of the trials. They found that the exogenous cue has a fast-acting effect on the performance. Thus, the peripheral cue was characterized as having a fast, transient response, and the central cue was characterized as having a slow, sustained response. More specifically, the central cues elicited a deliberate shift of attention that is characterized by a monotonic rise to an asymptote, while peripheral cues produce a quick rise and then fall to a lower asymptotic level (and, perhaps, inhibition of return at still longer intervals). Similar findings have been reported by (Cheal and Lyon, 1991; Nakayama and Mackeben, 1989). Thus, on the basis of these studies, it may be concluded that both endogenous and exogenous cues affect early stage of processing, but that the effects of the respective cues depend on the cue-to-target intervals.

The effects of orienting attention to selective attributes of mental representations have also been investigated using ERP and functional magnetic resonance imaging (fMRI) measures. The studies measuring ERPs have revealed that the N1 and P1 components were larger for stimuli presented at the cued location compared to the uncued location and the effects were independent of the paradigms used (Hopfinger and Mangun, 1998; Mangun and Hillyard, 1990; 1991; Mangun et al., 1993; van der Lubbe and Woestenberg, 1997). The differences between exogenous and endogenous cueing effects have also been supported by ERP data (Hopfinger and Mangun, 1998; Mangun and Hillyard, 1991; van der Lubbe and Woestenberg, 1997). ERP studies using exogenous cue showed enhanced P1 component for cued location compared to uncued location at short SOAs (less than 300 ms: Hopfinger and Mangun 1998; van der Lubbe and Woestenberg, 1997). At larger SOAs, the effect was reversed with smaller P1 amplitude for cued location compared to uncued location with exogenous cue. With results from ERP studies, it has been argued that the P1 effect reflects inhibitory activity at the unattended location (or distracter suppression) and the N1 effect reflects a limited capacity discrimination mechanism (or target enhancement) applied to the attended location (Luck, 1995).

Behaviourally, attentional benefits associated with informative retrocues and precues have been found to be similar (Griffin and Nobre, 2003), at least when the number or objects or features in the array fall within the limits of working memory capacity (Wheeler and Treisman, 2002). In spite of a high degree of overlapping activations, some selectivity was observed in the right inferior parietal cortex, which was specifically activated by spatial precues whereas the prefrontal regions were more engaged by spatial retrocues (Nobre et al., 2004). The spatial attention task may require maintenance of directionally specific expectancy and thus a spatial focus for the precues, which was mediated by the inferior parietal areas. In comparison, the prefrontal involvement with the retrocues could reflect top-down biasing signals during zooming of the focus of spatial attention in a working memory context that required selection and maintenance of target information and inhibition of irrelevant distracters (Nobre, 2004).

The ERP waveforms elicited by the spatially informative precues and retrocues relative to neutral precues and retrocues were comprised of early posterior and later frontal potentials (Griffin and Nobre, 2003) which were consistent with previous studies of visual spatial orienting (Hopf and Mangun, 2000). Mainly the fronto-parietal network in the brain has emerged as the neural system responsible for orientation functions of spatial attention in tasks where attention is directed towards peripheral stimuli, both during cued detection at several locations (Corbetta et al., 1993).

A major difference between exogenous and endogenous cuing effect is shown with different effects obtained at longer SOAs. With exogenous cuing, a reversal of the cueing effect has been reported when cue-to-target asynchrony is longer than 300ms i.e., RTs were faster at the invalidly cued location compared to the validly cued location (Posner and Cohen, 1984). However, with endogenous cue the optimal effect has been reported with longer cue-to-target interval (> 400 ms). This phenomenon has been termed as inhibition of return (IOR). IOR can last up to 3-4 seconds and typically operates with environmental co-ordinates rather than retinal co-ordinates (Vecera and Luck, 2000). IOR magnitude decreases with increasing distance of the target from the original cued location.

Inhibition of already visited locations and objects becomes increasingly critical when we are performing a search in the environment. IOR has been investigated to some extent using ERP measures to show that reflexive attention mechanisms can modulate visual processing (McDonald et al., 1999). The P1 component which appears at early stages of processing at about 120 ms after stimulus onset shows reduced amplitudes for targets presented at cued locations (at long SOAs). The P1 activity has been related to inhibitory perceptual processes. For example, P1 amplitude was reduced whenever stimulus was presented at an unattended location indicating suppression of information from locations that are marked to be irrelevant (Luck, 1995). Yet there is no clear consensus regarding the psycho-physiological correlates of IOR as P1 effects do not always coincide with behavioural IOR (Eimer, 1994a; 1994b). The other effects seen in ERP components include a reduced positivity in the latency range of 200-300 ms and increased P3 amplitudes for targets at cued location (McDonald et al., 1999). Another component reported to show differences with cueing conditions is termed as "negative difference" or Nd obtained by subtracting ERP of uncued condition (with reduced negativity) from cued ERP (Eimer, 1994a).

4.2. Object-Based Attention

When information is selected on the basis of object-based properties, then it is referred to as object-based selection. A significant number of studies have provided support for space-based attention but studies have also provided evidence for object-based attention as well. Questions have been raised about object-based selection since an object is located in a particular location in space. It has been argued that an object cannot be selected without selecting the spatial region in which it is located. On the contrary, other studies have argued that object-based selection may exist without considering spatial information (Chun and Wolfe 2001; Duncan, 1984).

To demonstrate object-based attention, Duncan (1984) used a display containing two superimposed objects. He found that participants were most accurate in reporting two features from the same object than reporting one feature from each of the two objects. Duncan (1984) argued that the shift of attention is purely object-based as both objects were overlapped at the same location. In a recent study by Blaser et al. (2000), two superimposed circular striped Gabor patches were presented as objects, which changed smoothly (drifted) along their feature dimensions (orientation, spatial frequency, and color). Even though, both patches were at the same location, observers could attend to one of them and track as the object changed in feature-space. Similar to Duncan's (1984) study, the performance was better with features from the same Gabor patch compared to when each feature was from a different Gabor patch.

To investigate the interaction between location-based selection and object-based selection, Egly et al. (1994) have conducted a study by showing a display consisting two rectangular bars with fixation cross in the centre. Participants were asked to attend to one end of the two rectangles in a display by flashing a light at the end of the rectangles. The flashing light was presumed to engage stimulus-driven attentional mechanisms. Participants were then asked to detect the onset of the target that appears at the cued location, in opposite end (i.e. uncued end) of the cued object, or in the other uncued object. There was an object advantage: participants were faster to detect targets in the uncued but same object compared to the uncued but different object. This was true even when those locations were equidistant from the cued location. Thus objects in the scene play an important role in guidance of the selective attention.

Lavie and Driver (1996) used a variation of Duncan's (1984) paradigm to examine potential limitations of object-based attention. They had participants view two lines (with dots and dashes) to compare the length of the two target elements that differ from each other in colour (the target were either dot or a dash in a line). They found that participants made this comparison more quickly when each element was on the same line than when each element was on a different line, revealing a typical object advantage. However, they also investigated whether stimulus driven attention might modulate the object advantage. The object advantage did not occur when spatial attention was narrowly focused. In particular, the object advantage did not appear to extend beyond the focus of attention. The result suggested that the object-based attention has a limited role in space and object advantage is reduced or eliminated for the portions of a scene that are outside the focus of attention.

The issue was further investigated by Abrams and Law (2000) by conducting a series of experiments. In their study, they presented two bars having a notch at the end and participants' task was to discriminate the size of the notch presented at the bars. The task was preceded by four conditions: the valid cue near condition (the notch appeared at the cued side of the display and on different bars), the invalid near condition (the notch appeared at the uncued side or opposite side of cue and on different bars), the far condition (the notch appeared on different bars and on the other side of the display), the object condition (the notch appeared on the other side of the display and on the same bar). They concluded that spatially directed attention and object-based attention can coexist, and object-based attention was observed in the absence or presence of focused spatial attention. They showed that participants were faster and more accurate in detecting the relative size of notch bars objects when both notches were presented on the cued side of the display than when they were presented on the uncued side of the display. Importantly, the participants were faster in comparing notch size when the notches were both on the bars of the same object than when they were a similar distance apart but on different objects, demonstrating an object-based effect. It has been suggested that once a part of an object is selected, it helps to process the whole object even if the other part is irrelevant to the current task or out of the focus of attention (Abrams and Law, 2000).

Similar to IOR for locations discussed in the earlier section, IOR has also been observed with several object-based properties such as colour, shape, and size (Pratt et al., 1997; Riggio et al., 2004). Using cueing task, Tipper (1991) have shown object-based IOR at longer intervals between cue and target (greater than 400 ms), IOR was observed for the cued object (object-based IOR) rather than the cued location. Other evidence for modulation of IOR based on object-based information includes greater IOR for face versus non-face targets (Theeuwes and van der Stigchel, 2006).

IOR has also been shown for spatial frequencies (Brown, 2009; Srinivasan and Brown, 1997). Greater IOR has been found with conditions that favour processing the P-ventral pathway and lesser IOR has been found with conditions that favour processing in the M-dorsal pathway (Brown, 2009). Inhibition has been found with both cued objects and locations, suggesting the existence of object-based and location-based components of IOR that reflect the operation of two partially separate inhibitory mechanisms (Abrams and Pratt, 2000).

Evidence for object-based selection also comes from ERP and neuroimaging studies (Arrington et al., 1999; He et al., 2004; O'Craven et al., 1999; Valdes-Sosa et al., 1998). Valdes-Sosa et al. (1998) have investigated the neural correlates of object-based attention with ERPs and they found that the larger posterior P1 and N1 components of the motion-onset ERPs were associated with the attended set of rotating dots (object) compared to the unattended set (object). Evidence for feature based selection was provided by a classic experiment using PET in which subjects were asked to pay attention to different features of the same visual arrays (the colour, shape or speed of the motion of the element) (Rees et al., 1997). Different regions in extrastriate cortex were active when subjects attended to different features of the same arrays (Rees et al., 1997). This result cannot be accounted for in terms of either space-based or object-based selection because all of the visual features were present in the same location and all were properties of the same objects. Instead, it argues strongly that attention can directly affect the extraction and/or representation of specific visual features.

A fMRI study by O'Craven et al. (1999) has revealed evidence for both feature-based and object-based attention. In the study, two stimuli, a face and a house, were superimposed transparently in the same location. On each trial, either the face or the house oscillated back and forth along one of the four axes. Observers were required to report the three different attributes (face, house or the direction of motion or the position of the stationary stimulus, which was slightly displayed from the centre). The neural activity in each of the three cortical areas responsible for processing the specific visual attribute was higher when the corresponding visual attribute was the irrelevant property of an attended object than when it was a property of an unattended object. For example, the signal was higher in MT or MST for the moving face than the house if the face was attended, even though all the features were present in the same location and motion was completely irrelevant to the task. These results strongly indicate that objects function as the unit of attentional selection and directing attention to one attribute of an object would produce not only enhancement of the activation in the cortical region coding that attribute, but also enhancement of the other features associated with that object.

An ERP study by He et al. (2000) demonstrated that object-based processing affects the sensory areas in the posterior region and space-based processing affects the anterior regions in the brain. Furthermore, Arrington et al. (1999) found greater activation for the bounded region selection than unbounded region especially in the left hemisphere in an object-based attention study using event-related fMRI. The results discussed so far indicate that object information plays a significant role in selection and both types of attention are implemented in different circuits and areas in the brain.

The results from facilitation and inhibition studies indicate that both spatially directed attention and object-based attention can operate simultaneously and may employ different mechanisms. He et al. (2004) investigated object-based attention by manipulating cue validity using the same paradigm used in the Egly et al. (1994) study. They found effects for both types of selection in the high cue validity condition whereas the object-based attention effects were present only in the low cue-validity condition. Furthermore, the ERP results have demonstrated object-based effects at sensory level over the posterior areas of brain, and space-based effects over the anterior region (He et al., 2004). On the basis of these results, they suggested that space-based and object-based components of attention reflect mainly voluntary and reflexive mechanisms, respectively (He et al., 2004).

The selection of objects have been explained through a biased competition model of attention according to which activations from multiple objects in a visual display compete with each other with only the winner moving on to awareness or working memory (Desimone and Duncan, 1995). According to the biased competition model, attentional selection is an emergent effect of competition between neural representations in multiple brain systems. Further, the bias can be affected by top-down processing that helps in selecting relevant information or bottom up processing that influences the probability of selecting (capturing) particular information. Evidence for biased competition in the brain comes from a fMRI study in which visual stimuli were presented simultaneously as well as in succession (Kastner et al., 1998). They found that the neural activations in V4 are reduced in the simultaneous condition compared to the successive condition indicating competition between visual representations in the brain. While both location-based and object-based selection can be based on one unit at a time, it is possible for the system to select more than one object or location. Below we discuss the issues associated with selection when multiple items need to be processed.

5. Divided Attention

How many objects/tasks can be attended and performed at the same time? Most of the studies have examined the attentional processes involved in focusing on a single object or location i.e. more focused attention compared to divided attention. Our everyday experiences show that we may have to process information about more than one object at a time, but the efficiency of the performance depends on the nature of the objects/tasks to be performed. This section deals with various issues related with divided attention on more than one location/object over space and time. Studies have examined our ability to divide attention between multiple objects and focus attention on target(s) rather than distractors (Awh and Pashler, 2000; Cavanagh and Alvarez, 2005; Raymond et al., 1992). Studies using visual search and attentional blink have been extensively used to study divided attention in space and time (Duncan et al., 1994; Treisman and Gelade, 1980; Wolfe, 1994).

5.1. Visual Search

A number of accounts have been put forth to show how attention modulates the amount of time required for visually searching for objects in the environment (Treisman and Gelade,

1980; Wolfe, 1994). In a visual search task, participants look for one or more targets if they are present (among distractors) by saying "yes" or "no" in a present/absent task or reporting the designated target from the *n*-alternative targets' in a forced-choice search task. The number of items presented in the display is called display set size (Pashler, 1998). The efficiency of visual search can be assessed by observing changes in performance: reaction times and accuracy as a function of set-size or number of items in the display. The search is faster if the target is defined by a single feature (feature search) than by two or more features mainly because greater amount of attentional resources are required for the latter or the separate features have to be bound together (conjunction search; Treisman and Gelade, 1980). For instance, in feature search the target differs from the distractors in a single feature only such as finding a red X among green Os. In conjunction search, target shares features from distractors in more than one dimension such finding a green T among green Os and red Ts. In general, the search times are dependent on the number of items present in a display in a conjunction search but not in the feature search. This is because the target tends to "pop out" in a feature display because of its uniqueness, whereas attention needs to be shifted from one item to another until the target is found in a conjunction display.

Using ERP, studies have shown the N2pc component (a negative going voltage deflection that is typically observed in a latency range of 200-300 ms) after the onset of the visual search array. It is largest over the visual cortex in the hemisphere contralateral to the location of the attended object with in the visual search array (Woodman and Luck, 1999). The N2pc activity is indicative of rapid shifts of attention towards objects before they are recognized (Luck and Hillyard 1994a; 1994b) which typically occurs during visual search.

Why is feature search more efficient than conjunction search? To explain this dichotomy, Treisman proposed the feature integration theory (Treisman and Gelade, 1980). She has argued that visual search is a two-stage process: pre-attentive and attentive. Further, she has suggested that an efficient feature search entails parallel (pre-attentive) and conjunction search involves serial (attentive) processing (Treisman and Gelade, 1980) as attention is required to conjoin or bind multiple features into a single object. Hence, conjunction searches are serial (Treisman and Gelade, 1980), and withdrawing attention produced errors for binding features, known as "illusory conjunctions" (Treisman and Schmidt, 1982). A more recent model is the guided search model (Wolfe, 1994) that argues that visual search can be guided by the information from any number of preattentive parallel processes (Wolfe et al,. 1989). They have suggested that preattentive processes guide the spotlight of attention towards a likely target (Wolfe, 1994). Guided search model is a two stage model from spatially parallel to serial enabling computation of more complex features of objects. The efficiency of the search is a function quality of the guidance provided by the parallel processing (Wolfe et al., 1989). Both top-down and bottom-up processing is used to rank the items present in a visual display according to attentional priority that guides visual search.

5.2. Attentional Blink (AB)

Studies on visual search have suggested that shift of attention is relatively fast and on an average we spend very few milliseconds on each search items to search the target items (Johnson and Proctor, 2004; Treisman and Gelade, 1980; Wolfe, 1994). On the contrary, other studies have suggested that visual attention is not a high-speed switching mechanism as suggested by estimates from visual search tasks (Ward et al., 1996; Duncan et al., 1994). Generally, the speed of shift of attention from one object or location to another is studied using attentional dwell time or attentional blink paradigms. Attentional blink and attentional dwell time paradigms have been effectively used to study the temporal aspects of attention (Duncan et al., 1994; Raymond et al., 1992; Ward et al., 1996). Attentional blink is a method of displaying an array of items in rapid succession at a fixed location. The items consist of two targets presented at different lags along with distractors. Participants' task is to identify both the targets and report at the end of the trial. Similarly, attentional dwell time or the two target paradigm is a method of displaying two targets at two different locations with variable SOA. The task is to identify both the targets and report at the end of the trial.

The studies suggest that when stimuli are presented serially, the first target interferes highly with the processing of the second target when both were separated by 300-500 ms approximately (Duncan, 1994; Moore et al., 1996; Ward et al., 1996) and this interference reduces as the temporal separation between the two stimuli increases (Nieuwenstein et al., submitted). Thus we normally require at least 200-300 ms to shift our attention from one location to another location or from one object to another object. The attentional dwell time has also been found to be the same for targets present at same or different locations (Nieuwenstein et al., in press; Ward et al., 1996).

It has been argued that the AB reflects the limitation of attentional resources shared by the competing targets for access to the visual short term memory (VSTM) for eventual identification and action (Chun and Potter, 1995; Duncan and Humphreys, 1999; Duncan, Ward, and Shapiro, 1994). Attentional resource has been argued to play a crucial role in shifts of visual attention and necessary for processing all the perceptual stimuli.

Using fMRI, Marois, Chun, and Gore (2000) have supported the limited-capacity account and showed activations in right intraparietal and frontal cortex that are responsible for capacity-limited processes. Further, they have suggested a significant role for the right parieto-frontal network (implicated in attentional control and enhancement) in capacity-limited processing responsible for AB (Marois et al., 2000). Marcantoni et al. (2003) have shown activations in inferotemporal and posterior parietal cortex, lateral frontal and cerebellum in short delay condition (shorts lags < 300 ms) suggesting the role of these brain areas in decisions under masking and interference and support the notion of limited-capacity processing.

Using event related potentials (ERP), studies have shown suppressed P3 ERP component for target items presented at short lags (<300ms), indicating the difficulties in consolidation in working memory (Kranczioch et al., 2003). Kranczioch et al. (2003) have shown P3 ERP component during (i.e. at short lags when AB occur) and after AB period, indicating that the target presented during AB period can reach to working memory and does not support the limited-capacity view of AB. But it should be noted that the P3 ERP component at short lag (lag 2) was smaller than larger lags (lag 7), indicating different degrees of overlapping and interference.

Recently, Olivers et al. (2006) have argued that the impaired performance on the second target is due to the overinvestment of attentional resources on the AB task. They have manipulated the sharing of attentional resource with an additional task to the AB main task and reported a better performance with the second target compared to other AB studies (Duncan, et al., 1994; 1996; Raymond et al., 1992). In addition, they have also presented positive, negative and sad emotions in addition to the AB task and found improved performance on the second target when participants were shown positive information compared to negative or neutral information. They have suggested that the sharing of attentional resources improves the performance on AB task for second target. Though, they have argued in terms of over investment of attentional resource allocated centrally to the AB task, it can also be argued that an additional task enables distribution of attentional resources and increases the scope of attention which in turn results in better performance on the second target compared to when attention was narrowly focused on the main task.

6. FOCUSED AND DISTRIBUTED ATTENTION

Attention can be distributed over space or narrowly focused on an object or region in space. Studies on focused attention have indicated that focused attention is limited to a small number of regions/items (typically 4) in space (Awh and Pashler, 2000; Cavanagh and Alvarez, 2005). In a study with moving targets among distractors that had to be tracked, only up to four targets could be tracked indicating the limited capacity for focused attention (Cavanagh and Alvarez, 2005). In change detection tasks, changes between an initial display and a subsequent display can be detected efficiently only up to four squares (Luck and Vogel 1997). The number of items that can be selected depend on the detailed nature of the representations with selection of more items possible when coarse representations are formed (Franconeri et al., 2007).

Treisman (2006) have argued that the two different types of attentional deployment (focused or distributed) lead to differences in attentional processing with focused attention enabling detailed analysis of specific features/objects and distributed attention facilitating global/whole registration of objects/group of objects' properties. Distributed attention compared to focused attention benefits the computation of statistical properties in a scene (Atchley and Andersen, 1995; Dakin and Watt, 1997; Parkes et al., 2001) and may also benefit shifts of visual attention (Olivers and Nieuwenhuis, 2006).

The statistical properties include many mean judgments in various stimulus dimensions including orientation (Dakin and Watt, 1997; Parkes et al., 2001), motion speed (Atchley and Andersen, 1995), and motion direction (Williams and Sekuler, 1984). For instance, the accurate estimation of average tilt of oriented signals in a crowded visual display requires attention to be distributed over space compared to focus on an orientation of individual items (Parkes et al., 2001). Statistical estimation is found to remain good under difficult perceptual conditions, such as brief set exposure duration, or the insertion of a delay between two sets that need to be compared based on mean estimation (Chong and Treisman, 2003). In addition, performance in mean judgment tasks has been found to be independent of set size (Ariely 2001; Chong and Treisman, 2005a). Extraction of statistical properties can be modulated by features that are attended previously and therefore are not considered as an automatic process (de Fockert and Marchant, 2008).

An alternative account for the lack of set size effects on computing statistical information is the sub-sampling strategy, which has been used as an alternate explanation for the findings of mean judgment of size (Myczek and Simons, 2008). A number of simulations were performed where subsets were selected at random and on average those subsets had a mean size similar to that of the entire set. As a result, simulations based on subset-averaging of one or two items were very similar to the performance of participants who were instructed to average the entire set. However, the strategy of subset-averaging may not explain all the findings on distributed attention based tasks of mean computation. For example, in dual task conditions, the task of mean judgment benefited from tasks requiring distributed or global attention (pop-out search) compared to focused attention task (conjunction search) (Chong and Treisman, 2005b). This emphasizes parallel processing mechanisms for distributed attention contrary to serial processing mechanisms for focused attention. The parallel accounts for distributed attention were also confirmed when an advantage in mean judgment was observed with successive presentation compared to simultaneous presentation of sets (Chong and Treisman, 2005b).

All of these observations have demonstrated a distributed deployment of attention based on task demands. However, attention can be deployed differently based on object properties as well. The next section focuses on the distributed and focused deployment of attention and their potential relationship with emotional and hierarchical stimuli.

6.1. Emotions and Scope of Attention

Control of attention has been shown to be influenced by the current affective state of the observer (Hasher et al., 2007). Emotional stimuli have been found to interact with the scope of attention. Negative states are associated with a constriction of attentional focus (Derryberry and Reed, 1998) at the expense of encoding peripheral details (Christianson and Loftus, 1990). Positive states are associated with the broad scope of attention (broaden-and-build theory: Fredrickson, 2004; Fredrickson and Branigan, 2005; Wadlinger and Issacowitz, 2006) and reduction in selectivity (Fenske and Eastwood, 2003; Rowe et al., 2005).

It has been found that positive emotion increases the thought-action repertoires due to the enhancement of the global scope of cognitive processing (Fredrickson, 2004; Frederickson and Branigan, 2005). In addition, identification of positive emotions has been linked to global processing and identification of negative emotions has been linked to local processing (Srinivasan and Hanif, 2010). Recognition memory for happy or sad happy emotional faces has been shown to interact with load and the scope of attention (Srinivasan and Gupta, 2010). Moreover, positive and global information necessitates distributed attention and negative and local information requires more focused attentional processing (Srinivasan and Hanif, 2010; Srinivasan and Gupta, 2010; Srivastava and Srinivasan, 2010).

Srivastava and Srinivasan (2010) have investigated the role of emotional stimuli in shifts of visual attention between objects. In their study, participants were presented with happy and sad stimuli using attentional dwell time paradigm. Two experiments were conducted by manipulating emotional faces as T1 or T2 in separate experiments. In the first experiment, emotional stimuli (T1) were followed by the neutral target (T2). The results showed less impairment for neutral T2 performance when it was preceded by the happy faces compared to the sad faces. This could be due to lesser attentional resources or broadening of attention associated with happy faces. To investigate whether happy stimuli demand less attentional resources, a second experiment was conducted using emotional stimuli as T2 preceded by a neutral T1. Results showed better identification of the happy faces than sad faces indicating less attentional demand for happy stimuli compared to sad face. These findings support theories that argue for reciprocal links between emotions and the scope of attention and indicate that broadening of attention requires less attention (Srivastava and Srinivasan, 2010) and therefore, interferes less with the subsequent target processing.

6.2. Global-Local Processing and Scope of Attention

Processing complex objects such as faces or scenes that are composed of parts and sub-parts involve perceptual parsing or perceptual grouping for object identification and can be thoughts in terms of global or local levels of processing (Kimchi, 1992; Navon, 1977; Robertson et al., 1993). Most studies have used compound/hierarchical stimuli (letters/shapes) to understand the hierarchical nature of perceptual processing (Navon, 1977). A global hierarchical stimulus is composed of either identical or different smaller local stimuli. For instance, a global letter "S" is made up of local letters "S" or "H". Participants' task is to identify the letters or digits either at the local or global level. If the level of the non-target letter is different or same from the target level then stimulus is considered inconsistent (incongruent) or consistent (congruent) respectively. Perceptual processing is considered to either start from global processing followed by local processing (Navon, 1977) or hierarchically build up from processing local information towards global processing (Kimchi, 1992). Therefore, it can be concluded that the bias in attending to global or local percept is dependent on the properties of local items.

Irrespective of the global/local precedence, it is important to note that one can shift attention from one level to another. Using compound stimuli, Robertson et al. (1993) have shown a relationship between global-local form processing and scope of attention. They found that the selected/attended area could be contracted or expanded by attending to local or global information respectively (Robertson et al., 1993). Robertson and colleagues (1988; 1993) have shown that global processing was faster when attention was distributed over space and local processing was faster when attention was narrowly focused to the central region. Similarly, a large number of studies have shown faster and better understanding of gist of scene when attention was distributed over space compared to being narrowly focused on the local items (Greene and Oliva, 2009; Tatler, Gilchrist, and Risted, 2003). These studies indicate that processing of global patterns is linked to increased scope of attention compared to local processing.

Recently, a study conducted by Srivastava and Srinivasan (2009; 2010), investigated the time course of shift of visual attention manipulated by differences in scope of attention. In one of their experiments, participants were shown compound letters and digits as T1 and T2 to vary the scope of attention using attentional dwell time paradigm. They found an advantage in covert shifts of attention from one object to another attended at the global level involving distributed attention compared to when both or either of the objects were attended at local level entailing focused attention. Most importantly, they found efficient identification of both the objects attended at global level compared to when both or either of the objects were attended at local level indicating that distributed attention facilitates global processing over space.

7. MODELS OF ATTENTION

Many computational models have been proposed for simulating various attentional processes in the past thirty years (Bundesen, 1990; 1998; Bundesen et al., 2005; Heinke and Humphreys, 2003; Logan, 1996; Rolls and Milward, 2000; Wallis and Rolls, 1997). Models include mathematical models of attention as well as connectionist models for space- and object-based selection.

7.1. Connectionist Models

Many models have been proposed to explain findings from selective attention tasks (Heinke and Humphreys, 2003; Rolls and Milward, 2000). They include VisNet and Selective attention for identification model (SAIM) that have been proposed for object recognition and visual selective attention. VisNet was proposed by Wallis and Rolls (1997) and modified by Rolls and Milward (2000), which is based on cortical architecture. Rolls and colleagues tried to study various aspects of invariant object-recognition in the primate visual system and investigated the way neurons learn to respond similarly to the gradually transforming inputs they receive over the short term to the same object. VisNet is a four-layered feed-forward network in which each layer converges from a small region of the preceding layer with a competition between the neurons within the layers. They used a trace learning rule for training the model for invariance transformation. Trace rule was a modified Hebbian rule that modifies synaptic weights according to both the current firing rates and the firing rates to recently seen stimuli enabling the network to learn invariances in object as they get transformed.

VisNet was trained with simple stimuli such as T, L, and +, or more complex stimuli such as faces. Relatively few different stimuli such as faces were used for training. The network was trained by presenting a stimulus in one training location, calculating the activation of the neuron, then its firing rate, then updating its synaptic weights. The sequence of location changes in sets of five locations. They applied more biological realistic activation functions for the neurons by replacing the power function with a sigmoid function. The sigmoid function models not only the threshold nonlinearity of neurons, but also the fact that eventually the firing rate of neuron saturates at high rate. In addition, it also provides good performance with sparseness than were possible with a power activation function.

SAIM (Heinke and Humphreys, 2003) is a model for visual selective attention and object-recognition that was proposed to simulate and explain findings from visual attention studies on object recognition. They proposed that attention can help to produce translation invariance pattern recognition by serving as a 'window' through which activation from a retinal field must pass before being transmitted through to recognition system. They suggested that window could be shifted around to different spatial positions; same units will modulate recognition, irrespective of lateral positions of the objects in the retinal field. Consequently, recognition units do not need to be coded for every location in the visual field, since the same input will occur across different retinal positions, normalized by shifting the attentional window. The shifting of an attentional window across space is enabled with two sets of units: one set of units map the contents in the visual field into the attention window or focus of attention (FOA), performed by content network (Olshausen et al., 1993). The second sets of units control the shifting of the FOA, by selection network. The third one is the knowledge network that has stored knowledge in the form of template units and recognizes the contents of the FOA via simple template matching. These three components operate in parallel. The location map stores location information for the attended objects and prevent these locations from being selected again, which determines attention switching behaviour. The results from SAIM have been shown to fit with findings from many studies (e.g., Egly et al., 1994).

7.2. Mathematical Models

One way to elucidate the mechanisms involved in selective attention is by manipulating perceptual discriminability using noise (Lu and Dosher, 2005). They have proposed a perceptual template model based approach to see whether the mechanisms involved in attention are amplification, external noise reduction, and multiplicative noise reduction. Using the PTM, Lu and Dosher (2005) predicted that stimulus enhancement would improve performance only in the low-noise conditions and external noise reduction would improve performance mostly in the high-noise conditions. They found evidence for these two mechanisms but not for multiplicative noise reduction.

Detailed models of visual attention have also been proposed by Bundesen, Logan and their colleagues (Bundesen, 1990; 1998; Bundesen et al., 2005; Logan, 1996). Bundesen's theory of visual attention (TVA) models perceptual categorization as a function of visual selective attention. The perceptual categorization is determined by $v(x,i)$ which is determined by

$$v(x,i) = \eta(x,i)\beta_i \frac{w_x}{\sum_{z \in S} w_z}$$

where $\eta(x,i)$ is the instantaneous strength of the sensory evidence that element x belongs to category i, β_i is a perceptual decision bias associated with category i, S is the set of all elements in the visual field, and w_x and w_z are attentional weights of elements x and z respectively. The weight of an element an element x is dependent on the processing priority indicates the importance of attending to the elements that belong to a given category. The weight of an element x is given by

$$w_x = \sum_{j \in R} \eta(x,j)\pi_j$$

where R is the set of all perceptual categories, $\eta(x,j)$ is the instantaneous strength of the sensory evidence that element x belongs to category j and π_j is the perceptual processing priority of category j. TVA uses two mechanisms, filtering and pigeon holing (Bundesen, 1990; 1998). Filtering depends on attentional weights, which are determined by perceptual processing priorities. Pigeon holing works through perceptual bias. Bundesen's model has been successfully applied to many tasks including selection from multielement displays and other visual search tasks.

The code theory of visual attention (CTVA) combines space-based and object-based selection (Bundesen, 1998; Logan, 1996). CODE theory models a perceptual grouping process based on assuming that each unit in a given location produces a Laplacian distribution centred at that location. The distributions are added to give a CODE surface. To obtain a perceptual group, a threshold is applied to the CODE surface and the regions above the threshold are cut. Different groupings can be obtained by manipulating the threshold. The CODE is linked to TVA using the concepts of spatial focusing and feature catch. Spatial focusing determines the spatial scope of attention and hence determines the attended elements in the visual display. It is possible to process information from more than one unit simultaneously if the focus of attention contains more than one unit. Feature information is obtained from the CODE surface after the application of the threshold, which will be used as input to the Bundesen's TVA (Bundesen, 1998).

CONCLUSION

Attention has been a popular topic of research in psychology, cognitive science and the neurosciences for the past fifty years. The studies using multiple methodologies have enabled us to gain a better understanding of selective attention in terms of the nature of selection processes and its consequence for purposeful and efficient behaviour. Findings from cognitive neuroscience have enabled us to verify theoretical proposals on the locus of selection and the mechanisms involved in selection. Efforts are being made to link mathematical models of attention to findings from cognitive neuroscience (Bundesen et al., 2005).

ACKNOWLEDGMENTS

We are thankful to Rashmi Gupta for her comments.

REFERENCES

Abrams, R.A., and Law, M.B. (2000). Object-based visual attention with endogenous orienting. *Perception and Psychophysics, 62,* 818-833.

Abrams, R.A., and Law, M.B. (2000). Object-based visual attention with endogenous orienting. *Perception and Psychophysics, 62,* 818-833.

Abrams, R.A., and Pratt, J. (2000). Oculocentric coding of inhibited eye movements to recently attended locations. *Journal of Experimental Psychology: Human Perception and Performance, 26,* 766-788.

Ariely, D. (2001). Seeing sets: Representation by statistical properties. *Psychological Science, 12,* 157–162.

Arrington, C.M., Rao, S.M., Mayer, A.R., and Carr, T.H. (1999). Brain mechanisms underlying object-based attention: How do we select a bounded region of space? *Journal of Cognitive Neuroscience, Supplement S, 68.*

Atchley, P., and Andersen, G. (1995). Discrimination of speed distributions: Sensitivity to statistical properties. *Vision Research, 35,* 3131–3144.

Averbach, E., and Coriell, A.S. (1961). Short-term memory in vision. *Bell Systems Technical Journal, 40,* 309-328.

Awh, E., and Pashler, H. (2000) Evidence for split attentional foci. *Journal of Experimental Psychology: Human Perception and Performance, 26,* 834–846.

Blaser, E., Pylyshyn, Z.W., and Holcombe, A. (2000). Tracking an object through feature-space. *Nature, 408,* 196-199.

Broadbent, D. (1958). *Perception and Communication.* New York: Pergamon Press.

Brown, J. (2009). Visual streams and selective attention. In N. Srinivasan (Ed.) *Progress in Brain Research: Attention, 176,* 47-64.

Bundesen, C. (1990). A theory of visual-attention. *Psychological Review, 97,* 523-547.

Bundesen, C. (1998). A computational theory of visual attention. *Philosophical Transactions of the Royal Society of London, Series B, 353,* 1271–1281.

Bundesen, C., Habekost, T., and Kyllingsbæk, S. (2005). A neural theory of visual attention: Bridging cognition and neurophysiology. *Psychological Review, 112,* 291-328.

Cavanagh, P., and Alvarez, G.A. (2005). Tracking multiple targets with multifocal attention. *Trends in Cognitive Sciences, 9,* 349–354.

Cheal, M., and Lyon, D. R. (1991). Central and peripheral precueing of forced-choice discrimination. *The Quarterly Journal of Experimental Psychology, 43A,* 859-880.

Cherry, E.C. (1953). Some experiments on the recognition of speech, with one and two ears. *Journal of the Acoustical Society of America, 25,* 975–979.

Chong, S. C., and Treisman, A. (2005a). Statistical processing Computing the average size in perceptual groups. *Vision Research, 45,* 891–900.

Chong, S.C., and Treisman, A. (2003). Representation of statistical properties. *Vision Research, 43,* 393–404.

Chong, S.C., and Treisman, A. (2005b). Attentional spread in the statistical processing of visual displays. *Perception and Psychophysics, 67,* 1–13.

Christianson, S.A., and Loftus, E. (1990). Some characteristics of people's traumatic memories. *Bulletin of the Psychonomic Society, 28,* 195–198.

Chun, M. M., and Potter, M. C. (1995). A two-stage model for multiple target detection in rapid serial visual presentation. *Journal of Experimental Psychology: Human Perception and Performance, 21,* 109-127.

Chun, M., and Wolfe, J. M. (2001). Visual Attention. In B. Goldstein (Ed.), *Blackwell Handbook of Perception.* Oxford, UK: Blackwell Publishers Ltd.

Corbetta, M., Miezin, F., Shulman, G., and Petersen, S. (1993). A PET study of visuospatial attention. *Journal of Neuroscience, 13,* 1202–1226.

Dakin, S., and Watt, R.J. (1997). The computation of orientation statistics from visual texture. *Vision Research, 37,* 3181–3192.

de Fockert, J.W., and Marchant, A.P. (2008). Attention modulates set representation by statistical properties. *Perception and Psychophysics, 70,* 789–794.

Derryberry, D., and Reed, M.A. (1998). Anxiety and attentional focusing: Trait, state and hemispheric influences. *Personality and Individual Differences, 25,* 745–761

Duncan, J. (1980). The locus of interference in the perception of simultaneous stimuli. *Psychological Review, 87,* 272-300.

Duncan, J. (1984). Visual attention. In B. Goldstein (Ed.), *Blackwell Handbook of Perception.* Oxford, UK: Blackwell Publishers Ltd.

Duncan, J., and Humphreys, G. (1989). Visual search and stimulus similarity. *Psychological Review, 96,* 433-458.

Duncan, J., Ward, R., and Shapiro, K. (1994). Direct measurement of attention dwell time in human vision. *Nature, 369,* 313-315.

Egly, R., Driver, J., and Rafal, R.D. (1994). Shifting Visual attention between objects and locations: Evidences from normal and parietal lesions subjects. *Journal of Experimental Psychology: General, 123,* 161-177.

Eimer, M. (1994a). An ERP study on visual spatial priming with peripheral onsets. *Psychophysiology, 31,* 154-163.

Eimer, M. (1994b). "Sensory gating" as a mechanism for visuospatial orienting: Electrophysiological evidence from trial by trial cuing experiments. *Perception and Psychophysics, 55,* 667-675.

Ericksen, C.W., Hoffman, J.E. (1973). The extent of processing of noise elements during selective encoding from visual display. *Perception and Psychophysics, 14,* 155-160.

Fenske M.J., and Eastwood, J.D. (2003). Modulation of focused attention by faces expressing emotion: evidence from flanker tasks. *Emotion, 3,* 327-343.

Franconeri, S.L., Alvarez, G.A., and Enns, J.T. (2007). How many locations can be selected at once? *Journal of Experimental Psychology: Human Perception and Performance, 33,* 1003-1012.

Fredrickson, B.L. (2004). The broaden and build theory of positive emotion. *Philosophical Transactions: Biological Sciences (The Royal Society of London), 359,* 1367-1377.

Fredrickson, B.L., and Branigan, C. (2005). Positive emotions broaden the scope of attention and thought-action repertoires. *Cognition and Emotion, 19,* 313-332.

Green, M.R., and Oliva, A. (2009). Recognition of natural scenes from global properties: Seeing the forest without representing the trees. *Cognitive Psychology, 58,* 137-176

Griffin, I.C. and Nobre, A.C. (2003). Orienting attention to locations in internal representations. *Journal of Cognitive Neuroscience, 15,* 1176-1194.

Handy, T.C., and Mangun, G.R. (2000). Attention and spatial selection: Electrophysiological evidence for modulation by perceptual load. *Perception and Psychophysics, 62,* 175-186.

Hasher, L., Lustig, C., and Zacks, R.T. (2007). Inhibitory mechanisms and the control of attention. In A. Conway, C. Jarrold, M. Kane, A. Miyake, and J. Towse (Eds.), *Variation in Working Memory* (pp. 227–249). New York: Oxford University Press.

He, X., Fan, S., Zhou, K., and Chen, L. (2004). Cue Validity and Object-Based Attention. *Journal of Cognitive Neuroscience, 16,* 1085-1095.

Heinke, D., and Humphreys, W.G. (2003). Computational models of visual selective attention: A review. In G. Houghton (Ed.), *Connectionist Models in Psychology* (pp. 273-312), Psychology Press.

Heinz, H.J., Luck, S.J., Mangun, G.R., and Hillyard, S.A., (1990) Visual event-related potentials index focused attention within bilateral arrays. I. Evidence for early selection. *Electroencephalography and Clinical Neurophysiology, 75,* 511-527.

Hopf, J.M., and Mangun, G.R. (2000). Shifting of visual attention in space: an electrophysiological analysis using high spatial resolution mapping. *Clinical Neurophysiology, 40(Suppl.)*, 61-67.

Hopfinger, G.B., and Mangun, G.R. (1998) reflexive attention modulates processing of visual stimuli in human extrastriate cortex. *Psychological Science, 9,* 441-447.

Johnson, A. and Proctor, R.W. (2004). *Attention: Theory and Practice.* USA: Sage Publications.

Jonides, J. (1981). Voluntary versus automatic control over the mind's eye's movement. In J.B. Long and A.D. Baddeley (Eds.). *Attention and performance IX* (pp.187-203). Hillsdale. NJ: Lawrence Erlbaum Associates Inc.

Kastner, S., DeWeerd, P., Desimone, R., and Ungerleider, L.C. (1998). Mechanisms of directed attention in the human extrastriate cortext as revealed by functional MRI. *Science, 282,* 108-111.

Kimchi, R. (1992). Primacy of wholistic processing and global/local paradigm: a critical review. *Psychological Bulletin, 112,* 24-38.

Kranczioch, M., Debener, S., and Engel, A.K. (2003). Event-related potential correlates of the attentional blink phenomenon. *Cognitive Brain Research, 17,* 177-187.

LaBerge, D., 1983. Spatial extent of attention to letters and words. *Journal of Experimental Psychology: Human Perception and Performance, 9,* 371-379.

Lavie, N. (2001). Capacity limits in selective attention: Behavioural evidence and implications for neutral activity. In J. Braun, C. Koch, and J.L. Davis (Eds.). *Visual Attention and Cortical Circuits* (pp. 49-68). Cambridge, MA: MIT Press.

Lavie, N., and Driver, J. (1996). On the Spatial extent of attention in object-based visual selection. *Perception and Psychophysics, 58,* 1238-1251.

Lavie, N., and Tsal, Y. (1994). Perceptual load as a major determinant of the locus of selection in visual attention. *Perception and Psychophysics, 56,* 183-197.

Logan, G.D. (1996). The CODE theory of visual attention: An integration of space-based and object-based attention. *Psychological Review, 103,* 603-649.

Lu, Z., and Dosher, B.A. (2005). External noise distinguishes mechanisms of attention. In L. Itti, G. Rees, and J.K. Tsotsos (Eds.) *Neurobiology of Attention* (pp. 448-453), Elsevier Academic Press.

Luck, S.J. (1995). Multiple mechanisms of visual-spatial attention: Recent evidence from human electrophysiology. *Behavioural Brain Research, 71,* 113-123.

Luck, S.J. (2006). The operation of attention-millisecond by millisecond-over the first half second. In H. Ogmen, and B.H. Breitmeyer, (Eds.), *The First Half Second: The Microgenesis and Temporal Dynamics of Unconscious and Conscious Visual Processes* (pp. 187-206). London: The MIT Press.

Luck, S.J., and Hillyard, S.A. (1994a). Electrophysiological correlates of feature analysis during visual search. *Psychophysiology, 31,* 291-308.

Luck, S.J., and Hillyard, S.A. (1994b). Spatial filtering during visual search: Evidence from human electrophysiology. *Journal of Experimental Psychology: Human Perception and Performance, 20,* 1000-1014.

Luck, S.J., and Vogel, E.K. (1997). The capacity of visual working memory for features and conjunctions. *Nature, 390,* 279-281.

Luck, S.J., Hillyard, S.A., Mouloua, M., and Hawkins, H.L. (1996). Mechanisms of visual-spatial attention: Resource allocation or uncertainty reduction? *Journal of Experimental Psychology: Human Perception and Performance, 22,* 725-737.

Mangun, G.R. and Hillyard, S.A. (1990). Allocation of visual attention to spatial location: Event-related brain potentials and detection performance. *Perception and Psychophysics, 47,* 532-550.

Mangun, G.R. and Hillyard, S.A. (1991). Modulations of sensory evoked brain potentials indicate changes in perceptual processing during visual-spatial priming. *Journal of Experimental Psychology: Human Perception and Performance, 17,* 1057-1074.

Mangun, G.R., Hillyard, S.A. and Luck S.J. (1993). Electrocortical substrates of visual selective attention. In D. Meyer and S. Kornblum (Eds.), *Attention and Performance XIV,* (pp. 219-243), Cambridge, MA: MIT Press.

Marcantoni, W. S., Lepage, M., Beaudoin, G., Bourgouin, P., and Richer, F. O. (2003). Neural correlates of dual task interference in rapid visual streams: An fMRI study Brain and *Cognition, 53,* 318-321.

Marois, R., Chun, M.M., Gore, J.C. (2000). Neural Correlates of the Attentional Blink. *Neuron, 28,* 299-308.

McDonald, J.J., Ward, L.M., and Kiehl, K.A. (1999). An event-related brain potential study of inhibition of return. *Perception and Psychophysics, 61,* 1411-1423.

Moore, C.M., Egeth, H., Berglam, L.R., and Luck, S.J. (1996). Are attentional dwell times consistent with serial visual search? *Psychonomic Bulletin and Review, 3,* 360-365.

Moray, N. (1959). Attention in dichotic listening: affective cues and the influence of instructions. *Quarterly Journal of Experimental Psychology, 27,* 56-60.

Muller, H.J., and Rabbitt, P.M.A. (1989). Reflexive and voluntary orienting of visual attention: Time course of activation and resistance to interruption. *Journal of Experimental Psychology: Human Perception and Performance, 15,* 315-330.

Myczek, K., and Simons, D.J. (2008). Better than average: Alternatives to statistical summary representations for rapid judgments of average size. *Perception and Psychophysics, 70,* 772–788.

Nakayama. K., and Mackeben, M. (1989). Sustained and transient components of focal visual attention. *Vision Research, 29,* 1631-1647.

Navon, D. (1977). Forest before trees: the precedence of global features in visual perception. *Cognitive Psychology, 9,* 353-383.

Nieuwenstein, M.R., Vlaskamp, B.N.S., Hooge, I.T.C., Van der Lubbe, R.H.J., and Johnson, A. Object-based attentional dwelling. Manuscript submitted for publication.

Nobre, A.C. (2004). Probing the flexibility of attentional orienting in the human brain. In Posner, M. I. (Ed.) *Cognitive Neuroscience of Attention* (pp. 157-179), New York: The Guilford Press.

Nobre, A.C., Coull, J., Vandeberghe, R., Maquet, P., Frith, C.D., and Mesulam, M.M. (2004). Directing attention to locations in perceptual versus mental representations. *Journal of Cognitive Neuroscience, 16,* 363-373.

O'Craven, K., Downing, P., and Kanwisher, N. (1999). fMRI Evidence for objects as the units of attentional selection. *Nature, 401,* 584-587.

Olhausen, B.A., Anderson, C. H., and Van Essen, D. C. (1993). A neuro-biological model of visual attention and invariant pattern recognition based on dynamic routing of information. *Journal of Neuroscience, 13,* 4700–4719.

Olivers, C.N.L., and Nieuwenhuis, S. (2006). The beneficial effects of additional task load, positive affect, and instruction on the attentional blink. *Journal of Experimental Psychology: Human Perception and Performance, 32*, 364–379.

Parkes, L., Lund, J., Angelucci, A., Solomon, J., and Morgan, M. (2001). Compulsory averaging of crowded orientation signals in human vision. *Nature Neuroscience, 4*, 739–744.

Pashler, H. (1998). *The Psychology of Attention*. Cambridge, MA: MIT Press.

Posner, M I., and Cohen, Y.A. (1984). Components of visual orienting. In H. Bouma and D. G. Bouwhuis (Eds.), *Attention and Performance X* (pp. 531-556). Hillsdale, NJ: Erlbaum.

Posner, M. I. (1980). Orienting of attention. *Quarterly Journal of Experimental Psychology, 32*, 3-25.

Pratt, J., Kingstone, A., and Khoe, W. (1997). Inhibition of return in location- and identity-based choice decision tasks. *Perception and Psychophysics, 59*, 1241-1254.

Raymond, J.E., Shapiro, K.L., and Arnell, K.M. (1992). Temporary suppression of visual processing in an RSVP task: An attentional blink? *Journal of Experimental Psychology: Human Perception and Performance, 18*, 849-860.

Rees, G., Frith, C. D. and Lavie, N. (1997). Modulating irrelevant motion perception by varying attentional load in an related task. *Science, 28*, 1616–1619.

Riggio, L., Patteri, I., and Umilta, C. (2004). Location and shape in inhibition of return. *Psychological Research, 68*, 41-54.

Robertson, L. C., Egly, R., Lamb, M. R., and Kerth, L. (1993). Spatial attentional and cueing to global and local levels of hierarchical structure. *Journal of Experimental Psychology, 19*, 471-487.

Robertson, L.C., Lamb, M.R., and Knight, R.T. (1988). Effects of temporal-parietal junction on perceptual and attentional processing in humans. *Journal of Neuroscience, 8*, 3757-3769.

Rolls, T. E., and Milward T. (2000). A Model of invariant object recognition in the visual system: Learning rules, activation functions, lateral inhibition, and information-based performance measures. *Neural Computation, 12*, 2547-2572.

Rowe, G., Hirsh, J.B, and Anderson, A.K. (2007). Positive affect increases the breadth of attentional selection. *Proceedings of National Academy of Sciences, 104*, 383-388.

Srinivasan, N., and Brown, J.M. (1997). Spatial frequency based inhibition of return. *Investigative Ophthalmology and Visual Science* (Suppl.), 38, S371.

Srinivasan, N., and Gupta, R. (2010). Emotion-attention interactions in recognition memory for distractor faces. *Emotion, 10*, 207-215.

Srinivasan, N., and Hanif, A. (2010). Global-happy and local-sad: Perceptual processing affects emotion identification. *Cognition and Emotion, 24*, 1062-1069.

Srinivasan, N., Srivastava, P., Lohani, M., and Baijal, S. (2009). Focused and Distributed Attention. In N. Srinivasan (Ed.), *Progress in Brain Research: Attention, 176*, 87-100.

Srivastava, P. and Srinivasan, N. (2009) Dynamics of Attentional Shifts with Hierarchical Stimuli. Proceeding of 25th Annual Meeting of the International Society of Psychophysics, Galway, Ireland.

Srivastava, P., and Srinivasan, N. (2010). Time course of visual attention with emotional faces. *Attention, Perception, and Psychophysics, 72*, 369-377.

Tatler, B., Gilchrist, I., and Risted, J. (2003). The time course of abstract visual representation. *Perception, 32*, 579–593.

Theeuwes, J. and van der Stigchel, S. (2006). Faces capture attention: evidence from inhibition of return. *Visual Cognition, 13,* 657-665.

Tipper, S.P., Driver, J., and Weaver, B. (1991). Object-centered inhibition of return of visual attention. *Quarterly Journal of Experimental Psychology, 43A,* 289-298.

Treisman, A. and Gelade, G. (1980). A feature-integration theory of attention. *Cognitive Psychology, 12,* 97-136.

Treisman, A. (2006). How deployment of attention determines what we see. *Visual Cognition, 14,* 411–443.

Treisman, A., and Schmidt, H. (1982). Illusory conjunctions in the perception of objects. *Cognitive Psychology, 14,* 107-141.

Treisman, A.M. (1964). Selective attention in man. *British Medical Bulletin, 20,* 12-16.

Valdes-Sosa, M., Bobes, M.A., Rodriguez, V., and Pinilla, T. (1998). Switching attention without shifting the spotlight object-based attentional modulation of brain potentials. *Journal of Cognitive Neuroscience, 10,* 137–151.

Van der Lubbe, R.H.J., and Woestenburg, J.C. (1997). Modulation of early ERP components with peripheral cues: A trend analysis. *Biological Psychology, 45,* 143-158.

Vecera, P.S. and Luck, S.J. (2000). Attention. *Encyclopedia of the Human Brain, 1,* 269-284.

Wadlinger, H.A. and Issacowitz, D.M. (2006). Positive mood broadens visual attention to positive stimuli. *Motivation and Emotion, 30,* 89-101.

Wallis, G., and Rolls, E.T. (1997). A model of invariant object recognition in the visual system. *Progress in Neurobiology, 51,* 167–194.

Ward, L.M. (1982). Determinants of attention to local and global features of visual forms. *Journal of Experimental Psychology: Human Perception and Performance, 8,* 562-581.

Ward, R., Duncan, J., and Shapiro, K. (1996). The slow time-course of visual attention. *Cognitive Psychology, 30,* 79–109.

Wheeler, M.E., and Treisman, A.M. (2002). Binding in short-term visual memory. *Journal of Experimental Psychology: General, 131,* 48–64.

Williams, D. W., and Sekuler, R. (1984). Coherent global motion percepts from stochastic local motions. *Vision Research, 24,* 55–62.

Wolfe, J.M. (1994). Guided Search 2.0: A revised model of visual search. *Psychonomic Bulletin and Review, 1,* 202-238.

Wolfe, J.M., Cave, K.R., and Franzel, S.L. (1989). Guided search: An alternative to the feature integration model for visual search. *Journal of Experimental Psychology: Human Perception and Performance, 15,* 419-433.

Woodman, G.F. and Luck, S J. (1999). Electrophysiological measurement of attention during visual search. *Nature, 400,* 867-869.

Yantis, S., and Jonides, J. (1984). Abrupt visual onsets and selective attention: Evidence from visual search. *Journal of Experimental Psychology: Human Perception and Performance, 10,* 601- 621.

Yeshurun, Y., and Carrasco, M. (1998). Attention improves or impairs visual performance by enhancing spatial resolution. *Nature, 396,* 72-75.

In: Expanding Horizions of the Mind Science(s)　　　　ISBN: 978-1-62808-705-5
Editors: P.N. Tandon, R.C. Tripathi and N. Srinivasan　©2013 Nova Science Publishers, Inc.

Chapter 8

CELLULAR AND MOLECULAR MECHANISMS OF MEMORY

Shiv K. Sharma

National Brain Research Centre, Manesar, Haryana, India

ABSTRACT

Memory, a higher order function of the brain is fundamental to our ability to appropriately deal with the present and plan for the future. As is true for many other abilities, importance of memory is better recognized when we loose it. Considerable effort is directed towards understanding how we store information and retrieve it at a later time. Studies from several different model systems using a variety of approaches have enormously contributed to our present day understanding of how memory is formed. In this chapter, I discuss the synaptic and molecular processes that play important roles in the formation of memory.

1. INTRODUCTION

James McGaugh in his book "Memory and Emotion" has rightly pointed out that "a life without memory would be no life at all." The ability to store information in the form of memory, and retrieve it at a later time is crucial for our proper functioning on a day-to-day basis. It is also necessary for suitable adaptation in an environment which changes constantly. In some sense, memory is vital to our very existence. It links us with our past, and with the future. Taking a clue from the past events, we modify our behavior in the present. And, we plan our future based on the previous knowledge and logical expectations.

Memory was traditionally studied in psychology. Several decades ago researchers started investigating memory using a variety of approaches including molecular biology, electrophysiology and computational techniques. Slowly, it became apparent that we can learn a lot about memory with the help of available techniques used to study other biological processes. In this chapter, I discuss different aspects of memory and its cellular and molecular mechanisms. This chapter is not meant to be comprehensive, but to provide the readers a flavor of the levels of research that hopes to eventually understand the mechanisms that are involved in making memories.

2. MEMORY

Learning is the process of acquiring new information, and memory is the process by which that information is retained. Memory can be broadly divided into two types: declarative or explicit memory and non-declarative or implicit memory (Squire and Zola, 1996). Memories of facts (semantic memory) and events (episodic memory) fall in the declarative memory category, whereas memories such as sensitization and conditioning are classified in the non-declarative category. An important role for hippocampus, a sea horse-shaped structure in the brain, in declarative memory was shown by studies of Brenda Milner on the patient famously known as H.M. (Henry Gustav Molaison; his full name was revealed only after his death) who underwent bilateral medial temporal lobe resection to help cure epileptic seizures. Several later studies have supported the central role of hippocampus in different kinds of memories. After the surgery, H.M. could not remember the events that happened sometime before the surgery. And he was unable to make long-term declarative memory, although he was good at some other tasks. After Milner, H.M. was studied by many investigators. Indeed, H.M. has helped us learn a lot about memory formation and storage. The inability to recall previous memories is referred as retrograde amnesia, and not being able to make new memories is referred as anterograde amnesia. Memory can be divided into at least two phases: short-term memory and long-term memory. The distinction between short- and long-term memories is based on temporal profile and mechanistic features.

After encoding, memory needs to undergo a process of consolidation that makes it relatively stable. Several molecular events are involved in the consolidation of memory (Alberini, 2005). These include molecular signaling, protein synthesis and transcription. In addition to other targets, signaling pathways converge on transcriptional and protein synthesis machinery that eventually contribute to the consolidation process. When memory is recalled, it again becomes susceptible to disruption. After recall, memory undergoes reconsolidation that makes it stable (Alberini, 2005). There are several similarities between the processes of consolidation and reconsolidation, but there are some differences also. The important point is that memory undergoes time-dependent stabilization processes.

The ultimate goal of memory research is to understand the mechanisms involved in the formation and retrieval of memory in humans. However, for obvious reasons, several different model systems are used to study behavioral, cellular and molecular aspects of memory. These include Aplysia, Drosophila, Caenorhabditis elegans, crab, chicks and rodents. Aplysia, a marine mollusk, displays simple forms of memory such as habituation, dishabituation and sensitization. This animal shows also relatively complex forms of memory such as classical and operant conditioning. Aplysia has been widely studied for sensitization, a form of memory in which the response of the animal to a non-harmful stimulus increases after it has been exposed to a noxious stimulus. The large size of its neurons has made them accessible to different kinds of analyses. The available genetic tools have made the fruit fly, Drosophila, an attractive system to examine the genes involved in olfactory conditioning. Conditioning of the proboscis-extension response has extensively been studied in the honey bee. The nematode, Caenorhabditis elegans has been used for study of habituation. In the rodents, several different kinds of memories are studied. The most common tasks used in the laboratory are Morris water maze, conditioning tasks, avoidance task and object recognition task. The Morris water maze task is a spatial memory task in which the animal, using spatial cues, tries to locate a platform hidden in water. Spatial memory task is dependent on hippocampus. In conditioning, an association is made between two stimuli, the conditioned stimulus (CS) and the unconditioned stimulus (US). The response of the animal after conditioning is referred as the conditioned response (CR). In the example of Ivan Pavlov's experiment, the dog makes an association between sound of the bell (CS) and food (US). And, the salivation response is the CR. In fear conditioning task, the animal associates an electric shock to a cue (cued fear conditioning) or a context (contextual fear conditioning). The cued fear conditioning requires amygdala whereas contextual fear conditioning requires both hippocampus and amygdala. The object recognition task is based on the natural exploratory behavior of the rodents and it tests the memory for an object. Collectively, studies using different systems have greatly advanced our understanding of cellular and molecular processes that are involved in memory formation and its maintenance.

Although in the laboratory, different animal model systems are used to study properties of memory and its mechanisms, some features of memory are conserved across different systems. These include: (1) Memory can be divided in at least two phases, short-term memory and long-term memory, (2) Whereas short-term memory is independent of new protein and RNA synthesis, long-term memory critically requires these events, and (3) Massed training is less effective in inducing long-term memory than the spaced training. In a multi-trial task, memory retention is more robust when the trials during training are distributed over time (spaced training) than when the trials are presented with little or no intervening rest interval (massed training). This effect is commonly referred as the "spacing effect". It is interesting that for the massed and spaced training regimens, the amount of training is constant and only the time between training trials differs. Additionally, the spacing effect has a pronounced impact on longer-lasting memories than the memories that last for a short time.

3. SYNAPTIC PLASTICITY

The nervous system has remarkable ability to change in response to the external signal. The ability of a synapse (the region of communication between the pre- and postsynaptic neurons) to change its strength in response to stimuli is referred as synaptic plasticity. Activity-dependent synaptic plasticity plays important roles during the nervous system development. But its role is not limited to the developmental stages. In the adults also synaptic plasticity plays a major role in the proper functioning of the nervous system and adaptation to the environment.

Donald Hebb suggested that memory may involve strengthening of synapses due to coordinated firing of the pre- and postsynaptic cells. Hebb's postulate that came to be known as "fire together wire together," has been a guiding principle in the investigation on mechanisms of memory. In the mammalian system, two kinds of synaptic plasticity have mostly been studied. These are called long-term potentiation (LTP) and long-term depression (LTD). In LTP, the synaptic strength increases whereas the opposite happens in LTD. It was in 1973, when working on the hippocampus Timothy Bliss and Terje Lømo presented a detailed description of long-term potentiation. Since the publication of that seminal paper by Bliss and Lømo (1973) on LTP, various groups have studied it as a prominent candidate for cellular basis of memory, and investigated its properties and molecular mechanisms. Similar to memory, LTP has at least two phases that can be distinguished based on the molecular requirement. Whereas the early phase of LTP does not require gene expression and protein synthesis, the late phase of LTP has a critical requirement for these events. LTP has been observed *in vitro as well as in vivo.*

In vitro, LTP is mostly induced by tetanic stimulation, the theta burst protocol and the pairing protocol. In the tetanic protocol, high frequency stimulation is used to stimulate the presynaptic pathway. The theta burst protocol, considered physiologically more relevant, is based on the theta rhythm which is seen in the hippocampus when the animal is exploring the environment. In the pairing protocol, the postsynaptic depolarization is paired with the presynaptic stimulation. In the hippocampus, LTP has been studied at the synapses between the entorhinal cortex and dentate gyrus, between dentate gyrus and CA3 and between CA3-CA1 regions. The pathway between the entorhinal cortex and dentate gyrus is called the perforant pathway, the pathway between the dentate gyrus and CA3 is called mossy fibre pathway, and the pathway between the CA3 and CA1 regions is called the schaffer collateral pathway.

The LTP is commonly studied in the schaffer-collateral pathway (CA3-CA1 synapses). The *N*-methyl *D*-aspartate (NMDA) receptor, a type of glutamate receptor is known to play important role in the induction of LTP in this pathway. It requires the presence of glutamate and postsynaptic depolarization for the influx of calcium ions. The presence of glutamate is required for the binding to the receptor, and the postsynaptic depolarization removes the magnesium block normally present in this ion channel. Thus, the NMDA receptor functions as a co-incident detector for the presynaptic glutamate release and postsynaptic activity. The activation of the NMDA receptor leads to calcium influx which then starts the signaling mechanisms that are required for the development of LTP. LTP shows input specificity, associativity and cooperativity. Input specificity says that out of several synapses made to the postsynaptic cell, only the stimulated synapses show LTP. When strong stimulation to one pathway is paired with a weak stimulation to another pathway, both synapsing to the same postsynaptic cell, LTP can be induced in the weakly stimulated pathway also. This is called associativity. Cooperativity says that a critical number of presynaptic fibres need to be stimulated to induce LTP (both events- glutamate release and postsynaptic depolarization are required for the induction of LTP).

Although LTP is mostly studied in the hippocampus, it has been observed in several other brain regions where it is thought to play important role in information processing. NMDA receptor is critical for the induction of LTP in other brain regions also. LTP is not one phenomenon; its mechanisms differ in different brain regions, and depending on the experimental conditions. LTP at the dentate gyrus-CA3 synapses is NMDA receptor-independent. The LTP in the CA3-CA1 synapses can be induced in an NMDA receptor-independent manner also.

The importance of NMDA receptor has been studied using specific inhibitors as well as genetic strategies. Both pharmacological as well as genetic manipulation experiments show that NMDA receptor activity is critical for LTP and memory (Morris et al., 1986; Tsien et al., 1996; Tang et al., 1999).

Another kind of synaptic plasticity referred as synaptic facilitation of the sensory and motor neuron synapses has extensively been studied in Aplysia (Kandel, 2001). Synaptic facilitation is considered a cellular analog of memory for sensitization. It is induced by serotonin, a modulatory neurotransmitter. Synaptic facilitation also has different phases that can be distinguished based on their temporal profile and mechanistic requirements.

4. MOLECULAR MECHANISMS OF SYNAPTIC PLASTICITY AND MEMORY

Over the years, studies aimed at examining the molecular mechanisms of synaptic plasticity and memory have identified several protein kinases, transcription factors and other proteins that play important roles in these processes (Lynch, 2004). These include calcium/calmodulin-dependent kinase, protein kinase A, protein kinase C, tyrosine kinases, extracellular signal-regulated kinase, cAMP response element binding protein, CREB binding protein and C/EBP. Both pharmacological as well as genetic approaches have contributed to the identification of different molecules that are critical for the processes of synaptic plasticity and memory in different systems. I will discuss some of them in more detail.

4.1. Calcium/Calmodulin-Dependent Kinase II

As mentioned earlier, at the CA3-CA1 synapses, calcium entry through the NMDA receptors plays a critical role in the induction of LTP. This led to the identification of signaling enzymes that are activated by calcium and are important for the establishment of LTP. One such enzyme is the calcium/calmodulin dependent kinase II (CaMKII), a major constituent of the postsynaptic density. LTP induction leads to activation of CaMKII. Although the initial activation of CaMKII is calcium-dependent, once it gets autophosphorylated, it becomes independent of calcium and remains activated for a longer time. Studies have demonstrated that CaMKII plays crucial roles in LTP and memory (Silva et al., 1992a; 1992b).

4.2. Protein Kinase A

The levels of cAMP are increased after LTP. Protein kinase A (PKA) is one of the main targets of cAMP, although the activity of other molecules could also be modulated by cAMP. PKA is activated by LTP inducing stimuli. Several studies have shown that PKA signaling is critical for LTP and memory (Elgersma and Silva, 1999). Kandel and colleagues showed that in transgenic mice when PKA activity is reduced by expression of a regulatory subunit, the late phase of LTP is reduced (Abel et al., 1997). Moreover, these transgenic mice show deficits in long-term memory. Importantly, the early phase of LTP and short-term memory are unaffected in these mice. Genetic studies have identified important role of the cAMP-PKA pathway in memory formation in Drosophila (Davis, 1996). In Aplysia, PKA plays important role in serotonin-induced facilitation of the sensory-motor synapses. Thus, PKA plays important roles in synaptic plasticity and memory in different systems.

4.3. Extracellular Signal-Regulated Kinase

The extracellular signal-regulated kinase (ERK) is activated in the hippocampus in response to LTP inducing stimuli. Using pharmacological inhibitor, English and Sweatt (1997) showed that ERK activity is required for LTP in the hippocampus. In addition to the hippocampus, ERK is involved in LTP in other regions of the brain. The requirement of ERK activity is typically examined using the inhibitor for the upstream kinase, MEK, which phosphorylates and activates ERK. Several studies have shown that ERK plays a critical role in different kinds of memories (Atkins et al., 1998; Blum et al., 1999). In Aplysia, studies have shown that ERK is activated by serotonin, and that ERK activity is required for long-term synaptic facilitation as well as long-lasting memory for sensitization (Martin et al., 1997; Sharma et al., 2003a). ERK is a convergent point for several signaling pathways and may function as a "molecular node". In addition to other targets, ERK could modulate LTP and memory due to its effects on protein synthesis and transcription, events that are important for long-lasting synaptic plasticity and memory.

4.4. cAMP Response Element Binding Protein

The cAMP response element binding protein (CREB) is a transcription factor that binds to the cAMP response element (CRE; a small DNA sequence present in several genes) and activates transcription. Early studies in Aplysia showed that sequestration of CREB using CRE blocks serotonin-induced long-term facilitation (LTF) of sensory motor synapses (Dash et al., 1990). Conversely, injection of the phosphorylated and thus active CREB facilitates LTF induction. In Drosophila, CREB is involved in the associative memory formation (Yin et al., 1994). CREB is activated after tetanic stimulation that gives rise to LTP. CREB can be activated by CaMKIV and ERK. ERK phosphorylates and activates CREB through p90rsk. The CREB mutant mice have deficits in memory (Bourtchuladze et al., 1994). Using antisense RNA approach to reduce CREB levels, Guzowski and McGaugh (1997) found that this led to a deficit in long-term memory. Thus, it is clear that disruption of CREB activity interferes with synaptic plasticity and memory formation. In addition, increasing the levels of CREB facilitates memory (Josselyn et al., 2001). These studies show that CREB plays important role in synaptic plasticity and memory in invertebrates and vertebrates.

Several genes, including zif268 are targets for CREB. This immediate early gene (IEG) is a transcription factor and thus takes part in the transcriptional regulation of late response genes. Zif268 mutant mice show impairment in LTP and memory (Jones et al., 2001). Moreover, learning leads to an increase in the expression of this IEG. Zif268 contains CRE as well as serum response elements in its promoter region. Thus, the expression of zif268 may be controlled by both CREB as well as Elk transcription factors. Since both of these transcription factors are targets of ERK, the ERK pathway may contribute significantly to the regulation of zif268 expression.

4.5. Chromatin Modifications

The fact that new RNA synthesis is required for LTP and memory led to investigations examining the regulation of transcription factors in synaptic plasticity and memory. It is clear that in addition to transcription factors, the status of chromatin also regulates transcription. Nucleosome, which consists of DNA wrapped around the histone octamer, is the repeating unit of chromatin. Both histones and DNA undergo modifications, and these modifications play a deterministic role in transcription. DNA methylation is well known to play a role in the regulation of gene expression during development. Sweatt and colleagues showed that DNA methylation plays critical roles in LTP and memory also (Levenson et al., 2006; Miller and Sweatt, 2007). Histones undergo several posttranslational modifications. One such modification, acetylation, is positively correlated with enhanced transcriptional activity. The acetylation level of a protein in a cell is determined by the activities of acetylases and deacetylases. It has become clear that histone acetylation plays important roles in synaptic plasticity and memory in systems ranging from invertebrates to vertebrates (Sharma, 2010). Increasing histone acetylation (and may be acetylation of other proteins) by inhibiting deacetylases facilitates synaptic plasticity and memory. Genetic manipulation studies also have shown that transcriptional co-activators with histone acetylase activity regulate LTP and memory. Thus, DNA methylation and histone acetylation also play important roles in the establishment of synaptic plasticity and memory.

4.6. Local Protein Synthesis

The formation of LTP and memory require the synthesis of new proteins. Although the neuronal cell body plays critical roles in the synthesis of proteins, it has become clear that some of the newly synthesized proteins that play a role in LTP may come from the pool of RNA that is present at the synaptic site and is synthesized "locally" in response to stimulation. Early studies showed that ribosomes are present in the dendrites. Since then, several studies have examined the regulation of components of protein synthesis machinery and protein synthesis itself in the dendrites. Studies examining functional significance of local protein synthesis have identified important role for local protein synthesis in synaptic plasticity (Sutton and Schuman, 2006). For example, Kandel and colleagues showed that local protein synthesis is essential for serotonin-induced synaptic facilitation of the sensory-motor synapses in Aplysia (Martin et al., 1996). In addition, local protein synthesis is required for LTP in the hippocampus (Bradshaw et al., 2003).

4.7. Molecular Constraints

From the above described studies, it is clear that several molecules positively regulate the development of synaptic plasticity and memory. This, however, is only part of the story. There are molecular events that negatively regulate these processes (Abel et al., 1998). For example, protein phosphatase 1 acts as a constraint on memory (Genoux et al., 2002). Similarly, protein phosphatase 2B (calcineurin) negatively regulates LTP and memory (Malleret et al., 2001; Sharma et al., 2003b). The overall phosphorylation status of a protein in a cell is dependent upon the relative activities of the kinase and the phosphatase. Thus, whereas blocking kinase activity inhibits LTP and memory, blocking phosphatase activity actually facilitates LTP and memory. Apart from the phosphatases, a recent study (Guan et al., 2009) showed that a histone deacetylase, HDAC2, functions as negative constraint on memory: overexpression of HDAC2 inhibits memory formation, but HDAC2 deficiency facilitates memory formation. Thus, relative activities of positive and negative regulators determine the induction of synaptic plasticity and memory.

4.8. Maintenance and Expression

While a lot is known about the induction of LTP, relatively less is known about how LTP is maintained and expressed. PKMzeta is an atypical form of protein kinase C which does not contain the autoinhibitory domain and is thus constitutively active. Studies have shown that PKM zeta is involved in the maintenance of LTP (Ling et al., 2002). Furthermore, PKM zeta plays important role in the maintenance of memory (Pastalkova et al., 2006). Some synapses contain NMDA receptors but lack α-amino-3-hydroxyl-5-methyl-4-isoxazole-propionate (AMPA) receptors. Thus, these synapses display NMDA receptor-mediated response, but no AMPA receptor-mediated response. These are called silent synapses. During LTP, these synapses get "un-silenced" due to AMPA receptor insertion. Thus, AMPA receptor trafficking plays important role in the expression of LTP (Malinow, 2003). In Aplysia, PKA is known to play critical role in the expression of memory (Sutton et al., 2001). Thus, protein phosphorylation is involved in the maintenance of synaptic plasticity and memory.

CONCLUDING REMARKS

Elucidation of molecular and cellular mechanisms of memory formation and retrieval remains one of the most challenging problems in neuroscience. It is important to understand memory mechanisms not only from intellectually stimulating exercise point of view, but also for treatment of the disorders that manifest in memory impairment, such as Alzheimer's disease. Tremendous progress has been made in past several decades. But our understanding of mechanisms of memory formation is far from complete. LTP has been widely studied as a candidate cellular mechanism for at least some kinds of memories. Although proving the causal relationship between LTP and memory is a daunting task, it is clear from the literature that there exists a great deal of similarity in the molecular mechanisms involved in LTP and memory. Several studies have shown that pharmacological treatments that inhibit LTP also inhibit memory. In addition, genetic manipulation studies also suggest that LTP and memory may be related. That is not to say that LTP indeed underlies memory formation. There are studies where dissociation between LTP and memory has been observed. Although this issue is not yet settled, a study by Bear and colleagues showed that memory training can indeed induce LTP in some synapses of the hippocampus (Whitlock et al., 2006). This is an important development in the study of cellular basis of memory formation.

Modern memory research has benefited from several disciplines of science including genetics and molecular biology. The development of new tools will obviously help in furthering this research. It is hoped that the continued pace of research in this field will provide a comprehensive picture about the processes that go into making memories.

REFERENCES

Abel, T., Martin, K.C., Bartsch, D., and Kandel, E.R. (1998). Memory suppressor genes: inhibitory constraints on the storage of long-term memory. *Science, 279*, 338-341.

Abel, T., Nguyen, P.V., Barad, M., Deuel, T.A., Kandel, E.R., and Bourtchouladze, R. (1997). Genetic demonstration of a role for PKA in the late phase of LTP and in hippocampus-based long-term memory. *Cell, 88*, 615-626.

Alberini, C.M. (2005). Mechanisms of memory stabilization: are consolidation and reconsolidation similar or distinct processes? *Trends in Neurosciences, 28*, 51-56.

Atkins, C.M., Selcher, J.C., Petraitis, J.J., Trzaskos, J.M., and Sweatt, J.D. (1998). The MAPK cascade is required for mammalian associative learning. *Nature Neuroscience, 1*, 602-609.

Bliss, T.V., and Lomo, T. (1973). Long-lasting potentiation of synaptic transmission in the dentate area of the anaesthetized rabbit following stimulation of the perforant path. *Journal of Physiology, 232*, 331-356.

Blum, S., Moore, A.N., Adams, F., and Dash, P.K, (1999). A mitogen-activated protein kinase cascade in the CA1/CA2 subfield of the dorsal hippocampus is essential for long-term spatial memory. *Journal of Neuroscience, 19*, 3535-3544.

Bourtchuladze, R., Frenguelli, B., Blendy, J., Cioffi, D., Schutz, G., and Silva, A.J. (1994). Deficient long-term memory in mice with a targeted mutation of the cAMP-responsive element-binding protein. *Cell, 79*, 59-68.

Bradshaw, K.D., Emptage, N.J., and Bliss, T.V. (2003). A role for dendritic protein synthesis in hippocampal late LTP. *European Journal of Neuroscience, 18*, 3150-3152.

Dash, P.K., Hochner, B., and Kandel, E.R. (1990). Injection of the cAMP-responsive element into the nucleus of Aplysia sensory neurons blocks long-term facilitation. *Nature, 345*, 718-721.

Davis, R.L. (1996). Physiology and biochemistry of Drosophila learning mutants. *Physiological Review, 76*, 299-317.

Elgersma, Y., and Silva, A.J. (1999). Molecular mechanisms of synaptic plasticity and memory. *Current Opinion in Neurobiology, 9*, 209-213.

English, J.D., and Sweatt, J.D. (1997). A requirement for the mitogen-activated protein kinase cascade in hippocampal long term potentiation. *Journal of Biological Chemistry, 272*, 19103-19106.

Genoux, D., Haditsch, U., Knobloch, M., Michalon, A., Storm, D., and Mansuy, I.M. (2002). Protein phosphatase 1 is a molecular constraint on learning and memory. *Nature, 418*, 970-975.

Guan, J.S., Haggarty, S.J., Giacometti, E., Dannenberg, J.H., Joseph, N., Gao, J., Nieland, T.J., Zhou, Y., Wang, X., Mazitschek, R., Bradner, J.E., DePinho, R.A., Jaenisch, R., and Tsai, L.H. (2009). HDAC2 negatively regulates memory formation and synaptic plasticity. *Nature, 459*, 55-60.

Guzowski, J.F., and McGaugh, J.L. (1997). Antisense oligodeoxynucleotide-mediated disruption of hippocampal cAMP response element binding protein levels impairs consolidation of memory for water maze training. *Proceedings of the National Academy Science USA, 94*, 2693-2698.

Jones, M.W., Errington, M.L., French, P.J., Fine, A., Bliss, T.V., Garel, S., Charnay, P., Bozon, B., Laroche, S., and Davis, S. (2001). A requirement for the immediate early gene Zif268 in the expression of late LTP and long-term memories. *Nature Neuroscience, 4*, 289-296.

Josselyn, S.A., Shi, C., Carlezon, W.A. Jr, Neve, R.L., Nestler, E.J., and Davis, M. (2001). Long-term memory is facilitated by cAMP response element-binding protein overexpression in the amygdala. *Journal of Neuroscience, 21*, 2404-2412.

Kandel, E.R. (2001). The molecular biology of memory storage: a dialogue between genes and synapses. *Science, 294*, 1030-1038.

Levenson, J.M., Roth, T.L, Lubin, F.D., Miller, C.A., Huang, I.C., Desai, P., Malone, L.M., and Sweatt, J.D. (2006). Evidence that DNA (cytosine-5) methyltransferase regulates synaptic plasticity in the hippocampus. *Journal of Biological Chemistry, 281*, 15763-15773.

Ling, D.S., Benardo, L.S., Serrano, P.A., Blace, N., Kelly, M.T., Crary, J.F., and Sacktor, T.C. (2002). Protein kinase Mzeta is necessary and sufficient for LTP maintenance. *Nature Neuroscience, 5*, 295-296.

Lynch, M.A. (2004). Long-term potentiation and memory. *Physiological Review, 84*, 87-136.

Malinow, R. (2003). AMPA receptor trafficking and long-term potentiation. *Philosophical Transactions of Royal Society B Biological Science, 358*, 707-714.

Malleret, G., Haditsch, U., Genoux, D., Jones, M.W., Bliss, T.V., Vanhoose, A.M., Weitlauf, C., Kandel, E.R., Winder, D.G. and Mansuy, I.M. (2001) Inducible and reversible enhancement of learning, memory, and long-term potentiation by genetic inhibition of calcineurin. *Cell, 104*, 675-686.

Martin, K.C., Casadio, A., Zhu, H., Yaping, E., Rose, J.C., Chen, M., Bailey, C.H., and Kandel, E.R. (1996). Synapse-specific, long-term facilitation of aplysia sensory to motor synapses: a function for local protein synthesis in memory storage. *Cell, 91*, 927-938.

Martin, K.C., Michael, D., Rose, J.C., Barad, M., Casadio, A., Zhu, H., and Kandel, E.R. (1997). MAP kinase translocates into the nucleus of the presynaptic cell and is required for long-term facilitation in Aplysia. *Neuron, 18*, 899-912.

Miller, C.A., and Sweatt, J.D. (2007). Covalent modification of DNA regulates memory formation. *Neuron, 53*, 857-869.

Morris, R.G., Anderson, E., Lynch, G.S., and Baudry, M. (1986). Selective impairment of learning and blockade of long-term potentiation by an N-methyl-D-aspartate receptor antagonist, AP5. *Nature, 319*, 774-776.

Pastalkova, E., Serrano, P., Pinkhasova, D., Wallace, E., Fenton, A.A., and Sacktor, T.C. (2006). Storage of spatial information by the maintenance mechanism of LTP. *Science, 313*, 1141-1144.

Sharma, S.K. (2010). Protein acetylation in synaptic plasticity and memory. *Neuroscience and Biobehavioral Reviews*. doi:10:1016/j.neubiorev.2010.02.2009

Sharma, S.K., Sherff, C.M., Shobe, J., Bagnall, M.W., Sutton, M.A., and Carew, T.J. (2003a). Differential role of mitogen-activated protein kinase in three distinct phases of memory for sensitization in Aplysia. *Journal of Neuroscience, 23*, 3899-3907.

Sharma, S.K., Bagnall, M.W., Sutton, M.A., and Carew, T.J. (2003b). Inhibition of calcineurin facilitates the induction of memory for sensitization in Aplysia: requirement of mitogen-activated protein kinase. *Proceedings of the National Academy Sciences USA, 100*, 4861-4866.

Silva, A.J., Stevens, C.F., Tonegawa, S., and Wang, Y. (1992a). Deficient hippocampal long-term potentiation in alpha-calcium-calmodulin kinase II mutant mice. *Science, 257*, 201-206.

Silva, A.J., Paylor, R., Wehner, J.M., and Tonegawa, S. (1992b). Impaired spatial learning in alpha-calcium-calmodulin kinase II mutant mice. *Science, 257*, 206-211.

Squire, L.R., and Zola, S.M. (1996). Structure and function of declarative and nondeclarative memory systems. *Proceedings of the National Academy Sciences USA, 93*, 13515-13522.

Sutton, M.A., and Schuman, E.M. (2006) Dendritic protein synthesis, synaptic plasticity, and memory. *Cell, 127*. 49-58.

Sutton, M.A., Masters, S.E., Bagnall, M.W., and Carew, T.J. (2001). Molecular mechanisms underlying a unique intermediate phase of memory in aplysia. *Neuron, 31*, 143-154.

Tang, Y.P., Shimizu, E., Dube, G.R., Rampon, C., Kerchner, G.A., Zhuo, M., Liu, G., and Tsien, J.Z. (1999). Genetic enhancement of learning and memory in mice. *Nature, 401*, 63-69.

Tsien, J.Z., Huerta, P.T., and Tonegawa, S. (1996). The essential role of hippocampal CA1 NMDA receptor-dependent synaptic plasticity in spatial memory. *Cell, 87*, 1327-1338.

Whitlock, J.R., Heynen, A.J., Shuler, M.G., and Bear, M.F. (2006). Learning induces long-term potentiation in the hippocampus. *Science, 313*, 1093-1097.

Yin, J.C., Wallach, J.S., Del Vecchio, M., Wilder, E.L., Zhou, H., Quinn, W.G., and Tully, T. (1994). Induction of a dominant negative CREB transgene specifically blocks long-term memory in Drosophila. *Cell*. 79:49-58.

In: Expanding Horizions of the Mind Science(s)   ISBN: 978-1-62808-705-5
Editors: P.N. Tandon, R.C. Tripathi and N. Srinivasan   ©2013 Nova Science Publishers, Inc.

Chapter 9

EFFECTS OF STRESS ON COGNITIVE VERSUS EMOTIONAL FUNCTION, CELLS, CIRCUITS AND BEHAVIOR

Aparna Suvrathan, Rajnish P. Rao and Sumantra Chattarji
National centre for Biological Sciences, GKVK Campus, Bellary Road, Bangalore, India

ABSTRACT

Although we think of memories as being rooted in the past, they have a profound influence on how we respond to experiences in the future. Memories come in many different flavors, some more potent than others. Explicit memories of facts or events, encoded by a brain structure called the hippocampus, tend to fade away with time. In contrast, unconscious emotional memories of stressful experiences, formed in a brain structure called the amygdala, appear to leave an indelible mark that may last for a lifetime. The rapid and efficient encoding of fear memories by the amygdala help us cope with threatening stimuli in the future, but it also comes at a high price. These emotional memories etched into the amygdalar circuitry can also become maladaptive. For example, high anxiety and mood lability are cardinal symptoms of many stress disorders. Here we review recent findings that provide insights into the cellular and synaptic mechanisms underlying the cognitive and emotional effects of stress.

ABBREVIATIONS

HPA axis	Hypothalamo-Pituitary-Adrenal Axis
PTSD	Post Traumatic Stress Disorder
BLA	Basolateral Amygdala
GR	glucocorticoid receptor
MR	mineralocorticoid receptor
NMDA	N-methyl-D-aspartate
AMPA	alpha-amino-3-hydroxy-5-methyl-4-isoxazolepropionic acid
EPSC	Excitatory Post Synaptic Current
IPSC	Inhibitory Post Synaptic Current

1. INTRODUCTION

Some experiences are memorable, others forgettable. How does a particular experience leave its mark as a memory in the brain? For more than a century, the search for a biological basis of memory formation has centered on the synapse, the junction where information is passed from one brain cell to another. The remarkable ability of synapses to change in response to experience, a property described as "synaptic plasticity", is believed to mediate long-term storage of information in the brain. Although common cellular and molecular mechanisms underlying synaptic plasticity and memory have been identified over the past few decades, we know little about why some memories last for a long time while others fade away. Specifically, why are memories of emotional events often more powerful and persistent? Why do war veterans or victims of severe stress continue to have vivid flashbacks of traumatic events from their past, while their cognitive abilities diminish? Stress disorders bring these questions into sharp focus because prolonged stress has contrasting effects on different types of memories. Stress impairs memories of facts and events, a key aspect of cognitive function, which depend on a brain structure called the hippocampus. In contrast, stress greatly amplifies emotional memories, which are processed by another structure called the amygdala. But little is known about the neural basis for this contrast. Gaining mechanistic insights into the cellular underpinning of this contrast is also critically important for effective therapeutic interventions against stress-related psychiatric disorders, such depression, chronic anxiety, etc.

Clinical studies of human patients suffering from various stress disorders have provided a valuable foundation and direction for animal models that can use a range of neurobiological tools and models to analyze the effects of stress at multiple levels of neural organization – from molecular and cellular correlates at one end to a behavioral output at the other. Such basic research then leads to an intellectual framework for understanding the mechanistic basis for such disorders, thereby bridging the gap to clinical research. Within this context, this review focuses on a detailed description of a range of observations from animal models, on the effects of stress on the two key brain areas involved in cognitive and emotional function respectively – the hippocampus and amygdala - at the synaptic, cellular, molecular and behavioral levels. We attempt to synthesize these findings into a cellular model of stress-induced modulation of synaptic structure and function that may mediate the short and long term effects of stress in the hippocampus and amygdala.

2. EFFECTS OF STRESS ON THE HIPPOCAMPUS AND AMYGDALA

A range of animal models of stress have been used to analyze specific cellular and molecular changes in brain areas that are known to be involved in learning and memory. Critically, both the hippocampus and the amygdala play a pivotal role in feedback regulation of the stress response via the hypothalamic– pituitary–adrenal (HPA) axis (Herman et al., 1989; Jacobson and Sapolsky, 1991; Sapolsky et al., 1991; Herman and Cullinan, 1997). Yet this regulation works in diametrically opposite directions; the hippocampus negatively regulates the HPA axis, whereas the amygdala has a positive feedback to it. In addition, their behavioral outputs, in turn, are differentially modulated by stress. Hippocampus dependent declarative memory formation is impaired by stress, while fear learning, which depends on the amygdala, is enhanced. Such a differential modulation of different forms of memory by behavioral stress is of clear adaptive benefit to an animal, and is widely observed in the enhanced memory for events with emotional relevance.

Anatomical studies have shown that hippocampal inputs on to the paraventricular nucleus (PVN) of the hypothalamus and GABAergic inhibitory neurons in the hypothalamus are excitatory, and thus enhance GABAergic tone. On the other hand, those from the amygdala are inhibitory and reduce GABAergic tone. (Herman et al., 1989; Jacobson and Sapolsky, 1991; Sapolsky et al., 1991; Pitkanen and Amaral, 1994; Herman and Cullinan, 1997). This is how enhanced hippocampal input, by enhancing inhibition in the thalamus, suppresses the HPA axis, whereas the amygdala does the opposite. At the behavioral level, stress facilitates aversive learning but impairs spatial learning in rodents (Shors et al., 1992; Luine et al., 1994).

The hippocampus has long been known to be essential for the formation of memories. In stress-induced disorders such as post-traumatic stress disorder (PTSD), reduction in hippocampal volumes have been reported by many neuroimaging studies (Rauch et al., 2006; Bremner, 2007) (Gilbertson et al., 2002; Bremner et al., 2003) and these are believed to underlie the associated functional deficits. Abnormalities in hippocampal structure and function are particularly relevant in light of the central role played by this structure in both the neuroendocrine stress response and memory deficits. These studies have been extensively reviewed elsewhere (McEwen, 1999).

Unlike the hippocampus, amygdala activation is reported to be positively correlated with the severity of PTSD symptoms (Rauch et al., 1996; Shin et al., 2004; Armony et al., 2005; Protopopescu et al., 2005). The amygdala, as mentioned earlier, is involved in the formation of memories with emotional salience and decades of animal research demonstrates a pivotal role for the amygdala in Pavlovian fear learning (LeDoux, 2000). Fear conditioning is a well established learning paradigm in which animals are trained to associate a previously neutral sensory stimulus (the conditioned stimulus, or CS) with a coincident aversive stimulus (unconditioned stimulus, or US) – such as an electric shock. The US always elicits a response, such as freezing in rodents in response to an electric shock; and this response is known as an unconditioned response (UCR). When the animal is re-exposed to the CS alone, a conditioned response (CR) is elicited. Measurement of the CR provides a measure by which the strength of the association that was formed between the CS and the US can be tested. The role of the amgydala in the formation of this association has been studied for many years, and it is believed that thalamic afferent synapses to the lateral amygdala (LA) are strengthened in the initial stages of fear conditioning (Lamprecht and LeDoux, 2004; Maren and Quirk, 2004). The amygdala is a medial temporal lobe structure, organized into about 13 distinct nuclei. There is precise information flow between these nuclei, with connections having been mapped extensively. However, the lateral nucleus of the amygdala serves as the input nucleus, with heavy projections out to other nuclei. There is a convergence of inputs via sensory thalamic and cortical pathways onto the cells of the LA, and the information that exits from this nucleus is likely to be substantially processed (McDonald, 1982, 1992; Pitkänen, 2000; Sah et al., 2003).

We first discuss the major findings from rodent models of chronic or prolonged forms of behavioral stress and their effects on the cells and synapses of the hippocampus, which is involved in the formation of episodic or declarative memories, followed by a summary of the effects of stress on a brain area that has an opposing influence on the HPA axis, and mediates the formation of fear memories –the amygdala.

3. STRESS-INDUCED STRUCTURAL PLASTICITY

The dendrites of a neuron are capable of remodeling and change, even in the adult brain. Through its modulation of postsynaptic dendritic surface area, such dendritic remodeling has a profound impact on the availability of synaptic inputs, and thereby synaptic plasticity. Excitatory synaptic contacts in the brain lie on dendritic spines, which are micron-scale projections from the neuronal dendrite. It has long been hypothesized that morphological and numerical alterations in dendritic spines, underlie long-term structural encoding of behavioral experiences. In the context of learning and memory, structural plasticity of spines may have different implications. The possibility we discuss extensively in the following sections envisages that spine synapses are the primary site of plasticity elicited by stressful experiences.

3.1. Chronic Stress and the Hippocampus

Analysis of the morphology of neurons in the rat hippocampus shows that 21 days (6 hours per day) of repeated restraint stress produces significant dendritic remodeling in CA3 pyramidal neurons (Watanabe et al., 1992; Magarinos and McEwen, 1995b, 1995a; McEwen, 1999; Sousa et al., 2000). This morphological change is specifically in the form of a shortening and debranching of apical dendrites (Conrad et al., 1999) and is mediated by mechanisms involving high levels of glucocorticoid secretion, glutamate, and serotonin (Magarinos and McEwen, 1995b). Later work has shown that even shorter durations of immobilization stress (2 hours per day, for 10 days) are capable of causing significant atrophy of both apical and basal dendrites on CA3 pyramidal cells (Vyas et al., 2002). The hippocampal volume loss seen in stress-triggered disorders such as PTSD may potentially be explained by these findings in rodent models of stress-induced neuronal atrophy (Gurvits et al., 1996; Bremner et al., 2003).

However, it has been seen that repeated application of the same stressor can lead to habituation in the stress response (Melia et al., 1994). This raises the possibility that although chronic stress triggers dendritic remodeling, it may eventually set in motion adaptive changes that counter the initial effects of stress on dendritic morphology. Such homeostatic mechanisms, triggered by prolonged stress, could be mediated by changing the number of spines, thereby regulating the overall synaptic connectivity in the affected area. This possibility finds support in the observation that dendritic atrophy in hippocampal CA3 pyramidal neurons, caused by repeated restrained stress, is accompanied by a numerical increase in spines (Sunanda et al., 1995). This is indicative of an adaptive mechanism that may compensate for the loss of dendritic area for synaptic inputs to terminate. Interesting questions about the relationship between such spine and dendritic plasticity are raised by these findings. It remains to be seen whether they always go hand in hand, have a particular temporal evolution with one kind of change always preceding the other, or even whether they mutually work for the maintenance of homoeostasis. The rules governing these changes may also be different in different brain areas.

3.2. Stress and the Amygdala

As discussed earlier in this review, the amygdala and the hippocampus have opposing feedback onto the stress response system of the HPA axis. Although repeated stress that produces dendritic atrophy in the CA3 region impairs hippocampal-dependent learning (Conrad et al., 1996), the basolateral amygdala (BLA) has been shown to be essential for stress-induced facilitation of aversive learning (Shors and Mathew, 1998). These observations emphasize the need to study the effects of stress on the amygdala at the cellular level.

Using the same analytical tools of morphometry as in the hippocampus, the first cellular evidence for contrasting patterns of stress-induced structural plasticity was found in the basolateral amygdala (BLA) (Vyas et al., 2002). Here, chronic immobilization stress (2 hours per day, for 10 days) induced dendritic atrophy and debranching in CA3 pyramidal neurons of the hippocampus, which is consistent with earlier reports using other forms of chronic stress. On the other hand, principal neurons in the basolateral complex of the amygdala (BLA) exhibited *enhanced* dendritic arborization in response to the same chronic stress. This stress-induced enhancement in dendritic arborization was restricted only to BLA pyramidal and stellate neurons, which are presumably excitatory projection neurons (McDonald, 1982) (McDonald, 1992). The behavioral relevance of such stress-induced structural change was seen in the enhancement of anxiety after ten days of stress. Only chronic immobilization stress caused a significant increase in anxiety-like behavior (Vyas and Chattarji, 2004). On the other hand, chronic unpredictable stress, which failed to enhance anxiety, did not cause any dendritic remodeling of BLA principal neurons (Vyas et al., 2002).

This correlation between stress-induced dendritic hypertrophy in the BLA and behavioral anxiety is strengthened by another study where animals are kept for 21 days in stress free conditions after chronic immobilization stress. Even after the recovery period, animals continue to exhibit enhanced anxiety (Vyas and Chattarji, 2004; Vyas et al., 2004). Strikingly, at the cellular level, the stress-induced increase in dendritic arborization of BLA projection neurons was also as persistent as enhanced anxiety after 21 days of recovery.

Further emphasizing the differences in the amygdala and the hippocampus, hippocampal CA3 atrophy is reversible in this same recovery period (Vyas et al., 2004). Following a 10-day exposure to immobilization stress, dendritic hypertrophy is accompanied by an equally robust increase in spine-density that spreads across both primary and secondary dendrites of BLA pyramidal neurons (Mitra et al., 2005; Vyas et al., 2006). Thus, chronic stress leads to a significant increase in the structural basis of synaptic connectivity in the BLA.

To summarize, studies on the plasticity of neuronal structure that use tools of morphometric analysis of spines and dendritic arbor show that there are critical differences in the way the hippocampus and the amygdala respond to stress. This difference at the cellular level correlates with the way memories mediated by these two brain areas are affected differently by stress, and by the opposing way in which they feed back onto the primary mediator of the effects of stress – the HPA axis.

4. Synaptic and Molecular Correlates of Learning and Memory

The behavioral and morphometric studies of the hippocampus and amygdala discussed above are complemented and extended by bottom-up approaches that analyze these effects on a sub-cellular scale. The focus on spine and dendritic structure is believed to imply differences in synaptic connectivity and plasticity. Direct experimental evidence of physiological changes in synaptic transmission, and the molecular basis for these changes, is discussed in the following sections.

4.1. Adrenal Steroid Action

Upon exposure to stressful conditions, the organism responds suitably to new conditions, and to restore homeostasis. Stress triggers the activation of the HPA axis, which brings about its downstream actions by the classical stress hormones, i.e. the glucocorticoids. Also, there are other mediators such as epinephrine/norepinephrine, the parasympathetic and cytokine systems which help in the attempt to adapt and restore homeostasis (McEwen, 2007). Yet severe or prolonged stress results in damage to several brain areas due to the continued exposure to these same hormones. An example of such effects is ischemia and seizures generated in the hippocampus (McEwen, 2007).

In addition to the way mediators of the stress response interact with and modulate each other, adrenal steroids show an inverted U-shaped dose response curve (Diamond et al., 1992; Joels, 2006). This action involves the recruitment of the mineralocorticoid receptors (MRs) (Pascual-Le Tallec and Lombes, 2005) and the glucocorticoid receptors (GRs) (Zhou and Cidlowski, 2005). A large body of work has resulted in several mouse over expression (Ferguson and Sapolsky, 2007; Lai et al., 2007; Rozeboom et al., 2007) or knockout models (Gass et al., 2000; Berger et al., 2006) which have been of critical importance to delineate the roles played by these receptors in the stress response. The MRs, which have been shown to be involved in the stability of neuronal networks and hippocampal cell survival (Sloviter et al., 1989; McEwen et al., 1992; Macleod et al., 2003), are the first to be saturated by the rising glucocorticoid levels as they have a higher affinity to the hormone. With a further rise in hormone levels, the lower affinity GRs take over as the chief mediators of glucocorticoid action. It is also these GRs expressed in the brain that are key to the feedback inhibition of the HPA axis (Joels, 2006).

The regulatory mode of action of adrenal steroids is complicated by a mind-boggling diversity in transcriptional regulation. Upon ligand binding, the receptors translocate to the nucleus and either bind to the DNA as heterodimers (Karst et al., 2000) or interact with one of the several transcription factors (AP-1, NFκB, CREB, STATs) (De Bosscher et al., 2003) as monomers. Further, different brain regions not only express splice variants of the receptors, but also have different MR/GR ratios. Also, the local intracellular hormone concentrations can be regulated by the expression of modifying enzymes (11-β steroid dehydrogenase 1 and 2, 3-α and 5-α reductases) (Edwards et al., 1988; Funder et al., 1988; Seckl and Walker, 2001). Additionally, these brain regions also express various combinations of transcriptional co-activators and co-repressors and can therefore result in differential gene expression.

This wide array of factors may partly explain the differential effects of stress on the hippocampus and the amygdala. While low levels of glucocorticoids have been reported to have trophic effects in the hippocampus (Gould et al., 1990), repeated stress and high dose corticosteroid treatment promotes dendritic remodeling (McEwen, 1999) by the downstream actions of excitatory amino acids and NMDA receptors. In the amygdala, adrenal steroids cause dendritic lengthening and an associated increase in anxiety-like behaviour (Mitra and Sapolsky, 2008). Yet, at the same time, priming with adrenal steroids is also implicated in the prevention of stress-induced anxiety and induction of excitatory spine synapses (Rao et al., 2007). The actions of adrenal steroids therefore are very much dependent on their concentration and time of action, and also on the activities of other mediators such as excitatory amino acids and serotonin. For example, adrenal steroids are required for serotonin to exert a gating effect on synaptic transmission the BLA (Stutzmann et al., 1998), and to increase excitability of BLA neurons and in the reduction of inhibitory tone (Duvarci and Pare, 2007).

4.2. Effects on Excitatory Synaptic Transmission

The effects of stress on spine density have been of great interest because excitatory synapses lie on these structures. The impact of stress on transmission and plasticity at excitatory synapses in the hippocampus and the amygdala is likely to be significantly different. One common effect of stress on both the hippocampus and the amygdala lies in the rise in extracellular levels of the excitatory neurotransmitter glutamate (Lowy et al., 1993; Reznikov et al., 2007). The immediate synaptic effect of stress therefore seems to be common in both brain areas, which subsequently exhibit contrasting patterns of structural plasticity of dendrites and spines. This parallels the initial increase in glucocorticoid stress hormones which is also elevated in both areas following stress. The cellular mechanisms that result in different changes in dendritic arbor, in response to stress, must therefore lie downstream of a common pathway in both brain areas. This common pathway includes the rise in glucocorticoid hormomes, and an increase in extracellular glutamate. These results implicate the effect of glucocorticoids on different glutamate receptor subtypes, and synaptic plasticity mechanisms mediated by them.

In the hippocampus, where the effects of stress on excitatory synapses have been much more extensively studied than in the amygdala, a major target is the N-methyl-D-aspartate (NMDA) receptor (Bartanusz et al., 1995; Weiland et al., 1997). There is also evidence that the effects of stress on dendritic remodeling lie downstream of the activation of this receptor. In response to stress, signaling through the NMDAR is increased, partly through an increase in deactivation time of the NMDAR-mediated EPSCs in the CA3 region of the hippocampus (Kole et al., 2002). Following upon the widespread activation of NMDA receptors, the increased levels of intracellular calcium may make the dendritic cytoskeleton become depolymerized or undergo proteolysis (McEwen, 1999).

Stress hormones require the activation of the NMDA receptor for their action on dendritic arbor in the hippocampus. This has been experimentally shown using antagonists to the NMDA receptor, which block stress-induced hippocampal plasticity. In addition to the effect of stress on the morphology of dendrites, stress causes an impairment of neurogenesis in the dentate gyrus and this too is dependent on the NMDA receptor (Magarinos and McEwen,

1995b; Kim et al., 1996; Gould et al., 1997; Baker and Kim, 2002). Further proof comes from studies where direct application of the stress hormone corticosterone in-vitro enhances NMDA currents (Takahashi et al., 2002), and glucocorticoids increase NR2A and NR2B subunit expression in the hippocampus after chronic administration (Weiland et al., 1997).

Fear learning, critically dependent on the amygdala, is enhanced after stress. This enhancement is blocked if the NMDA receptor is blocked during stress, showing that the same dependence on NMDA is true for the amygdala (Shors et al., 1992). The NMDA receptor, therefore, acts as a critical initial step in the action of stress responses on both the amygdala and hippocampus – resulting in both amygdala- and hippocampus-dependent behavioral impairments, as well as synaptic plasticity in both brain areas.

In addition to regulating NMDA receptor currents, stress and glucocorticoids also regulate excitatory amino acid transporters in the hippocampus (Reagan et al., 2004; Autry et al., 2006). Chronic stress increases Glt1a and Glt1b expression in the CA3 region of the hippocampus, but adrenal steroids appear to suppress the expression of Glt1a, as shown by studies with adrenalectomy and adrenal steroid replacement (Autry et al., 2006). Adrenal steroids therefore appear to act in a suppressive manner similar to inflammatory responses, and regulate transporters in the opposite direction (McEwen, 2007).

Synaptic plasticity has long been the focus of studies on the neural substrates of learning and memory. The capacity of synapses to change in strength is believed to underlie the storage of information in the brain. As such, the role of stress in modifying plasticity in the hippocampus and the amygdala would give mechanistic insight into the behavioral and learning impairments that are observed. Direct evidence on the modulation of synaptic plasticity by adrenal steroids is provided by in-vitro work on assays of long-term potentiation (LTP) (Bliss and Lomo, 1973). Correlating with the well-established phenotype of stress-induced dendritic atrophy in the hippocampus, acute stress or elevation of glucocorticoids has been shown to impair LTP, or a similar form of plasticity - primed-burst potentiation (PBP) in both the hippocampal CA1 region and the dentate gyrus. Further, there is a U-shaped dose response curve, with low levels of corticosterone facilitating PBP and high levels inhibiting PBP in the CA1 region (McEwen, 2007). Glucocorticoids also lower the threshold for NMDA-dependent long-term synaptic depression (LTD) in the hippocampus, which is in turn dependent on voltage-gated calcium channels (Kerr et al., 1992). They act on neurons in both the hippocampus and the basolateral amygdala (BLA) to increase the amplitude of high-voltage activated calcium currents, without changing any of their passive properties (Karst et al., 2002). However, glucocorticoids have also been shown to cause an increase in the excitability of lateral amygdala neurons (Duvarci and Pare, 2007).

The effect of stress on synaptic plasticity is not restricted merely to chronic forms of stress. A single acute episode of stress also impairs LTP and enhances LTD (Shors and Thompson, 1992; Garcia et al., 1997; Shors et al., 1997; Chaouloff et al., 2007). This defect in synaptic plasticity correlates with the stress-induced impairment of hippocampus-dependent memory (Baker and Kim, 2002). Stress interferes with performance of hippocampus-dependent tasks, but facilitates tasks such as eyeblink conditioning, in both rats (Shors et al., 1992) and humans (Spence and Taylor, 1951). The difference between the effects of short or prolonged periods of stress lies in how long memory continues to be impaired. Hippocampus-dependent spatial learning tasks are impaired by chronic stress and these changes stay well beyond the duration of the stressor (Bodnoff et al., 1995). Unlike chronic stress, however, the changes in plasticity caused by acute stress do not last long (Garcia et al., 1997).

While discussing the temporal dimension, the humandisorder of post-traumatic stress disorder (PTSD) becomes particularly relevant. In this disorder, behavioral impairments are seen long after the original stressor. Here, the amygdala is of particular importance because of the finding that there can be a delayed onset of anxiety like behavior and increase in BLA spine density after a single stressful experience (Mitra et al., 2005; Kohda et al., 2007). There is, however, no evidence of a delayed onset of aberrations in hippocampus-dependent spatial memory tasks. It is possible the amygdala, which does not recover rapidly from the effects of chronic stress (Vyas et al., 2004), may influence or modulate the hippocampus, where stress-induced changes do recover (Akirav and Richter-Levin, 1999).

These findings are particularly striking because the increase in anxiety and BLA spine-density is evident not immediately, but only in a delayed manner 10 days after the acute stress (Mitra et al., 2005). In contrast to the effects of chronic stress there is no dendritic remodeling and changes in spine density are localized proximal to the cell soma. This study suggests that BLA spinogenesis in itself may be adequate to increase behavioral anxiety, a correlation that is strengthened by another study where BLA spinogenesis, caused by transgenic over expression of BDNF, leads to enhanced anxiety (Govindarajan et al., 2006).

These hippocampal effects are known to be dependent on the action of glucocorticoids (Xu et al., 1998; Blank et al., 2002). Glucocorticoids are also known to bind to GRs in the BLA and enhance retention memory for learning tasks (Roozendaal and McGaugh, 1997; Karst et al., 2000; McGaugh and Roozendaal, 2002; Vouimba et al., 2007). While the stress effects are dependent on glucocorticoids, there is evidence that this is not sufficient to explain the behavioral enhancement of memory seen after stress, and that this enhancement may be selective for certain forms of memory (Shors, 2001; Beylin and Shors, 2003).

The action of stress, via glucocorticoids, impairs plasticity in the amygdala too (Maroun and Richter-Levin, 2003), even though amygdala-dependent fear conditioning learning is enhanced (Shors et al., 1992). The duration of stress, however, may be critically important as there is evidence that the populations of cells recruited by acute versus chronic stress may be different (Reznikov et al., 2008). Acute stress causes an increase in unit activity in the basolateral amygdala, and re-exposure to the stressful context itself causes an increase in activity (in an NMDA-dependent manner) (Shors, 1999).

4.4. Effects on Inhibitory Synaptic Transmission

Although the focus has largely been on excitatory neurotransmission, the link between stress-induced anxiety disorders and the GABAergic system in the amygdala is a well-established one (Rainnie et al., 2004; Shekhar et al., 2005). Stress and glucocorticoids reduce inhibitory currents in the basolateral amygdala (Duvarci and Pare, 2007; Suvrathan et al., 2007), as in the hippocampus (Gronli et al., 2007). Duvarci and Pare (2007) have shown the direct effect of in-vitro administration of glucocorticoids on intrinsic excitability and inhibitory synaptic transmission in the lateral amygdala.

We have seen that acute stress results in a delayed decrease in spontaneous miniature inhibitory post-synaptic currents (mIPSCs) (Suvrathan, 2008). Earlier work had shown, as mentioned earlier, that acute stress results in a delayed structural change in the form of increase in spine density in the LA, ten days later (Mitra et al., 2005). The chronic effects of stress in producing a similar increase in spines in the LA are also accompanied by a decrease in mIPSCs. Although the delayed structural effects of acute stress are less wide-spread than those of chronic stress (Vyas et al., 2002), they are similar in terms of modified amygdalar output. It is possible that there is an original period of lowered inhibition which sets off a cascade of synaptic and molecular events that result, ten days later, in the changes in both excitatory and inhibitory synapses in the LA.

The network within the amygdala is tightly controlled by inhibition (Ehrlich et al., 2009). Behaviorally, the actions of the benzodiazepine drugs as anxiolytics are known to be through an enhancement of inhibitory GABAergic currents (Rainnie et al., 2004; Shekhar et al., 2005; Suvrathan et al., 2006). One of the primary behavioral effects of stress is in the development of anxiety-like behaviors. Such anxiety and therapeutic strategies to combat it may therefore be tightly coupled to levels of inhibition in the amygdala and possibly in other brain areas. The shift to a disease state may result from the increase in excitability caused by the increase in NMDA currents (Weiland et al., 1997; Kole et al., 2002), AMPA binding (Tocco et al., 1991), and extracellular glutamate (Lowy et al., 1993; Reznikov et al., 2007) seen after stress or glucocorticoid application, along with a reduction in levels of inhibition. The change in the excitatory and inhibitory synapses, therefore, shifts the balance of the network further and further away from homeostasis. It is possible that the inhibitory system has a gating role in this cascade of events, and may be a key target for therapeutic intervention.

The direct role of adrenal steroids on inhibitory neurotransmission has also been studied. Low levels of corticosterone alter mRNA levels for specific subunits of $GABA_A$ receptors in hippocampal area CA3 and the dentate gyrus of ADX rats (Orchinik et al., 1994), whereas stress levels of corticosterone has produced different effects on $GABA_A$ receptor subunit mRNA levels and receptor binding in hippocampal subregions, including CA3 (Orchinik et al., 2001). In both the hippocampus and the amygdala (Duvarci and Pare, 2007), therefore, corticosterone regulates excitability via the inhibitory neurons.

CONCLUSION

A growing body of experimental and clinical evidence gathered over the past several decades suggests that, in addition to affecting endocrinologic and physiologic aspects of body and brain function, stress also has a profound impact on neural plasticity at multiple levels of neural organization. These effects on neuroplasticity mechanisms are manifested as behavioral effects on cognitive and emotional function at one end of the spectrum and down to the level of cells, synapses, and molecules at the other. Moreover, the rich database of hippocampal synaptic plasticity has provided a very useful foundation for analyzing the cellular and molecular effects of stress, and their functional implications in terms of cognitive deficits triggered by stress. Although much of the early evidence on stress-induced plasticity emerged from studies of the hippocampus, more recent work has taken into account the important role played by another brain structure, the amygdala, which is critical for processing emotional information. As described in this review, a combination of behavioral, morphological, electrophysiological, pharmacological, and molecular techniques have been used to explain how stress elicits global changes in a brain's output, on the basis of local changes in individual neurons in the hippocampal and amygdalar networks. Findings from these studies have given rise to an intellectual framework that enables us to investigate molecular and biochemical underpinnings of neuronal plasticity mechanisms, and how these mechanisms are adversely affected by stress and emotional disorders.

REFERENCES

Akirav, I., and Richter-Levin, G. (1999). Biphasic modulation of hippocampal plasticity by behavioral stress and basolateral amygdala stimulation in the rat. *Journal of Neuroscience, 19,* 10530-10535.

Armony, J.L., Corbo, V., Clement, M.H., and Brunet, A. (2005). Amygdala response in patients with acute PTSD to masked and unmasked emotional facial expressions. *The American Journal of Psychiatry, 162,* 1961-1963.

Autry, A.E., Grillo, C.A., Piroli, G.G., Rothstein, J.D., McEwen, B.S., and Reagan, L.P. (2006). Glucocorticoid regulation of GLT-1 glutamate transporter isoform expression in the rat hippocampus. *Neuroendocrinology, 83,* 371-379.

Baker, K.B., and Kim, J.J. (2002). Effects of stress and hippocampal NMDA receptor antagonism on recognition memory in rats. *Learning and Memory, 9,* 58-65.

Bartanusz, V., Aubry, J.M., Pagliusi, S., Jezova, D., Baffi, J., and Kiss, J.Z. (1995). Stress-induced changes in messenger RNA levels of N-methyl-D-aspartate and AMPA receptor subunits in selected regions of the rat hippocampus and hypothalamus. *Neuroscience, 66,* 247-252.

Berger, S., Wolfer, D.P., Selbach, O., Alter, H., Erdmann, G., Reichardt, H.M., Chepkova, A.N., Welzl, H., Haas, H.L., Lipp, H.P., and Schutz, G. (2006). Loss of the limbic mineralocorticoid receptor impairs behavioral plasticity. *Proceedings of the National Academy of Sciences of the United States of America, 103,* 195-200.

Beylin, A.V., and Shors, T.J. (2003). Glucocorticoids are necessary for enhancing the acquisition of associative memories after acute stressful experience. *Hormones and Behavior, 43,* 124-131.

Blank, T., Nijholt, I., Eckart, K., and Spiess, J. (2002). Priming of long-term potentiation in mouse hippocampus by corticotropin-releasing factor and acute stress, implications for hippocampus-dependent learning. *Journal of Neuroscience, 22,* 3788-3794.

Bliss T.V., and, Lomo, T. (1973). Long-lasting potentiation of synaptic transmission in the dentate area of the anaesthetized rabbit following stimulation of the perforant path. *Journal of Physiology, 232,* 331-356.

Bodnoff, S.R., Humphreys, A.G., Lehman, J.C., Diamond, D.M., Rose, G.M., and Meaney, M.J. (1995). Enduring effects of chronic corticosterone treatment on spatial learning, synaptic plasticity, and hippocampal neuropathology in young and mid-aged rats. *Journal of Neuroscience, 15,* 61-69.

Bremner, J.D. (2007). Functional neuroimaging in post-traumatic stress disorder. *Expert Review of Neurotherapeutics, 7,* 393-405.

Bremner, J.D., Vythilingam, M., Vermetten, E., Southwick, S.M., McGlashan, T., Nazeer, A., Khan, S., Vaccarino, L.V., Soufer, R., Garg, P.K., Ng, C.K., Staib, L.H., Duncan, J.S., and Charney, D.S. (2003). MRI and PET study of deficits in hippocampal structure and function in women with childhood sexual abuse and posttraumatic stress disorder. *The American Journal of Psychiatry, 160*, 924-932.

Chaouloff, F., Hemar, A., and Manzoni, O. (2007). Acute stress facilitates hippocampal CA1 metabotropic glutamate receptor-dependent long-term depression. *Journal of Neuroscience, 27*, 7130-7135.

Conrad, C.D., Galea, L.A., Kuroda, Y., and McEwen, B.S. (1996). Chronic stress impairs rat spatial memory on the Y maze, and this effect is blocked by tianeptine pretreatment. *Behavioral Neuroscience, 110*, 1321-1334.

Conrad, C.D., LeDoux, J.E., Magarinos, A.M., and McEwen, B.S. (1999). Repeated restraint stress facilitates fear conditioning independently of causing hippocampal CA3 dendritic atrophy. *Behavioral Neuroscience, 113*, 902-913.

DeBosscher, K., Vanden-Berghe W., and Haegeman, G. (2003). The interplay between the glucocorticoid receptor and nuclear factor-kappaB or activator protein-1, molecular mechanisms for gene repression. *Endocrine Reviews, 24*, 488-522.

Diamond, D.M., Bennett, M.C., Fleshner, M., and Rose, G.M. (1992). Inverted-U relationship between the level of peripheral corticosterone and the magnitude of hippocampal primed burst potentiation. *Hippocampus, 2*, 421-430.

Duvarci, S., and Pare, D. (2007). Glucocorticoids enhance the excitability of principal basolateral amygdala neurons. *Journal of Neuroscience, 27*, 4482-4491.

Edwards, C.R., Stewart, P.M., Burt, D., Brett, L., McIntyre, M.A., Sutanto, W.S., de Kloet E.R., Monder, C. (1988). Localisation of 11 beta-hydroxysteroid dehydrogenase--tissue specific protector of the mineralocorticoid receptor. *Lancet, 2*, 986-989.

Ehrlich, I., Humeau, Y., Grenier, F., Ciocchi, S., Herry, C., and Luthi, A. (2009). Amygdala inhibitory circuits and the control of fear memory. *Neuron, 62*, 757-771.

Ferguson, D., and Sapolsky, R. (2007). Mineralocorticoid receptor overexpression differentially modulates specific phases of spatial and nonspatial memory. *Journal of Neuroscience, 27*, 8046-8052.

Funder, J.W., Pearce, P.T., Smith, R., and Smith, A.I. (1988). Mineralocorticoid action, target tissue specificity is enzyme, not receptor, mediated. *Science, 242*, 583-585.

Garcia, R., Musleh, W., Tocco, G., Thompson, R.F., and Baudry, M. (1997). Time-dependent blockade of STP and LTP in hippocampal slices following acute stress in mice. *Neuroscience Letters, 233*, 41-44.

Gass, P., Kretz, O., Wolfer, D.P., Berger, S., Tronche, F., Reichardt, H.M., Kellendonk, C., Lipp, H.P., Schmid, W., and Schutz, G. (2000). Genetic disruption of mineralocorticoid receptor leads to impaired neurogenesis and granule cell degeneration in the hippocampus of adult mice. *EMBO Reports, 1*, 447-451.

Gilbertson, M.W., Shenton, M.E., Ciszewski, A., Kasai, K., Lasko, N.B., Orr, S.P., and Pitman, R.K. (2002). Smaller hippocampal volume predicts pathologic vulnerability to psychological trauma. *Nature Neuroscience, 5*, 1242-1247.

Gould, E., Woolley, C.S., McEwen, B.S. (1990). Short-term glucocorticoid manipulations affect neuronal morphology and survival in the adult dentate gyrus. *Neuroscience, 37*, 367-375.

Gould, E., McEwen, B.S., Tanapat, P., Galea, L.A., and Fuchs, E. (1997). Neurogenesis in the dentate gyrus of the adult tree shrew is regulated by psychosocial stress and NMDA receptor activation. *Journal of Neuroscience, 17*, 2492-2498.

Govindarajan, A., Rao, B.S., Nair, D., Trinh, M., Mawjee, N., Tonegawa, S., and Chattarji, S. (2006). Transgenic brain-derived neurotrophic factor expression causes both anxiogenic and antidepressant effects. *Proceedings of the National Academy of Sciences of the United States of America, 103,* 13208-13213.

Gronli, J., Fiske, E., Murison, R., Bjorvatn, B., Sorensen, E., Ursin, R., Portas, and C.M. (2007). Extracellular levels of serotonin and GABA in the hippocampus after chronic mild stress in rats. A microdialysis study in an animal model of depression. *Behavior and Brain Research, 181,* 42-51.

Gurvits, T.V., Shenton, M.E., Hokama, H., Ohta, H., Lasko, N.B., Gilbertson, M.W., Orr, S.P., Kikinis, R., Jolesz, F.A., McCarley, R.W., and Pitman, R.K. (1996). Magnetic resonance imaging study of hippocampal volume in chronic, combat-related posttraumatic stress disorder. *Biological Psychiatry, 40,* 1091-1099.

Herman, J.P., and Cullinan, W.E. (1997). Neurocircuitry of stress, central control of the hypothalamo-pituitary-adrenocortical axis. *Trends in Neuroscience, 20,* 78-84.

Herman, J.P., Schafer, M.K., Young, E.A., Thompson, R., Douglass, J., Akil, H., and Watson, S.J. (1989). Evidence for hippocampal regulation of neuroendocrine neurons of the hypothalamo-pituitary-adrenocortical axis. *Journal of Neuroscience, 9,* 3072-3082.

Jacobson, L., and Sapolsky, R. (1991). The role of the hippocampus in feedback regulation of the hypothalamic-pituitary-adrenocortical axis. *Endocrine Reviews, 12,* 118-134.

Joels, M. (2006). Corticosteroid effects in the brain, U-shape it. *Trends in Pharmacological Science, 27,* 244-250.

Karst, H., Karten, Y.J., Reichardt, H.M., de Kloet, E.R., Schutz, G., and Joels, M. (2000). Corticosteroid actions in hippocampus require DNA binding of glucocorticoid receptor homodimers. *Nature Neuroscience, 3,* 977-978.

Karst, H., Nair, S., Velzing, E., Rumpff-van Essen, L., Slagter, E., Shinnick-Gallagher, P., and Joels, M. (2002). Glucocorticoids alter calcium conductances and calcium channel subunit expression in basolateral amygdala neurons. *European Journal of Neuroscience, 16,* 1083-1089.

Kerr, D.S., Campbell, L.W., Thibault, O., and Landfield, P.W. (1992). Hippocampal glucocorticoid receptor activation enhances voltage-dependent Ca2+ conductances, relevance to brain aging. *Proceedings of the National Academy of Sciences of the United States of America, 89,* 8527-8531.

Kim, J.J., Foy, M.R., and Thompson, R.F. (1996). Behavioral stress modifies hippocampal plasticity through N-methyl-D-aspartate receptor activation. *Proceedings of the National Academy of Sciences of the United States of America, 93,* 4750-4753.

Kohda, K., Harada, K., Kato, K., Hoshino, A., Motohashi, J., Yamaji, T., Morinobu, S., Matsuoka, N., and Kato, N. (2007). Glucocorticoid receptor activation is involved in producing abnormal phenotypes of single-prolonged stress rats, a putative post traumatic stress disorder model. *Neuroscience, 148,* 22-33.

Kole, M.H., Swan, L., and Fuchs, E. (2002). The antidepressant tianeptine persistently modulates glutamate receptor currents of the hippocampal CA3 commissural associational synapse in chronically stressed rats. *Europian Journal of Neuroscience, 16,* 807-816.

Lai, M., Horsburgh, K., Bae, S.E., Carter, R.N., Stenvers, D.J., Fowler, J.H., Yau, J.L., Gomez-Sanchez, C.E., Holmes, M.C., Kenyon, C.J., Seckl, J.R., and Macleod, M.R. (2007). Forebrain mineralocorticoid receptor overexpression enhances memory, reduces anxiety and attenuates neuronal loss in cerebral ischaemia. *Europian Journal of Neuroscience, 25,* 1832-1842.

Lamprecht, R., and LeDoux, J. (2004). Structural plasticity and memory. *Nature Reveiws of Neuroscience, 5,* 45-54.

LeDoux, J.E. (2000) Emotion circuits in the brain. *Annual Reveiws of Neuroscience, 23,* 155-184.

Lowy, M.T., Gault, L., and Yamamoto, B.K. (1993). Adrenalectomy attenuates stress-induced elevations in extracellular glutamate concentrations in the hippocampus. *Journal of Neurochemistry, 61,* 1957-1960.

Luine, V., Villegas, M., Martinez, C., and McEwen, B.S. (1994). Repeated stress causes reversible impairments of spatial memory performance. *Brain Research, 639,* 167-170.

Macleod, M.R., Johansson, I.M., Soderstrom, I., Lai, M., Gido, G., Wieloch, T., Seckl, J.R., and Olsson, T. (2003). Mineralocorticoid receptor expression and increased survival following neuronal injury. *Europian Journal of Neuroscience, 17,* 1549-1555.

Magarinos, A.M., and McEwen, B.S. (1995a). Stress-induced atrophy of apical dendrites of hippocampal CA3c neurons, comparison of stressors. *Neuroscience, 69,* 83-88.

Magarinos, A.M., and McEwen, B.S. (1995b). Stress-induced atrophy of apical dendrites of hippocampal CA3c neurons, involvement of glucocorticoid secretion and excitatory amino acid receptors. *Neuroscience, 69,* 89-98.

Maren, S., and Quirk, G.J. (2004). Neuronal signalling of fear memory. *Nature Reveiws Neuroscience, 5,* 844-852.

Maroun, M., and Richter-Levin, G. (2003). Exposure to acute stress blocks the induction of long-term potentiation of the amygdala-prefrontal cortex pathway in vivo. *Journal of Neuroscience, 23,* 4406-4409.

McDonald, A.J. (1982). Neurons of the lateral and basolateral amygdaloid nuclei, a Golgi study in the rat. *Journal of Comparaive Neurology, 212,* 293-312.

McDonald, A.J. (1992). Cell types and intrinsic connections of the amygdala. New York, Wiley-Liss.

McEwen, B.S. (1999). Stress and hippocampal plasticity. *Annual Reveiws of Neuroscience, 22,* 105-122.

McEwen, B.S. (2007). Physiology and neurobiology of stress and adaptation, central role of the brain. *Physiological Reveiws, 87,* 873-904.

McEwen, B.S., Gould, E.A., and Sakai, R.R. (1992). The vulnerability of the hippocampus to protective and destructive effects of glucocorticoids in relation to stress. *The British Journal of Psychiatry. Supplement,* 18-23.

McGaugh, J.L., and Roozendaal, B. (2002). Role of adrenal stress hormones in forming lasting memories in the brain. *Current Opinions in Neurobiology, 12,* 205-210.

Melia, K.R., Ryabinin, A.E., Schroeder, R., Bloom, F.E., and Wilson, M.C. (1994). Induction and habituation of immediate early gene expression in rat brain by acute and repeated restraint stress. *Journal of Neuroscience, 14,* 5929-5938.

Mitra, R., and Sapolsky, R.M. (2008). Acute corticosterone treatment is sufficient to induce anxiety and amygdaloid dendritic hypertrophy. *Proceedings of the National Academy of Sciences of the United States of America, 105,* 5573-5578.

Mitra, R., Jadhav, S., McEwen, B.S., Vyas, A., and Chattarji, S. (2005). Stress duration modulates the spatiotemporal patterns of spine formation in the basolateral amygdala.

Proceedings of the National Academy of Sciences of the United States of America, 102, 9371-9376.

Orchinik, M., Weiland, N.G., and McEwen, B.S. (1994). Adrenalectomy selectively regulates GABAA receptor subunit expression in the hippocampus. *Molecular and Cellular Neuroscience, 5,* 451-458.

Orchinik, M., Carroll, S.S., Li, Y.H., McEwen, B.S., and Weiland, N.G. (2001). Heterogeneity of hippocampal GABA(A) receptors, regulation by corticosterone. *Journal of Neuroscience, 21,* 330-339.

Pascual-Le Tallec L., and Lombes, M. (2005). The mineralocorticoid receptor, a journey exploring its diversity and specificity of action. *Molecular Endocrinology, 19,* 2211-2221.

Pitkanen, A., and Amaral, D.G. (1994). The distribution of GABAergic cells, fibers, and terminals in the monkey amygdaloid complex, an immunohistochemical and in situ hybridization study. *Journal of Neuroscience, 14,* 2200-2224.

Pitkänen, A. (2000). Connectivity of the rat amygdaloid complex. In The Amygdala (Ed.) J. P. Aggleton, USA: Oxford University Press.

Protopopescu, X., Pan, H., Tuescher, O., Cloitre, M., Goldstein, M., Engelien, W., Epstein, J., Yang, Y., Gorman, J., LeDoux, J., Silbersweig, D., and Stern, E. (2005). Differential time courses and specificity of amygdala activity in posttraumatic stress disorder subjects and normal control subjects. *Biological Psychiatry, 57,* 464-473.

Rainnie, D.G., Bergeron, R., Sajdyk, T.J., Patil, M., Gehlert, and D.R., Shekhar, A. (2004). Corticotrophin releasing factor-induced synaptic plasticity in the amygdala translates stress into emotional disorders. *Journal of Neuroscience, 24,* 3471-3479.

Rao, R.P., Tomar, A., McEwen, B.S., and Chattarji, S. (2007). Multiple roles for corticosterone in the modulation of stress-induced anxiety.In, *Annual Meeting of the Society for Neuroscience,* p 840.821. San Diego.

Rauch, S.L., Shin, L.M., and Phelps, E.A. (2006). Neurocircuitry models of posttraumatic stress disorder and extinction, human neuroimaging research--past, present, and future. *Biological Psychiatry, 60,* 376-382.

Rauch, S.L., van der Kolk, B.A., Fisler, R.E., Alpert, N.M., Orr, S.P., Savage, C.R., Fischman, A.J., Jenike, M.A., Pitman, R.K. (1996). A symptom provocation study of posttraumatic stress disorder using positron emission tomography and script-driven imagery. *Archives of General Psychiatry, 53,* 380-387.

Reagan, L.P., Rosell, D.R., Wood, G.E., Spedding, M., Munoz, C., Rothstein, J., and McEwen, and B.S. (2004). Chronic restraint stress up-regulates GLT-1 mRNA and protein expression in the rat hippocampus, reversal by tianeptine. *Proceedings of the National Academy of Sciences of the United States of America, 101,* 2179-2184.

Reznikov, L.R., Reagan, L.P., and Fadel, J.R. (2008). Activation of phenotypically distinct neuronal subpopulations in the anterior subdivision of the rat basolateral amygdala following acute and repeated stress. *Journal of Comparative Neurology, 508,* 458-472.

Reznikov, L.R., Grillo, C.A., Piroli, G.G., Pasumarthi, R.K., Reagan, L.P., and Fadel, J. (2007). Acute stress-mediated increases in extracellular glutamate levels in the rat amygdala, differential effects of antidepressant treatment. *Europian Journal of Neuroscience, 25,* 3109-3114.

Roozendaal, B., and McGaugh, J.L. (1997). Glucocorticoid receptor agonist and antagonist administration into the basolateral but not central amygdala modulates memory storage. *Neurobiology of Learning and Memory, 67,* 176-179.

Rozeboom, A.M., Akil, H., and Seasholtz, A.F. (2007). Mineralocorticoid receptor overexpression in forebrain decreases anxiety-like behavior and alters the stress response in mice. *Proceedings of the National Academy of Sciences of the United States of America, 104,* 4688-4693.

Sah, P., Faber, E.S., Lopez De Armentia, M., and Power J. (2003). The amygdaloid complex, anatomy and physiology. *Physiological Reveiws, 83,* 803-834.

Sapolsky, R.M., Zola-Morgan, S., and Squire L.R. (1991). Inhibition of glucocorticoid secretion by the hippocampal formation in the primate. *Journal of Neuroscience, 11,* 3695-3704.

Seckl, J.R., and Walker, B.R. (2001). Minireview, 11beta-hydroxysteroid dehydrogenase type 1- a tissue-specific amplifier of glucocorticoid action. *Endocrinology, 142,* 1371-1376.

Shekhar, A., Truitt, W., Rainnie, D., and Sajdyk, T. (2005). Role of stress, corticotrophin releasing factor (CRF) and amygdala plasticity in chronic anxiety. *Stress, 8,* 209-219.

Shin, L.M., Orr, S.P., Carson, M.A., Rauch, S.L., Macklin, M.L., Lasko, N.B., Peters, P.M., Metzger, L.J., Dougherty, D.D., Cannistraro, P.A., Alpert N.M., Fischman, A.J., and Pitman, R.K. (2004). Regional cerebral blood flow in the amygdala and medial prefrontal cortex during traumatic imagery in male and female Vietnam veterans with PTSD. *Archives of General Psychiatry, 61,* 168-176.

Shors, T.J. (1999). Acute stress and re-exposure to the stressful context suppress spontaneous unit activity in the basolateral amygdala via NMDA receptor activation. *Neuroreport, 10,* 2811-2815.

Shors, T.J. (2001). Acute stress rapidly and persistently enhances memory formation in the male rat. *Neurobiology Learning and Memory, 75,* 10-29.

Shors, T.J., and Thompson, R.F. (1992). Acute stress impairs (or induces) synaptic long-term potentiation (LTP) but does not affect paired-pulse facilitation in the stratum radiatum of rat hippocampus. *Synapse, 11,* 262-265.

Shors, T.J., and Mathew, P.R. (1998). NMDA receptor antagonism in the lateral/basolateral but not central nucleus of the amygdala prevents the induction of facilitated learning in response to stress. *Learning and Memory, 5,* 220-230.

Shors, T.J., Weiss, C., and Thompson, R.F. (1992). Stress-induced facilitation of classical conditioning. *Science 257,* 537-539.

Shors, T.J., Gallegos, R.A., and Breindl, A. (1997). Transient and persistent consequences of acute stress on long-term potentiation (LTP), synaptic efficacy, theta rhythms and bursts in area CA1 of the hippocampus. *Synapse, 26,* 209-217.

Sloviter, R.S., Valiquette, G., Abrams, G.M., Ronk, E.C., Sollas, A.L., Paul, L.A., Neubort, S. (1989). Selective loss of hippocampal granule cells in the mature rat brain after adrenalectomy. *Science, 243,* 535-538.

Sousa, N., Lukoyanov, N.V., Madeira, M.D., Almeida, O.F., and Paula-Barbosa, M.M. (2000). Reorganization of the morphology of hippocampal neurites and synapses after stress-induced damage correlates with behavioral improvement. *Neuroscience, 97,* 253-266.

Spence, K.W., and Taylor, J. (1951). Anxiety and strength of the UCS as determiners of the amount of eyelid conditioning. *Journal of Experimental Psychology, 42,* 183-188.

Stutzmann, G.E., McEwen, B.S., and LeDoux, J.E. (1998). Serotonin modulation of sensory inputs to the lateral amygdala, dependency on corticosterone. *Journal of Neuroscience, 18*, 9529-9538.

Sunanda Rao, M.S., and Raju, T.R. (1995). Effect of chronic restraint stress on dendritic spines and excrescences of hippocampal CA3 pyramidal neurons--a quantitative study. *Brain Research, 694*, 312-317.

Suvrathan, A. (2008). In, *Society for Neuroscience*, p Submitted. Washington, D.C.

Suvrathan, A., Bennur, S., and Chattarji, S. (2006). Silent synapses speak up in the amygdala. In, *Annual Meeting of Society for Neuroscience*, p 370.314. Atlanta.

Suvrathan, A., Tomar, A., and Chattarji, S. (2007). Whats synaptic transmission in the amygdala got to do with PTSD? In, *Annual Meeting of Society for Neuroscience*, p 841.812. San Diego.

Takahashi T., Kimoto T., Tanabe N., Hattori T.A., Yasumatsu N., and Kawato S. (2002). Corticosterone acutely prolonged N-methyl-d-aspartate receptor-mediated Ca2+ elevation in cultured rat hippocampal neurons. *Journal of Neurochemistry, 83*, 1441-1451.

Tocco, G., Shors, T.J., Baudry, M., and Thompson, R.F. (1991). Selective increase of AMPA binding to the AMPA/quisqualate receptor in the hippocampus in response to acute stress. *Brain Research 559, 168-171.*

Vouimba, R.M., Yaniv, D., and Richter-Levin, G. (2007). Glucocorticoid receptors and beta-adrenoceptors in basolateral amygdala modulate synaptic plasticity in hippocampal dentate gyrus, but not in area CA1. *Neuropharmacology, 52*, 244-252.

Vyas, A., and Chattarji, S. (2004). Modulation of different states of anxiety-like behavior by chronic stress. *Behavioral Neuroscience, 118*, 1450-454.

Vyas, A., Pillai, A.G., and Chattarji, S. (2004). Recovery after chronic stress fails to reverse amygdaloid neuronal hypertrophy and enhanced anxiety-like behavior. *Neuroscience, 128*, 667-673.

Vyas, A., Jadhav, S., and Chattarji, S. (2006). Prolonged behavioral stress enhances synaptic connectivity in the basolateral amygdala. *Neuroscience, 143*, 387-393.

Vyas, A., Mitra, R., Shankaranarayana Rao, B.S., and Chattarji S (2002). Chronic stress induces contrasting patterns of dendritic remodeling in hippocampal and amygdaloid neurons. *Journal of Neuroscience, 22*, 6810-6818.

Watanabe, Y., Gould, E., Cameron, H.A., Daniels, D.C., and McEwen, B.S. (1992). Phenytoin prevents stress- and corticosterone-induced atrophy of CA3 pyramidal neurons. *Hippocampus, 2*, 431-435.

Weiland, N.G., Orchinik, M., and Tanapat, P. (1997). Chronic corticosterone treatment induces parallel changes in N-methyl-D-aspartate receptor subunit messenger RNA levels and antagonist binding sites in the hippocampus. *Neuroscience, 78*, 653-662.

Xu, L., Holscher, C., Anwyl, R., and Rowan, M.J. (1998). Glucocorticoid receptor and protein/RNA synthesis-dependent mechanisms underlie the control of synaptic plasticity by stress. *Proceedings of the National Academy of Sciences of the United States of America, 95*, 3204-3208.

Zhou, J., and Cidlowski, J.A. (2005). The human glucocorticoid receptor, one gene, multiple proteins and diverse responses. *Steroids, 70*, 407-417.

In: Expanding Horizions of the Mind Science(s)
Editors: P.N. Tandon, R.C. Tripathi and N. Srinivasan

ISBN: 978-1-62808-705-5
©2013 Nova Science Publishers, Inc.

Chapter 10

NEUROECONOMICS OF INDIVIDUAL DECISION MAKING AT MULTIPLE LEVELS: A REVIEW

V. S. Chandrasekhar Pammi[1] and Krishna P. Miyapuram[2]
[1]Centre of Behavioural and Cognitive Sciences, University of Allahabad, India
[2]Unilever R and D, Olivier van Noortlan, Vlaardingen, AT, The Netherlands

ABSTRACT

NeuroEconomics is a rapidly developing multidisciplinary research area that employs neuroscience techniques to explain economic theories of human behaviour and makes efforts to bring researchers from the disciplines of psychology, neuroscience, and economics to a common platform. The foundations of this science lie in the pioneering experiments of Kahneman and Tversky (1979). The past decade has seen a remarkable effort by neuroscientists and economists in applying neurophysiological (animal experiments recoding single cell behaviour) and neuroimaging (such as event related potentials, functional magnetic resonance imaging) techniques to understand brain activity while experimental tasks that involve economic decisions are being performed. In this review we would like to present studies from single cell recording to behavioural level while presenting various components related to the process of individual decisions. Along with the future scope of computational modelling, the existing models related to individual decision making will be presented. Taking further, the upcoming research area of collective decision making will be reviewed.

1. METHODS OF STUDYING BRAIN FUNCTION: A BRIEF INTRODUCTION

Neuroeconomics is a rapidly developing multidisciplinary research area that employs neuroscience techniques to explain economic theories of human behaviour. Understanding the function of the brain can be done at various levels from molecular level to studying the entire central nervous system (Churchland and Sejnowski, 1992). We review research on decision making from neuronal level to neuroimaging studies, including some evidence from lesion studies. In this section, we briefly describe methodological aspects of studying brain function.

The primary information processing cells in the brain are neurons; around 100 billion neurons can be found in the human brain. The other type of cells, the Glia, outnumber the neurons by at least 10 times, and are involved in maintaining the structural integrity of brain tissue and energy metabolism. The dendrites of the neurons gather inputs from other neurons through an electrochemical junction called the synapse. Information is processed by the neuron in the form of electrical impulse (action potential) that travels down its tail, the axon, it is then passed on to the next synapse by releasing a neurotransmitter (Kandel, Schwartz, Jessell, 2000; Gazzaniga, Ivry, Mangun, 2002). Excitatory transmitter (e.g., glutamate) increase firing in the post-synaptic neuron, while Inhibitory transmitters (e.g., GABA) decreases firing in the post-synaptic neuron. Neurophysiological recordings of neural responses can be done extra-cellularly while awake-animals engage in behavioural tasks. The invasiveness of neurophysiological recordings (implanting of electrodes) limits its use in humans. However, it would be possible to record from patients with electrodes implanted for therapeutic purpose (e.g. epileptic patients).

Brain imaging techniques provide an excellent tool for studying human brain from network level to systems level. It is possible to non-invasively image brain activity in healthy volunteers by measuring electrical/magnetic fields created by electrical activity of the brain. The electrical impulses sent out by the neurons propagate through the brain tissue to the scalp. Electrodes placed on the scalp can record this current. This is known as Electroencephalo-gram (EEG). The electrical field also creates a small magnetic field that can be measured using magneto-encephalogram (MEG). The advantage of EEG and MEG is high temporal resolution essentially being able to measure signals changing over milliseconds. While EEG/MEG can be regarded as relatively direct measures of brain activity, the main drawback of these techniques is the low spatial resolution. It is possible to localize brain activity within >1 cm. The various currents in the scalp are added up and there is no unique solution to localization of the signal source inverse problem. MEG is insensitive to signals from radially-oriented sources and hence does not record from deep brain regions (Volkow, Rosen, Farde, 1997).

The alternative approach to studying function of human brain is to measure blood flow, which is an indirect measure of neural activity. In a typical positron emission tomography (PET) study, subject is injected with a radioactive tracer that has a relatively short half-life (~120 seconds). The subject performs a behavioural task, while the photons emitted from radioactive decay are counted by the scanner for every location in the brain. Increased blood flow results in increased delivery of the radioactive tracer, which further results in a larger signal being emitted from active regions of the brain. PET can be used with radio-labelled neurotransmitters or receptor agonists, which allows imaging of the level of various neurotransmitters or receptors in selected regions of the brain. The disadvantages of the PET technique, apart from the subject being exposed to radioactive substance, are low spatial (~1cm) and temporal resolution (~2 min).

Functional magnetic resonance imaging (functional MRI or fMRI), has been widely popular as a brain imaging technique ever since Ogawa et al., (1990) discovered the level of blood oxygen was visible on certain types of MRI scans referred to as "blood oxygen-level dependent" (BOLD) contrast. The BOLD technique depends on the fact that deoxygenated haemoglobin is paramagnetic and becomes magnetic when placed in a magnetic field, while oxygenated haemoglobin does not exhibit such properties. Ogawa et al., (1992) applied the BOLD technique to image activity in visual cortex to alternating on/off visual stimulus. The fMRI technique has relatively high spatial resolution (~1mm) and low temporal resolution (1sec) (Logothetis et al., 2001; Heeger and Ress, 2002; Ugurbil, Toth and Kim, 2003; Logothetis, 2008). The disadvantage with fMRI technique is the artefacts in some areas of the brain near air-filled cavities such as medial orbitofrontal, inferior temporal cortex.

Electrical and haemodynamic techniques have different resolutions of imaging. While fMRI has high spatial resolution (~ 1 mm) and low temporal resolution (> 1 second), EEG/MEG has low spatial resolution (~ 1 cm) and high temporal resolution (~ 1 millisecond). Multimodal imaging allows integration of fMRI and MEG/EEG results to combine the advantages of these techniques (George et al., 1995; Dale, and Halgren, 2001; Shibasaki, 2008; Bandettini, 2009).

While brain imaging techniques demonstrate the involvement of a brain area during performance of a behavioural task these techniques cannot tell us whether a particular brain area is essential for the concerned behaviour (Chandrasekhar, 2005; Miyapuram, 2008). Lesions to particular brain areas allow determination of whether a particular brain region is necessary for a particular cognitive function. Lesion studies can be performed in animals to produce very anatomically precise lesions. In humans, patients with well defined lesions (for example, due to injury) can be studied, but is less precise due to the varied nature of lesions. With new technique of Transcranial Magnetic Stimulation (TMS), reversible lesions can be momentarily induced by passing electrical current in the cortex, thus allowing study function of human brain at systems level. Similarly, TMS can be used to stimulate specific region of the brain to elicit activation

2. ECONOMIC THEORIES OF HUMAN CHOICE BEHAVIOUR

Decision making is the process of choosing one out of several alternatives. A choice is made whenever an organism is confronted with more than one alternative for which an action is necessary to acquire or avoid these alternatives. The deliberative process that precedes the choice is called "decision". The study of human decision making has been a longstanding tradition in microeconomics. Traditional accounts in economics have a normative flavour which prescribe what the decisions should be, based on optimization of the payoffs. Recently, with the concepts of psychology, descriptive theories of decision making have evolved that describe the kinds of judgments and decisions people actually make in practice. Cognitive, motivational and affective factors limit the rational behaviour approach taken by normative theories. A description of the normative theories of decision making, followed by descriptive theories is presented next.

Typically analysis of choice behaviour is based on revealed preference: if an alternative A is chosen over alternative B, it means that the person actually prefers A over B, for whatever reason. The expected utility theory (Bernoulli, 1763/1958; Von Neumann and Morgenstern, 1944) attaches a subjective value or utility to each alternative and suggests that people make choices that maximize their (expected) utility. Hence, when a person prefers A over B, the utility of A is greater than that of B.

Many a time the outcomes are probabilistic and have some degree of uncertainty associated with them. Pascal, way back in 1670 (Pascal, 1670 /1958), conjectured that human choice behaviour could be understood by the expected value (product of probability and objective value i.e. magnitude of the outcome).

For example, consider a situation in which an individual is faced with a choice between two lotteries A and B. Lottery A has an equal chance (50% chance or probability) of winning $10 and $90 and lottery B has an equal chance of winning $40 and $60. The expected values of the two lotteries are

A: (10 x 50/100) + (90 x 50/100) = 50
B: (40 x 50/100) + (60 x 50/100) = 50

The expected values of both the lotteries A and B are the same. Hence, the individual must be indifferent to either of these two lotteries. Consider another situation in which the two lotteries are A: 25% probability of winning $10 and 75% probability of winning $90 and B: 25% probability of winning $40 and 75% probability of winning $60.

The expected values of the two lotteries are

A: (10 x 25/100) + (90 x 75/100) = 70
B: (40 x 25/100) + (60 x 75/100) = 55

Hence lottery A is expected to be chosen because it has higher expected value than B.

Consider a third situation in which the individual is again faced with two lotteries A: 50% probability of winning $10 or $90 and B: getting $50 for sure (100% probability). In this situation, most people would choose lottery B, even though the two lotteries have the same expected value. This is referred to as *risk aversion*, i.e. people always tend to choose certain outcomes and are aversive to choose the risky or uncertain (probability < 100%) outcome.

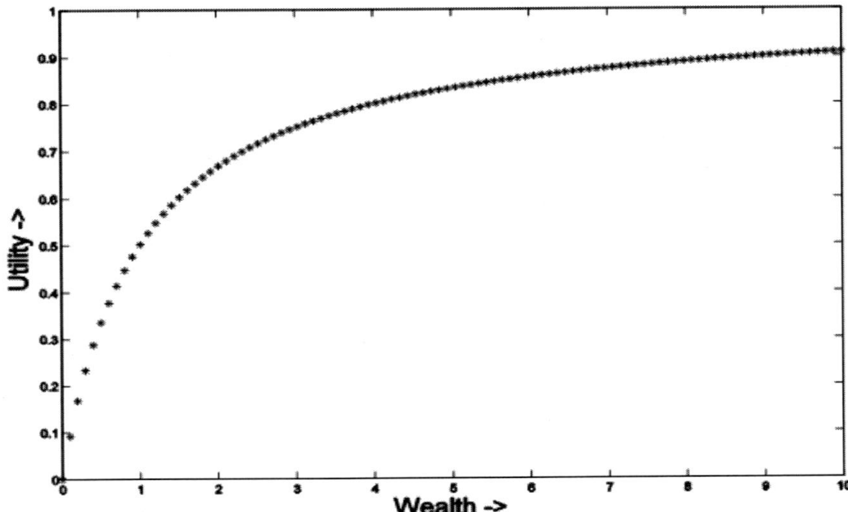

Figure 1. A hypothetical utility function depicting the relationship between wealth and utility (modified from Doya, 2008). The utility function f(r; c) = r/(r+c) where c=1 and r is the reward amount (wealth).

The *certainty equivalent* is the guaranteed payoff at which a person is indifferent between accepting the guaranteed payoff and a higher but uncertain payoff. Additionally, the difference between the certainty equivalent and the expected value of the lottery is known as the *risk premium*, which is the amount that the individual actually 'pays' in order to avoid risk. Choices made by an individual can be classified into risk averse, risk seeking or risk neutral if the certainty equivalent is less than, greater than or equal to the expected value of the lottery, respectively.

Daniel Bernoulli (1763/1958) proposed that a mathematical function should be used to correct the expected value to account for risk aversion. Bernoulli suggested that the subjective value or the utility that people assign to an outcome depends on the wealth of the assigning person and grows more slowly than its objective value (magnitude). Intuitively this means that an offer of $10 has more value (=utility) to somebody whose total wealth is $100 than to somebody richer, whose total wealth is $100,000. More specifically the principle of diminishing marginal rate of utility (Bernoulli, 1763/1958) states that the utility of each additional dollar decreases with increasing wealth. Bernoulli proposed that increase in magnitude is always accompanied by an increase in the utility, which follows a concave (more specifically, a logarithmic) function of magnitude (Figure 1). Hence, individuals behave as to maximise the expected utility, instead of the expected value. Extending the mathematical notion of expected value (Pascal, 1670 /1958), the expected utility of each alternative is calculated as the sum of utilities weighted (multiplied) by the associated probability. The alternative with highest expected utility is chosen.

The Expected value and Expected utility are defined as follows:

- Expected Value = sum (magnitude x probability)
- Expected Utility = sum (utility x probability)

Where utility is the subjective value that follows a concave function of magnitude for risk averse people and a convex function for risk seeking people.

Von Neumann and Morgenstern (1944) formalized the Expected Utility theorem describing under what conditions (or axioms) preferences can be (numerically) represented using a mathematical function. This should allow for a cardinal representation of preferences i.e. allowing quantification of how much an option is preferred over another one. The axioms viz., completeness, transitivity and continuity establish that an individual behaves rationally. An implicit assumption in expected utility theorem was independence of lotteries. The independence axiom was challenged by well-known paradoxes (Allais, 1953; Ellsberg, 1961).

The following is an example of Allais Paradox. Given a choice between two lottery A : 10% chance of winning $5 million and lottery B : an 11% chance of winning $1 million, people would tend to choose the lottery A which has higher expected value, but their preferences would reverse when an option of winning $1 million for sure is added to both the gambles. It looks like lotteries with guaranteed outcomes (high probability) are weighted differently by people. The Ellsberg paradox suggests that people prefer situations with known probabilities compared to unknown probabilities. This is known as *ambiguity aversion*. Depending on whether the probabilities are known or unknown, the uncertain situation can be classified as risky or ambiguous, respectively. Consider the example of an urn having a total of 90 balls of which it is known that 30 balls are red and the remaining 60 are either black or yellow in an unknown proportion. Consider a choice between two options A: You receive

$100 if you draw a red ball and B: You receive $100 if you draw a black ball; most people prefer to choose A. When the two options are modified such that A : You receive $100 if you draw a red ball or an yellow ball and B : You receive $100 if you draw a black ball or a yellow ball, most people reverse their preferences and choose option B. The Allais and Ellsberg paradoxes challenge the Expected utility theory, which is a normative theory.

The expected utility theory is based on two economic parameters, namely utility (subjective value) and probability. The third parameter that underlies many decision making situations is the delay, which can be defined as the waiting time until the outcome is realised. Many decisions in everyday life result in delayed outcomes. Decisions between outcomes that occur at different points in time are referred to as intertemporal decisions, and the process of systematically devaluating outcomes over time is called temporal discounting. Discounted utility model can be considered equivalent to the expected utility theory in time domain. The expected utility theory suggested that decision makers choose between options based on probability weighted sum of utilities. The discounted utility model posits that decisions are made as a weighted sum of utilities with temporal discount factors as weights. In addition to maximisation of utility rate, the discounted utility model assumes a constant discount rate (Samuelson, 1937). Constant discounting ensures that a preference between two delayed options is only dependent on the time delay between them. The order of preferences should be preserved at all possible time points (time consistency or stationarity). The exponential discount function has a constant discount rate and is adopted by discounted utility theory.

The expected utility and discounted utility models provide normative descriptions of human choice behaviour. Empirical research has identified many systematic deviations from the ideal economic situations. While the normative models tell us what the optimal decisions should be, several descriptive models have been proposed to adequately describe human choice behaviour. Below, we describe two such models — the prospect theory and the regret theory.

Two key assumptions of expected utility theory are: 1) preferences should not depend on procedure (the way in which they are elicited), description, and context i.e. the other available options, and 2) preferences should depend on impact of consequences on final states of wealth. Empirical research has identified a number of anomalies in choice behaviour. For example, most people prefer a gamble of winning $3000 for sure over a risky gamble yielding $4000 with 80% probability. However when the gamble is presented for losses, people have their preferences reversed and choose the risky option when gambling to avoid losses (Kahneman and Tversky, 1979). This is known as *reflection effect* i.e. the risk attitude reverses because of the reflection of prospects (options) around 0 (changing from gains to losses). Furthermore, Kahneman and Tversky (1979) demonstrated that the gains must be much larger than the losses in order that subjects would choose to play a mixed gamble involving both gains and losses. This is referred to as loss aversion suggesting that individuals are more sensitive to losses (steeper utility function) than to gains. Loss aversion is closely linked to the endowment effect, which refers to the tendency of people to value an object they possess more highly than they would if they did not possess it (Tversky and Kahneman, 1991). Lastly, the way the problem is defined determines the choice of the individual (Kahneman and Tversky, 1984; Kahneman and Tversky, 2000). For example, assume that a ship sailing with 600 people is going to be drowned, two options are available A: 200 people will be saved and B: there is a 1/3 probability that 600 people will be saved and a 2/3 probability that no one will be saved. Most people prefer option A. When the problem was framed negatively and people were asked to choose between two options C: 400 people will die and D: there is 1/3 probability that no one will die and 2/3 probability that 600 people will die, most people preferred option D. In all the four options, the expected value was identical (200 lives saved), but the way the problem was framed changed the preferences of people from risk averse (option A) to risk seeking (option D). This is the well-known *framing effect*.

The Prospect theory of Kahneman and Tversky (1979) suggests the value function to be defined in relation to a reference point (usually the current wealth) and follows a concave function for gains (values above the reference point) and a convex function for values lower than the reference point (losses). In addition, the function is steeper for losses than for gains. While the expected utility theory assumes that people weigh outcomes according to the actual probability of occurring, prospect theory suggests non-linear probability weighting. Individuals calculate the weighted sum of values of outcomes as a non-linear function of probability over-weighing small probabilities and under-weighing middle and large probabilities. According to this theory, the value of a prospect is defined as the product of value function and the probability weighting function (Figure 2). It is to be noted that the utility function is substituted with a subjective value function defined over gains and losses relative to a *status quo* (reference point) and is weighted by a probability weighting function.

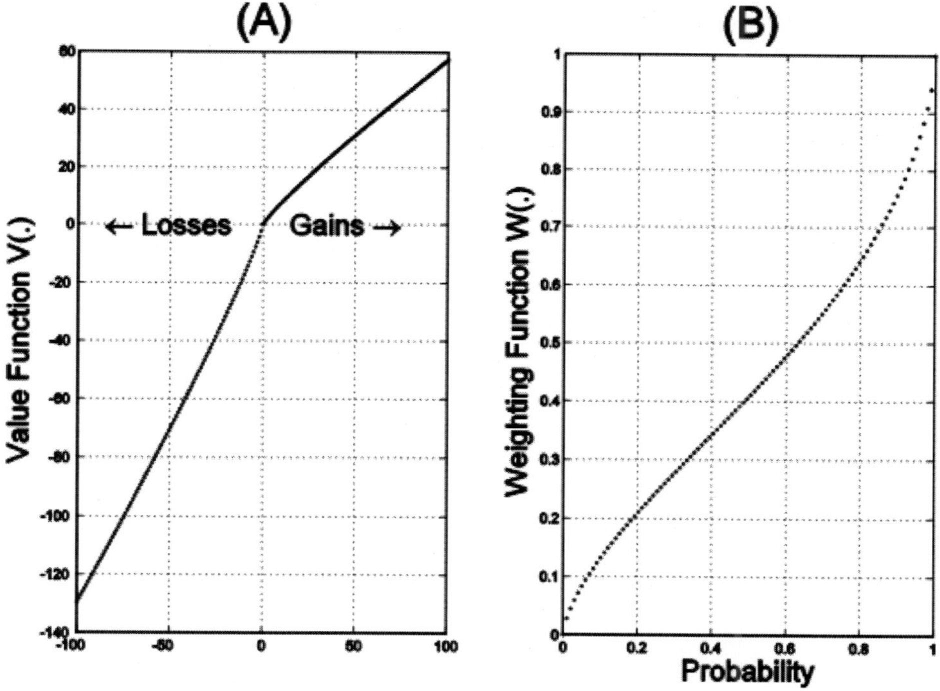

Figure 2. Value and probability weighting functions from prospect theory (adapted from Trepel, Fox and Poldrack, 2005). (A) Value Function V(.) is plotted as a function of gains and losses with respect to a reference point. $V(q) = q^{\alpha}$ where $q \geq 0$, and $V(q) = -\lambda(-q)^{\beta}$ where $q < 0$ (here, $\alpha = 088$, $\beta = 0.88$, $\lambda = 2.25$). The α, β measures the curvature of the value function and λ is the coefficient of loss aversion. (B) Probability weighting function W(.) for gains plotted as a function of the probability p of occurrence of an event. $W(p) = \zeta p^{\gamma} / (\zeta p^{\gamma} + (1-p)^{\gamma})$ where ζ represent the elevation of weighting function and γ measures its curvature (here, $\zeta = 0.69$, $\gamma = 0.69$).

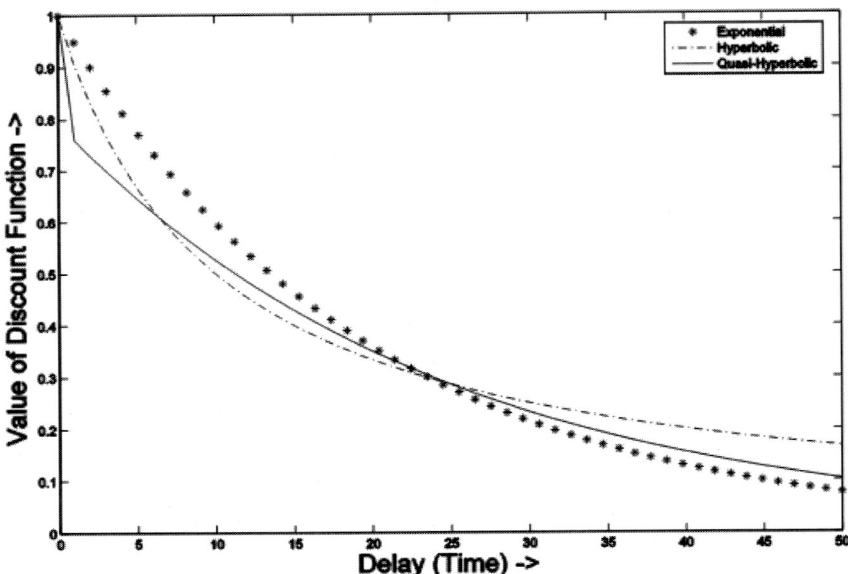

Figure 3. The exponential, hyperbolic and quasi-hyperbolic temporal discount functions (modified from Berns, 2007; Engelmann and Brooks, 2009) for 50 time points. A) The exponential discount function (γ^t) where γ (here we took, 0.95) is the discount rate and t is the time factor. B) The hyperbolic discount function ($1/(\alpha*t+1)$) where α=0.1. C) The Quasi-hyperbolic function is a piece-wise function similar to exponential discount function except for the first time unit (1, $\beta*\delta$, $\beta*\delta^2$,..., $\beta*\delta^t$) where β and δ are two-factors modelling the constant discount rate and discount rate in the standard exponential formula (here, β=0.792, δ= 0.96).

The regret theory is another popular descriptive model of human choice behaviour. An individual feels regret when he makes a choice that results in an outcome that is worse than what would have occurred had he made an alternative choice. A feeling of rejoice occurs when the option chosen yields a more favourable outcome than an alternative decision (Chandrasekhar et al., 2008). Interest in experimental economics had started with a theory developed by (Bell, 1982, 1983); Loomes and Sugden, (1982, 1987) which was used to explain the violations of the classical expected utility theory. The regret theory explained the Allais paradox (Allais, 1953) with an anticipatory regret signal that could avoid decisions that would lead to worse outcomes. The regret theory incorporated an anticipatory component which is emotional in nature (regret-rejoice function) into the expected utility framework. The experience of regret or rejoice at the outcome measures the amount of utility or disutility that the subjects experienced while comparing the obtained outcome with the alternative outcomes that could have been chosen (called as *counter factual* process). But importantly, the back propagated (in time) regret signal would be useful for the decision making process. Interest in behavioural economics started with this anticipatory regret signal which supported irrational behaviour while addressing the interaction of emotions with decision making under uncertainty.

Decisions are often encountered with temporal factors. The intertemporal choices are the decisions about the consequences that play out in time (Engelmann and Brooks, 2009) and are seen in most of the decisions in our daily lives such as decisions about spending money, purchase of health insurance, money investment and food intake (Berns, Laibson, Loewenstein, 2007). This temporal aspect in decision making process involves tradeoffs between immediate consumption and considerations about better future payoffs. This phenomenon was observed as either discounting of positive or negative payoffs or the preferences of payoffs at different points in time (Frederick, Loewenstein, and O'Donoghue, 2004). Earlier in 1937, Paul Samuelson proposed a model of standard discounted utility theory explaining how individuals weigh future utilities with the help of exponential discounting factor (Samuelson, 1937). This model of discounting was popular among economists investigating choice behaviour (Green and Myerson, 1996) where the present value is equal to the discounted value of a reward amount (the actual value) available after a delay (in units of time). Another competing model backed by psychologists was the hyperbolic discounting model. This model explained people sensitiveness to time delays associated with prospects and the subjective value of a prospect (in this model) depended on the ratio of amount to time. Another model consisting of quasi-hyperbolic discount function incorporating additional features in discounting function (Phelps and Pollack, 1968; Laibson, 1997) was proposed while explaining the choices among smaller, sooner payoffs and larger, later payoffs. The discount function is a two-parameter function (beta-delta preference) with beta (value ranging from 0 to 1 and is constant discount for all future time periods) and delta (value less than or equal to 1 and is discount rate in the standard exponential formula) factors. The equations along with the forms of discounting functions are shown in Figure 3.

Most of our decisions under uncertainty are either risky or ambiguous. Risky decisions are taken when the probabilities of uncertain outcomes are known whereas ambiguous decisions are taken when the probabilities of uncertain outcomes are unknown (O'Neill and Kobayashi, 2009). Behavioural evidence illustrating the distinction between risk and ambiguity in decision making was first depicted by Ellsberg (1961). He observed that the subjects responded differently to risky and ambiguous choices, and they mostly selected choices with risky outcomes. This demonstrates the existence of *ambiguous averse people* who would like to choose uncertain outcomes with only known information.

When people were given a choice between lotteries with sure smaller valued outcome versus an uncertain outcome with a larger value, some people chose the sure outcome whereas some selected the uncertainty outcome while risking themselves. The people who took sure certain outcomes were called *risk averse* while the people who preferred risky outcome (preferring a lottery) were called *risk seeking*. If individuals were indifferent between a lottery and its expected value, then they are called as *risk neutral* (Friedman and Savage, 1948; Kalenscher, 2009). Later, Tversky and Kahneman (1992) elegantly demonstrated that in low probability gain conditions, subjects were risk seeking, and in low probability loss conditions they were observed to be risk averse. In addition, under high probability gain conditions, subjects were risk averse and in high probability loss conditions, the subjects were observed to be risk seeking. This result explained real life behaviours such as investing, purchase of insurances, etc.

Every choice under uncertainty could be associated with a risk factor and that would reflect subjects' preference over the available options (uncertain in nature). The behaviour of subjects whether they are risk seeking or risk averse can be measured with the help of risk factor. The risk factor for a gamble is computed as the variance of the outcome of the selected options and the difference in risk computed among options could be used in the decision making (Engelmann and Tamir, 2009).

Risk and ambiguity are important components of decision making. Loewenstein and colleagues (2001) with risk-as-feelings hypothesis argued that feelings (which was dealt as separate component in the earlier decision making models) influence cognitive evaluations and also effect behaviour (which occurs after the cognitive evaluation). Their argument evolved into a framework consisting of bidirectional link between feelings and cognitive evaluation modules, and further influences behaviour while leading to outcomes.

3. NEUROSCIENCE OF DECISION MAKING: NEUROPHYSIOLOGICAL STUDIES

The central dogma is that all decisions originate from the brain. The human and primate brains have similar structure and have large brain relative to body-size when compared against other animals. Electrophysiological technique enables us to record activity of neurons, the information processing cells of the brain, in awake and behaving rodents, pigeons, and primates. In this section, an overview of primate brain areas followed by a selective review of some neurophysiological studies of decision making is presented.

The primate brain is separated into two almost symmetrical halves, the left and the right hemispheres. Each hemisphere is covered by the cerebral cortex, a thick layer of gray matter formed mostly by neuronal cell bodies. To increase the surface area, the cerebral cortex is heavily folded forming gyri (the raised portions) and sulci (the grooves, the deeper ones are called fissures). The higher cognitive function of primates is mostly attributed to the largely developed frontal lobe, which lies directly behind the forehead. The central sulcus separates the frontal lobe from the parietal lobe and forms the boundary between the motor and sensory areas, respectively. The sensori-motor cortices have a topographic map of the human body (homunculus) with larger portions of the brain devoted to more sensitive parts of the body (e.g. face). The temporal lobe is located laterally on each hemisphere and implicated in semantics and long-term memory. The occipital lobe at the back of the brain contains the visual cortex. Deeper inside the brain we can find the sub-cortical structures like basal ganglia (striatum), amygdala, hippocampus and other portions of the limbic system including the cingulate cortex. The cerebellum is situated at the bottom of the brain. It is presumed that various functions are organized in a modular fashion in the brain (functional specialisation).

A simple decision making task would involve integrating the sensory information and processing it before an action is made. Gold and Shadlen (2000, 2007) used a motion detection task in which the monkey had to disambiguate the dominant direction of movement among a randomly moving set of dots. Their results showed that monkeys weighed the evidence as a log likelihood ratio favouring one direction over the other. Assuming a constant rate of information accumulation, waiting for a longer time resulted in more accurate decision. Neurons in medial temporal area (MT) fire selectively to motion in a particular direction. The lateral intra-parietal (LIP) neurons integrate information across time representing opposing motion sensors in MT. When this information reaches a threshold, it gives rise to a decision. This is an example of perceptual decision making.

Value based decision making involves an additional evaluation step in the process of decision making, possibly transforming the choices into a common currency of reward value. Platt and Glimcher (1999) used a *probability matching task* in which the monkey needs to move its eyes towards the location of a probabilistically rewarding target. The fixation cue changed colour to denote which target will be rewarded and the monkey was allowed to freely choose between the two targets. It was found that the probability of a response and activity in the LIP matches the probability that it will be rewarded although this is less optimal than making the most rewarded response every time. More precisely, the ratio of choices between two alternatives will be the same as the ratio of probabilities of these alternatives being rewarded. This is in accordance with the well known matching law (Herrnstein, 1961). Sugrue et al. (2004) demonstrated that in a dynamic foraging environment, LIP neurons tracked the changing values of alternative choices through time. They concluded that matching provides a behavioural method of observing the internal representation of value. Recently Kiani and Shadlen (2009) demonstrated that the neurons in the parietal cortex represented the formation of decision direction and also the degree of certainty underlying the decision that was opted out.

Primate decision making is also reported to involve other areas of the brain such as the anterior cingulate cortex (Kennerley et al., 2006; Rushworth and Behrens, 2008), posterior cingulate cortex (McCoy and Platt, 2005), cingulate motor area (Shima and Tanji, 1998), dorsolateral prefrontal cortex (Lee and Seo, 2007), orbitofrontal cortex (Padoa-Schioppa and Assad, 2006; Tsujimoto, Genovesio, and Wise, 2009) and basal ganglia (Samejima et al., 2005; Samejima and Doya, 2007; Lau and Glimcher, 2008).

The basal ganglia, particularly the ventral striatum (and nucleus accumbens) have a critical role in reward, decision making, and motivated behaviour. Dopamine, a modulatory neurotransmitter, is linked to reward prediction (Doya, 2008). The dopaminergic system projects heavily to the striatum and prefrontal cortex. Dopamine neurons fire in response to unpredicted rewards and stimuli predicting rewards. The observation that dopamine neurons show suppressed activation from their baseline firing when a predicted reward does not occur lead to the hypothesis that dopamine neurons represent *reward prediction error* (Schultz et al., 1997; Daw and Doya, 2006). Dopamine appears to mediate the motivational aspects of a stimulus as blocking dopamine impairs wanting but not liking of reward (Berridge and Robinson, 1998). Schultz and colleagues have convincingly demonstrated that dopamine neurons represent economic parameters of decision such as magnitude, probability, expected value and uncertainty (Fiorillo, Tobler, and Schultz, 2003; Fiorillo, 2008; Schultz et al., 2008; Fiorillo, Newsome, and Schultz, 2008). When monkeys learned stimuli with varying levels of reward probability, an increase in sustained firing for stimuli with greater risk (closer to 0.5 reward likelihood) was observed during the delay period between stimulus and reward (Fiorillo et al., 2003).

Using single-cell recordings on pigeons Kalenscher et al. (2005) presented evidence that the brain region, which is functionally analogous to prefrontal cortex varied inversely with the length of delay in time during anticipation for reward and co-varied with the expected value of the outcome. Further this study revealed the reward preference shift from large to smaller rewards and validated the hyperbolical discount function incorporating amount of reward and time delay based on neural data obtained. In a recent study on three rhesus monkeys Kim et al. (2008) found that the prefrontal cortex (especially the dorsolateral prefrontal neurons) encoded the temporally discounted value of reward expected from a particular choice.

Neurophysiological studies on primates and pigeons have provided ample evidence that single neurons (and groups of them) can compute decision variables in the brain.

4. Neuroimaging Studies of Individual Decision Making

Although neurophysiological studies provide an unprecedented approach to study neural mechanisms at cellular level, the limited communication and cognitive capabilities of non-human primates restricts the investigation of decision making. Early neuroimaging studies have replicated the results from primates in human subjects and extended the putative view of neural structures involved in decision making. In this section, a selected review of neuroimaging studies investigating basic decision parameters, namely, magnitude, probability, delay, and uncertainty (risk and ambiguity) is provided. Particular emphasis is given on those studies that are relevant to economic theories of decision making.

A number of neuroimaging studies have found distinct neural systems processing reward (gains) and punishment (losses) information. The reward and punishment are also referred to as valence (i.e. positive and negative) of the outcomes. Delgado et al. (2000) asked participants to guess whether the value of the card was higher or lower than 5. Participants received monetary reward ($1.00) for a correct guess, punishment ($0.50) for an incorrect guess or neutral feedback when the outcome was equal to 5. While sustained activation in the dorsal striatum was observed following reward, a sharp decrease of response below baseline was observed after punishment. In another study by Knutson et al. (2001), cues signalled the potential reward ($0.20, $1.00, or $5.00), punishment ($0.20, $1.00, or $5.00) or no monetary outcome. Subjects performed a button-press task with a target to win or avoid losing money and the task difficulty was adjusted so that subjects should succeed on ~66% of target responses. Nucleus accumbens activity correlated with anticipation of reward and self-reported happiness, whereas activation in medial caudate increased in anticipation of both rewards and punishments. These early neuroimaging studies established that gains and losses were represented distinctly in the brain. Particularly, there has been ample evidence of medial-lateral distinction in the orbitofrontal cortex (OFC) representing rewards and punishments respectively (see Elliot, Dolan and Frith, 2000 for a review). In a visual reversal-learning task, the choice of a correct stimulus lead to a probabilistically determined monetary reward and the choice of an incorrect stimulus lead to a monetary loss. After a few trials, the correct and incorrect stimuli reversed their contingencies. Using this paradigm, O'Doherty et al. (2001) found a medial-lateral distinction for rewarding and punishing outcomes, respectively. In a subsequent study, O'Doherty et al. (2003) found that ventromedial and orbital prefrontal cortex (PFC) are not only involved in representing the valence of outcomes, but also signal subsequent behavioural choice. The anterior insula / caudolateral OFC was related to behavioural choice and was active in trials that required a switch in stimulus choice the subsequent trials.

Seymour, Singer and Dolan (2007a) reviewed studies in both animals and humans and concluded that gains (or, more generally, appetitive states) are encoded by medial orbitofrontal cortex, whereas losses (or aversive events) are represented by activity at anterior insula and lateral orbitofrontal cortex. All of these areas have reciprocal functional connections with both the striatum and amygdala, which are supposed to convey signals related to the value of a choice. This medio-lateral regional specialization has also been suggested by O'Doherty (2007) and the distinction between dorsal and ventral regions of the medial prefrontal cortex (MPFC) roles in the decision making under uncertainty was also recently suggested (Xue et al., 2009).

According to decision affect theory (Mellers et al., 1997), responses to a given outcome depend on counterfactual comparisons. Breiter et al. (2001) presented subjects with three outcomes in which subjects could win or loose money. Three kinds of prospects (good: $10, $2.50, $0, intermediate: $2.50, $0, -$1.50 and bad: $0, -$1.50, -$6) were used. Thus $0 on a good prospect will be experienced as a loss and the same outcome in a bad prospect would be experienced as a win. Partial evidence for this was observed clearly in time courses of nucleus accumbens and amygdala for the good and bad prospects, but not so for intermediate prospect. Haemodynamic responses in the amygdala and orbital gyrus tracked the expected values of the prospects.

DeMartino et al. (2006) found amygdala activity that correlated with the behavioural prediction of the framing effect (Kahneman and Tversky, 1979); its activity was enhanced when the choice was safe (as opposed to risky) in the gains domain, whereas the inverse happened in the losses domain. On the contrary, dorsal anterior cingulate cortex (ACC) activity followed the opposite pattern (higher activity for the risky option in the gains domain; higher activity for the safe option in the losses domain). In addition, medial orbitofrontal cortex (OFC) activity correlated with the susceptibility to framing effect. Poldrack and colleagues (Tom et al., 2007) tried to measure BOLD responses correlating with loss aversion. They offered binary (p=0.5) mixed gambles resulting in either a positive or a negative outcome. Typical areas of the reward system, and especially ventral striatum and ventromedial prefrontal cortex, were activated in correlation with both increasing gains and decreasing losses.

The neuroeconomic studies on regret and rejoice using functional MRI consistently reported medial orbitofrontal cortex (OFC) involvement with the experience of regret (Camille et al., 2004; Coricelli et al., 2005; Chandrasekhar et al., 2008). The experience of rejoice or relief signals were found in the cortical and subcortical brain regions such as ventral striatum and mid-brain (Chandrasekhar et al., 2008), and anterior ventrolateral prefrontal cortex (Fujiwara et al. 2009).

Decisions could also incorporate disappointment or elation signals. Disappointment is a reduction in utility from a bad outcome purely due to an unfavorable realization of a random variable. When the outcome obtained is lower than expected, subjects might feel disappointment (Loomes and Sugden, 1986; Roese, 1997). Elation is opposite to that of disappointment and analogous to rejoicing. The experimental conditions of these neuroeconomic studies (Camille et al., 2004; Coricelli et al., 2005) were designed based on partial feedback to investigate neural correlates of disappointment. In partial feedback conditions only the opted gamble output was shown but in their full feedback conditions, the outcome of decision along with the outcomes of alternatives were shown. Coricelli et al., (2005) showed the involvement of midbrain (periaqueductal gray matter), precentral gyrus (S2), subcallosal gyrus, middle temporal gyrus correlated with the magnitude of disappointment.

Dread is an important temporal component in the process of decision making. In a decision making study with the incentives being milder electrical shocks, Gregory Berns and colleagues at Emory University investigated neurobiological substrates for dread (Berns et al., 2006) and observed modulation in subjective experience of dread through BOLD responses measured in the pain network. This study illustrated the importance of functional MRI methodology to distinguish subjective experience (disutility) of dread while explaining the temporal aspect in decision making under uncertainty. This temporal aspect called as *intertemporal choice* often effects decisions (Berns, Laibson, Lowenstein, 2007). Recently, Carter and Krug (2009) pointed to the role of amygdala and anterior cingulate cortex in the anxiety related processes (such as dread) and suggested further work in establishing the circumstances in which the functional activity in anterior cingulate cortex (ACC) for diagnosis and treatment of dread related disorders.

Camerer and colleagues (Hsu et al., 2005) are one of the first people to show the neural correlates of risk and ambiguity associated with choices under uncertainty based on neuroimaging data from normal population and behavioural data from frontal lobe diseased patients .They found the dorsal striatum was more sensitive to risk than to ambiguity and the level of ambiguity was positively correlated with the activations in amygdala and lateral orbitofrontal cortex. Interestingly, they found that the ambiguity aversion correlated with the activations in the orbitofrontal cortex and behavioural results from patients with orbitofrontal lesions were risk- and ambiguity- neutral.

In another recent study, Huettel et al. (2006) used a modified paradigm with certain and uncertain choices to dissociate risk (uncertainty of known probabilities) and ambiguity (uncertainty of unknown probabilities) related brain regions. They demonstrated the involvement of lateral prefrontal cortex with ambiguity and the posterior parietal cortex with risk associated with the decision making process with monetary gambles. In summary, these two studies (Hsu et al., 2005; Huettel et al., 2006) demonstrated the involvement of frontal cortex and amygdala with ambiguity, and the posterior partietal cortex and straitum with risk associated with the decision making process. These studies categorically defined these two processes (ambiguity and risk) either by fully presenting or completely avoiding one of the two (O'Neill and Kobayashi, 2009).

In a recent study incorporating partial conditions of ambiguity using aversive outcomes (electrical shocks), Bach et al. (2009) varied ambiguity with the help of three cues; no ambiguity (risky cues), intermediate level of ambiguity and full ambiguity (that they called as ignorance cue). Their results show the involvement of posterior inferior frontal gyrus and posterior parietal cortex with the intermediate ambiguous cue compared with risky and ignorance cues. However, they could not observe any activation for risky or ignorance cues compared with the intermediate ambiguity cues. As an indication, the cognitive processes such as ambiguity or risk or ignorance is prevalent in any decision making process under uncertainty and thus should be given a thought while designing any study.

Engelmann and Tamir (2009) used choice between uncertain lotteries to investigate the neural correlates of subjective valuations involving risk factor. Their results show increased activations in anterior and posterior cingulate cortex, superior frontal gyrus, caudate nucleus, and substantia nigra as a function of risk levels. Their further analysis revealed significant correlations between risk-seeking attitudes and neural activity in superior and inferior frontal gyri, medial and lateral orbitofrontal cortex, and parahippocampal gyrus. The risk-aversive attitudes were observed to be correlated with the activations in caudate.

A recent study by Shultz and colleagues distinguished reward value coding from risk attitude related brain activations (Tobler et al., 2007). In this study they presented stimuli of varying expected values by manipulating magnitude and probability of rewards. Their results suggested the role of striatum in combining the magnitude and probability of rewards to produce expected value signal. Further they show the risk aversion choices were correlated with responses from the lateral orbitofrontal cortex and the risk seeking choices were sub-served by medial sites of the orbitofrontal cortex (OFC).

In an initial meta-analysis of decision-making on brain regions, Krain et al., (2006) argued for distinct neural systems sub-serving risk and ambiguity related choices. It was suggested that the orbitofrontal cortex and rostral parts of anterior cingulate cortex were associated with risky decisions, whereas the dorsal parts of anterior congulate cortex and dorsolateral prefrontal cortex were involved in the ambiguous decision making process.

With the help of quasi-hyperbolic discount function, in a neuroeconomic study on fourteen normal humans involving intertemporal choices McClure et al. (2004) found separate neural systems underlying immediate and delayed rewards. They examined neural correlates of time discounting while subjects made a series of choices between monetary reward options that varied by delay (same day to 6 weeks later) to delivery. They found ventral striatum, medial OFC, medial prefrontal cortex (PFC), posterior cingulate and left posterior hippocampus were related to choice of immediate rewards. In contrast, regions of lateral prefrontal cortex and posterior parietal cortex are engaged uniformly by inter-temporal choices irrespective of delay. Recently, McClure et al. (2007) used primary rewards (fruit juice or water) with time delay of minutes instead of weeks and found similar activation patterns as in their previous study. When the delivery of all rewards was offset by 10 min, there was no further differential activity in limbic reward-related areas; hence suggesting that time discounting is not a relative concept.

In the quasi-hyperbolic discounting model the beta and delta designations refer to the two valuation systems: the beta system, which is carried by emotion and overvalues immediate rewards, and the delta system, which discounts at a constant rate over time (Boettiger et al., 2007). With a modified intertemporal task Luhmann et al. (2008) dissociated the time-delay and the probability associated with the monetary outcomes. Their functional imaging results pointed out to a set of brain regions such as the posterior cingulate cortex, parahippocampal gyri, frontopolar cortex which were activated with the temporal aspect of intertemporal task. Their brain-behaviour analysis revealed the frontopolar cortex (Brodmann area 10) activation magnitudes are significantly influenced by the time delay.

Using passive and active decision making tasks with the help of aversive outcomes, the study by Berns and colleagues (2006) revealed modulations in the pain network with the subjective experience of dread. With the neuroimaging data, this study showed how dread factor in the anticipation to outcome, i.e. the temporal aspect would enter into the decision making process. The above mentioned studies with intertemporal choice (Loewenstein, 1987) still leave us with many open questions related to the intertemoporal preferences during choice and the underlying neuronal mechanisms (Engelmann and Brooks, 2009).

The occurrence of rewards is uncertain in the dynamic world. Both the environment and the behaviour of other agents render the rewards partly unpredictable. Uncertainty can be in the expected magnitude of the reward (characterized by the variance) or the probability (p) of the reward (maximum uncertainty at $p = 50\%$) or the time of delivery of the reward. The economic conception of uncertainty is formalized as risk or ambiguity. Outcomes with known probabilities are risky and outcomes with unknown probabilities are ambiguous.

Dreher et al. (2006) used three slot machines with two different reward values ($10, $20) and reward probabilities (0.25, 0.5), so that one pair of slot machines had the expected value matched. To avoid counterfactual comparison, a common outcome of no reward with a probability of 1 served as a fourth slot machine. They found that midbrain region responded transiently to higher reward probability at the time of cue and to lower reward probability at the time of reward outcome and in a sustained fashion to reward uncertainty during the delay period. A frontal network co-varied with the reward prediction error signal both at the time of the cue and at the time of the outcome. The ventral striatum showed sustained activation that co-varied with maximum reward uncertainty during reward anticipation. Their results suggest distinct functional networks encoding economic decision parameters.

Berns et al. (2001) delivered subjects with fruit juice and water in a temporally predictable or unpredictable manner. Unpredictability of rewards resulted in significant activity in nucleus accumbens and medial orbitofrontal cortex, while predictability resulted in activation predominantly in the superior temporal gyrus. Unlike reward learning, the source of prediction in Berns et al. (2001) was based on the sequence of stimuli. Reward learning literature has focused on the computational models such as the temporal difference (TD) model incorporating temporal prediction error. Using appetitive conditioning paradigm O'Doherty et al., (2003b) demonstrated activity in the ventral striatum and OFC with the error signal when taste reward was omitted or unexpectedly delivered in some of the trials. Signals predicted by the temporal difference models were found to correlate with activity in the ventral striatum and the anterior insula in a second-order pain learning task (Seymour et al., 2004). Seymour et al. (2007b) found that striatal activation reflected positively signed prediction error in anterior region for rewards. However, the activation was found in posterior regions for losses in a probabilistic Pavlovian task to compare winning / losing money in two conditions when the alternative was respectively winning / losing nothing, or losing / winning money. In the next section, we provide an overview of these computational models and suggest how these can be extended taking economic decision making into consideration.

Loss aversion is an important concept that originated from the prospect theory developed by Kahneman and Tversky (1979) using hypothetical risky gambles and was explained with the help of value function varying between losses and gains. They proposed that the value function was generally concave for gains and convex for losses. They also suggested that the reference point was important and the deviations (gains or losses) were dependant on this point. The important observation from their experimental results was that the value function was steeper for losses compared to gains. It lead to an important concept called loss aversion, which states that people exhibit greater sensitivity to losses than gains (Trepel, Fox and Poldrack, 2005) while making decisions under risk (or uncertainty). The loss aversion was studied experimentally using buying and selling of goods or losses and gains of monetary gambles. The important point to be noted while considering individual variability in decision making process was that some people are not loss averse (Abdellaoui, Bleichrodt and Paraschiv, 2007).

Two studies investigated the neural basis for loss aversion using functional MRI on normal humans. With the help of buying and selling of goods (MP3 songs), Weber et al. (2007) demonstrated loss aversion for goods and money. They suggested that activations in left amygdala and left caudate nucleus are associated with loss aversion for goods (while selling songs), while activations in right parahippocampal gyrus are associated with loss aversion for money (while buying songs). Tom et al. (2007) recently investigated loss aversion for monetary gambles. The subjects in their study either accepted or rejected a gamble offering 50/50 chance of winning/losing monetary value. Their results showed that the activity in a unique set of brain regions such as ventral striatum and ventromedial prefrontal cortex increased with gains and decreased with losses. Based on the slope of value function for losses, they further demonstrated that these two regions exhibit neural correlates of behavioural loss aversion (Dreher, 2007). These studies leave us with an open question as to whether loss aversion is due to framing effect of the experiment or is medium (goods or money) dependant.

5. INDIVIDUAL DECISION MAKING: COMPUTATIONAL MODELS

Economic decision making models suggest that among a number of alternatives or options available, the option yielding highest utility is chosen. The outcome of the choice is realized only after the choice is made. Invariably, decisions must be made on the predicted outcome. Prediction plays a central role in shaping the behaviour of an individual. Learning to make predictions is, thus, an integral process of individual decision making. Animal learning theories have proposed predictive learning models based on classical (Pavlovian) and instrumental conditioning phenomenon. While instrumental conditioning requires an action to be made by the animal and thus more relevant for decision making, the Pavlovian conditioning paradigm emphasizes on the associations made by the animal to predict an upcoming reward.

Fundamental to predictive learning models is the concept of prediction error, which is the difference between the predicted outcome and the actual (obtained) outcome. In the context of animal learning theory, the Rescorla-Wagner learning rule suggests that learning progresses in proportion to the prediction error (Rescorla and Wagner, 1972). The speed of learning is governed by a constant learning rate parameter. The prediction error acts as a feedback signal to update the predicted value of the outcome. Notably, the Rescorla-Wagner learning rule is useful when the outcome is immediate. For delayed outcomes, it is necessary to incorporate the element of time. A real-time extension of the Rescorla-Wagner model is the temporal difference (TD) model developed by Sutton and Barto (1981; 1990). The predicted value of a delayed outcome or reward is calculated as a temporally discounted sum of all future rewards. Rewards that occur closer in time have greater weight and those farther in time have a smaller weight. For example, using an exponential discounting function, with γ as discount factor, the reward predicted V_t at time t is given by

$$V_t \leftarrow \lambda_{t+1} + \gamma\lambda_{t+2} + \gamma^2\lambda_{t+3} + \gamma^3\lambda_{t+4} + \ldots$$

The following recursive relationship allows estimation of the current prediction and avoids the necessity to wait until all future rewards are received in that trial.

$$Vt \leftarrow \lambda t + \gamma Vt + 1$$

We can now define the temporal difference error that must approach zero with learning as

$$\delta t = \lambda t + 1 + \gamma Vt + 1 - Vt$$

and the learning is governed by

$$\Delta VA = \alpha A\beta(\lambda t + 1 + \gamma Vt + 1 - Vt)$$

The temporal difference model forms the class of computational models referred to as reinforcement learning. Implementations of reinforcement learning models such as the actor-critic architecture provide an account of choice behaviour. The actor/critic model has two components – Actor and Critic. The actor implements a policy for which actions should be chosen depending upon their predicted outcome. The critic provides an error signal comparing the predicted outcome to the actual or obtained outcome. An individual learns to achieve a goal (maximize reward) by navigating through the space of states (making decisions - actor) using the reinforcement signal (updating the value function - critic). In the temporal difference (TD) model, the TD error guides the updating of value function $V(S_t)$ when transitioning from state S_t to state S_{t+1}. Q-learning and its variants have offered estimation of value functions over state-action pairs, so that in a given state s, the organism chooses the action a that maximizes the value $Q(s,a)$. The updating of value function Q is done similar to the TD model (Watkins and Dayan, 1992).

Research in computational models incorporating the regret signal is becoming important (Cohen, 2008; Marchiori and Warglien, 2008). Most of the current models use prediction-error signal for learning the internal parameters of the system. Another class of error signal that originated from the regret theory, the fictive-error signal, (Chandrasekhar et al., 2008; Lohrenz et al., 2007) was argued to be essential in future learning algorithms. This error signal compares the obtained outcome with the expected values of the alternatives that could have been chosen. This error is different from the prediction error signal which compares only the obtained outcome with the expected (or desired) outcome.

Several considerations are needed for applying the predictive learning models mentioned above to economic decision making. First, the future reward value should be replaced by the expected utility of the future reward incorporating both the magnitude weighting and probability weighting. Second, a large body of literature has made a clear proposition that gains and losses are treated differently by individuals. Hence predictive learning models have to consider the relative weights of gains versus losses instead of a simple flipping of the sign to represent losses. Third, the temporal discounting function used conforms to the exponential discounted utility. The more general hyperbolic (or quasi-hyperbolic) discounting function needs to be incorporated for weighting the time points.

In the following section, we discuss the generalization of individual decision making to collective decision making.

6. Taking Further: Collective Decision Making

So far decision making investigations pointed to various components associated with the neurobiological underpinnings of individual decision making process. The outcomes or payoffs would only affect the individual of interest. However, collective decision making is the process of making decisions taking into consideration consensus among peers. The field investigating various components and the neuronal systems of collective decision making is an emerging research area under the umbrella of social psychology (Berns, 2008). The nascent area of collective decision making has potential implications for the betterment of society (such as good policy making, etc.).

Most of the research in decision making process in the social context began few years ago with experimental paradigms involving various components such as altruism, trust, empathy, cooperation and competition. One of the widely used experimental paradigm was *ultimatum game*. The classic version of an ultimatum game first proposed by Guth et al. (1982) involved two players in which one was a proposer (say, X) and the other was a receiver (say, Y). The player X is endowed with an amount with which he/she has to make an economic exchange with his/her peer. The player X (who is also called proposer) would offer part of the amount endowed to him/her to the receiver Y. The player Y can accept or reject the offer. If he/she accepts the proposed division amount, the economic exchange gets implemented as it is. If the player Y rejects the offer, both players get nothing.

There are also experimental paradigms with which the reciprocal interactions among individuals during economic exchange can be investigated. One example is a *trust game*, in which a player called investor will make an interaction with a partner called trustee. The investor decides how much to endow to invest with trustee and during the transfer to trustee the amount gets multiplied (say, thrice) and trustee would return some or all amount. This is an elegant paradigm to investigate the economic investment and exchange with trusted partners (Rilling, King-Casas, Sanfey, 2008).

It is interesting to note that some components associated with economic exchange such as envy, revenge, altruism, trust, empathy, cooperation, and competition between two players can be studied using the principles of these two paradigms. We believe that the modified two player game paradigms can be utilized to investigate the collective decision making process. There are also experimental paradigms involving strategic interactions (Camerer, 2004) and the equilibrium (Nash, 1950).

The results from important neuroeconomic studies using two-person economic games along with a possible methodology will be reviewed in this section. Montague, Berns et al. (2002) proposed a new methodology using simultaneous functional MRI called *hyperscanning* for recording neurobiological underpinnings during social interactions. Though there are many technical and conceptual bottlenecks to be solved, in future, this methodology might play an important role for investigation of collective decision making or interpersonal economic exchanges.

In one of the first neuroeconomic investigations using ultimatum game, Sanfey and colleagues (2003) depicted the neural correlates of emotional and cognitive processes. They found unfair offers activated anterior insula and dorsolateral prefrontal cortex and they further observed enhanced activity in anterior insula when the unfair offers were rejected by the responder. A subsequent study by the same group (Wout et al., 2005), using repetitive transcranial magnetic stimulation technique, demonstrated the causal role of right dorsolateral prefrontal cortex (DLPFC) in the strategic decision making process.

Using functional MRI with a modified multi-round trust game involving economic exchange, King-Casas et al. (2005) demonstrated the involvement of caudate (in dorsal striatum) with reciprocity involving trust. One of the important findings from this study was that the caudate registered the social prediction errors and eventually guided the reciprocity. Using a computer game to investigate the neural correlates of cooperation and competition, Decety et al. (2004) demonstrated the involvement of medial orbitofrontal cortex in the process of cooperation and that of inferior parietal and medial prefrontal cortices in the process of competition. Using an intranasal Oxytocin infusion (a neuropeptide) on human subjects playing a trust game, Ernst Fehr and colleagues (Kosfeld, 2005) demonstrated that

Oxytocin has a role in increasing initial money transfers by the investors and thus point to its key function in economic interaction with trust. A recent study by Montague and colleagues (Tomlin et al., 2006) investigated two-person economic exchange and found the role of cingulate cortex in agent-specific responses. Using emotional faces to demonstrate empathy, Carr et al. (2003) showed the important role played by the brain region Insula.

Though there are many studies in the literature, we have pointed to some important ones. The questions that remain open to neuroimaging methods are how the network activity, the connectivity among brain regions associated with the economic exchange, would predict the nature of interaction and how this would influence the partners in the economic exchange. The methodological issues related to hyperscanning should also be addressed.

CONCLUSION

In this paper, efforts were made to bring together various components of neuroeconomics of decision making under uncertainty. Methods for investigating brain functions, the various economic theories of behaviour of human choice, the neuronal mechanisms sub-serving at the cellular and the system levels, and the computational models relevant to the process of decision making were reviewed. The future research in collective decision making along with some open questions related to neuroeconomics were also presented.

ACKNOWLEDGMENTS

We are thankful for inspirational support from our current and past colleagues and mentors. We thank Drs. Raju Bapi, Kenji Doya and Ahmed. VSC Pammi is thankful to Narayanan Srinivasan, Ranganatha Sitaram, Debarati Banerjee, Gregory Berns, Giuseppe Pagnoni, Charles Noussair, Monica Capra, Jan Engelmann and Sara Moore. KPM would like to thank Yorgos Christopoulos, Philippe Tobler, Shunsuke Kobayashi and Wolfram Schultz.

REFERENCES

Abdellaoui, M., Bleichrodt, H., and Paraschiv, C. (2007). Loss aversion under prospect theory: a parameter-free measurement. *Management Science, 53(10)*, 1659-1674.

Allais, M. (1953). Le comportement de l'Homme rationnel devant le Risque : Critique des tostulats et axiomes de l'Ecole Americaine. *Econometrica, 21,* 503-546.

Bach, D. R., Seymour, B., and Dolan, R. J. (2009). Neural activity associated with the passive prediction of ambiguity and risk for aversive events. *The Journal of Neuroscience, 29,* 1648-1656.

Bell, D.E., 1983. Risk Premiums for Decision Regret. *Management Science, 29,* 1156-1166.

Bell, D.E., 1982. Regret in decision making under uncertainty. *Operations Research, 30,* 961-981.

Bandettini, P. A. (2009). What's new in neuroimaging methods? *Annals of the New York Academy of Sciences, 1156,* 260-293.

Berns, G. S., McClure, S. M., Pagnoni, G., and Montague, P. R. (2001). Predictability modulates human brain response to reward. *The Journal of Neuroscience, 21,* 2793–2798.

Berns G. S., Chappelow, J., Cekic, M., Zink, C. F., Pagnoni, G., and Martin-Skurski, M. E. (2006). Neurobiological substrates of dread. *Science, 312(5774),* 754-758.

Berns, G. S., Laibson, D., and Loewenstein, G. (2007). Intertemporal choice--toward an integrative framework. *Trends in Cognitive Science, 11,* 482-488.

Berns, G.S. (2008). Neuropolicy Center Confronts the Biological Basis of Collective Decision Making. *Press report,* Emory University, Atlanta, USA.

Berridge, K. C., and Robinson, T. E. (1998). What is the role of dopamine in reward: hedonic impact, reward learning, or incentive salience? *Brain Research Reviews, 28,* 309-369.

Bernoulli, D. (1763/1958). Exposition of a new theory on the measurement of risk. *Econometrica, 22,* 23-36.

Boettiger, C. A., Mitchell, J. M., Tavares, V. C., Robertson, M., Joslyn, G., D'Esposito, M., and Fields, H. L. (2007). Immediate reward bias in humans: Fronto-Parietal networks and a role for the Catechol-O-Methyltransferase 158Val/Val genotype. *The Journal of Neuroscience, 27,* 14383–14391.

Breiter, H. C., Aharon, I., Kahneman, D., Dale, A., Shizgal, P. (2001). Functional imaging of neural responses to expectancy and experience of monetary gains and losses. *Neuron, 30,* 619–639.

Camerer, C.F. (2004). Behavioral game theory: Predicting human behavior in strategic interactions. In C. F. Camerer, G. Loewenstein and M. Rubin (Eds.), *Advances in Behavioral Economics* (pp. 374-292). New York: Princeton University Press.

Camille, N., Coricelli, G., Sallet, J., Pradat-Diehl, P., Duhamel, J. R., and Sirigu, A. (2004). The involvement of the orbitofrontal cortex in the experience of regret. *Science, 304,* 1167-1170.

Carr, L., Iacoboni, M., Dubeau, M-C., Mazziotta, J. C., and Lenzi, G. L. (2003). Neural mechanisms of empathy in humans: A relay from neural systems for imitation to limbic areas. *Proceedings of National Academy of Sciences, USA, 100,* 5497-5502.

Carter, C. S., and Krug, M. K. (2009). The functional neuroanatomy of dread: Functional magnetic resonance imaging insights into generalized anxiety disorder and its treatment. *American Journal of Psychiatry, 166,* 263-265.

Chandrasekhar, P.V.S. (2005). *Functional MRI investigation of Complex Sequence Learning.* PhD Thesis, Department of Computer and Information Sciences, University of Hyderabad, India.

Chandrasekhar, P. V. S., Capra, C. M., Moore, S., Noussair, C., and Berns, G. S. (2008). Neurobiological regret and rejoice functions for aversive outcomes. *Neuroimage, 39),* 1472-1484.

Churchland, P.S., and Sejnowski, T. J. (1992). *The Computational Brain.* Bradford Book, MIT Press, Cambridge MA.

Cohen, M. D. (2008). Learning with Regret. *Science,* 319, 1052-1053.

Coricelli, G., Critchley, H. D., Joffily, M., O'Doherty, J. P., Sirigu, A., and Dolan, R. J. (2005). Regret and its avoidance: A neuroimaging study of choice behavior. *Nature Neuroscience, 8,* 1255-1262.

Dale, A. M., and Halgren, E. (2001). Spatiotemporal mapping of brain activity by integration of multiple imaging modalities. *Current Opinion in Neurobiology, 11,* 202-208.

Daw, N. D. Doya, K. (2006). The computational neurobiology of learning and reward. *Current Opinion in Neurobiology, 16,* 199-204.

Decety, J., Jackson, P. L., Sommerville, J. A., Chaminade, T., and Meltzoff, A. N. (2004). The neural bases of cooperation and competition: an fMRI investigation. *Neuroimage, 23,* 744-751.

Delgado, M. R., Nystrom, L. E., Fissell, C., Noll, D. C., and Fiez, J. A. (2000). Tracking the hemodynamic responses to reward and punishment in the striatum. *Journal of Neurophysiology, 84,* 3072–3077.

DeMartino, B., Kumaran, D., Seymour, B., and Dolan, R. (2006). Frames, biases, and rational decision-making in the human brain. *Science, 313,* 684–687.

Doya, K. (2008). Modulators of decision making. *Nature Neuroscience, 11(4),* 410-416.

Dreher, J-C., Kohn, P., and Berman, K. F. (2006). Neural coding of distinct statistical properties of reward information in humans, *Cerebral Cortex, 16,* 561-573.

Dreher, J-C. (2007). Sensitivity of the brain to loss aversion during risky gambles. *Trends in Cognitive Sciences, 11,* 270-272.

Elliott, R., Dolan, R. J., and Frith, C. D. (2000). Dissociable functions in the medial and lateral orbitofrontal cortex: Evidence from human neuroimaging studies. *Cerebral Cortex, 10,* 308-317.

Ellsberg, D. (1961). Risk, Ambiguity, and the Savage Axioms, *Quarterly Journal of Economics, 75* (4), 643–669.

Engelmann, J. B., and Brooks, A.M. (2009). Behavioral and neural effects of delays during Intertemporal choice are independent of probability. *The Journal of Neuroscience, 29(19),* 6055-6057.

Engelmann, J. B., and Tamir, D. (2009). Individual Differences in Risk Preference Predict Neural Responses during Financial Decision-Making. *Brain Research,* 1290, 28-51.

Fiorillo, C. D., Tobler, P. N., and Schultz, W. (2003). Discrete coding of reward probability and uncertainty by dopamine neurons. *Science, 299,* 1898-902.

Fiorillo, C. D. (2008). Towards a general theory of neural computation based on prediction by single neurons. *PLoS One, 3(10),* e3298.

Fiorillo, C. D., Newsome, W. T., and Schultz, W. (2008). The temporal precision of reward prediction in dopamine neurons. *Nature Neuroscience, 11,* 966 – 973.

Frederick, S., Loewenstein, G., and O'Donoghue, T. (2004). Time discounting and time preference: A critical review. *Journal of Economic Literature, 40,* 351-401.

Friedman, M. and Savage, L. J. (1948). The utility analysis of choices involving risk. *The Journal of Political Economy, 56,* 279-304.

Fujiwara, J., Tobler, P. N., Taira, M., Iijima, T.,and Tsutsuia, K-I. (2009). A parametric relief signal in human ventrolateral prefrontal cortex. *NeuroImage, 44,* 1163-1170.

Gazzaniga, M. S., Ivry, R. B., and Mangun, G. R. (2002). Cognitive Neuroscience (Second Edition), W. W. Norton and Company, New York.

George, J. S., Aine, C. J., Mosher, J. C., Schmidt, D. M., Ranken, D. M., Schlitt, H. A., Wood, C. C., Lewine, J. D., Sanders, J. A., and Belliveau, J. W. (1995). Mapping function in the human brain with magnetoencephalography, anatomical magnetic resonance imaging, and functional magnetic resonance imaging. *Journal of Clinical Neurophysiology, 2,* 406-431.

Gold, J. I., and Shadlen, M. N. (2000). Representation of a perceptual decision in developing oculomotor commands. *Nature, 404,* 390-394.

Gold, J. I., and Shadlen, M. N. (2007). The neural basis of decision making. *Annual Review of Neuroscience, 30,* 535-574.

Green, L., and Myerson, J. (1996). Exponential versus hyperbolic discounting of delayed outcomes: risk and waiting time. *American Zoologist, 36,* 496-505.

Guth, W., Schmittberger, R., and Schwarze, B. (1982). An experimental analysis of ultimatum bargaining. *Journal of Economic Behavior and Organization, 3,* 367-388.

Heeger, D. J. and Ress, D. (2002). What does fMRI tell us about neural activity? *Nature Reviews, 3,* 142–150.

Herrnstein, R. J. (1961). Relative and absolute strength of responses as a function of frequency of reinforcement. *Journal of the Experimental Analysis of Behavior, 4,* 267-272.

Hsu, M., Bhatt, M., Adolphs, R., Tranel, D., and Camerer, C. F. (2005). Neural systems responding to degree of uncertainty in human decision making. *Science, 310,* 1680-1683.

Huettel, S. A., Stowe, C. J., Gordon, E. M., Warner, B. T., and Platt, M. L. (2006). Neural signatures of Economic preferences for risk and ambiguity. *Neuron, 49,* 765–775.

Kahneman, D. and Tversky, A. (2000). Choices, Values and Frames. Cambridge University Press.

Kahneman, D. and Tversky, A. (1984). Choices, Values and Frames. *American Psychologist, 39,* 341-350.

Kahneman, D., and Tversky, A. (1979). Prospect theory: an analysis of decision under risk. *Econometrica, 47(2),* 263-291.

Kalenscher, T. (2009). *Decision-making and Neuroeconomics.* In Encyclopedia of Life Sciences (ELS), John Wiley and Sons Ltd., Chichester.

Kalenscher,O., Windmann, S., Diekamp, B., Rose, J., Guentuerkuen, O., and Colombo, M. (2005). Single units in the Pigeon brain integrate reward amount and time-to-reward in an impulsive choice task. *Current Biology, 15(7),* 594-602.

Kandel, E. R., Schwartz, J. H., and Jessell, T. M. (2000). Principles of Neural Science. McGraw-Hill, Health Professions Division, New York.

Kennerley, T. W., Walton, M. E., Behrens, T. E. J., Buckley, M. J., and Rushworth, M. F. S. (2006). Optimal decision making and the anterior cingulate cortex. *Nature Neuroscience, 9,* 940 – 947.

Kiani, R,, and Shadlen, M. N. (2009). Representation of confidence associated with a decision by neurons in the parietal cortex. *Science, 324(5928),* 759-64.

Kim, S., Hwang, J., and Lee, D. (2008). Prefrontal coding of Temporally discounted values during intertemporal choice. *Neuron, 59,* 161-172.

King-Casas, B., Tomlin, D., Anen, C., Camerer, C. F., Quartz, S. R., and Montague, P. R. (2005). Getting to know you: reputation and trust in a two-person economic exchange. *Science, 308,* 78-83.

Knutson, B., Adams, C. M., Fong, G. W., and Hommer, D. (2001). Anticipation of increasing monetary reward selectively recruits nucleus accumbens. *The Journal of Neuroscience, 21,* RC159.

Kosfeld, M., Heinrichs, M., Zak, P. J., Fischbacher, U., and Fehr, E. (2005). Oxytocin increases trust in humans. *Nature, 435(7042),* 673-676.

Krain, A.L., Wilson, A. M., Arbuckle, R., Castellanos, F. X., and Milham, M. P. (2006). Distinct neural mechanisms of risk and ambiguity: a meta-analysis of decision-making. *NeuroImage, 32,* 477-484.

Laibson, D., (1997). Golden eggs and hyperbolic discounting. *Quarterly Journal of Economics, 112*, 443–477.

Lau, B., and Glimcher, P. W. (2008). Value Representations in the Primate Striatum during Matching Behavior. *Neuron, 58*, 451-463.

Lee, D., Seo, H. (2007). Mechanisms of reinforcement learning and decision making in the primate dorsolateral prefrontal cortex. *Annals of the New York Academy of Sciences, 1104*, 108-122.

Loewenstein, G. F., Weber, E. U., Hsee, C. K., and Welch, N. (2001). Risk as feelings. *Psychological Bulletin, 127(2)*, 267-286.

Loewenstein, G. (1987). Anticipation and the valuation of delayed consumption. *The Economic Journal, 97*, 666-684.

Logothetis, N. K. (2008). What we can do and what we cannot do with fMRI. *Nature, 453(7197)*, 869-878.

Logothetis, N. K., Pauls, J., Augath, M., Trinath, T., and Oeltermann, A. (2001). Neurophysiological investigation of the basis of the fMRI signal. *Nature, 412*, 150-157.

Lohrenz, T., McCabe, K., Camerer, C. F., and Montague, P. R. (2007). Neural signature of fictive learning signals in a sequential investment task. *Proceedings of the National Academy of Science, USA, 104(*, 9493-9498.

Loomes, G., and Sugden, R., 1982. Regret theory: an alternative theory of rational choice under uncertainty. *The Economic Journal, 92*, 805-824.

Loomes, G., and Sugden, R., 1986. Disappointment and dynamic consistency in choice under uncertainty. *The Review of Economic Studies, 53*, 271–282.

Loomes, G., and Sugden, R., 1987. Some implications of a more general form of regret theory. *Journal of Economic Theory, 41*, 270-287.

Luhmann,C. L., Chun, M. M., Yi, D-J., Lee, D., and Wang, X-J. (2008). Neural dissociation of delay and uncertainty in Intertemporal choice. *The Journal of Neuroscience, 28(53)*, 14459-14466.

Marchiori, D., and Warglien, M. (2008). Predicting human interactive learning by regret-driven neural networks. *Science, 319*, 1111-1113.

McClure, S. M., Laibson, D. I., Loewenstein, G., and Cohen, J. D. (2004). Separate neural systems value immediate and delayed monetary rewards. *Science, 306(5695)*, 503-507.

McClure, S.M., Ericson, K.M., Laibson, D.I., Loewenstein, G., and Cohen, J.D. (2007). Time discounting for primary rewards. *The Journal of Neuroscience, 27*, 5796-5804.

McCoy, A. N., Platt, M. L. (2005).Risk-sensitive neurons in macaque posterior cingulate cortex. *Nature Neuroscience, 8*, 1220-1227.

Mellers, B.A., Schwartz, A., Ho, K., and Ritov, I. (1997). Decision affect theory: emotional reactions to the outcomes of risky options. *Psychological Science, 8*, 423–429.

Miyapuram, K.P. (2008) Human neuroimaging of visual presentation and imagination of reward. Ph.D. Thesis, University of Cambridge, UK.

Montague, P.R., Berns, G.S., Cohen, J.D., McClure, S.M., Pagnoni, G., Dhamala, M., Wiest, M.C., Karpov, I., King, R.D., Apple, N., and Fisher, R.E. (2002). Hyperscanning: Simultaneous fMRI during linked social interactions. *Neuroimage, 16*, 1159-1164.

Nash, J. (1950). Equilibrium points in N-person games. *Proceedings of National Academy of Sciences, USA, 36*, 48-49.

Ogawa, S., Lee, T. M., Kay, A. R., and Tank, D. W. (1990). Brain magnetic resonance imaging with contrast dependent on blood oxygenation. *Proceedings of the National Academy of Sciences, USA, 87*, 9868–9872.

Ogawa, S., Tank, D. W., Menon, R., Ellermann, J. M., Kim, S., Merkle, H., and Ugurbil, K. (1992). Intrinsic signal changes accompanying sensory stimulation: Functional brain mapping with magnetic resonance imaging. *Proceedings of the National Academy of Sciences, USA, 89*, 5951–5955.

O'Doherty, J.P., Kringelbach, M. L., Rolls, E. T., Hornak, J., and Andrews, C. (2001). Abstract reward and punishment representations in the human orbitofrontal cortex. *Nature Neuroscience, 4*, 95–102.

O'Doherty, J. P., Critchley, H., Deichmann, R., and Dolan, R. J. (2003a). Dissociating valence of outcome from behavioral control in human orbital and ventral prefrontal cortices. *The Journal of Neuroscience, 23*, 7931-7939.

O'Doherty, J., Dayan, P. Friston, K.J., Critchley, H.D., and Dolan, R.J. (2003b). Temporal difference models and reward-related learning in the human brain. *Neuron, 38(2)*, 329-337.

O'Doherty J.P. (2007) Lights, Camembert, Action! The role of human orbitofrontal cortex in encoding stimuli, rewards and choices. *Annals of the New York Academy of Sciences, 1121*, 254-272.

O'Neill, M., and Kobayashi, S. (2009). Risky business: Disambiguating ambiguity - related responses in the brain. *Journal of Neurophysiology, 102*, 645-647.

Padoa-Schioppa, C., and Assad, J. A. (2006). Neurons in the orbitofrontal cortex encode economic value. *Nature, 441*, 223-226.

Pascal, B. (1670/1958). *Pascal's Pensées*. E.P. Dutton and Co., Inc.: New York, NY.

Phelps, E. S., and Pollack, R. A. (1968). On second-best national saving and game-equilibrium growth. *Review of Economic Studies, 35*, 185–199.

Platt, M. L., and Glimcher, P. W. (1999). Neural correlates of decision variables in parietal cortex. *Nature, 400,* 233-238.

Rescorla, R. A., Wagner, A. R. (1972). A theory of Pavlovian conditioning: Variations in the effectiveness of reinforcement and nonreinforcement. In: Classical Conditioning II: Current Research and Theory (Black AH, Prokasy WF, Eds) New York: Appleton Century Crofts, pp. 64-99.

Rilling, J., King-Casas, B., Sanfey, A. G. (2008). The neurobiology of social decision-making. *Current Opinion in Neurobiology, 18,* 159-165.

Roese, N.J., 1997. Counterfactual thinking. *Psychological Bulletin, 121,* 133–148.

Rushworth, M. F. S., and Behrens, T. E. J. (2008). Choice, uncertainty and value in prefrontal and cingulate cortex. *Nature Neuroscience, 11,* 389 – 397.

Samejima, K., Ueda, Y., Doya, K., and Kimura, M. (2005). Representation of action-specific reward values in the striatum. *Science, 310,* 1337–1340.

Samejima, K., and Doya, K. (2007). Multiple representations of belief states and action values in corticobasal ganglia loops. *Annals of the New York Academy of Sciences, 1104,* 213-228

Samuelson, P. (1937). A Note on measurement of utility. *The Review of Economic Studies, 4(2),* 155–61.

Sanfey, A. G., Rilling, J. K., Aronson, J. A., Nystrom, L. E., and Cohen, J. D. (2003). The neural basis of economic decision-making in the Ultimatum Game. *Science, 300(5626),* 1755-1758.

Shibasaki, H.(2008). Human brain mapping: hemodynamic response and electrophysiology. *Clinical Neurophysiology, 119,* 731-743.

Shima, K., and Tanji, J. (1998). Role for Cingulate Motor Area Cells in Voluntary Movement Selection Based on Reward. *Science, 282(5392),* 1335 – 1338.

Schultz, W., Preuschoff, K., Camerer, C., Hsu, M., Fiorillo, C. D., Tobler, P. N., and Bossaerts, P. (2008). Explicit neural signals reflecting reward uncertainty. *Philosophical Transactions of the Royal Society B: Biological Sciences, 363(1511),* 3801-11.

Schultz, W., Dayan, P., and Montague, P. R. (1997). A Neural Substrate of Prediction and Reward. *Science, 275(5306),* 1593 – 1599.

Sugrue, L. P., Corrado, G. S., and Newsome, W. T. (2004). Matching behavior and the representation of value in the parietal cortex. *Science, 304(5678),* 1782-1787.

Seymour, B., O'Doherty, J.P., Dayan, P., Koltzenburg, M., Jones, A., Dolan, D., Friston, K. J., and Frackowiak, R. S. J. (2004). Temporal difference models describe higher-order learning in humans. *Nature, 429,* 664-667.

Seymour, B., Singer, T., and Dolan, R. (2007a). The neurobiology of punishment. *Nature Reviews: Neuroscience, 8,* 300-311.

Seymour, B., Daw, N., Dayan, P., Singer, T., and Dolan, R. (2007b). Differential encoding of losses and gains in the human striatum. *The Journal of Neuroscience,* 27(18), 4826-4831.

Sutton, R.S., and Barto, A.G. (1981). Toward a modern theory of adaptive networks: expectation and prediction. *Psychological Review, 88,* 135–170.

Sutton, R.S., and Barto, A. G. (1990). Time-derivative models of Pavlovian reinforcement. In: Learning and computational neuroscience: foundations of adaptive networks. In Gabriel M, Moore J, (Eds.), pp 497–537. Cambridge, MA: MIT.

Tobler, P. N., O'Doherty, J. P., Dolan, R. J., and Shultz, W. (2007). Reward value coding distinct from risk-attitude related uncertainty coding in human reward systems. *The Journal of Neurophysiology, 97,* 1621-1632.

Tom, S. M., Fox, C. R., Trepel, C., and Poldrack, R. A. (2007). The neural basis of loss aversion in decision-making under risk. *Science, 315,* 515-518.

Tomlin, D., Kayali, A. M., King-Casas, B., Anen, C., Camerer, C. F., Quartz, S. R., and Montague, P. R. (2006). Agent-specific responses in the cingulate cortex during economic exchanges. *Science, 312(5776),* 1047-1050.

Trepel, C., Fox, C. R., and Poldrack, R. A. (2005). Prospect theory on the brain? Towards a cognitive neuroscience of decision under risk. *Cognitive Brain Research, 23,* 34-50.

Tsujimoto, S., Genovesio, A., and Wise, S. P. (2009). Monkey orbitofrontal cortex encodes response choices near feedback time. *The Journal of Neuroscience, 29(8),* 2569-74.

Tversky, A., and Kahneman, D. (1992). Advances in prospect theory: cumulative representation of uncertainty. *Journal of Risk and Uncertainty, 5,* 297-323.

Tversky, A. and Kahneman, D. (1991). Loss Aversion in Riskless Choice: A Reference Dependent Model. *Quarterly Journal of Economics, 106,* 1039-1061.

Ugurbil, K., Toth, L., and Kim, D. S. (2003). How accurate is magnetic resonance imaging of brain function? *Trends in Neuroscience, 26,* 108–114.

Von neumann, J., and Morgenstern, O. (1944). *Theory of games and Economic behavior.* Princeton university press. Princeton.

Volkow, N. D., Rosen, B., and Farde, L. (1997). Imaging the living human brain: Magnetic resonance imaging and positron emission tomography. *Proceedings of National Academy of Sciences, USA, 94*, 2787–2788.

Watkins, C. J. C. H., and Dayan, P. (1992). Q-learning, *Machine Learning, 8*, 279-292.

Weber, B., Aholt, A., Neuhaus, C., Trautner, P., Elger, C. E., and Teichert, T. (2007). Neural evidence for reference-dependence in real-marker-transactions. *NeuroImage, 35*, 441-447.

Wout, M. V., Kahn, R. S., Sanfey, A. G., and Alerman, A. (2005). Repetitive transcranial magnetic stimulation over the right dorsolateral prefrontal cortex affects strategic decision-making. *Neuroreport, 16*, 1849-1852.

Xue, G., Lu, Z., Levin, I. P., Weller, J. A., Li, X., and Bechara, A. (2009). Functional dissociations of risk and reward processing in the medial prefrontal cortex. *Cerebral Cortex, 19*, 1019-27.

In: Expanding Horizions of the Mind Science(s) ISBN: 978-1-62808-705-5
Editors: P.N. Tandon, R.C. Tripathi and N. Srinivasan ©2013 Nova Science Publishers, Inc.

Chapter 11

HUMAN COGNITION AND THE FRONTAL LOBE

P. N. Tandon

AIIMS and National Brain Research Centre, Society

"In the 21st century, we are discovering more and more about the brain and the role of emotions, and challenging old ideas about how we learn, make decisions, act and remember".

Chris Frith (2008)

ABSTRACT

The chapter discusses the role of frontal lobe in human cognition. Frontal lobe is highly developed in humans and consists of different areas subserving potentially different higher cognitive functions. The chapter discusses the results of studies on the dorsolateral prefrontal cortex and the medial prefrontal cortex. The dorsolateral prefrontal cortex has been implicated in working memory and executive functions and the evidence for the role of dorsolateral prefrontal cortex from lesion, neurophysiological and imaging studies are presented. The chapter also discusses the role of medial prefrontal cortex in reward and emotional processing. Finally the chapter presents findings on theory of mind and also the role played by orbitofrontal cortex.

1. COGNITION

As currently used, the term, "cognition" refers to many different processes by which organisms understand and make sense of the world. Perception, attention, memory, problem solving and action planning would all be examples of cognition (Frith, 2008). To this may be added emotions, thoughts, self awareness, judgement and even social setting depends upon our ability to understand the emotions, intentions and beliefs of others i.e. ability to "read other person's mind". There are no doubt other mental functions besides those mentioned here like speech and language, empathy, body image, imagery etc. but for the purpose of this essay it is intended to restrict primarily to those which have a predominant involvement of the frontal lobes particularly the Prefrontal Cortex (PFC).

2. FRONTAL LOBES

Evolution's latest addition to mammalian brain – the prefrontal cortex – is undoubtedly the most highly developed in humans. It is the largest part of the human brain. Though long recognized as the part of the brain which endows the qualities that differentiate human being from all other animals, it is surprising that for the clinicians it had been considered "a silent" area of the brain till recently. Anatomically the frontal lobes consist of the part of the brain in front of the central sulcus and thus include the pre central gyrus which is primarily concerned with motor function. Broadly speaking it may be subdivided into a dorso-lateral, a ventromedial and a fronto-basal or orbitofrontal components based on cytoarchitectural, grossanatomical, configuration, connectivity and functional specialization. No doubt it includes not only these cortical areas but also sub-cortical white matter as also some of the deeper grey matter i.e. the basal ganglia. For the purpose of this essay, the focus would be on the three cortical subdivisions mentioned above excluding the motor strip and the supplementary motor area; as also the subcortical elements. At the outset it may be clarified that unlike the sensory-motor cortex or the Broca's and Wernicke's areas or for that matter that visual cortex, the cognitive functions are not strictly localized to a particular well defined area of the cortex, nor does any particular part of the cortex is totally limited to a single cognitive function. Different regions of the frontal cortex are found to be involved in more than one cognitive functions and a single cognitive function often involves diverse regions of the brain, no doubt interconnected to each other through neural networks. Barbas (2000) summarized the connections underlying the synthesis of cognitive functions in the primate prefrontal cortices as follows. Distinct domains of the prefrontal cortex in primates have a set of connections suggesting that they have different roles in cognition, memory, and emotions. Caudal lateral prefrontal areas (area 8 and 46) receive projections from cortices representing early stages in visual or auditory processing, and from intraparietal and posterior cingulate areas associated with oculomotor guidance and attentional processes. Cortical input to areas 46 and 8 is complemented by projections from the thalamic multiform and parvicellular sectors of the mediodorsal nucleus associated with oculomotor functions and working memory. In contrast, caudal orbitofrontal areas receive diverse input from cortices resenting late stages of processing within every unimodel sensory cortical system. In addition, orbitofrontal and caudal medial (limbic) prefrontal cortices receive robust projections (from amygdala, associated with emotional memory, and from medial temporal and thalamic structures associated with long-term memory. Prefrontal cortices are linked with motor control structures related to their specific roles in central executive functions. Caudal lateral prefrontal areas project to brainstem oculomotor structures, and are connected with premotor cortices effecting head, limb and body movements.

In contrast, medial prefrontal and orbitofrontal limbic cortices project to hypothalamic visceromotor centers for the expression of emotions. Lateral, orbitofrontal, and medial prefrontal cortices are robustly interconnected, suggesting that they participate in concert in central executive functions. Through their widespread connections, prefrontal limbic cortices may exercise a tonic influence on lateral prefrontal cortices, inextricably linking areas associated with cognitive and emotional processes. The integration of cognitive, mnemonic and emotional processes is likely to be disrupted in psychiatric and neurodegenerative diseases which preferentially affect limbic cortices and consequently disconnect major feedback pathways to the neuraxis.

On the other hand Duncan and Owen (2000) summarized the common regions of the human frontal lobe recruited by diverse cognitive functions. This article considers the contribution of recent functional imaging data to this state of affairs. Indeed, although these data show strong evidence for regional differentiation within prefrontal cortex, they also help to explain the difficulty of defining component frontal functions at the cognitive level. For many different cognitive demands, there is joint recruitment of three frontal regions: mid-dorsolateral; mid-ventrolateral, extending along the frontal operculum to the anterior insula; and the dorsal part of the anterior cingulate. To this extent, cognitive activation studies give results that are very regionally specific within prefrontal cortex. This recruitment, however, is extremely similar from one cognitive demand to another, suggesting a specific network of prefrontal regions recruited to solve diverse cognitive problems.

Miller and Cohen (2001) in a detailed review proposed an integrative theory of prefrontal cortex function. The prefrontal cortex has long been suspected to play an important role in cognitive control, in the ablity to orchestrate thought and action in accordance with internal goals. Its neural basis, however, has remained a mystery. They proposed that cognitive control stems from the active maintenance of patterns of activity in the prefrontal cortex that represent goals and the means to achieve them. They provide bias signals to other brain structures whose net effect is to guide the flow of activity along neural pathways that establish the proper mappings between inputs, internal states, and outputs needed to perform a given task. We review neurophysiological, neurobiological, neuroimaging and computational studies that support this theory and discuss its implications as well as further issues to be addressed.

Cabeza and Nyberbg (2000) confirmed these observations following a review of over 250 PET and fMRI studies. In the context of regional activations observed across cognitive domains, it must also be noted that activation of one and the same region in two distinct domains need not imply that the region has the same functional role in both cases. Rather, it has been argued that the functional role of a brain region depends, at least in part, on its neural context. Neural context refers to the pattern of interactions among brain regions. More generally, this perspective suggests that brain regions are not committed to specific functions, but also may play a role in a variety of cognitive and other operations. The data presented in here help to identify network components for various cognitive operations. These extensive observations are systematically tabulated in Table 11 by them.

3. MODES OF EXPLORING HUMAN COGNITION

The earliest information on human cognition was the result of careful psychological and neural deficits associated with well defined lesions of the brain. Beginning with the identification of speech and language related areas by Broca and Wernicke, a large number of clinical syndromes like Korsakoff Psychosis, Anton Syndrome, Capgras delusion, Pain

Asymbolia, Blind-Sight etc. provided increasing insights about the neural basis of human cognition. Similarly electrical stimulation of the brain in conscious, co-operative individuals undergoing surgery for epilepsy provided a wealth of information on human cognition (Penfield and Jasper, 1954). Based on the information gained from the patients, investigators created animal models of some of these conditions. Earlier attempts to use Electroencephalography (EEG) and single neuron electrophysiology for this purpose did not prove to be very useful till recently when the techniques of ERP (Event Related Potentials) and Transcranial Magnetic Stimulation have been used. However, it is with the advent of Positron Emission Tomography (PET), Single Photon Emission Computerized Tomography (SPECT) and above all functional Magnetic Resonance Imaging (fMRI) that has made it possible to study increasingly complex cognitive functions in normal and diseased human beings. In an empirical review of 275 PET and fMRI studies Cabeza and Nybert provided an extensive account of "Imaging of Cognition". For perception and imagery, activation patterns included primary and secondary regions in the dorsal and ventral pathways. For attention and working memory, activations were usually found in prefrontal and parietal regions. For language and semantic memory retrieval, typical regions included left prefrontal and temporal regions. For episodic memory encoding, consistently activated regions included left prefrontal, medial-temporal, and posterior midline regions. For priming, deactivations in prefrontal (conceptual) or extrastriate (perceptual) regions were consistently seen. For procedural memory, activations were found in motor as well as in non-motor brain areas. Analysis of regional activations across cognitive domains suggested that several brain regions, including the cerebellum, are engaged by a variety of cognitive challenges.

4. FUNCTIONS OF THE DIFFERENT PARTS OF THE FRONTAL LOBE

The motor functions of the precentral gyrus, supplementary motor area and the frontal eye-field and also the role of Broca's area for speech and language have been extensively elaborated for nearly a century and hence would not be discussed here. Notwithstanding the fact that some of these functions play a role in higher cognitive functions, the major emphasis in this review will be primarily on complex mental functions like memory, thought, emotions, judgement and social behaviour about which a lot of new knowledge has accumulated in recent years, utilizing the evolving imaging technologies mentioned above. It is important to recognize that these technologies indicate the association of a region of the brain with a specific function. It does not necessarily imply that this association signify causality i.e. structure-function correlation. Nevertheless this information is of vital importance in our understanding of the overall organization of brain functions and their significance for clinical diagnostic and therapeutic purposes. While there are no absolute boundaries between the different regions of the frontal lobe mentioned above for the sake of simplification the following account is divided into the three components of the frontal lobe e.g. Dorsolateral Prefrontal cortex (DLPFC), Ventromedical Prefrontal cortex (VMPFC), and basifrontal or orbitofrontal cortex (OFC). Some investigators have further subdivided these regions into smaller functioning units, while others have considered the functional units as networks, including other regions outside the frontal lobe also (Whittle et al. 2006, Cabeza and Nyberg 2000, Miller et al. 2002)

5. DORSOLATERAL PREFRONTAL CORTEX (DLPFC)

This is part of the frontal lobe anterior to the motor strip to the frontal pole, hence also referred to as prefrontal cortex. It consists of the lateral surface of the brain from the interhemispheric fissure above to the Sylvian fissure below. Evidence from neuropsychological, electrophysiological and functional neuroimaging research suggests a critical role for the lateral prefrontal cortex in "executive control" for goal directed behaviour (Cummings, 1993).

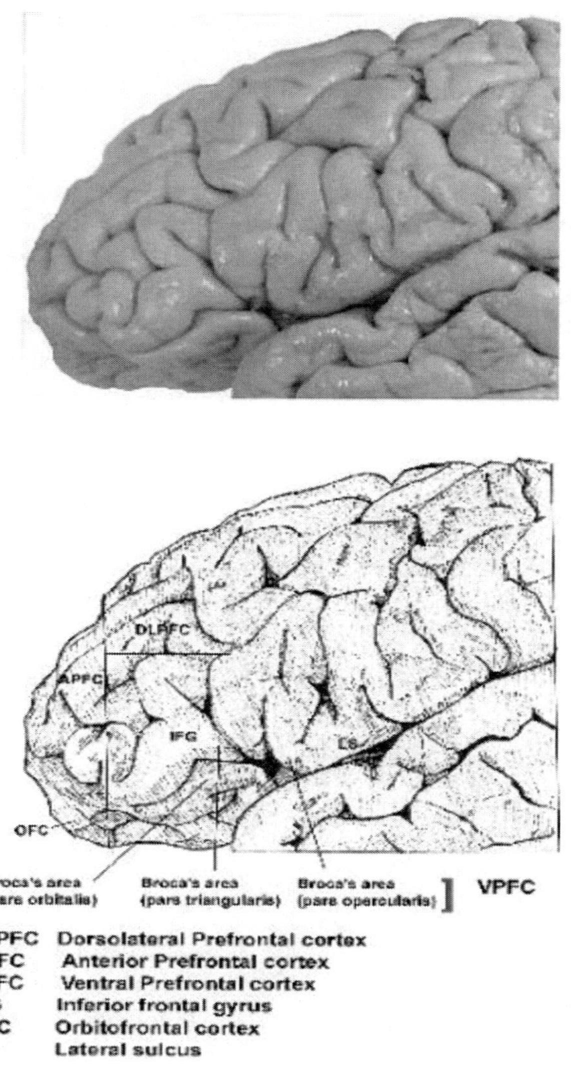

Figure 1. Vental prefrontal cortex.

This term implies a wide range of cognitive processes such as focused and sustained attention, fluency and flexibility of thoughts for planning and regulating adaptive and goal directed behavior (Damasio and Anderson, 1993; Stuss and Benson, 1986). Mental flexibility implies ability to rapidly alter behavior according to the requirement of the task at hand.

ERP studies have revealed that even prior to initiation of any motor activity, electrical potentials are observed in the prefrontal motor cortex. These are called "Readiness Potentials", which imply the role of this region in planning for a voluntary motor activity.

The role of DLPFC cortex in memory function especially for short term or working memory has long been established. Functional imaging studies of normal subjects performing tasks of episodic encoding, retrieval and recognition support the importance of frontal lobe function in the memory process. However, recent studies have demonstrated the role of the region in strategies used in learning new information and systematically searching memory, activation of remote memories (Mega, 2003). It has been suggested that left prefrontal activation is associated with encoding and right prefrontal activation is associated with retrieval tasks, whether the tasks are verbal or non-verbal (Buckner et al., 1998; Kapur et al., 1995b; Wagner et al., 1998a). Imaging studies revealed that prefrontal activations were particularly prominent during working memory and memory retrieval (episodic and semantic) which included monitoring, organization, and planning (Goldman-Rakic, 1996). Semantic memory retrieval and episodic memory encoding were usually localized to left PFC, while those during sustained attention and episodic retrieval were mostly right-lateralized and those during working memory were typically bilateral (Cabeza and Nyberg, 2000). Miller et al. (2002) observed that the dorsolateral PFC including Broadmann's areas 9,46,9/46 and the ventrolateral PFC including Broadmann's areas 44,45,47 subserved different aspects of memory functions. The former was activated in non-spatial working memory tasks while the latter was activated in tasks demanding active retrieval of object as well as spatial information.

Functional neuroimaging studies have revealed involvement of left PFC in verbal encoding, especially during semantic compared to non-semantic encoding. Wagner et al. (1998a) claimed that neural signatures during encoding differ from events subsequently remembered and events subsequently forgotten. Parahippocampal and prefrontal regions were found to act independently to promote encoding of events attributes important for conscious remembrance. Furthermore, neuroimaging studies of episodic recognition memory consistently demonstrated retrieval – associated activation in right anterior and DLPFC. These findings were corroborated by PET studies. However, recognition resulted in greater activation in bilateral frontal opercular cortex, medial prefrontal cortex (near anterior cingulate) and bilateral posterior inferior frontal gyrus (Wagner et al., 1998b).

Prefrontal cortex is known to perform executive functions. Given its role in organizing complex behavior and its extensive connections the PFC seems to be a likely region where integration of diverse sensory information could take place for knitting together behaviorally relevant associations (Miller, 1999). The superior dorsolateral PFC (area 9 and10) is the centre for an executive cognition network that links the posterior parietal lobe and anterior cingulate gyrus. Its functions include the ability to organize behavioral response to solve a complex problem (including the strategies used in learning new information and systematically searching memory), generating motor programs, and using verbal skill to guide behavior (Mega, 2003).

The lateral PFC has an important role in executive control of goal directed behavior which involves a wide range of cognitive processes such as focused and sustained attention, fluency and flexibility of thoughts in the generation of solution to novel problems and planning and regulating adaptive behavior (Damasio and Anderson, 1993, Knight and D'Esposito, 2003)

The extensive reciprocal connections between the DLPFC to virtually all cortical and sub-cortical structures places it in a unique neuroanatomical position to monitor and manipulate diverse cognitive functions (Knight and D'Esposito, 2003).

In addition to its involvement in memory and executive functions, PFC has been found to be involved in a variety of other behavior e.g. moral judgement (Green et al., 2001), Laughter and appreciation of jokes (Boynton, 2002), evaluating reward and judging its fairness or otherwise (Crockett et al., 2008; McClure et al., 2004; Richmond et al., 2003), Self-awareness (Ehrsson et al., 2004), temperament (Whittle et al., 2006), meditation (Newberg et al., 2001), and introception (Craig, 2009).

Neuroimaging investigations by Crochett et al. (2008) during psychological studies of social behavior implicated both DLPFC and ventral prefrontal cortex (VPFC) in regulating reactions to unfair offers. Disrupting DLPFC function with transcranial magnetic stimulation (TMS) led to decreased rejection of unfair offers. However, individuals with VPFC damage rejected a higher proportion of unfair offers than the controls (Crockett et al., 2008).

Investigating neural basis of distributive justice Hsu et al. (2008) demonstrated the important role of insular cortex, nearby paralimbic regions including septal-subgenual area involved in fairness and empathy, altruism and social attachment and risk perception.

Frontal lobe lesions in humans are associated with a variety of memory impairments, including deficits in short-term memory, free recall, memory for temporal information etc. Patients with DLPFC lesions were found to be impaired on free recall tests of remote memory for both public events and famous faces. The pattern of these deficits suggests an impairment in the ability to organize effective retrieval strategy. Yet in clinical setting the routine psychological tests e.g. Wechsler memory test fail to reveal a functional deficit (Mangels et al., 1996). It has been suggested that their memory deficits are not due to anterograde amnesia but probably due to an impairment in strategic processes associated with organization and monitoring at encoding and retrieval. The deficit in working memory evident in Parkinson's disease and Schizophrenia have been postulated to arise from a deficiency in the function of the PFC. Frontal lobe lesions within the premotor cortex (area 9 and 10) have also been found to be associated with limb – akinetic apraxia, impairment in learning manual sequences and in the organization of spontaneous gestures (Grafton, 2003).

Recent investigations have surprisingly revealed the role of PFC – including DLPFC in regulating emotions. The involvement of amygdala, the parahippocampal gyrus and hypothalamus limbic system in control of emotions has been known for a long time. However, it is only as a result of recent neuro-imaging studies that the role of PFC in this respect has been discovered. While the medial PFC is more important even the DLPFC is seen to participate in some of the emotional functions. According to Pessoa (2008) while PFC is primarily involved in cognition, recent functional studies of the LPFC provide evidence that cognition and emotion are integrated in this region. While LPFC has a role in maintaining and manipulating information, it can also integrate this content with both affective and motivational information. (further details of neural basis of emotions will be discussed in sections under MPF and OFC which no doubt play a much greater role than the DLPFC).

6. MEDIAL PREFRONTAL CORTEX

This includes cortex on the medial aspect of the frontal lobe. Consisting of the cingulate gyrus and areas above and below it (i.e. Broadmann's areas 24, 23, 8, 9, 10, 11, 25, 31, 32). As a matter of fact parts of the MFC specially the anterior cingulate cortex (ACC) constute an intrinsic component of the "Limbic System" or part of the Papez's circuit. The cingulate region is roughly divided into an anterior (Broadmann's areas 32, 24), a central (areas 23, 31) and posterior (retrosplenial area: not really a part of the PFC). Major reciprocal connections with the ACC exist with DLPFC (areas 8, 9, 10, 46). These are responsible for directing attentional systems. ACC activation was observed in working memory, semantic generation and episodic memory tasks. Imaging studies indicated that ACC is involved in "attention to action", initiating appropriate responses and suppress inappropriate ones (Cabeza and Nyberg, 2000; Paus et al., 1993). Furthermore, recent neuroimaging studies have revealed that the ACC is active when human participants engage in social interaction (Rudebeck et al., 2006). Damage to this area is reported to result in acquired psychopathy and akinetic mutism.

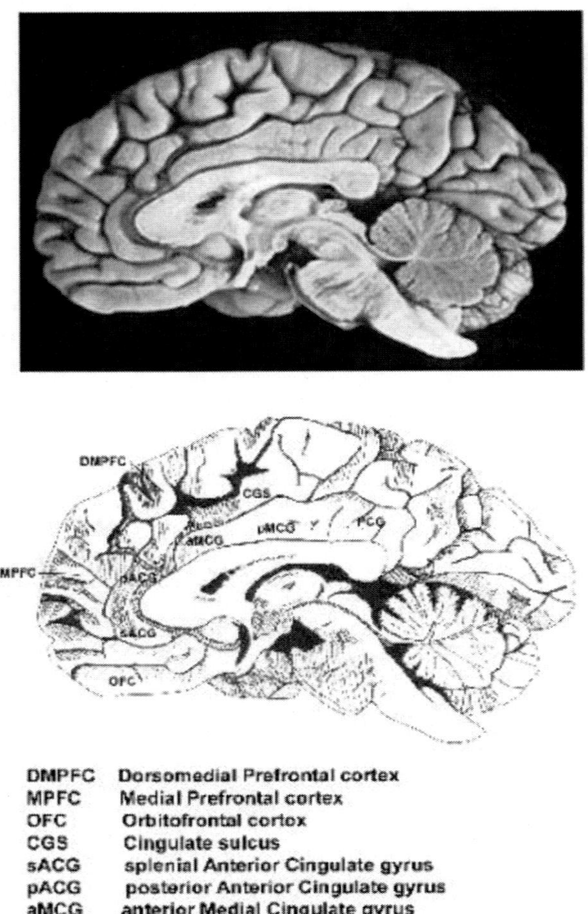

Figure 2. Medial prefrontal cortex.

Ventromedial PFC – particularly rostral ACC alongwith subgenual anterior cingulate play an important role in emotional regulation. It is believed to suppress the emotional information received from amygdala (Blair, 2008). There are bidirectional connections between amygdala and MPFC as also OFC (Ghashghaei and Barbas, 2002).

MPFC along with amygdala and hippocampus is implicated in fear conditioning and extinction. Fear neurons in the amygdala receive inputs from the hippocampus and project to the MPFC. Extinction neurons by contrast are connected to the MPFC in both direction (Sah and Westbrook, 2008).

Nucleus accumbens, on the medial surface of the frontal lobe, below the genu of the corpus callosum is known to be an important part of the reward circuit. The normal function of this circuit is to ensure that behaviors such as eating and sex are perceived as rewarding. It is not surprising then that the same circuit is involved in craving for addictive drugs (Abbott, 2002). Ramamurthi and his colleagues at Neurological Institute at Madras used stereotactic lesion in the cingulum for treatment of addiction. According to Richmond et al. (2003) the MPFC particularly the ACC is important in judging a reward's value. Both functional imaging in humans and single neuron recording in monkeys have shown that MPFC, including the cingulate cortex are activated during periods of expectancy about the imminence and probability of a reward associated with strong emotional component.

7. THEORY OF MIND

Recent years have witnessed the emergence of a new field of cognition i.e. Social Cognition. When we interact with another person, we assume that they have minds like ours and try to predict their behavior on the basis of contents of their minds; their beliefs and desire (Frith and Singer, 2008). This has been called, "Theory of Mind". Discovery of mirror neurons, (initially in the brains of monkeys) and contributions of modern neuroimaging have led to accumulation of vast literature on the subject (Gallagher and Frith, 2003; Ochsner et al., 2004; Zimmer, 2003).

Frith and colleagues in a series of publications (Frith and Frith, 1999; Frith and Singer, 2008; Gallagher and Frith, 2003) found that in spite of the wide range of investigations, a small set of brain regions is activated in all these studies. These include the MPFC, posterior superior temporal cortex and temporal poles. MPFC was one of the regions consistently activated when subjects performed tasks in which they have to think about the mind states of others.

Ochsner et al. (2004) and Pronin (2008) using fMRI compared the neural processes supporting inference about one's own and other individual's emotional states. They found activation of the MPFC when individuals judged some aspects of their own or someone else's mental states.

Zimmer (2003) in his studies on the theory of mind found anterior paracingulate neurons to be active. However, in addition to MPFC, superior temporal sulcus, the temporal pole and the amygdala were found to play important role.

Those interested in details may refer to a special issue of the Philosophical Transactions of Royal Society B Social Intelligence: from Brain to Culture, Volume 362, 2007. However, for the limited purpose of this presentation mention is made of the role of MPFC and associated areas of the brain subserving this function.

It is therefore not surprising that complex human social interaction is disrupted when the frontal lobe is damaged. This has been confirmed by lesion studies in monkey involving anterior cingulate and the orbito frontal cortices (Rudebeck et al., 2006).

8. BASIFRONTAL OR ORBITO – FRONTAL CORTEX

The basofrontal region of the frontal lobe, also called Orbito-Frontal Cortex (OFC) consists of the gyrus rectus (Broadmann's area 14) and what was initially labelled as area II by Broadmann. This was later further divided into areas 12,13 and 47 by Walker (1940), Barbas and Pandya (1989) and Carmichael and Price (1994).

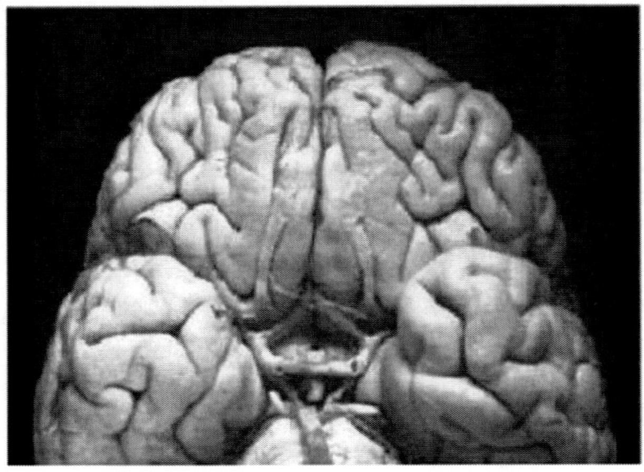

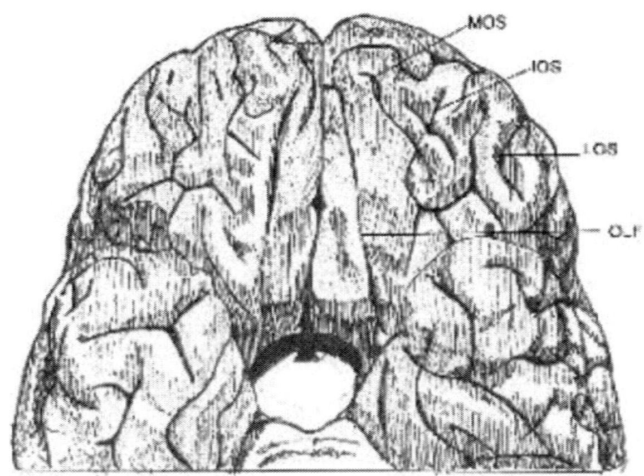

IOS	Intermediate Orbital Sulcus
LOS	Lateral Orbital Sulcus
MOS	Medial Orbital Sulcus
OLF	Olfactory Sulcus

Figure 3. Basifrontal or Orbito – Frontal Cortex.

Robust bidirectional connections have been observed between OFC, MPFC and amygdala. The OFC axons terminated densely in a narrow band around the borders of the magnocellular basi lateral nucleus of the amygdala. This suggests that the OFC exercises control on the internal processing of the amygdala which no doubt is the most important brain region controlling emotions (Ghashghaei and Barbas, 2002). A meta-analysis of 106 PET and fMRI studies of human emotions indicated that the limbic system alone may not be the only neural substrate of all emotions. Lateral OFC activity was reported in a higher proportion of studies of anger relative to other emotions. Focal lesions in this region was specifically associated with changes in aggressive behavior in humans (Murphy et al., 2003).

In addition to emotional processing, reward and decision making OFC region has also been shown to be involved in the analysis of visual information and in visual memory. OFC has been found to be involved in object recognition and encoding information. Bare et al. (2006) using magnetoencephalography observed differential activity in the left OFC (in addition to the temporal region) during an object recognition task.

In addition, Frey and Petrides (2000) in their PET studies observed that the rostral OFC (area 11), which is primarily linked with the anterior medial temporal limbic region and LPFC, is involved in the process of encoding of new information specially on the right side for novel abstract visual stimuli. Bilateral lesions of the OFC in non human primates result in a severe recognition memory impairment. Hirstein (2005) in his book, "Brain Fiction, Self-Deception and the Riddle of Confabulation", provided evidence that damage to OFC, especially on the right side results in confabulation. He suggested that, "Confabulation occurs when a perceptional or mnemonic process fails and the failure is not detected by frontal processes."

In a meta analysis of data from neuroimaging and neurophysiology Kringelbach and Rolls (2004) observed that functionally OFC can be divided into a medial and a lateral subdivision. Whereas medial OFC activity was related to monitoring the reward value of different reinforcers. The lateral OFC was related to the evaluation of punishers which way lead to change in ongoing behaviour. They provide a detailed description of the neuroanatomy, cyto-architecture of the OFC along with its functional correlation.

CONCLUDING REMARKS

Prefrontal lobe that is, most of the frontal lobe in front of the precentral sulcus is the largest and most highly evolved, part of the human brain. On the basis of gross anatomy, cytoarchitecture and functional attributes it can be divided into three major components – the dorsolateral, ventromedial and basofrontal. Each of these has been further subdivided into smaller functional units. Theoretically it also consists of insular opercula and deeply situated some of the basal ganglia.

Recent advances in brain imaging and psychophysics have now been able to assign a variety of cognitive functions to different parts of the extensive prefrontal cortex. This review is primarily restricted to the three main subdivisions i.e. DLPFC, MPFC, OFC. It cannot be over emphasized that none of these regions are solely restricted to one or more cognitive function, nor do they act as independent entities. Though some parts are predominantly involved in one or more cognitive modality but the actual function is dependent upon dynamic coalitions of networks of brain areas (Barbas, 2000). Thus, while the hypothalamus was one of the first regions linked to emotions, it was the description of the Papez circuit that advanced the network theory. To this was later added the amygdala and hippocampus and now as recorded above we know the role played by parts of DLPFC, MPFC and OFC in a variety of emotions like fear, anger, aggression etc (Cordoso, 2002; Duncan and Owen, 2000; Ghashghaei and Barbas, 2002; Miller and Cohen, 2001; Sah et al., 2008). Not surprisingly complex behaviours like moral judgement and distributive justice involve regions as diverse as DLPFC, Septal subgenual part of MPFC, the insular cortex and the striatum (Crockett et al., 2008; Greene et al., 2001; Hsu et al., 2008; McClure et al., 2004). Similarly, Newberg et al. (2001) reported involvement of DLPFC, left superior parietal lobe and thalamus during the complex cognitive task of meditation. Ehrsson et al. (2004) and Craig (2009) provided evidence of involvement of frontal lobe in human awareness and feeling of self.

These observations are not just of academic interest but help to explain the various psychological and psychiatric disorders (Whittle et al., 2006). At the same time this knowledge is already being utilized for therapeutic purpose like deep brain stimulation for depression and obsessive compulsive disorders (Miller, 2009), transcranial magnetic stimulation, and cognitive and behavioural therapy.

ACKNOWLEDGMENT

Thanks to Dr. Soumya Iyengar, Associate Professor, National Brain Research Centre, Manesar for providing the three photographs and labeled figures from a brain specimen in her lab.

REFERENCES

Abbott, A. (2002). Addicted. *Nature, 419*, 872-874.

Barbas, H., and Pandya, D.H. (1989). Architecture and intrinsic connections of the prefrontal cortex in the rhesus monkey. *Journal of Comparative Neurology, 286*, 353-375.

Barbas, H. (2000). Connections underlying the synthesis of cognition, memory and emotion in primate prefrontal cortices. *Brain Research Bulletin, 52*, 319-330.

Blair, R.J.R. (2008). The amygdala and ventromedial prefrontal cortex: Functional contributions and dysfunction in psychopathy. *Philosophical Transactions of the Royal Society B, 363*, 2557-2565.

Boynton, P. (2002). It's no laughing matter. *New Scientist*, pp 48-51.

Buckner, R., and Koustaal, W. (1998). Functional neuroimaging studies of encoding, priming, and explicit memory retrieval. *Procedings of the National Academy of Sciences USA, 95*, 891-898.

Cabeza, R., and Nyberg, L. (2000). Imaging Cognition II: An empirical review of 275 PET and fMRI studies. *Journal of Cognitive Neuroscience, 12*, 1-47.

Carmichael, S.T., and Price, J.L. (1994). Architectonic subdivision of the orbital and medial prefrontal cortex in the macaque monkey. *Journal of Comparative Neurology, 346*, 366-402.

Cordoso, S. (2002). It's no laughing matter. *New Scientist*, 48-57.

Craig, A.D. (2009). How do you feel – now? The anterior insula and human awareness. *Nature Reviews Neuroscience, 10*, 59-70.

Crockett, M.J., Clark, L., Tabibnia, G., Lieberman, M. D., and Robbins, T. W. (2008). Serotonin modulates behavioural reactions to unfairness. *Science, 320*, 1092-1095,

Cummings, J.L. (1993). Frontal-subcortical circuits and human behaviour. *Archives of Neurology, 50*, 873-880.

Damasio, A.S., and Anderson, S.W. The Frontal lobes. In K.M. Heilman and E. Valestein, *Clinical Neurophysiology*. 3rd ed (pp. 409-460), New York: Oxford University Press.

Duncan, J., and Owen, A.M. (2000). Common regions of the human frontal lobe recruited by diverse cognitive demands. *Trends in Neurosciences, 23*, 475-483.

Duncan. J., and Owen, A.M. (2000). Common regions of the human frontal lobe recruited by diverse cognitive demands. *Trends in Neurosciences, 23*, 475-493.

Ehrsson, H.H., Spence, C., and Passingham, R.E. (2004). That's my hand! Activity in premotor cortex reflects feelings of ownership of a limb. *Science, 302*, 875-877.

Frey. S., and Petrides, M. (2000). Orbitofrontal cortex: A key prefrontal region for encoding information. *Procedings of the National Academy of Sciences USA, 97*, 8723-8727.

Frith, C. (2006). No one really uses reason. *New Scientist*, p 45.

Frith, C. (2008). Social Cognition. *Philosophical Transactions of the Royal Society B, 363*, 2033-2034.

Frith, C.D., and Singer, T. (2008). The role of social cognition in decision making. *Philosophical Transactions of the Royal Society B, 363*, 3875-3886.

Galagher, H.L., and Frith, C.D. (2003). Functional imaging of "theory of mind". *Trends in Cognitive Sciences, 7*, 77-83.

Ghashghaei, H.T., and Barbas, H. (2002). Pathways for emotion: interactions of prefrontal and anterior temporal pathways in the amygdala of the rhesus monkey. *Neuroscience, 115*, 1261-1279.

Grafton, S. (2003). Frontal lobe apraxia: In Mark D'Esposito (Ed) *Neurological Foundation of Cognitive Neuroscience* (p 245), Cambridge, Mass: MIT Press.

Hirstein, W. (2005). *Brain Fiction: Self-Deception and the Riddle of Confabulation*. MIT Press, Cambridge, MA.

Hsu, M., Anen, C., and Quartz, S.R. (2008). The right and the good: distributive justice and neural encoding of equity and efficiency. *Science, 320*, 1092-1095.

Kapur, S., Craig, F.I.M., Jones, C., Brown, G.M., Houle, S., and Tulving, E. (1995). Functional role of the prefrontal cortex in retrieval of memories: a PET study. *Neuroreport, 6*, 1880-1884.

Knight, R.T., and D'Esposito, M. (2003). Lateral prefrontal syndrome. A disorder of executive control. In *Neurological Foundation of Cognitive Neuroscience*, (Eds) Mark D'Esposito. The MIT Press, Cambridge, Mass.

Kringelbach, M.L., and Rolls, E.T. (2004). The functional neuroanatomy of the human orbitofrontal cortex: evidence from neuroimaging and neurophysiology. *Progress in Neurobiology, 72*, 341-372.

Mangels, J.A., Gershberg, F.B., and Shimanmuna, A.P. (1996). Impaired retrieval from remote memory in patients with frontal lobe damage. *Neurophychology, 10*, 32-41.

McClure, S.M., Laibson, S.I., Loewenstein, G. and Cohen, J.D. (2004). Separate neural systems value immediate and delayed monetary rewards. *Science, 306*, 503-507.

Mega, M.S. (2003). Amnesia: A disorder of episodic memory. In *Neurological Foundations of Cognitive Neuroscience* (Ed) Mark D'Esposito, The MIT Press Cambridge, Mass, pp 41.54.

Miller, E.K. and Cohen, J.D. (2001). An integrative theory of pre-frontal cortex function. *Annual Review of Neuroscience, 24*, 167-202.

Miller, E.K. (1999). The Prefrontal cortex: Complex neural properties for complex behaviour. *Neuron, 22*, 15-17.

Miller, G. (2009). Rewiring faulty circuits in the brain. *Science, 323*, 1554-1556.

Miller, N.G., Machado, L. and Knight, R.T. (2002). Contributions of subregions of the prefrontal cortex to working memory: Evidence from brain lesions in humans. *Journal of Cognitive Neuroscience, 14*, 673-686.

Murphy, F.C., Nimmo-Smith, I., Lawrence, A.D. (2003). Functional neuroanatomy of emotions: A meta-analysis. *Cognitive, Affective, Behavioral Neuroscience, 3*, 207-233.

Newberg, A., Alavi, A., Baime, M. et al. (2001). The measurement of regional cerebral blood flow during the complex cognitive task of meditation; a preliminary SPECT study. *Psychiatry Research, 106*, 113-122.

Ochsner, K.N., Knierim, K., Ludlow, D.H. et al. (2004). Reflecting upon feelings: An fMRI study of neural systems supporting the attribution of emotions to self and other. *Journal of Cognitive Neuroscience, 16*, 1746-1772.

Paus, T., Petrides, M., Evans, A.C., and Meyer, E. (1993). Role of human anterior cingulate cortex in oculomotor, manual and speed responses: A positron emission tomography study. *Journal of Neurophysiology, 70*, 453-469.

Penfield, W.P., and Jasper, H. (1954). *Epilepsy and Functional anatomy of the Human Brain.* Little Brown and Co. Boston.

Pessoa, L. (2008). On the relationship between emotion and cognition. *Nature Reviews Neuroscience, 9*, 148-158.

Pronin, E. (1940). How we see ourselves and how we see others. *Journal of Comparative Neurology, 73*, 59-86.

Richmond, B.J., Liu, Z., and Shidara, M. (2003). Predicting future rewards. *Science, 301*,179-180.

Rudebeck, P.H., Buckley, M.J., Walton, M.E., and Rushworth, M.F.S. (2006). A role for the Macaque anterior cingulate gyrus in social valuation. *Science, 31*, 1310-1312.

Sah, P., and Westbrook, F. (2008). The Circuit of fear. *Nature, 454*, 589-590.

Stuss, D.T., and Benson, D.F. (1986). *The Frontal Lobes.* Raven Press, New York.

Wagner, A.D., Desmond, J.E., Glover, G.H., and Gabrieli, J.D.E. (1998). Prefrontal cortex and recognition memory: Functional – MRI evidence for context-dependent retrieval processes. *Brain, 121*, 1985-2002.

Wagner, A.D., Schater, D.L., Rotte, M. et al. (1998). Building memories: Remembering and forgetting of verbal experiences as predicted by brain activity. *Science, 281*,1188-1191.

Walker, E.A. (2000). Quoted by Frey and Petrides. *Proceedings of the National Academy of Sciences, 97*, 5723-8727.

Whittle, S., Allen, N.B., Lubman, D.I., and Yiicel, M. (2006).The neurobiological basis of temperament: Towards a better understanding of psychopathology. *Neuroscience and Biobehavioral Reviews, 30*, 511-525.

Zimmer, C. (2003). How the mind reads other minds? *Science, 300*, 1079-1080.

In: Expanding Horizions of the Mind Science(s)   ISBN: 978-1-62808-705-5
Editors: P.N. Tandon, R.C. Tripathi and N. Srinivasan   ©2013 Nova Science Publishers, Inc.

Chapter 12

LANGUAGE PROCESSING IN BILINGUALS AND BILITERATES

Shyamala K. Chengappa[1] and Prema K.S. Rao[2]

[1]Department Speech Language Pathology & OC(DHLS), All India Institute of Speech
and Hearing, Manasagangothri, Mysore, India
[2]Department of Special Education, All India Institute of Speech and Hearing,
Manasagangothri, Mysore, India

ABSTRACT

This chapter addresses language processing in bilinguals and biliterates and discusses issues that have been raised with respect to such studies. Specific issues on language processing in bilinguals such as cortical representation, neuroanatomical and functional organization, evidence from bilingual aphasia, psycho linguistic accounts of bilingualism, types of bilingualism, neurolinguistic models on bilingualism, models of bilingual memory, cognitive advantages and disadvantages of bilingualism and other crucial factors are discussed with research support. Available research findings on the cross language transfers of one language over the other during the period of acquisition of biliteracy like influence of structure and orthography on processing of biliteracy are deliberated in detail. Suggestions are proposed for incorporating empirical database in making policy decisions in the educational sector and remedial teaching practices to facilitate effective bi/multiliteracy in the multilingual context of India.

1. INTRODUCTION

Language is a system of arbitrary symbols by means of which members of a society interact with one another. It is the means by which messages are conveyed either through speech or through written codes used in a given language. There are four language skills-listening, speaking, reading and writing that are grouped into receptive and expressive skills. While listening and reading are considered as the receptive skills, speaking and writing are considered to be the expressive skills. Listening and speaking are considered primary while reading and writing are considered secondary language skills. Often it is believed that an individual begins with the listening skills before acquiring speaking skills and reading skills before acquiring writing skills. It can be said that 'secondary language' comprises of 'reading and writing skills' which are invariably the components of 'literacy skills'. Therefore, there is considerable relationship among the components of language and literacy either at the processing stage or at the production stage. The complexities of language and literacy skills make an interesting plane for study in the special population of bilinguals/multilinguals who are most often (while sometimes not) biliterates and multiliterates[1]. Sometimes, they may also be only bilinguals and not biliterates as seen commonly in certain contexts of immigration for example in the labour class. It appears to be advantageous to take a relook at these two contexts to establish the levels of association/dissociation between the two primary and the two secondary language skills, as it would clarify and impinge on the cognitive issues of processing in bilinguals and biliterates, eitherway.

Bilingualism has defied definition ever since it is conceptualized. There are as many characteristics and definitions of bilingualism as researchers in the field. This is more due to multifaceted nature of phenomenon rather than the ambiguity of the term, if any. Different researchers have their own theoretical and methodological perspectives with specific issues, contexts and bilingual population at hand. In a not so exhaustive review, several neurolinguistic, psycholinguistic and cognitive research database is extrapolated in the present paper. Many basic issues beginning from definition to variety of factors influencing, bilingual and biliterate phenomena like mixing and switching, interference, cognitive advantages and disadvantages , language structures, crosslinguistic transfer, orthographic issues etc are deliberated upon to highlight their impact on the phenomenon of cognitive processing in bilinguals and / biliterates.

1.1. Definitions of Bilingualism

Websters dictionary (1961) defines a bilingual as having or using two languages especially as spoken with the fluency characteristics of a native speaker; a person using two languages habitually with control like that of a native speaker and bilingualism as the constant oral use of two languages. Diebold (1961) however gives a minimal definition of bilingualism when he uses the term incipient bilingualism to characterize the initial stages of contact between the two languages. Thus any passive knowledge of the written language or any contact with a second language and the ability to use it in the environment of the native language was considered bilingualism by him.

Bilingualism refers to knowledge and use of two languages and an ability to make a meaningful utterance in either of the languages (Harding, Edith and Riley, 1986). If a speaker is fluent in two languages, then he is said to be a bilingual. A bilingual person is generally brought up in a culture where he/she is exposed to two languages from birth. On the basis of the way in which the two languages are learned, three types of bilingualism – the compound, coordinate, and sub-coordinate- are proposed by Weinrich (1953).They can also be learnt in

---

[1] Readers may note that the term 'bilingual/multilingual' and biliterate/multiliterate' are used synonymously for the sake of convenience in this paper.

simultaneous /successive context. For a review of these terminological issues readers are referred to Thirumalai and Chengappa (1986) and Harini and Chengappa (2010).

A functional or wholistic view of bilingualism takes into consideration those individuals who learn and use each of their languages for different purposes and in different communication contexts. Therefore, a functional view of bilingualism recognizes that a bilingual may become more competent in one language in certain communication contexts and competent in the other languages for other contexts. From the functional perspective, then it becomes important to ask why the languages were acquired, how they were acquired, whom they were acquired with and how they were used. Thus, the level of fluency in a language depends on the need for and use of a particular language in a particular situation or context at any given point of time.

Most of the definitions describe bilinguals as ideal speaker hearers of two languages and adhere to Chomskyan view point. In contrast, MacNamara (1967) proposed that a bilingual is anyone who possesses a minimal competence in any one of the four language skills: listening, comprehension, speaking, reading and writing in a language other than his mother tongue. Some researchers describe bilinguals as individuals with native like competence in second language whereas others limit it to minimal proficiency. These definitions suggest that the concept of bilingualism does not seem to be as simple as one would think. Paradis (1986) like Hamers (1981) suggested that bilingualism should be defined on a multidimensional continuum, considering an only linguistic structure and language skill which is probably a restricted view.

Grosjean (1997) defined it using complementarity principle that emphasizes the functionality of language. This is also termed as the wholistic or functional view of bilingualism. Grosjean (1989) delineated different ways in which bilingualism has been viewed. Two views of bilingualism are; the monolingual or fractional view which holds that the bilingual is (or should be) two monolinguals in one person, and the bilingual or wholistic view which states that the coexistence of two languages in the bilingual has produced a unique and specific speaker-hearer. The monolingual view sees a person as ideal bilingual, who has good knowledge of both languages and the rest, who in fact represent the vast majority of people, are not really bilinguals or are special types of bilinguals. The fact is that most of the people are not ideal bilinguals so the monolingual view has many demerits. A major consequence of the monolingual view is that language skills in bilinguals are usually appraised in terms of monolingual standards. Another consequence is that the contact of the bilingual's two languages is seen as accidental or anomalous and because of language interference. Thus, bilinguals rarely evaluate their language competencies as adequate, often amplify the monolingual view and hence criticize their own language competence.

A bilingual (or wholistic) view of bilingualism proposes that the bilingual is an integrated whole which cannot be easily decomposed into two separate parts. The co existence and constant interaction of two languages in a bilingual has produced different but complete linguistic entity. The bilingual uses the two languages separately or together for different purposes, in different domains of life, with different people. Because the needs and uses of two languages are usually quite different, the bilingual is rarely equally or completely fluent in his two language. Levels of fluency in a language will depend on the need for that language and will be domain specific.

By convention the language learnt first is called L1 and the language learnt second is called L2. Bialystok and Hakuta (1994) made a distinction between simultaneous (L1 and L2 learned about the same time), early sequential (L1 learned first and L2 relatively early in childhood) and late (from adolescence onwards) bilingualism. In addition, the term second language is often used to mean a language that is learned after the first or the native language is relatively well established. However, in the present era of changing global scenario, it is quite likely that the language learnt first would become the secondary language of use in later life and therefore a rigid definition of terminology is not advisable. However, these terminologies and the way a language is learnt and used have their implications on our understanding of the processing mechanisms employed while acquiring language and literacy by bilinguals and biliterates.

Thus, communicative competence of bilingual's cannot be evaluated through only one language; it must be studied instead through the bilingual's total language repertoire as it is used in his or her everyday life. How brain processes the acquisition/learning of the two/three languages in a bilingual/multilingual, however, is still an enigma.

1.2. Cognitive Advantages and Disadvantages of Bilingualism

Earlier studies have provided support for the negative effects of bilingualism on cognitive development. Darcy (1953) had concluded from the review of relevant studies that bilinguals suffered from a language handicap when measured by verbal tests of intelligence. It was proposed that bilinguals never reached comparable levels of linguistic proficiency as did monolinguals. Researchers, theoreticians and professionals alike often viewed the simultaneous acquisition of two languages with apprehension .Starting with the 1960s; the propensity has changed towards the positive advantages of L2 use. These studies denunciate earlier studies since they depended largely on L2 users who differed in many factors other than knowing a second language. One major limitation was that these studies did not control for socioeconomic status between the bilingual and monolingual subjects.

As McCarthy (1930) argued, bilingualism in America was confounded with SES since more than half of the children classified as bilinguals in early studies belonged to families from the unskilled labor group. There was also a failure to adequately assess and consider differences in degree of bilingualism. There was a lack of clarity in how researchers defined and evaluated the bilingual or monolingual status of their subjects. Hakuta (1986) claim that early psychologists used a societal definition of bilingualism in determining language proficiency. Such methods thus do not hold up to the present day scrutiny and may have thus resulted in biased results favoring monolinguals. Peal and Lambert's (1962) research suggested cognitive advantages to being bilingual, calling into question the validity of earlier studies and supporting the claims linguists had been making for years. With control in place for socio-economic variables many studies showed bilingual children as showing better metalinguistic skills across different language pairs. Ben-zee (1972) showed that bilinguals score lower in vocabulary but higher in verbal material. It was shown that high score in verbal learning is a sign of cognitive development. Lanco-worrall (1972) noted that bilinguals show more metalinguistic awareness in terms of language forms and properties Landry (1973) pinpointed that bilinguals are those who study a second language and they are better in diverse thinking than monolinguals. Powers and Lopez (1985) showed that bilinguals are better than monolinguals in the complex and perceptual motor coordination in the brain.

Furthermore, Galambos (1982) found that El Salvadoran children proficient in English and Spanish demonstrated a stronger syntactic orientation when judging grammatically correct and incorrect sentences in both languages. Various studies in last two decades have shown the influence of bilingualism on word awareness leading to better reading skills Bialystok and Herman, (1999); Bruck and Genesee (1993). These benefits were not just restricted to balanced bilinguals but significant benefits were observed for children whose contact with a second language was restricted Yellend, (1993). Research appears to suggest a positive relationship between bilingualism and a wide range of other cognitive measures, including enhanced ability to restructure perceptual solutions Balkan (1970), stronger performances in rule discovery tasks Bain, (1975), greater verbal ability and verbal originality, and precocious levels of divergent thinking and creativity Cummins and Gulutsan, (1974). Petiti and Holowka (2002) rejected the assumption that early language exposure to different language will lead to delay in language acquisition, and they specifically emphasized that the child's semantic concept and image would not be tarnished. A growing number of studies have reported advantages in nonverbal executive control tasks for bilingual children (Bialystok, 2001; Carlson and Meltzoff, 2008) and adults (Bialystok, Craik, Klein, and Viswanathan, 2004; Bialystok, Craik, and Ryan, 2006).

In one set of studies, bilingualism has been shown to accelerate the development of executive control in children (Bialystok, 2001) using nonverbal control tasks such as the flanker task (Mezzacappa, 2004), perceptual analysis and rule switching (Bialystok and Martin, 2004) .These effects persist into adulthood and appear to protect bilingual older adults against the decline of those processes in older age (Bialystok, Craik, Klein, and Viswanathan, 2004). The difference in executive control between monolinguals and bilinguals is larger in older age because the normal decline of these processes with aging is attenuated for bilinguals (Rajasudhakar and Chengappa, 2008). Across the lifespan, therefore, bilingualism boosts the development and postpones the decline of executive control on a variety of tasks both non verbal as well as verbal.

Researchers have proposed various theories to explain this positive relationship.

1.2.1. Objectification Theory

This theory claims that by acquiring two languages, bilinguals learn more about the forms as well as the functions of language in general, which affects various cognitive processes. Experience with two language systems may also enable bilinguals to have a precocious understanding of the arbitrariness.

The ability to objectify language is linked to a capacity Piaget (1929) termed non-syncretism, is the awareness that attributes of an object do not transfer to the word itself.

Edwards and Christophersen (1988) found that bilinguals may have an enhanced level of such understanding, and researchers such as Olson (1977) have shown such capacity to be linked to literacy. Lastly, by learning that two words can exist for a single referent, bilinguals may develop not only increased knowledge of their L1 and L2, but of language in general as a symbolic system. Thus, such children may process concepts through higher levels of symbolic and abstract thinking.

1 2.2. Codeswitching Theory

Because bilinguals are able to move rather easily from verbal production in one language to that in another, they may have an added flexibility. Peal and Lambert (1962) theorized that the ability to code-switch provides bilinguals with an added mental flexibility when solving cognitive tasks.

1.2.3. Verbal Mediation Theory

Bilinguals are believed to have an enhanced use of self regulatory functions of language as a tool of thought guiding inner speech or verbal thinking .Language may be a more effective tool for bilinguals in approaching cognitive tasks.

In order to understand the influences of bilingualism, it is important to take the unique features of participants, tasks and the relationships of those tasks to the constructs into consideration. The education level and the language of clients and their family, the literacy environment, the extent of the child's/adults proficiency in the first language, the purpose of learning the second language and the degree of community support are a few variables to be controlled.

2. PSYCHOLINGUISTIC ACCOUNTS OF BILINGUALISM

Bilingual cognitive representation is a debatable issue. How do bilinguals organize two or more languages for use in their brain? The concept of shared and separate memory had been critically evaluated by researchers by evaluating the important theoretical considerations underlying these concepts. The shared and the separate memory hypothesis are perhaps the most influential views of bilingual memory representation (Ijalba, Obler and Chengappa, 2004).According to this proposal, bilinguals either organize their two languages into shared memory store or into two separate memory systems, where each language is organized independently. However there were many mixed results obtained based on the above concepts and this has led researchers to conclude that even if experimental findings supported both hypothesis ,the shared memory model explained some aspects of language, while the separate model was more appropriate for other aspects. Moreover, it was suggested that bilinguals had neither separate nor shared memories,because some information was restricted to the language of encoding, while some was accessible to both linguistic systems (Durgunoglu and Roediger,1987;Kolers and Gonzalez,1980).

Neural support for the languages in polyglot speakers may be associated with different memory systems, a theory proposed by the Declarative/Procedural model. Effective use of the known languages by bilingual and polyglot speakers may in turn be determined by control over inhibition or activation of each language, a theory posited by the Inhibition Control model (Ijalba, Obler and Chengappa, 2004).

2.1. Declarative/Procedural –Memory Model

The learning of language seems to be mediated by two different memory mechanisms as they interact with age of acquisition (Paradis, 1994; Ullman, 2001).This model explains the importance procedural memory system which is responsible for the acquisition of motor and cognitive functions at an early stage. This system is mediated by frontal and basal ganglia structures with contributions from inferior parietal regions in the left hemisphere (Ijalba,

Obler and Chengappa, 2004). The lexical processing relies heavily on this system. The declarative memory system mediates semantic and episodic memory and is supported by medial and temporo-parietal structures in both left and right hemispheres. Second language learning, heavily relies on this system. Therefore later exposed languages tend to rely more on declarative memory, are often learned in more structured and less natural settings, such as school, and are more exclusively represented in the cerebral cortex than the first language. The role of what neuropsychologists call executive processes influencing inhibition may also play a role in language choice. See Ijalba, Obler and Chengappa (2004) for more detailed review of these models.

2.2. The Inhibitory Control Model

Polyglots are able to control and determine when to use each language, or when one language should be activated and the other suppressed or inhibited. The rules determining language –choice and language –switching behaviors in normal speakers allow great flexibility, from intrasentential language switches to intersential switches that may mark a shift in the language used (Ijalba, Obler and Chengappa, 2004). However, a general rule that is always observed by bilingual and polyglot speakers is that language switching takes place only when the listener shares the language codes being used. This ability to switch language codes only when contextually appropriate is a skill that has been reported in bilingual speakers as young as two years old (Leopold, 1949).

There are three main components in bilingual and polyglot speaker's language systems that must be accounted for regulation in the use of two or more languages, according to Green (1986).Any language system must first address the issue of control. How are normal bilingual speakers with access to both language systems able to control which language they produce? If a bilingual speaker were not able to effectively separate language systems, code switching would violate many of the rules observed by normal bilinguals in alternating between languages. Successful control of the language systems primarily implies the avoidance of error, such as blending two words in one language or across languages. In normal speakers, failure to exercise full control over an intact language system does occur and may be due to a variety of reasons, such as temporary distraction, stress, fatigue, or the influence of toxic substances. In the aphasic speaker .however brain damage itself provides stressful situation that may bring about failure to exercise full control over two or more language systems, even in otherwise non-stressful conditions.

The second component in the language system of bilingual and polyglot speakers involves the issue of activation determined by the internal representation of words from the known languages. Word frequency may be influential in determining activation or inhibition at both semantic and phonological levels of word production. More frequent words will generate larger activation patterns in lexical access than words that are less frequent or familiar. Finally, the third component described by Green involves the issue of resources linked to processes of control and regulation. Language representation systems in polyglot speakers must possess excitatory and inhibitory resources in the selection or deselect ion of languages when speaking.

Green (1998) explains that there are multiple levels of control in the language processing of bilingual speakers. Accordingly, he elaborated his inhibitory control (IC) model by specifying one level of control involving language task schemas that compete to control output; the locus of word selection is the lemma level and involves the use of language tags; and control at the lemma level is inhibitory and reactive. In these papers Green delineates a functional-control circuit with three basic loci of control: an executive locus that is in charge of supervisory attention and is critical in establishing and maintaining the speaker's language goals; a locus at the level of language-task schemas that the speaker selects; and a locus within the bilingual lexicosemantic system itself where lemmas are established.

The bilingual speaker possesses a conceptualizer that builds conceptual representations from long-term memory and is driven by a goal to communicate through language. The bilingual speaker possesses a conceptualizer that builds conceptual representations from long term memory and is driven by a goal to communicate through language. The speaker's communicative and planning intention is mediated by a supervisory attentional system (SAS) and components of the language system itself, such as lexico-semantic system and language task schemas. These language task schemas (e.g. translation or word production schemas) are in competition to control output (production) from the lexico-semantic system. The speaker's selection of a word is determined through communication between the SAS and the task schemas and between the SAS and the conceptualizer. Another important aspect of this system is that once a speaker specifies a task goal in order to avoid competing task schemas becoming active and producing error behavior.

The bilingual speaker must therefore first establish language task schemas to determine when to speak in one language or when to translate between languages. These task schemas determine access to the lexico-semantic system within each language when selecting responses in one language or another. Therefore the lexico-semantic system and a set of language task schemas (such as translation or word production) both compete to control output from the lexico-semantic system.

Language task schemas regulate responses from the lexico-semantic system by altering the activation levels of representation within the system and by inhibiting responses in the language that is not needed. In other words, when the speaker is responding in one of the languages, other languages are inhibited and cannot interfere by also introducing responses in the output. Cases of paradoxical translation in bilingual aphasics are a good example of how language task schemas can be disrupted. In cases of paradoxical translation the patient translates more efficiently from L1 to L2, where as normal bilinguals typically translate best when going from the less proficient to the more proficient language, usually L2 to L1.A failure to inhibit one language over another can be explained as a breakdown in the supervisory attentional system and in the control of language task schemas.

Code switching between languages in bilingual and polyglot speakers becomes a pathological behavior when it is inappropriately used within a context where speakers do not share both language codes. Increased use of switching between languages by aphasics may also denote linguistic deficits in both languages and the need to rely on all available language codes to communicate more effectively. Munoz, Marquardt, and Copeland (1999) compared discourse samples from four aphasic and four neurologically normal Hispanic bilinguals in monolingual-English, monolingual-Spanish, and bilingual contexts, finding consistent matching of the language context by both the aphasic and normal subjects. The aphasic subjects demonstrated increased code switching not evident in the speech samples of the normal subjects, indicating increased dependence on both languages for communication subsequent to neurological impairment. A comparison of code switching in Malayalam-

English and Kannada –English bilingual aphasics with normal's has also revealed that there is only increase in the quantity of code switching and no difference in type of switches (Bhat and Chengappa,2003;Chengappa and Krupa,2003).

Language choice is a third area of phenomena that the IC model can account for. Cases of alternating antagonism refer to a patient's alternating ability to control which language to use. Thus, the patient may be able to communicate in one language but not in another on some days and alternate to only using another language at a different time. Several of the phenomena observed in bilingual aphasia (namely, pathological switching between languages and translation disorders) are consistent with IC model that Green has developed (Ijalba, Obler and Chengappa, 2004).These models however, need to be verified in more number of studies both in the western and Indian contexts.

3. EVIDENCE FOR THE NEURAL ORGANIZATION

Neurological constraints of bilingualism can be explained on the basis of three important evidence available to date. For a detailed review of these aspects, refer to Chengappa and Bhat (2004). The first is the evidence available from the studies on cortical maturation, critical period for language acquisition and the cortical Lateralization for language. The second is the realization that levels or components of language are differentially distributed in the brain structurally and functionally and that there is to an amount of modality specificity. The third is the evidence from the brain damaged individuals, that there is little difference in the representation of languages in brain but with a few but extremely interesting exceptions (Whitaker 1978). The former two lines of evidence concern the normative evidence while the latter focuses on brain damaged.

3.1. Cortical Representation in Bilingualism

There have been numerous studies which suggest that bilingual's two languages are represented bilaterally, one language in each hemisphere, and some studies that suggest that one language is represented in the left and the other bilaterally (Albert andObler,1978).

The notion that each language is stored in a different location in brain evoked a wide controversy. Potzl (1925) denied the existence of a separate centre for each language. Minkowski (1927) insisted that different languages were represented in a common area of the cortex, Veyrac (1931) and Ombredane (1951) accepted that there was no special locus in the brain devoted to foreign languages.

Penfield (1965) attributed the lack of confusion between the two languages in a bilingual's language behaviour to a switch mechanism that enables the bilingual to shift easily from one language to the other without confusion, without translation and without a mother tongue accent.

Potzl (1925) had hypothetically located the switch mechanism in the Inferior parietal region of the dominant hemisphere. It was reasoned that, if this part of the brain was injured, the bilingual could no longer shift freely from one language to the other; the bilingual either perseverated in either of the two languages or got them confused.

Penfield (1953) and Penfield and Roberts (1959) categorically denied the existence of separate neural mechanism for each language of a bilingual. They also maintained that there was a bilateral representation, that early bilingualism enlarged the cortical areas devoted to language and that the brain mechanism was the same whether one, two or many languages were learned.

A couple of studies asserted that there was no evidence for anatomical separation of languages within the brain (Minkowski 1963, Gloning and Gloning 1965), although Lebrun (1971) still held the opinion that different languages or languages acquired differently tended to be subserved by cerebral structures that were to some extent different. Paradis (1977) denies the need for postulating a localized switch mechanism. In support of this view, Albert and Obler (1978) argued that a continuously operating monitor system utilizing many parts of the brain controls language processing in a bilingual typical or brain damaged.

3.2. Language Subservience in the Brain Damaged

Penfield's (1953) and Gloning and Gloning's (1965) observation that language is represented more bilaterally in bilinguals than in monolinguals found support in the clinical evidence on aphasia in bilinguals. There is reported to be a higher incidence of aphasia following right sided damage in bilinguals than that in monolinguals (Ovachrova, Raichev and Geleva 1968; Albert and Obler 1978). This suggested that probably language in polyglots had a broader, more dynamic and less lateralized organization.

It is known that, typically, all the languages of a bilingual or a multilingual are similarly represented in brain as in a monolingual speaker. There appears to be no need to hypothesize any special anatomical structure or function in the brain of bilingual as differentiated from a monolingual. The bilingual does not need any different mechanism to enable him to choose one language over the other at a given time from the one he needs to speak at all (Whitaker 1978). The non-dominant hemisphere, to reiterate Paradis (1977), may play a role in the language of some bilinguals, but it is capable of doing so in monolinguals as well.

In the few interesting cases where right hemisphere dominance is suggested it could be noticed that one or more of the following situations play the lead: left handedness or ambidexterity, early damage to the left hemisphere or post pubertal acquisition of the language in question.

While none of these situations could ensure a shift in hemispheric Lateralization of one or the other language, right hemisphere dominance is always associated with these in the studies reviewed. According to Whitaker (1978), even if language dominance shifts to the right hemisphere or if there is one language in the left and the other in the right hemisphere, the same areas corresponding to those in left hemisphere subserve the language functions. This again is only speculation, and does not reduce the right hemisphere involvement in bilinguals.

Green (1986) rightly pointed out that, whichever of this hypothesis is adopted; we still need to characterize how items of each language are selected. The selection mechanism between languages need not be different from that within each language. One major controversy is whether bilinguals possess neural mechanisms of a nature different from those possessed by unilinguals (Bhat and Chengappa, 2004). Some authors assume language organization to be qualitatively different in bilinguals whereas others assume that no mechanism exists in a bilingual that is not already operative in unilinguals and that cerebral organization in bilinguals does not qualitatively differ from that in unilinguals. Bilinguals are thus viewed as making use of the same cerebral mechanisms available to unilinguals, but to different extents in order to compensate for gaps in their linguistic competence by relying more heavily on the other available systems like pragmatics (Paradis,2001 b; Ijalba, Obler, Chengappa,2004).

3.3. Functional Neuroanatomical Studies

Over the past few years studies with the advent of advanced functional neuroanatomical techniques such as Positron emission tomography (PET) and functional magnetic resonance imaging (fMRI) investigations into language representation in the brain of bilingual subjects has become easy and quite frequent. Some of the studies reviewed here indicate differential brain organization and processing in bilinguals.

The earliest neuroimagiung study on bilinguals was made by Klein, Zatorre, Milner, Meyer and Evans, (1995) and PET revealed a greater activation of the left putamen during a word generation task in the less well-known language. This was attributed to increased articulatory demands imposed by speaking a language learnt later in life. A similar study was conducted by Klein, Milner, Zatorre, Zhao and Nikelski, (1999) on English Chinese bilinguals. Inspite of the fact subjects had learned English later in life, the two languages showed activation in the same cerebral structures (left inferior frontal, dorsalateral, temporal and parietal cortices and right cerebellum). Authors did not find any difference in the activation of the two languages in the basal ganglia and their results, suggested a common macroscopic cortical representation for L1 and L2.

Using a different technique of fMRI, Illes et al. (1999) investigated brain activation during semantic judgment tasks (concrete/abstract/judgment) in eight late English-Spanish bilinguals. They also found similar results as above and concluded the existence of similar microscopic processing for semantic stimulation of L1 and L2. Fiebach, Christian, Hahne and Friederici, (2003) utilized fMRI in order to dissociate semantic from syntactic aspects of auditory sentence processing. Their findings indicated that syntactic processing in second language users relies on similar cerebral areas as those utilized by native speakers. They interpret their findings as showing that the manner in which highly proficient bilinguals process semantic information is similar regardless of which language they speak. The more proficient a speaker is in his/her L2 more the pattern of activation of activation resembles that seen for L1 processing.

Albert and Obler (1978) found in an extensive review of clinical studies, experimental evidence for differential cortical Lateralization in bilinguals. They formulated the following hypotheses for further research:

- Language organization in the brain of an average bilingual is more bilateral than in a monolingual.
- Patterns of cerebral dominance may be different for each language in the brain of bilingual.
- Differential cerebral Lateralization for each language is not random but determined by many factors like age, manner and modality of second language acquisition.

- Cerebral dominance for language in the bilingual is not a rigid, predetermined, easily predicted phenomenon. It is rather a dynamic process, subject to variation throughout life and sensitive to environmental and educational influences.

These cerebral hemispheres too sub serve languages of bi/multilinguals in a differential manner. While there is no consensus even here, the various research database suggests hints at the directions of such language processing by the cortical structures.

3.4. A Neurolinguistic Model of Bilingualism

Galloway (1981) presents a neurolinguistic model of bilingualism based on a review of studies. She includes the following observations:

1. The right hemisphere does not appear to be more involved in L2 than the L1 in the early stages of second language learning, neither during informal naturalistic acquisition or classroom L2 learning.
2. The right hemisphere may be more involved during the early stages of learning to read a different orthographic system of the L2 (e.g., English Vs. Hebrew).
3. The L2 may appear more left lateralized than the L1 in readers of an L2 and/or classroom L2 performer without communicative experience in the L2.
4. The separate languages of the advanced L2 speaker or fluent bilingual do not appear to be differentially lateralized with respect to each other.
5. Language in advanced L2 performers in an L2 environment and/or balanced bilinguals may be more bilaterally represented than language in the monolingual.
6. Balanced bilinguals may develop different hemispheric processing strategies for language as a result of age of initial exposure to the L2 (before age 6 years versus after puberty).
7. Adult L2 speakers may preferentially adopt right (gestalt, inductive) or left (sequential, deductive) cognitive styles for L2 classroom performance.
8. Mental grammars (grammatical competence) of certain, specific languages do not appear to be governed more by the right hemisphere (than the other languages).
9. Certain socio-ethnic bilingual groups may appear less left-lateralized for language.

In the following section, the third evidence is considered.

3.5. Evidence from Aphasia

Cases have been reported in which individuals after cerebral trauma exhibited different patterns of recovery of their premorbidly used languages. The ployglot patient might be limited to speech in only one of his languages, might switch languages inappropriately, or might speak in an incomprehensible jumble made up on words and idioms belonging to different languages.

For instance, deficient writing in one of the languages may be seen whereas writing in other language may be intact: reading ability in one language may be lost whereas in the other languages it may be retained; oral comprehension of one language may be preserved as against the comprehension in the other, etc.

Penfield's (1965) hypothesized that, in early bilinguals, language tended to be represented bilaterally was favored by many. Gloning and Gloning (1965) observed that right brain damage caused aphasia more frequently in polyglots than in monoglots. Moreover they observed that the polyglotic linguistic behavior was not always associated with the lesion of the lower parietal lobe thus denying the existence of a bilingual switch mechanism.

3.6. Language Impairment and Recovery

Typically in a balanced bilingual aphasic, both the languages are equally affected if the languages involved allow a linguistic comparison. That is to say that if the variables mentioned (to be elaborated later) allow a comparison, then the extent of language functioning affected will be equal in both the languages.

This probably points to a better establishment of the notion that a bilingual's cortical organization is no different from that of a monolingual. But there are some significant instances, i.e., the case reports which deviate from the usual, where the languages involved are differentially impaired and are selectively recovered and this form an extremely interesting part of the study.

These instances occur in a significant minority and suggest the possibility of involvement of other cortical structures either in the same hemisphere or in the right hemisphere.

Various theories of aphasia in bilinguals and polyglots have been put forward. These theories deal mainly with the impairment and recovery of languages after cerebral insult. In the following section only a few theories that have been recently tested and interpreted are mentioned.

Galloway (1978) tested the following two theories in the case of a heptalingual who presented a differential linguistic picture after a massive cerebrovascular accident: Pitres (1895) suggested a theory that language most frequently and extensively used prior to cerebral insult would recover first. Ribot's (1906) theory suggested that the earliest acquired language(s) would be most resistant to aphasic impairment and that languages acquired later would be most impaired. Galloway found that in the case exemplified, Pitres theory was applicable and that the language that was extensively used prior to the insult was the one that was recovered first.

Since the measures of language disability could also be measures of language recovery to a certain extent especially in the immediate postmorbid period, the types of recovery or the patterns of recovery form an interesting area of research in a bilingual. Paradis (1977) identified five common modes of restitution.

1. *Synergistic*: Recovery in all the languages of a multilingual is simultaneous, this kind of restitution may be of two types.
   Parallel: All languages are similarly impaired and similarly recovered to a similar degree.

Differential: The languages are differently impaired; recovery occurs either at the same or at a different rate. The degree of recovery may be at the same rate for all languages.

2. *Antagonistic:* One language recovers to a certain extent first and it starts regressing when the other language begins to recover.

3. *Successive:* One language begins to recover after the other one has been recovered almost fully.

4. *Selective:* One or more languages do not recover at all or remain severely impaired.

5. *Mixed:* A mixed language or mutual interference between the languages is seen in the process of recovery.

Of these types, the synergistic recovery pattern seems to be the more common, and of which the parallel pattern is most common. Antagonistic pattern of recovery is seen to be least common. Successive recovery can also be seen in cases of selective recovery except that some languages do not recover at all. Whitaker (1978) classifies successive recovery as a type of synergistic differential group, the difference being apparently in the temporal order. The two types of synergistic recovery and the successive recovery can be explained along similar lines of natural recovery in a monolingual aphasic which includes type, extent and location of bilingual acquisition.

3.7. Types of Bilingualism and Aphasia Recovery

Various types of bilingualism have been identified and discussed since long. Among these, the coordinate, compound and subordinate trichotomy has been the most discussed (Weinreich 1953), especially so in the context of bilingual aphasics and their neurocognition.

Weinreich (1953) maintained that a bilingual need not be completely coordinate one or a completely compound one. The bilingual's two languages may stand in a variety of relationship with each other, ranging from an idealized perfect coordinateness to total compoundness or subordinateness with real individuals situated somewhere on the continuum, sharing in various degrees, parts of their linguistic systems with all three types.

Not so in contrast with these findings, Albert and other (1978) thought it was more likely that certain systems (e.g. voice production, phonological perception, deep semantics) would be compound for all bilinguals, while other systems (e.g. lexicon and syntax) would be coordinate to a greater or lesser extent. It is useful to retain the theoretical distinction between coordinate, compound and subordinate bilingualism and it should be possible to ascertain the extent to which and possibly the areas for which each bilingual is coordinate, compound or subordinate, by experimental investigations.

Jakobovits (1968) introduced the concept of degree of bilingualism into the coordinate-compound distinction, the proficiency level which is more a function of the amount of competence in each language than the type of bilingualism.

Lambert and Fillenbaum (1959) stated that coordinate bilingualism would posit more functionally separated neural structures underlying the two languages than would compound bilingualism. This suggested that languages which are learnt differently may be to some extent subserved by different neural substrates. Accordingly, after cerebral insult, languages which have been acquired in a compound way should be equally affected in aphasia, while languages which have been acquired in a coordinate way should be selectively impaired (Lebrun 1971).

According to Paradis's (1977) types of restitution, it was supposed that parallel recovery or equal restitution of both the languages in bilingual aphasics, proportionate to the degree of proficiency in each language, was consistent with the notion of compound bilingualism as well as coordinate bilinguals. On the first line of reasoning, if the basic linguistic rules are affected the two lexical systems are also affected and on the second line of reasoning, there is equal disturbance because of the supposed single anatomic representation for both the languages (Whitaker 1978). In the same frame posited, the cases of differential recovery and selective recovery would support the notion of coordinate bilingualism (Wald 1974). The subordinate type is difficult to be explained on the type and amount of available evidence. There is no way for that matter, at present to prove (or disprove) any of these theories except by way of large collection of case studies (Chengappa, 2001).

3.8. Interference and Bilingual Aphasia

Mutual inter language interference is a common phenomenon observed in normal bilinguals, especially in those who are yet to master the second language (Chengappa,2009). Interference is the use of features belonging to one language while speaking or writing another (Mackey 1970). The characteristics of degree of proficiency, function and alternation between languages are seminal to the type and amount of interference.

Weinreich (1953) observed that language mixing or interference occurred mainly at the semantic level and it was proposed to be the distinctive factor in the three types of bilingualism. Interference from first language to the second language was reported by many, due to bilingual imbalance in exposure and use. Thirumalai and Chengappa (1986) found that early language proficiency and mastery were crucial in interference with the less proficient demonstrating more interference than the proficient bilinguals.

Paradis (1978) reaffirmed that the interference occurred at the various strata along lines of stratificational grammar, viz., conceptual (concepts and their relationships), lexemic (grammatical elements like subject, verb, etc.), morphemic (radical, affixes, etc.), phonemic (phonemic elements like stress, clusters, etc.), phonetic (voicing, nasalisation, etc.) stratum.

Aphasiologists have observed that mixing or interference is a feature of bilingual aphasia as well. Selective mixing or compounding at various levels of language has been cited from the available literature by Paradis (1978). The aphasia in polyglots is reported to have presented language mixing at the levels of words or lexemes, phonemic level, prosody characteristics, graphemic, morphemic and semantic levels. Language mixing after cerebral insult in a bilingual is observed to be different from the picture presented by a normal bilingual. Bhat and Chengappa (2004) found quantitative differences only in code mixing and switching seen in bilingual aphasics compared to the normal bilinguals. However, this has not been elaborated upon in the scanty literature available on the topic (Chengappa, Bhat and Padakannaya, 2004).

3.9. Factors Influencing Language Disability and Recovery Patterns in Bilingual Aphasics

Several factors may contribute to the language disability and recovery pattern in bilingual aphasics (Galloway, 1981)

   i.   Cerebral dominance for language
   ii. Locus of the lesion
   iii. Extent and severity of the lesion
   iv. Learning contexts (manner of acquisition)
   v.   Age of language acquisition and the age of cerebral insult
   vi. Order of acquisition of the two languages
   vii. Primacy (mother tongue or not) of the language
   viii. Structural and functional differences between the languages
   ix. Visual element of the languages (as to whether it could be read and/or written)
   x.   Degree of premorbid proficiency in each language
   xi. Premorbid frequency of use of each language
   xii. Premorbid affective attitudes towards each language
   xiii. Appropriateness (to the patient) of a language
   xiv. Language processing styles
   xv. The language of the hospital staff
   xvi. The language employed in speech therapy

It may be noted that factors like age and manner of acquisition of language and their proficiency are crucial to typical as well as atypical bilingual performance (Bhat and Chengappa, 2005). Paradis (1977) found no consistent correlation among some of these factors and the patterns of restitution, thus convincing the obvious inadequacy of the present day research.

Several studies have tried to interpret the different recovery patterns observed in multilingual aphasics (Paradis 1977, 2001). The main issues at stake are the reasons why recovery patterns differ so much across patients and why a language recovers better than the other(s).

Pitres (1895) and other neurologists in the past century claimed that failure of a language to recover during the intermediate and late phases was not due to its loss, but rather to pathophysiological inhibitory effects caused by the lesion. Pitres had drawn this conclusion on the basis of a general assumption and some empirical studies. The general assumption was supported by several neurologists and presupposed that all languages of a bilingual or a polyglot subject were localized in common language areas (Paradis, 1978, 2001).

In conclusion, our knowledge of the organization of the bilingual brain is certainly more extensive as compared to the first monograph published by Pitres more than 100 years ago. However, many aspects of the organization of the bilingual brain are still unknown and further clinical research based on an appropriate methodology and finer experimental studies, at both the neuroanatomical and the physiological levels, will hopefully lead to more advanced understanding of language organization in the multilingual brain.

Grosjean (1998) identifies methodological and conceptual issues in studying bilinguals, first explaining the issues, namely bilingual participants, language mode, stimuli, tasks, and the models of bilingual processing, then discussing the problems caused by these issues and finally proposing tentative solutions to those problems.

It is imperative that we design our studies in such a way that above mentioned variables be controlled to the maximum possible extent to generate reliable results.

While this section has largely dealt with the processing of primary language skills i.e. speaking and listening, it has its repercussions on the acquisition of literacy too.

4. LANGUAGE PROCESSING IN BILITERATES

One of the issues debated in the literature on bilinguals and biliterates is the influence of one language over the other during the period of acquisition of biliteracy. While some children acquire two or more languages simultaneously in the same environment and later on learn other languages in different environments such as in schools, it is argued that the influence of L2 can lead to cognitive disadvantage for a child learning literacy skills in L1. This cognitive disadvantage is attributed by a few to the acquisition of second language that has writing system different from that of the first language. Empirical studies to investigate the effect of language and script structure on the processing mechanisms during acquisition of literacy are quite sparse to date. An attempt is made in the following section to present the available research findings and discuss its relevance from the perspective of the contribution of the structure of language(s) and script(s) to cognitive and literacy skill. Suggestions are proposed for incorporating empirical findings in making policy decisions in the educational sector and teaching practices to facilitate bi/multiliteracy in a multilingual country such as India.

4.1. Literacy and Biliteracy

Literacy is traditionally defined as the ability to read and write. A child develops literacy skills with the help of meaning of a given language. It often begins early, long before children encounter any formal school instruction in reading and writing, the stage generally termed as emergent literacy or early literacy. While reading is defined as the ability to obtain meaning from print (Heath, 1980), writing is the ability of an individual to use print to communicate with others. Multiple skills at the cognitive as well as linguistic levels are needed to be a literate as literacy involves more than mere learning the alphabet, learning how to form letters and spell words, and/or learning how to decode print that are typically taught in elementary school. An individual can be literate in one language or in more than one language. The term biliteracy is used to describe an individual's competencies in two written languages, developed at varying degrees, either simultaneously or successively (Dworin, 2003).When an individual gains the mastery of the fundamentals of speaking, reading and writing in two linguistic systems, he/she is considered to be a biliterate (Reyes, 2001). The unique cognitive and language skills in biliterate/multiliterate children has fascinated many researchers to explore and understand the underlying processing mechanisms involved in the acquisition of biliteracy/multiliteracy.

The writing system that a language uses affects children's acquisition of literacy because each system is based on a different set of symbolic relations and requires different cognitive skills (Coulmas, 1989). Recent studies have pointed out that the structural characteristics of different languages and their writing systems influence the relationships between orthography, phonology, morphology and meaning in processing written language. A few factors that are investigated in bilingual-biliterate children as against monolingual-monoliterate children are phonological processing and phonological awareness skills in monolingual-monoliterates, phonological awareness skills in different languages and scripts in bilingual-biliterates, script specific issues in relation to cognitive processing, structural differences in language and its influence on acquisition of literacy. Empirical studies are very few in number on the above owing to the fact that the documented reports are from only those countries where bi/multilingualism, bi/multiliteracy is most prevalent. However, these reports open a vista of interesting findings to suggest that cognitive mechanism for processing different languages and scripts are different from that reported for monolingual-monoliterates. A few empirical studies on the above issues and its relevance for cognitive processing in biliterates as against monoliterates are discussed in the following section.

4.2. Language Structure vs. Phonological Processing in Biliterates

Processing of language for literacy purposes is said to depend on the finer constituents of language such as the phonemes. Therefore, ability of an individual to be sensitive to and have awareness of the phonological constituents of a given language has been extensively studied with reference to biliteracy. Since phonemes are the smallest units of speech sounds, ability to be sensitive to, to be aware of and to manipulate these small units requires not only linguistic knowledge but also cognitive skills such as awareness, identification, discrimination, manipulation and such other skills to process these units. Phonological processing, phonological sensitivity, phonological naming and phonological memory are a few such skills addressed in research in the context of literacy and biliteracy (Wagner and Torgesen, 1987).

While *phonological processing* refers to the activities that require sensitivity to, manipulation of, or use of the sounds in words, *phonological awareness* is the conscious awareness of the sound structure of language and the ability to manipulate phonological segments in speech. The term phonological awareness is therefore viewed under the broader domain of phonological processing abilities. *Phonological sensitivity* on the other hand refers to the ability of a child to detect and manipulate the sound structure of oral language. For example, ability to identify words that rhyme, blend spoken syllables or phonemes together to form a word, delete syllables or phonemes from spoken words to form a new word, or count the number of phonemes in a spoken word. Progression from sensitivity to large and concrete units of sound (words and syllables) to subsyllabic units of onset (initial consonant or consonant cluster in a syllable) and rime (vowel and final consonant or consonant cluster in a syllable) to small and abstract units of speech sounds (phonemes) promotes the development of decoding skills in alphabetic languages since the graphemes in its written language correspond to speech sounds at the level of phonemes. Efficiency in *Phonological memory* (short term memory for sound-based information), *Phonological naming* (efficiency of retrieval of phonological information from permanent memory), *Rapid verbal naming skills* (the speed with which an individual performs on a phonological task) enables children to maintain an accurate representation of the phonemes associated with the letters of a word while decoding and therefore, devote more cognitive resources to decoding and comprehension processes. The pattern and the direction with which children progress during the development of phonological processing skills is said to be a reflection of the linguistic and orthographic structure of the language(s) that the child is learning to read and write. Alternatively stated, a child may follow a bottom-up or a top-down approach to process a given language and script depending on the nature of these two during the course of learning to be literate (Shanbal, in progress).

Given the wealth of literature on phonological awareness in alphabetic orthography such as English, there was a general notion that the phonological principles that apply to alphabetic orthography apply to other orthographies as well (such as logographic, syllabaries and semi-syllabaries). Apart from awareness to the phonological constituents of speech sounds, the characteristics of a writing system are also found to influence a child's sensitivity to phonological processing. As a result, in non-alphabetic languages, the progression from phonological sensitivity to phoneme awareness as described earlier is questioned. Therefore, influence of the nature of orthography (script specific features per se) has been the focus of investigation particularly in the interest of understanding the underlying mechanisms of phonological sensitivity in children learning two different types of scripts. Studies on non-alphabetic orthographies such as Chinese, Japanese and Indian languages (Rao, Shanbal and Khurana, 2010) have suggested that the phonological awareness that is so crucial for acquisition of reading in alphabetic orthographies is not so crucial for non-alphabetic orthographies. On the contrary, it is reported that the specific phonological skills develop as a consequence of exposure to the alphabetic scripts in the course of learning to read English (alphabetic script) in schools provided one of the languages of the biliterate child happens to be English or any other language with alphabetic script.

The above premise has paved way to cross-linguistic literature on contrastive phonology which highlights that there exists distinct differences in the phonological features of languages, the most important being the temporal (timing) and the frequency composition of speech sounds. Accordingly, the phonological properties of speech sounds that are assessed in the continuum of phonological process measures need to be examined taking into consideration the features of the language in question. The inherent differences in the phonological properties of speech sounds of different languages suggests that the measures that are employed for one language (say, English) cannot be directly applied to another language (say, Hindi, an Indian language). Given the innumerable languages of the world, there is a need to devise measures of phonological processing that cut across the barriers of language specific features in order to facilitate uniformity across research. Such an attempt would help in making meaningful comparisons of reports of several studies in this direction. Despite these limitations, the studies have suggested that the acquisition of literacy depends on the processing abilities of children which are bound to be different in monolingual children compared to bilingual children. These differences are also attributed to the orthographic depth hypothesis (Katz and Feldman, 1983) i.e., transparency of letter to phonology correspondence that is reported to play a major role in deciphering the script.

Support for this premise may be derived from a few Indian studies that examined the phonological processing abilities of bilingual-biliterate children. Given the fact that the majority of children in India are, by virtue of their linguistic and cultural diversity bilingual or multilingual before their entry into school, a spate of research on acquisition of literacy in a few Indian languages and scripts are reported in the 1990's (Karanth, 1992; Karanth and Prakash, 1996; Prema, 1998; Prema, 2006; Prema et al., 2010 among others). Children who learn literacy in shallow as well as deep orthography (for example, Indian children who learn two different types of scripts - English which is alphabetic; Indian languages that are syllabic/semi-syllabic), are likely to use differential processing strategies in learning to read besides exercising differential cognitive skills in order to be good readers. Since children in India are compelled to be literate in two or three languages with differing scripts, it is very interesting to examine the differential skills that either facilitate or interfere in the acquisition of biliteracy.

Gokani (1992) studied phonological awareness in relation to orthographic factors in reading in children from both English medium and Gujarati medium. She reported a significant difference in phoneme deletion task between English and Gujarati medium children in favor of English medium children. Savithri, Prema and Shilpashree (2004) investigated ability of monolingual (Kannada) and bilingual (Kannada-English, with native language being Kannada) children to identify and name time-warped phoneme/syllable. Their results indicated significant difference between Kannada-English bilingual-biliterate children studying in English medium and monolingual children from Kannada medium with the former faring better than the latter in the identification and naming of both the deleted phonemes and syllables suggesting that the ability to identify and name phonemes was better in the bilingual than in the monolingual children.

Prema (1998) examined the performance of certain linguistic operations that make use of information about the speech sound structure of a given language in Kannada speaking children studying in Kannada medium with English as a second language. She reported that phonological awareness appeared to emerge with exposure to alphabetic script in the later grades of upper primary level. A series of studies were conducted by Prema (1998-Kannada language), Akhila and Prema (2000-Tamil language), Swaroopa and Prema (2001) and Seetha and Prema (2002) (Malayalam language) to examine the influence of script specific features of alphabetic languages such as those of English on Kannada, Tamil and Malayalam – the three south Indian Dravidian languages with semi-syllabic script. Akhila and Prema (2000) reported that the development of rhyming skills in Tamil was not found to parallel with syllable deletion as seen for Kannada language (Prema, 1998). On the other hand, Swaroopa and Prema (2001), and Seetha and Prema (2002) found that rhyming and alliteration were potential indicators for adequate reading skills in Malayalam language. The results of the three studies conducted in series indicate that the literacy related factors are influenced by the underlying script system. On the basis of the three studies Prema (2006) proposed that the sub-classification of languages should take into consideration not only the linguistic features but also the script specific features.

The above studies reported that the phonemic awareness emerges as a consequence of learning alphabetic script (say, English) and not a necessary skill for literacy acquisition in semi-syllabic script. Alternatively stated, literacy in a semi-syllabic system is insufficient for the development of phonemic awareness. Hence, children undertaking bilingual education, particularly in languages with widely differing features are likely to exercise differential skills depending on the linguistic and script features because of which it is extremely important to sensitize language teachers to the possibility of differential underlying skills as well as the need to focus on training/enhancing the cognitive resources that are necessary for children learning two or three languages/scripts. This, however, calls for intensive efforts to empower teachers to meet the needs of bilingual children.

4.3. Influence of Orthographic Structure on Script Processing in Biliterates

The writing system that a language uses as mentioned earlier is found to influence children's acquisition of literacy because each system is based on a different set of symbolic relations and requires different cognitive skills (Coulmas, 1989). These relations place different demands on children's analysis of spoken language and their recording of the language in print. The task of learning to read in different writing systems is influenced by the knowledge of the language in question and also the type of writing system employed in the target language (Bialystok, Luk and Kuwan, 2005). Further, understanding these systems with different symbolic relations will need different processing mechanisms in order to acquire literacy in those languages.

Writing system employs scripts, the basic unit size for the mapping of graphic units to language units while a grapheme refers to the letter or combination of letters that represent a phoneme. Written language systems or scripts around the world are generally grouped into alphabetic scripts in which the basic unit represented by a grapheme is essentially a phoneme (in languages such as English, this relation can be one-to-many in both directions i.e., a phoneme can be realized by different graphemes and a grapheme can be realized by many different phonemes); Consonantal scripts (where not all sounds are represented. For example, vowels are not written down in Arabic and Hebrew); Syllabic scripts (in syllabic scripts the written units represent syllables for example, the Indian scripts of Hindi and Kannada); Logographic scripts where each symbol is equivalent to a morpheme (for example, Chinese and *the* Japanese script Kanji) (Shanbal, in progress).

All writing systems represent spoken language, but differences in the mappings between orthography, phonology and semantics give rise to small differences in their transparency (Goswami, Ziegler, Dalton and Schneider, 2001; Wimmer and Hummer, 1990). Ziegler and Goswami (2005) proposed the psycholingusitic grain size theory which suggests that differences in reading accuracy across languages reflect fundamental differences in the phonology of the languages and the reading strategies that are developed in response to orthography of that particular language. The orthographically consistent languages may rely more on grapheme-phoneme recoding strategies because grapheme-phoneme correspondences are relatively consistent, whereas in English, children cannot use smaller grain sizes as easily because inconsistency is found to be much higher for smaller grapheme units than for larger units (Treiman et al., 1995). As a result, for English, they need to use a variety of strategies supplementing grapheme-phoneme conversion strategies which can aid them in reading. Ziegler and Goswami (2005) believed that languages vary in the consistency with which phonology is represented in their orthography. According to them, the degree of inconsistency between languages and orthographic units can result in developmental differences in the grain size of lexical representations and accompanying differences in development of reading strategies in children across orthographies. Finally, difficulty at the *granularity level* reflects that there are many more orthographic units to learn when access to the phonological system is based on bigger grain sizes as opposed to smaller grain sizes. That is, there are more words than there are syllables, more syllables than there are rimes, more rimes than there are graphemes, and more graphemes than there are letters.

The differences in the nature of orthography and the script structure have led researchers to propose theories to explain reading and writing to understand the processing mechanisms of different aspects of literacy in bilingual-biliterate children. Geva and Wang (2001) propose that the cognitive processes that underlie development of literacy in first language also apply to the development of literacy in an individual's second language. Geva and colleagues (Geva and Siegel, 2000; Gholamain and Geva, 1999) proposed that the main theoretical positions can be reduced to two competing perspectives as the script dependent hypothesis and central processing hypothesis. The script dependent hypothesis posits that reading development should vary with the transparency of a particular orthography. While the transparent orthographies permit a simple direct one-to-one correspondence between letters and sounds of words, the less transparent orthographies, however, use more complex relationships between letters and sounds. Therefore, accurate word recognition skills develop more slowly in less transparent orthographies than they do in more transparent orthographies. The central processing hypothesis, on the other hand, assumes a universal approach to literacy acquisition. It proposes that reading development is not contingent upon the type and the nature of the orthography. Rather, common underlying linguistic and cognitive processes (such as working memory, verbal ability, naming and phonological skills) influence the development of reading across all languages. However, considering that the languages differ in terms of the regularity between written symbols (letters/graphemes) and sounds/phoneme i.e., the transparency of orthography is different for different languages, an individual may require specific cognitive processes for learning the scripts of different languages. In a nutshell, it can be said that the central processing hypothesis does embrace the script dependent hypothesis to explain the underlying cognitive processes that are required to learn to be literate in a given language/script. There is, however, an equivocal thought on the influence of one language/script over the other in a bilingual-biliterate child learning to read and write two different types of languages and scripts. As a result, the issue of cross-language transfer of literacy skills has also been investigated in this special population.

4.4. Cross Language Transfer of Literacy Skills in Biliterate Children

In the recent years there is a considerable amount of research on bilingual children developing biliteracy skills suggesting cross-language transfer of some of the phonological skills (Cossu, Shankweiler, Liberman, Katz and Tola, 1988 in Hispanic-English), (Stuart-Smith and Martin, 1999 in Punjabi-English; Geva and Wang, 2001 in Turkish-English; Shanbal and Prema i2006a; 2006b; 2006c in Indian languages) as well as positive transfer of literacy skills across languages (e.g., Bialystok, Luk and Kwan, 2005; Geva and Siegel, 2000; Geva, Wade-Woolley, and Shany, 1997; Oller and Eilers, 2002).

Investigations on bilingual-biliterate children to examine the issue of cross-language transfer of literacy skills in India with wide linguistic and cultural diversity are, however, very sparse. Shanbal and Prema conducted a series of studies (2006a; 2006b; 2006c; in press) in which, they investigated cross-language transfer of phonological skills in bilingual-biliterate children in three South Indian languages [Tamil-Kannada (Ta-K), Telugu-Kannada (Te-K) and Malayalam-Kannada (M-K) and Kannada-Kannada (K-K) group]. The results revealed a significant difference in the performance of children between the M-K group and K-K group but no significant difference between the Ta-K and K-K group or Te-K and K-K group. The results suggest better cross-language transfer between the languages close to each other than those that are distant with reference to language/script structure (Prema, 2006). Shanbal and Prema (2006c) further investigated the effect of cross-language transfer of semantic skills across cognate words on two groups of school going children (Kannada monolingual and Telugu-Kannada bilingual). The results revealed that while both the groups showed similar performance in accuracy on the morpho-syntactic task, the bilingual group was found to take a longer time compared to the monolingual group in learning to read their non-native language suggesting the likelihood of the process of cross-language transfer from native language to non-native language leading to delay in response latency. In a subsequent study (Shanbal and Prema, 2007) the relationship between phonological awareness and word reading in children learning to be literate in Kannada (semi syllabic) and English (alphabetic) was investigated. The results revealed cross-language facilitation with phonological awareness in Kannada (L1) contributing to word reading in English (L2). On the contrary, there was also cross-language facilitation from phonological awareness in English to word and non-word reading in Kannada. This suggests that there is a reciprocal relationship for cross language facilitation of phonological awareness in L1 to reading in L2 and phonological awareness in L2 to reading in L1 even though L2 is only the language for literacy function.

The results of the series of studies in a few Indian languages indicate that both facilitation and interference effects are evident in bilingual children learning two different languages. Despite the dissimilarity of script features, transfer of skills by way of facilitation effect seen at the phonological level is, no doubt, an advantage to bilingual children in India. However, in cognate languages, there is found to be interference effect in deciphering the meaning of words thus slowing down the processing speed. Large scale studies of this kind in pairs of school languages of India would be quite helpful for language planners and educational policy makers. Teaching practice with bilingual children can be more effective provided the subtle characteristics of languages and scripts are taken into consideration by the language teachers.

The studies reviewed so far emphasized the relation among language, script and acquisition of literacy. It is widely known that bilingual children generally have less proficiency in the non-native language and therefore, encounter difficulties in cognitive subjects such as mathematics and science. In order to plan and evolve adequate methods for educational practice with bilingual children, it is necessary to understand the underlying linguistic components in cognitive subjects. In the recent years cross-language studies and investigations on bilinguals/multilingual children have suggested that the efficiency in processing of digits and number words differ from language to language. The variations are accounted for the structural aspects of a given language(s). Literature reveals that number word processing in L1 is better than L2 in bilinguals and this is attributed to differences in cognitive functions involved in learning L1 and L2.

Studies on digits and number word processing indicate that the differences in processing are not only at the feature level but also seen at the higher level of language. For example, majority of the studies have reported differences in cognitive linguistic abilities for processing digits and number words (Ellis and Hennelly, 1980; Stigler, Lee and Stenvenson, 1986; Brysbaert and Dhondt, 1991; Neath, 1998; Bernado, 2001). These studies suggest that there are differences in the processing at the semantic level between the monolingual and the bilingual children. Such differences have their impact on learning the cognitive subjects such as science and mathematics. A few small scale studies employing digits and number words examined reaction time (Dheepa, Sreedevi and Prema, 2005), digit span, (Anjali, Savitha and Prema, 2006; Anitha, Anjali, Savitha and Prema, 2007) and semantic aspects in statement problems (Sumitha, Shwetha and Prema, 2005) involved in processing by Kannada-English bilingual children.

In the above studies the Kannada-English bilingual children showed better span for digits compared to number words but longer reaction time. However, there was a significant difference in the reaction time for processing digits and number words in each of the languages i.e., Kannada and English with the performance of bilingual children being better for number words in English than for Kannada number words. Although the bilinguals are second language learners of English, they performed better on number words in English (L2). This may be attributed to the linguistic complexity of Kannada language. Since some of the Kannada number words have longer syllable length than English (number 'nine' in English is a monosyllabic word, while /ombattu/ in Kannada meaning 'nine' is a trisyllabic word), the processing of number words by way of language dependent route takes a few milliseconds more than the processing of digits by way of direct route (Campbell, 1994; Dheepa, Sreedevi and Prema, 2005; Frenck-Mestre and Vaid, 1993; Marsh and Maki, 1976; McClain and Huang, 1982). The Kannada number words appear to have imposed a greater load on working memory (Baddeley and Hitch, 1974; Baddeley, 2000; Anjali, Savitha and Prema, 2006). Bilingual children who use more than one verbal code for number processing are likely to have a differential load on their working memory. While the numerical notation is trans-linguistic and is independent of language, the number words seem to be language dependent. The fact that the bilinguals who are second language learners of English performed better on L2 tasks i.e., on English number words suggests that the nature of language processing and the underlying skills required for processing should be taken into consideration while teaching bilingual children cognitive subjects such as mathematics.

The differential skills of monolingual and bilingual children in number word processing also have its impact on higher level mathematical problem solving such as that of statement problems (word problems). Sumitha, Prakash and Prema (2005) examined comprehension of statement problems in bilingual children (with mother tongue being one of the Indian languages) from grade IV and V studying in English medium. Statement problems were prepared using homophonous words and non-homophonous words keeping as reference the curriculum of language and mathematics of IV and V grade. The selected words were incorporated in the statement problems. (For example, 'some' and 'sum'; 'tense' and 'tens). The bilingual children fared poorly on the homophonous words compared to non-homophonous words. The bilingual children found it difficult to extract the right sense of the homophonous words suggesting a high possibility of interaction between the language vocabulary and math vocabulary an important factor to be considered by both language teachers and teachers of cognitive subjects.

The studies on the phonological awareness and phonological processing, cross language transfer of phonological and semantic skills, processing of numbers and digits by monolingual and bilingual-biliterate children suggest that it is necessary to give due considerations to the structural differences of the language(s) of instruction as well as the proficiency of children in L2 in designing teaching strategies for classroom practice with bilingual children. More studies in this direction with other Indian languages would enlighten the educators to re-frame their teaching practices.

The issue of bilingualism and biliteracy has received much attention in the recent decades all over the world. Many monolingual countries invest a great deal in order to develop into 'bilingual' nations (Reed, 2003). India is fortunate enough to inherit language diversity in its population and therefore, sincere efforts should be made to nurture bilingualism. In the unique situation of the country where many languages and dialects exist, bilingualism and biliteracy in school set-up gets further enriched in an additive manner (when students add a second language to their intellectual tool-kit while continuing to develop conceptually and academically in their first language, Cummins, 2000) by virtue of the national education policies such as the Three Language Formula (TLF) for school education. Therefore, instead of the term 'bliteracy', it seems appropriate to embrace the term 'multiple literacies' to address most of the questions and concerns raised in this chapter with reference to the complexity of language(s), the distinctness of the script features, the medium of instruction and the national policies for education. In order to nurture bilingualism and biliteracy, there is a need to make a comprehensive analysis of the educational system, national policies and the existing educational practice with bi/multilingual children in India with a strong empirical support from research studies.

The general notion among educationists is that bilingual children will not fare well in academic subject material. But, language acquisition studies have repeatedly emphasized the cognitive advantage that is available for bilingual/multilingual children. Since cognition, language and literacy go hand-in-hand, there is bound to be advantages to bilingual/multilingual children over monolingual children in the development of cognition and literacy. While the monolingual countries try to build up bilingual community by way of literacy activities through programs such as immersion in L2, children in India are immersed in two or more languages right from their early childhood by virtue of its cultural and linguistic diversity. Therefore, such notions may not entirely apply in the Indian context.

The proficiency in languages in which literacy skills are acquired is also highly debated from the theoretical point of view. Studies carried out on predominantly monolingual children from the West state that a child must attain a certain level of proficiency in both the native and second language in order to accrue the beneficial aspects of bilingualism. On the contrary, those carried out in countries with bilingual children state a positive association between additive bilingualism and students' linguistic, cognitive, or academic growth on French-English bilinguals, Irish-English bilinguals, and Ukrainian-English bilinguals (Cummins, 1979). Some of the studies reported on Indian children substantiate that bilingualism and multilingualism offers linguistic, academic or cognitive growth. The educational implication of these findings is that the development of literacy in two or more languages entails linguistic and academic benefits.

A comparative research at the All India Institute of Speech and Hearing, Mysore, India, on children exposed predominantly to one language (monolinguals) with those exposed to multiple languages (multilingual) has revealed that multilingual children fair significantly better in their cognitive skills and comprehension of language than the monolingual children (Prema and Geetha, 2005). The differences in linguistic structures and features of speech sounds of Indian languages and English (the language that is learnt in school) invariably turns out advantageous to the children since the children exercise different types of metalinguistic and metaphonological skills while learning two or more languages. Exposing children to

English right from their preschool years and or much before, therefore, should be viewed positively for success in literacy (Prema, 2006).

In our small scale studies cited in this chapter, significant findings were observed in the performance of bi/multilingual children on constructs such as phonological processing, phonological awareness, digits and number word processing, cross-language transfer of literacy related skills and processing of cognate languages. These constructs influence literacy learning in the preschool and school years and therefore, should be given the necessary weightage at the time of evolving national policies for education as well as framing national curriculum for children. Further, it is very crucial to sensitize all the pre service and in service teachers to the subtleties and nuances of literacy related constructs so that optimal education could be imparted to the bi/multilingual children.

The bilingual instructional programs should be integrated with language instruction so that students acquire the academic content of science, math, social studies, art and other related courses. Language proficiency should be fostered by developing critical awareness for language by encouraging students to compare and contrast their languages (e.g. phonics conventions, grammar, cognates, etc.), by providing students with extensive opportunities to carry out projects investigating their own and their community's language use, practices, and assumptions. In summary, it can be said that instructions in a bilingual/multilingual program should be geared to provide adequate input on the content, the language in question besides providing opportunities to use all the languages offered in the school environment for both communication and literacy. At this juncture, large scale empirical studies on bilingual children in classrooms in the Indian context appear very crucial. Such studies would be helpful in our understanding of the processing mechanisms of bilingual children in India learning to be literate in different languages with differences in linguistic structures and script features. Research in this direction is bound to make an impact on the role of cognitive skills in the acquisition of biliteracy, a potential area of study to be considered as an initiative of cognitive science.

CONCLUSION

Language processing in bilinguals and bilterates is amazing, awe-inspiring, unique and unparalleled. The phenomenon as complex as bilingualism is widely researched upon and much written about extensively resulting in a voluminous body of literature on the topic. In spite of this, there is no single consensus on any one aspect of the issue. This is primarily because of the heterogeneity of the bilingualism and its speakers along multidimensional aspects of proficiency in bilingualism, degree of bilingualism, language levels (phonological, morphological, lexical, semantic, discourse and its use pragmatics) in different language skills of understanding, speaking, reading and writing at any given point of time and modalities (auditory, visual, gestural etc) during these comprehension and production tasks. It is promising that a great deal of research though on a smaller scale and population(s) is going on in Indian languages. More research with very specific hypotheses, tasks and strategies may be expected to throw light on the matters and issues of bilingual and biliterate processing.

ACKNOWLEDGMENTS

We are grateful to Dr. Vijayalakshmi Basavaraj, Director, All India Institute of Speech and Hearing , Manasagangothri, Mysore for her initiative and continuous encouragement for the preparation of this paper.

REFERENCES

Akhila, P., and Prema K.S. (2000). Phonological awareness and orthographic skills in Tamil speaking children. Unpublished Master's dissertation submitted to the University of Mysore, Mysore.

Albert, M., and Obler, L., (1978). *The bilingual brain: Neuropsychological and neurolinguistic aspects of bilingualism.* New York: Academic Press.

Anjali., G., Savitha., S., and Prema, K.S. (2006). Digit span: Does language play a role? Paper presented at the National Conference organized by the Indian Speech and Hearing Association (ISHA 38), 3-5 February, Ahmedabad, India.

Anitha, T., Savitha., Anjali, G., and Prema, K.S. (2007). Phonological processing in bilinguals. In 6th international symposium on bilingualism conference abstracts, May 30-June 2,University of Hamburg, pp.317-318.

Baddeley, A.D., Thomson, N., and Buchanan, M. (1975). Word length and the structure of short-term memory. *Journal of Verbal Learning and Verbal Behavior, 14,* 575–589.

Baddeley, A. D. (2000). The episodic buffer: a new component of working memory? *Trends in Cognitive Sciences, 4,* 417-423.

Baddeley, A., and Hitch, G. (1974). Working memory. In Bower, G., (ed.) *The Psychology of Learning and Motivation,* pp. 47–89. Academic Press.

Balkan, L. (1970). In Bi/Multilingualism and Issues in Management of communication disorders with emphasis on Indian perspectives: *Language in India,* vol.9: 3, August 2009.

Ben-zee., S. (1972). In Bi/Multilingualism and Issues in Management of communication disorders with emphasis on Indian perspectives: *Language in India,* vol.9: 3, August 2009.

Bernado, A.B. (2001). Asymmetric activation of number codes in bilinguals: further evidence for the encoding complex model of number processing. *Memory and Cognition, 7,* 968–978.

Bhat, S., and Chengappa, S. (2003). *Code switching in normal Hindi-English and Kannada-English bilinguals.* Paper presented in national seminar on multilingualism in India. Osmania University, Hyderabad, India.

Bhat, S., and Chengappa, S. (2004). *Code switching and Code mixing in bilingual aphasics.* Unpublished doctoral thesis, University of Mysore, Mysore, India.

Bhat., S., and Chengappa, S. (2005). Code switching in normal and aphasic Kannada-English bilinguals. Proceedings of 4th International Symposium on Bilingualism. Edited by Cohen, J., McAlister, K., Rolstad, K. and McSwan,J. pp. 306-316. Somerville, MA: Cascadilla Press.

Bialystok, E., and Hakuta, K. (1994). *In other words: The psychology of science of second language acquisition.* New York: Basic Books.

Bialystok, E., and Herman,J. (1999). In Bi/Multilingualism and Issues in Management of communication disorders with emphasis on Indian perspectives: *Language in India,* vol.9: 3, August 2009.

Bialystok, E (2001).In Bi/Multilingualism and Issues in Management of communication disorders with emphasis on Indian perspectives: *Language in India,* vol.9: 3, August 2009.

Bialystok, E, Craik, F.I, Klein, R and Viswanathan, M .(2004). In Bi/Multilingualism and Issues in Management of communication disorders with emphasis on Indian perspectives: *Language in India,* vol.9: 3, August 2009.

Bialystok, E., Luk, G., and Kuwan, E. (2005). Bilingualism, biliteracy, and learning to read: interactions among languages and writing systems. *Scientific Studies of Reading, 9,* 43-61.

Bialystok,E., Craik,F.I., and Ryan,J.(2006). In Bi/Multilingualism and Issues in Management of communication disorders with emphasis on Indian perspectives: *Language in India,* vol.9: 3, August 2009.

Bain,B. (1975). In Bi/Multilingualism and Issues in Management of communication disorders with emphasis on Indian perspectives. *Language in India,* vol.9: 3, August 2009.

Brysbaert, M., and Dhondt, A. (1991). 'The syllable-length effect in number processing is Campbell, R. N. (1994). *Korean/English Bilingual Immersion Project: Title VII* Cognition, *10,* 591-596.

Bruck,M and Genesee,F. (1993). In Bi/Multilingualism and Issues in Management of communication disorders with emphasis on Indian perspectives: *Language in India,* vol.9: 3, August 2009.

Carlson., S.M and Meltzoff,A.N. (2008). In Bi/Multilingualism and Issues in Management of communication disorders with emphasis on Indian perspectives: *Language in India,* vol.9: 3, August 2009.

Chengappa.,S.(2001).Speech Language Pathology in the multilingual context of India. *Language in India,* vol.1,3. May 2001.

Chengappa,S. and Krupa (2003). Language mixing and switching in Malayalam-English bilingual aphasics. *Asia Pacific Rehabilitation Journal, 15,* 68-76.

Chengappa, S., and Bhat, S. (2004). The neurolinguistic perspectives of language acquisition, Lateralization and bilingualism: A review. *Language in India,* vol.4: 3, March 2004.

Chengappa.,S, Bhat, S., and Padakannaya, P. (2004). Reading and writing skills in multilingual and multiliterate aphasics. *Reading and writing, 17,* 121-132.

Chengappa, S (2009). Bi/Multilingualism and Issues in Management of communication disorders with emphasis on Indian perspectives. Key note address delivered at the 7th International symposium on bilingualism, Utrecht, Netherlands2009.

Cossu, G., Shankweiler, D., Liberman, I. S., Katz, L., andTola, G. (1988). Awareness of phonological segments and reading ability in Italian children. *Applied Psycholinguistics, 9,* 1-16.

Coulmas, F. (1989) .*Writing systems of the world.* Oxford, England: Basil Blackwell.

Cummins, J., and Gulutsan, M. (1974). In Bi/Multilingualism and Issues in Management of communication disorders with emphasis on Indian perspectives: *Language in India,* vol.9: 3, August 2009.

Cummins, J. (1979). Linguistic interdependence and the educational development of bilingual children. *Review of Educational Research, 49,* 222–251.

Cummins, J. (1979). Linguistic interdependence and the educational development of bilingual children. *Review of Educational Research, 49,* 222–251.

Darcy, N.T. (1953). In Bi/Multilingualism and Issues in Management of communication disorders with emphasis on Indian perspectives: *Language in India,* vol.9: 3, August 2009.

Dheepa. D, Sreedevi, M., and Prema K.S. (2005). Digit vs. number word processing in bilinguals. Paper presented at the International Conference of South Asian Languages, Tamil Nadu, India.

Diebold, A.R. (1961). Incipient bilingualism. *Language, 37,* 97-112.

Durgunoglu, A.Y., and Roediger, H.L (1987). Neurology In Bhatia.T.K and Ritchie,W.C (eds), *Handbook of Bilingualism,* pp. 70-89:Blackwell publishing.

Dworin, J.E. (2003). Insights into biliteracy development: Toward a bidirectional theory of bilingual pedagogy. *Journal of Hispanic Higher Education, 2,* 171–186.

Edwards, D., and Christophersen, H. (1988). In Bi/Multilingualism and Issues in Management of communication disorders with emphasis on Indian perspectives. *Language in India,* vol.9: 3, August 2009.

Ellis, N.C., and Hennelly, R.A. (1980). A Bilingual word-length effect: Implications for intelligence testing and the relative ease of mental calculations in Welsh and English, *British Journal of Psychology, 71,* 43–52.

Fiebach, S.A., Chriatian, J., Hahm, S., and Friedrici, A. (2003). *Syntax Vs semantics. A neuroanatomical distinction in bilingual brain rieschemeyer.* Paper presented in 4th International symposium on bilingualism. Arizona State University, Arizona, USA.

Galambos ,L. (1982). In Bi/Multilingualism and Issues in Management of communication disorders with emphasis on Indian perspectives: *Language in India,* vol.9: 3, August 2009.

Galloway., L. (1978). Cerebral organization in bilingualism and second language. Paper presented at the third Los Angeles Second Language Research Forum, USC.

Galloway., L (1981).*Contribution of the right cerebral hemisphere to language and communication: Issues in cerebral dominance with special emphasis on bilingualism, second language acquisition, sex differences and certain ethnic groups.* Unpublished Ph.D. dissertation. Department of Linguistics, UCLA.

Geva, E., and Wang, M. (2001). The development of basic reading skills in children: A cross language perspective. *Annual Review of Applied Linguistics, 21,* 182-204.

Geva, E., and Siegel, L. (2000). Orthographic and cognitive factors in the concurrent development of basic reading skills in two languages. *Reading and Writing, 12,* 1–30.

Geva, E., Wade-Woolley, L., and Shany, M. (1997). Development of reading efficiency in first and second language. *Scientific Studies of Reading, 1,* 119–144.

Gloning, I., and Gloning, K. (1965). Aphasien bei polyglotten. Beitrag zur dynamik des sprachabbaus sowie zur lokalisationsfrage dieser storungen. *Wiener Zeitschrift fur Nervenheilkunde, 22,* 362-397.

Gokani, V.P. (1992). The orthographic factor in phonological awareness with relation to reading. In M.Jayaram and S.R.Savithri (eds.). Research at AIISH, Dissertation abstracts: Volume 3, D 236, (47-48), AIISH, Mysore, India.

Goswami, U., Ziegler, J.C., Dalton, L. and Schneider, W. (2001). Pseudohomophone effects and phonological recoding procedures in reading development in English and German.*Journal of Memory and Language*, *45*, 648–664.

Green, D.W. (1986). Neurology. In Bhatia.T.K and Ritchie,W.C (eds), *Handbook of Bilingualism*, pp. 70-89: Blackwell publishing.

Green, D.W. (1998). Mental control of the bilingual lexico-semantic system. *Bilingualism: Language and cognition, 1*, 67-81.

Grosjean, F. (1989). Neurolinguists, beware! The bilingual is not two monolinguals in one person. *Brain and Language, 36*, 3-15.

Grosjean, F. (1997). The bilingual individual. *Interpreting: International Journal of Research and Practice in Interpreting, 2*, 163-191.

Grosjean, F.(1998). In Bi/Multilingualism and Issues in Management of communication disorders with emphasis on Indian perspectives: Online Journal www.languageinindia. com, vol.9: 3, August 2009.

Hakuta, K. (1986). Cognitive development of bilingual children. Centre for Language Education and Research Educational Report Series, No.3.UCLA.

Hamers, J.F.(1981). Psychological approaches to the development of bilinguality. In H. Baetens Beardsmore (Ed.) *Elements of bilingual theory* (pp.120-135).Free university of Brussels.

Harding-Esch, E., and Riley, P. (Eds.) (1986). *The Bilingual Family: A Handbook for Parents.* Cambridge: Cambridge University Press.

Harini, N.V., and Chengappa, S.K. (2010). Code mixing and code switching in simultaneous Vs successive bilingual Children. Student Research at AIISH, Vol.V111, Part-B-Speech –Language Pathology, 97-115.

Heath, S. B. (1980). The functions and uses of literacy. *Journal of Communication, 30*, 123-133.

Ijalba, E., Obler, L. K., and Chengappa, S (2004).Neurology: Bilingual aphasia. In W.C Ritchie and T. K. Bhatia (ed) *Handbook of Bilingualism* (pp 70-89). Black well Publishing.

Illes, J., Francis, W.S., Desmond, J.E., Gabrieli, J.D.E., Glover, G.H., Poldrack, R., Lee, C.J., and Wagner A.D. (1999).Convergent cortical representation of semantic processing in Bilinguals. *Brain and Language, 70*, 347-363.

Karanth, P. (1992). Developmental dyslexia in bilingual-biliterates. *Reading and Writing: An Interdisciplinary Journal, 4*, 297-306.

Karanth, P., and Prakash, P. (1996). A developmental investigation of onset, progress and stages in the acquisition of literacy, Project funded by NCERT, India.

Katz, L., and Feldman, L. B. (1983). Relation between pronunciation and recognition of printed words in deep and shallow orthographies. *Journal of Experimental Psychology: Learning, Memory and Cognition, 9,* 157-166.

Kolers, P., and Gonzalez, E. (1980). Neurology In Bhatia.T.K and Ritchie,W.C (eds), *Handbook of Bilingualism*, pp. 70-89: Blackwell publishing.

Jakobovits, L.A. (1968). Dimensionality of compound coordinate bilingualism. *Language learning.* Special issue No.3, 29-49.

Klein, D., Zatorre, R. J., Milner, B., Meyer. E., and Evans, E.C. (1995). The neural substrates of bilingual language processing: Evidence from PET. In M. Paradis (Eds.), *Aspects of bilingual aphasia* (pp.23-36), Oxford pergamon press.

Klein, D., Milner,B., Zatorre R.J., Zhao,V., and NikelskiJ., (1999).Cerebral organization in bilinguals:A PET study of Chineese-English Verb generation, *Neuroreport, 10*, 2841-28746.

Lambert, W., and Fillernbaum, S. (1959). A pilot study of aphasia among bilinguals. *Canadian Journal of Psychology, 13*, 28-34.

Lanco-worrall, A.D. (1972). In Bi/Multilingualism and Issues in Management of communication disorders with emphasis on Indian perspectives: *Language in India,* vol.9: 3, August 2009.

Landry, R. (1973). In Bi/Multilingualism and Issues in Management of communication disorders with emphasis on Indian perspectives: *Language in India,* vol.9: 3, August 2009.

Lebrun, Y. (1971). The neurology of bilingualism. *Word, 27*, 179-186.

MacNamara, J. (1967). The bilinguals linguistic performance-A psychological overview. *Journal of Social Issues*, 23, 58-77.

McCarthy, D.A. (1930). In Bi/Multilingualism and Issues in Management of communication disorders with emphasis on Indian perspectives: *Language in India,* vol.9: 3, August 2009.

Marsh, L., and Maki, R. (1976) Efficiency of arithmetic operations in bilinguals as a function of language. *Memory and Cognition, 4*, 459-464.

McClain, L., and Huang, J. (1982). Speed of simple arithmetic in bilinguals. Memory and Cognition, 10, 591-596.

Mezzacappa, E. (2004). In Bi/Multilingualism and Issues in Management of communication disorders with emphasis on Indian perspectives: *Language in India,* vol.9: 3, August 2009.

Minkowski, M. (1927). Klinischer beitrag zur aphasie bei polyglotten, speziell in hinblick aufs schweizer-*deutsche Schweizer Archiv fur Neurologie and Psychiatrie* 21, P.43-72.

Minkowski, M. (1963). On aphasia in polyglots. In L. Halpern (ed.), *Problems of dynamic neurology*. Jerusalem: Hebrew University.

Munoz, M.L., Marquardt,T.P., and Copeland,G.(1999). Neurology. In Bhatia.T.K and Ritchie, W.C (eds), *Handbook of Bilingualism*, pp. 70-89:Blackwell publishing.

Neath, I. (1998). 'Human Memory: An introduction to research, data and theory. Pacific Grove, CA: Brooks / Cole.

Oller, D. K., and Eilers, R. E. (Eds.). (2002). *Language and literacy in bilingual children.* Clevedon:Multilingual Matters.

Ombredane, A. (1951). *L:aphasie et l `e' laboration de la pensee explicite.* Paris: presses universitaires de France.

Ovachrova, P., Raicheve, R. and Geleva, T. (1968). Afazia u poligloti. *Nevrologia, Psikhiatriia I Nevrohirurgiia* 7, 183-190.

Paradis, M. (1977). Bilingualism and aphasia. In H. Whitaker and H. A. Whitaker (eds.), *Studies in Neurolinguistics*, 3. New York: Academic Press.

Paradis, M. (1978). The stratification of bilingualism. In M. Paradis (ed.) *Aspects of Bilingualism*. Columbia: Hornbeam Press, Incorporated.

Paradis, M. (1986).Henry Hecaen's contribution to neurolingusitics. *Journal of Neurolinguistics, 2*, 1-14.

Paradis, M. (1994). Neurology In Bhatia.T.K and Ritchie,W.C (eds), *Handbook of Bilingualism*, pp. 70-89: Blackwell publishing.

Paradis, M. (2001). *Bilingual and polyglot aphasia: Handbook of neuropsychology*. Oxford: Elsevier Science.

Peal, E., and Lambert,W.E. (1962). In Bi/Multilingualism and Issues in Management of communication disorders with emphasis on Indian perspectives: *Language in India*, vol.9: 3, August 2009.

Penfield, W. (1953). A consideration of the neurophysiological mechanisms of speech and some educational consequences. *Proceedings of the American Academy of Arts and Sciences*. 82, p. 199-214.

Penfield, W. and Roberts, L. (1959). *Speech and Brain mechanisms*. Princeton, New Jersey: Princeton University Press.

Penfield, W. (1965). Conditioning the uncommitted cortex for language learning. *Brain, 88*, 787-798.

Petiti, L.A and Holowka,S. (2002). In Bi/Multilingualism and Issues in Management of communication disorders with emphasis on Indian perspectives: *Language in India*, vol.9: 3, August 2009.

Piaget, J. (1929). In Bi/Multilingualism and Issues in Management of communication disorders with emphasis on Indian perspectives: *Language in India*, vol.9: 3, August 2009.

Pitres, A. (1895). Etude sur l' aphasie chez les polyglottes, *Revue de Medicine* 15, 873-899.

Potzl, O. (1925). Uber die parietal bedingte aphasie undihren einfluss auf das sprechen mehrer sprachen. *Zeitschrift fur die gesamte neurology und psychiatrie*, 96, 100-1124.

Powers.,L and Lopez.,R. (1985). In Bi/Multilingualism and Issues in Management of communication disorders with emphasis on Indian perspectives: *Language in India*, vol.9: 3, August 2009.ISSN 1930-2940.

Prema K. S. Rao., Shanbal J.C., and Khurana S. (2010). Bilingualism and Biliteracy in India: Implications for Education. In *International Perspectives on Bilingual Education Policy, Practice and Controversy*. Petrovic, J. E. (ed.) A volume in the series: International Perspectives on Educational Policy, Research and Practice *Series Editor(s): Kathryn M. Borman, Information Age Publishing Series, NC, USA*.

Prema, K.S., and Geetha, Y.V. (2005). Language Acquisition in Multilingual Children. Project Report. Project funded by AIISH Research Fund, the All India Institute of Speech and Hearing, Mysore, India.

Prema, K. S. (2006). Learn English and Kannada. In Deccan Herald National daily, India.

Prema, K. S. (2006). Reading Acquisition in Dravidian Languages, *International Journal of Dravidian Linguistics*, *35*, 111-126.

Prema, K.S. (1998). Reading acquisition profile in Kannada, Unpublished doctoral thesis submitted to University of Mysore.

Prema, K.S. (2008). What concrete challenges EFA Global Monitoring Report, UNESCO.org/EFA-GMR.htm.

Olson, D. (1977). In Bi/Multilingualism and Issues in Management of communication disorders with emphasis on Indian perspectives: *Online Journal* www.languageinindia. com, vol.9: 3, August 2009.ISSN 1930-2940.

Rajasudhakar, R., and Chengappa, S. (2008).Effects of age, gender and Bilingualism on cognitive-linguistic performance. Student Research at AIISH, vol.111, *Part-B-Speech-language Pathology*, 127-146.

Reed, T. (2003). The literacy acquisition of Black and Asian 'English as additional language' (EAL) learners: Anti racist assessment and intervention challenges. In L.Peer and Reid G. (eds.) *Multilingualism, literacy and dyslexia: A challenge for educators* (pp. 111-119), David Fulton publishers, London.

Reyes, M.L. (2001). Unleashing possibilities: Biliteracy in the primary grades. In M. Reyes de la Luz and J.J. Halcón (Eds.), *Best for our children: Critical perspectives on literacy for Latino students* (pp. 96–121). New York: Teachers College Press.

Ribot, T. (1906). *Diseases of Memory.* London: Paul, Tranch, Trubner.

Savithri, S. R., Prema K. S., and Shilpashre H. N. (2004). Performance of Learning Disabled Children on Time-warped Words in Phoneme/Syllable Stripped Condition, Frontiers of Research on Speech and Music (FRSM), ITC Sangeet Research Academy, Kolkata and Center for Advanced Study in Linguistics (CASL), Annamalai University, Tamil Nadu.

Seetha, L., and Prema, K.S. (2002). Reading acquisition in Malayalam: A profile of secondary graders. Unpublished Master's dissertation submitted to the University of Mysore, Mysore.

Shanbal, J. C., and Prema K. S. (2006a). Phonological skills in bilingual-biliterate children: Cross-language transfer within Dravidian languages of India, *Language Forum, 32,* 125-145.

Shanbal, J.C., and Prema, K.S. (2006b). Word length and orthographic complexity in bilingual-biliterate children: An investigation in reading Kannada Presented at 34th All India Conference for Dravidian Linguists, Trivandrum, India.

Shanbal, J.C., and Prema, K.S. (2006c). Cross-language transfer in bilingual children – Effect on literacy in a second language? Presented at 2006 Second Language Research Forum Conference, University of Washington, Seattle, USA.

Shanbal, J.C., and Prema, K.S. (2007). Phonological awareness and reading in bilingual-biliterate children. Presented at the 35th All India Conference for Dravidian Linguists, Mysore, India.

Shanbal, J.C. (in progress). Acquisition of Biliteracy in Children. Doctoral thesis in preparation, University of Mysore, Mysore.

Stigler, J. W., Lees, S-Y., and Stevenson, H. W. (1990). Mathematical knowledge: Mathematical knowledge of Japanese, Chinese, and American elementary school children. Reston, VA: National Council of Teachers of Mathematics.

Stuart-Smith, J., and Martin, D. (1999). Developing assessment procedures for phonological awareness for use with Punjabi-English bilingual children. *The International Journal of Bilingualism, 3,* 1, 55-80.

Stuart-Smith, J., and Martin, D. (1999).Developing assessment procedures for phonological awareness for use with Punjabi-English bilingual children. *The International Journal of Bilingualism, 3,* 55-80.

Sumitha, M.M., Shwetha, P., and Prema K.S. (2005). Language of mathematics or mathematics of language? Paper presented at the National Conference of Indian Speech and Hearing Association, India.

Swaroopa, K. P. and Prema, K.S. (2001). Checklist for Screening Language based Reading Disabilities (Che-SLR). Unpublished Master's dissertation submitted to the University of Mysore, Mysore.

Thirumalai, M. S. and Chengappa,S. (1986).*Simultaneous acquisition of two languages: An overview.* Mysore: Central institute of Indian languages press.

Treiman, R., Mullennix, J., Bijeljac-Babic, R. and Richmond-Welty, E.D. (1995). The special role of rimes in the description, use, and acquisition of English orthography. *Journal of Experimental Psychology: General, 124*, 107–136.

Ullman, M.T.(2001). Neurology In Bhatia.T.K and Ritchie,W.C (eds), *Handbook of Bilingualism*, pp. 70-89: Blackwell publishing.

Veyrac, G. J. (1931). Etude de l' aphasie chez les subjects polyglottes. These pour doctorat en medicine. Universite de Paris.

Wagner, R.K. and Torgesen, J.K. (1987). The nature of phonological processing and its causal role in the acquisition of reading skills. *Psychological Bulletin, 101*, 192–212.

Wald, B. (1974). Bilingualism. In B. J. Siegal, A. R. Beals and S. A. Tyler (ed), *Annual Review of Anthropology* 3. Palo Alto: Annual review.

Webster, A. M. (1961). *Websters sixth new collegiate dictionary.* Massachusetts: G.C. Marriam Company publishers.

Weinreich, U. (1953). *Languages in Contact. Findings and Problems.* Mouton, The Hague, N.Y. USA.

Weinreich, U. (1953). *Languages in contact,* New York: Publication of the linguistic circle of New York.

Whitaker, H. A. (1978). Bilingualism: A neurolinguistics perspective. In W. C. Ritchie (ed.). *Second language acquisition research: Issues and implications,* New York: Academic Press.

Wimmer, H. and Hummer, P. (1990). How German-speaking first graders read and spell: Doubts on the importance of the logographic stage. *Applied Psycholinguistics, 11*, 349-368.

Yellend,G.W. (1993).In Bi/Multilingualism and Issues in Management of communication disorders with emphasis on Indian perspectives. *Language in India*, vol.9: 3, August 2009.ISSN 1930-2940.

Ziegler, J. C., and Goswami, U. (2005). Reading acquisition, developmental dyslexia, and skilled reading across languages: A Psycholinguistic grain size theory. *Psychological Bulletin, 131*, 1, 3-29.

In: Expanding Horizions of the Mind Science(s)   ISBN: 978-1-62808-705-5
Editors: P.N. Tandon, R.C. Tripathi and N. Srinivasan   ©2013 Nova Science Publishers, Inc.

Chapter 13

BELIEF: A SCIENTIFIC PERSPECTIVE

P. N. Tandon

National Brain Research Centre, Manesar

"Man is made up of faith, as is his faith, so is he……"

Bhagvad Gita 17.3.28

"Lord, I believe, help thou mine unbelief"
Bible 9 : 24

"A man with a grain of faith in God never loses hope, because he ever believes in the
ultimate triumph of Truth".

M.K. Gandhi

ABSTRACT

The chapter discusses the study of belief from a scientific point of view. Major focus is on religious beliefs and the neuroscience of religion. Evolution and development of religious beliefs over the course of evolution in societies across the world are discussed. Studies on the nature of religious beliefs held by people across the world tell us the prevalence of beliefs about the existence of god. The chapter discusses neural findings associated with religious belief followed by a discussion on brain activity associated with meditation. The recent attempts to study religious belief indicate us the way beliefs are created in our brain through evolution and experience.

1. INTRODUCTION

The study of belief in all its forms has recently become a very hot topic among scientists particularly cognitive scientists and philosophers alike. According to the Progressive English

Dictionary – belief a noun of the verb believe means to have trust or to feel sure of the existence of. Often used synonymous to 'faith' it differs from it. It is said that faith is something that you know, while belief is what you trust. Prof. G.C. Pande, a distinguished philosopher and historian, commented that, in many cultures there are three kinds of beliefs since antiquity i.e., beyond body there is 'soul', death is not end of life, and existence of superhuman beings. Robin Dunbar, an evolutionary psychologist at the University of Liverpool, UK, and an author of a recent book, 'The Human Story', described religious belief as a part of human nature. He attempted to answer two questions about its origin: Why did it evolve, and at what stage did our ancestors start believing in God?

In this chapter it is proposed to discuss belief as it pertains to religion, spirituality, soul, God. A number of recent books have dwelt on this and related aspects of belief. To mention a few: "Memory, Brain and Belief" by Elaine Scarry (2001), "Born to Believe: God; Science, and the Origin of Ordinary and Extraordinary Beliefs' by Andrew Newberg and Mark Robert Waldman (2007), 'The Soul in the Brain: The Cerebral Basis of Language, Art, and Belief' by Michael R. Trimble (2007), 'The Psychological Roots of Religious Belief: Searching for Angels and The Parent-God', by M.D. Faber (2004), 'The Biology of Belief: Unleashing the Power of Consciousness, Matter and Miracle' by Bruce H Lipton (2008), 'The Reason for God: Belief in an Age of Skepticism': Timothy Keller (2008), 'Why God Wont Go Away: Brain Science and the Biology of Belief' by Andrew Newberg, Eugene D'Aquill and Vince Rowse (2001). Roger W Sperry a recipient of Nobel Prize for his work on neuroscience, contributed a detailed account on "A Search for Beliefs to live by consistent with Science" in 1995.

Voluminous literature on the subject, mostly by philosophers and religious sages, extending over centuries exist in India. However, no major contribution on the subject by Indian natural scientists, psychologists or cognitive scientist has been forthcoming in recent years.

Van Inwagen (2006) listed some of the fairly common beliefs about human beings

- Human beings have free will
- Human beings do not come to an end with death
- There is a supernatural order – that the natural world is not all there is, but rather exists within a 'surround' of personal and invisible powers

Belief in the Spirit or Soul or *Atman,* Cosmic self, or in God or a supreme being (i.e. religion) pervades through virtually all societies and cultures from the beginning of the primitive civilization till today. Such beliefs are a source of personal strength and a bonding force of a society. The large numbers of temples, mosques, synagogues, churches and other places of worship that span the length and breadth of our planet, and the crowds of people at these places – illiterate, scholars, scientists, poor and rich – bear testimony to the immeasurable force of this human trait. According to Graham Green, "one must distinguish between faith and belief. I have faith, but less and less belief in existence of God. I have a continuing faith that I am wrong not to believe and that my lack of belief stems from my own faults and my failure to love". Why do people believe at all? Among the many answers to this difficult question probably one by Dennett (2006) summarizes it best. According to him religion (by implication belief) has three purposes: to comfort us in our suffering and allay our fear of death, to explain things, we can not otherwise explain and to encourage group cooperation in the face of trails and enemies. Commenting of Dennett's views, Ayala (2006) thought these to focus on the negative aspects of religion and talked about the positive aspects, thus, "This includes how belief in after life help people by relieving their existential anxiety and how religious beliefs provide purpose and meaning to life and are a source of individual fulfilment and group cohesion. These have led to the emergence of systems of mortality and moral codes advocated by all major religions. In other words beliefs are the basis of all religion, spiritual movements, mystic faith and divinity.

2. DEFINITIONS

This brings me to consider definitions of some of the terms to be frequently used in this essay. There may not be unanimity about these, however, to avoid confusion, the definitions given below are the ones adopted in this essay.

i) *Science:* is an organised body of knowledge based on facts and principles gained by systematic study.

ii) *Religion:* Faith in the existence of God or a superhuman. Controlling power and intellect; Belief in God as creator and controller of Universe, a system of faith and worship based on such a belief.

iii) *Spirit:* Something which is non-physical, transcends space and time, infinite, indestructible, immortal, in nature often used synonymously soul.

iv) *Spiritual:* Adjective of Spirit, of the soul, of religion, non material things.

v) *Spirituality:* According to Karl Pribram (2002), the essence of spirituality is the losing of self within some larger domain of experience. According to Vedic Seers, spirituality "is an expression of the spiritual dimensions inside us" (Swami Harshanand (2002). Some people consider spirituality not only restricted to religious experience but find aesthetic beauty in art and music as forms of spirituality.

vi) *Soul:* Non-material or spiritual part of a human being which is believed to exist for-ever. Variously defined as something other than body and mind, synonymous with spirit, transcends material existence. According to Michael R. Trimble (2007) it is variously defined as, "psyche, inspiration, energy, a vital force that inspires, energises and stimulates us".

vii) *Consciousness:* (There is no unanimity on its definition). But for the limited purpose of this essay we may accept it to refer to having of perceptions, thoughts, feeling, and awareness of one's self and environment. Awareness of self and environment, attention, cognition, discrimination, memory, responsiveness and volition constitute consciousness. (Tandon, 1993)

viii) *Mind:* A complex information processing system which receives, stores, retrieves, transforms, transmits information; the faculty by which we think, includes thoughts, feelings, sentiments and intention. To some it is the whole spiritual nature. This is too limited a definition of mind. According to Eccles (1988): "I do not define mind, I only talk about it". The mind is infinitely complex _ _ _ _(fill the blank space)All our experiences – loving, hating, creating, imagining – everything is in the mind, and we experience it directly. Now we can work scientifically on this.

ix) *Mystic Experience*: feeling of closeness to God, mingling with Him, feeling a sense of extreme bliss, to connect with something divine, absorption of self into something larger. William James the most celebrated psychologist of the 19th century acknowledged the mystic state and recognized four dimensions of it. His Gilford Lectures (1792) dealt with in great details "Variety of Religious Experiences"

x) *Meditation:* is of several different types e.g., Yogic, Transcendental, Zen etc. The Tibetan word for meditation, gom, literally means "becoming familiar with "and Buddhist meditation practice is really about becoming familiar with the nature of your own mind".

3. Origin of Belief / Religion

"All religions are based on faith belief".
Faith is the foundation of what you hope and firm belief of what you can't see, and I think that's its essence".

Daute Alighieri: An Italian Thinker

In most societies and cultures, since time immemorial, people believe in something larger than themselves, all powerful and all pervading call it simply God or else give it a name like, Ram, Rahim, Allah, Christ. This is embodied in individual religions. The subject has been discussed and debated by theologians, philosophers and more recently by psychologists, cognitive and neuroscientists. The scientific studies on the subject have been given the dignified name of Neurotheology. Daniel Dennett, a philosopher and self proclaimed atheist in his book, "Breaking the Spell: Religion as a Natural Phenomenon, proposed that (end of inverted comma) "religions are part of human culture and should thus be subjected to historical, psychological and anthropological research. He thus seeks to bring religion within the fold of "a forthright, scientific, no-hold-barred investigation" as "one natural phenomenon among many" (Ayala, 2006). Nicolas Wade (2009) in his book, "The Faith Instinct: How Religion Evolved and Why it Endures?" argues that people have a genetic urge to worship and that the instinct is engraved in the brain's neural circuits because of the tremendous advantage religion confirmed on early societies.

Starting from the stand point that religious belief is a part of human nature Robin Dunbar, an evolutionary biologist at the University of Liverpool, UK, addressed two questions about its origin: why did it evolve and at what stage did our ancestors start believing in God. All men are convinced of the existence of the God declared Aristotle. Allison Brooks of the George Washington University, Washington, DC in his talk, "What is Human?" Archaeological Perspective on the Humanness to the Pontifical Academy concluded, "Capacity for some of the most human qualities: creativity, empathy, reverence, spirituality, aesthetic appreciation, abstract thought and problem solving (rationality) were already evident soon after the emergence of our species". John Cacioppo of the University of Chicago, leader of a team of University researchers on effect of spirituality on health commented, "People are born with the capacity for spirituality".

Until now, the co-evolution of religious ritual and society was an assumption. A study (reported in PNAS DOI: 10.1073/pnas.0408551102) carried out in Mexico by Joyce Marcus and Kent Flannery, using carbon dating sites presumed to be used for initiation rites dates these to 8600 years. It claims that "Temple sanctification and an increase in bloodletting and human sacrifice emerged around 3100 years ago". Some archaeologists believe it began as far back as 200,000 years ago with Neanderthals. This has been based on looking for evidence of grave goods in burials, since these atleast unequivocally imply belief in after – life. Such burials imply a sophisticated theology. At some stage in our evolutionary past, our ancestors began living in groups that were too large for social grooming (observed in monkeys and apes) to provide for effective glue. Religion allowed larger groups to bond. Anthropologists agree in recognising that human beings have practiced some form of religious activity ever since their first appearance on the horizon (Sorondo, 2007).

Religion is not just about rituals, it also has an important cognitive component – its theology. To create a theology our ancestors needed to evolve cognitive abilities that far exceed those found in any other animal species (Dunbar, 2006). Lawrence Krauss, Director for Education and Research in Cosmology and Astrophysics at the Case Western Research University, Cleveland, Ohio, in his book, "Hiding in the Mirror" states, 'Spirituality and its religious faith is deeply ingrained in human culture, and many people rely on their religious convictions to make sense of life. Whatever ones personal views about religion, it is undeniable that scientific understanding alone does not encompass the range of the human intellectual experience."

3.1. Evolution of Religion / Belief in God or Superhuman Divine Power

Dean H Hamer, a geneticist at the National Cancer Institute, USA in his book, "The God Gene" went to the extent of declaring that he has discovered the God Gene, and he claimed that psychologists, neurologists, and even evolutionary biologists have offered insights how spiritual behaviours and beliefs emerge from the brain.

Pierre Teilhard de Chardin (1881-1955) a Teleologist in his book, "The Phenomenon of Man" attributed the belief in divinity to evolution; "Evolution is pushing man towards a higher goal, an omega point, which can be described as collective divinity. A cosmic divine manifestation is in the making". Some of the writings of Sri Aurobindo also point in this direction. "Nature has evolved beyond Matter and manifested Life, beyond Life and manifested Mind, so she must evolve beyond Mind,_ _ _ _ _ _ _ _ _ _ and able to develop the power and perfection of the Spirit". At yet another place he propounded, "Mind founded in life developed intellect _ _ _ _ _ _ _ _ _ _ _ _ till it reached spiritual perception". And furthermore, "As an evolving principle it will mark a stage in the human ascent and evolve a new type of human being: thus development must carry in it an ascending humanity which will embody more and more the turn towards spirituality_ _ _ _ a climb towards a divinised manhood and the divine life." It is interesting to note that a Western Scholar in his book, "Space, Time and Deity" had echoed similar thoughts, "The cosmic process has now reached the human level and man is looking forward to the next higher quality of deity is a stage in time beyond the human".

It is therefore reasonable to assume that throughout human existence there have been an all pervading belief about the existence of God or a superhuman divine power. This constituted the guiding force of all religions as propounded in their scriptures. This was specially deliberated in extreme details in the Vedic literature. It was equally strongly supported by the writings of the philosophers centuries ago. It is therefore important to consider the views of ancient philosophers and other scholars. Plato said, "Man is both a terrestrial tree and a celestial plant that inhabits two worlds – the physical and the spiritual". Emerson claimed, "Humanity is our actuality but divinity is our potentiality". According to Voltaire, "Faith consists in believing when it is beyond the power of reason to believe".

JB Priestly while extolling science commented, "without science we are helpless children. But without a deep religion we are blundering fools, _ _ _ _ ". And one of our most outstanding scientists of yesteryear Satyendra Nath Bose pointed out, "People forget that science is culture, a child must learn from science many things, a true scientific attitude not only engenders objectivity but also spiritual value". In this connection it is revealing to recall the words of Einstein – the greatest scientist of the twentieth century, "I do not arrive at my understanding of the fundamental laws of the universe through my rational mind _ _ _ _ _ _ _ _. The cosmic religious experience is the strongest and noblest mainspring of scientific research". Roger W Sperry proposed that religiously deep-rooted beliefs or faiths are no longer illusions. They are sometimes overwhelmingly functional: "Subjective belief is no longer a mere impotent epiphenomenon of brain activity. It becomes a powerful impelling force in its own right. I no longer need to keep my religion and my science separate. As thing stand _ _ _ I no longer need to believe, as a scientist, that I and my world are governed solely from below up ward through the 'fundamental forces of physics' in a totally mindless and purposeless cosmos, indifferent to human concerns. Quoted by Amit Goswami: The Self-Aware Universe, Bantam, New York, 195, p271

In 1997, University of Georgia history professor Edward Larson and journalist Larry Witham on the basis of a survey concluded that 39% of scientists believe in God. (Larson and Witham : Nature 386, 435-36, 1997) (Larson and Witham : Nature 394, 313, 1998) (Nature 460, 310-11, 2009). A 9 July 2009, survey by AAAS and Pew Research Centre found that figure to be 33%. But this dropped to 7% of respondents among those who were Members of the US National Academy of Sciences. More than half of those who responded to a USA Today/ABC News/Stanford University Medical Centre poll said they use prayer to control pain. Of those, 90 percent said it worked well and 51 percent say "very well".

Surprisingly this is the same figure (40%) reported by James H Leuba, a psychologist, who surveyed American biological and physical scientists in 1914 and again in 1933. Yet another survey by Scientific American found that notwithstanding the phenomenal discoveries in Science the same to be true in 1997-1998. However, this was not true among the "elite" group of scientists – the members of the US National Academy of Science. While 95 percent of the biologists evincinced atheism and agnosticism, surprisingly one in six mathematicians expressed belief in personal God. It is worth mentioning that such great scientists like Copernicus, Kepler and Newton were profoundly and personally religious.

Astronomer Jocelyn B. Burnell found a place for both science and religion in her life stated, "I don't think God created the world in any physical form, but that's not to say there isn't a God". Emblazened on the cover of Newsweek in 1998, was "Science finds God". This statement was motivated by the deliberations of a major symposium organised at the University of California, Berkley, "Science and the Spiritual Quest". More than 20 scientists, including a physics Nobel Laureate, testified that science either led them to God or was not an obstacle to faith". Charles Townes NL, inventor of Lasers acknowledged that, "God was a 'source of strength' during these historic discoveries aiding him at times to overcome self-doubt".

An International Survey: Worldviews and Opinions of Scientists in June 2008 by Ariela Keysar and Barry A Kosmin, in collaboration with N. Innaiah from India for Institute for the Study of Secularism in Society and Culture, of Trinity College in Hartford, Connecticut surveyed 1,100 participants from 130 Indian Universities and Research Institutes in June 2008 found the following:-

One fourth (26%) had no doubts about the existence of God, another 15% had occasionally some doubts though they generally believed in God. Another 30% said they don't believe in a personal God but believed in a higher power. Nearly 29% believed in Karma (Sins and Deeds of Past Life), 26% in life after death and 20% in reincarnation. The majority of these scientists think of themselves as "Spiritual". The term "spiritual" meant different thing to different persons but two thirds opted for either "commitment to higher human ideals, such as peace, harmony or well-being" (34%) or "a higher level of human consciousness or awareness" (31%).

Ever since Immanuel Kant devastatingly criticised all proofs of existence of a creator in the 18th Century, theologians and philosophers have been embarrassed about God. Keith Ward, Professor of Divinity at Gresham College, London is reported to indicate, "Today, however, there is no need to be defensive for modern physicists have pushed the pendulum the other way. Indeed, modern physics makes belief in God more than it has been since Kant. Some quantum physicists postulate mentality and consciousness as an irreducible property at the quantum level and propose an understanding of consciousness at the level of brain functioning through the concept of the collapse of quantum wave function. Martin Luther King Jr. lamented, "Our scientific power has over run over spiritual power. We have guided missiles and misguided human mind".

4. NEUROBIOLOGY OF BELIEF

On the basis of the rapidly growing knowledge of evolution, structure and function of the brain, Vittorio Marcozi (1992), an anthropologist acclaimed that it is now being generally recognised that the functions of the brain could be classified as follows :-

- Physiological : Sense perception, voluntary motor activity
- Psychological : Memory, Emotion, Language
- Intellective : Ideas, Concepts
- Spiritual ; Psychic

The important question is therefore, "What is going on inside our brains when we believe"? How does that trigger physical changes in body?", or in short, "What is the biological basis of belief?" According to Ramachandran, who has been studying "disorders of belief", it is a fascinating question and poorly studied" quoted by Matluk (2006). However, Zubieta (2005) acknowledged that faith in God is almost as ubiquitous as faith in medicine. There is no doubt that it is a real and powerful force. The evidence suggests that belief is a conscious, rational process. This is supported by behavioural studies, thus hinting that it is amenable to scientific studies.

According to Dean Hamer of NIH, author of 'The God Gene', "Belief changes the tenor of brain and is mediated by the same neurotransmitters for example, dopamine and serotonin – that mediate other emotions" (Hamer, 2004). He found that there is a variant of gene called VMAT2 that may be associated with greater spirituality. The VMAT2 protein seems to control the flow in the brain of monoamines like dopamine and serotonin. Based on certain beliefs religion acts as a kind of glue that holds society together because they exploit a whole suite of rituals that trigger the release of endorphins, creating a mild 'high'. Endorphins also 'tune up' the immune system, which probably explains why religious people are healthier (Dunbar, 2006).

Dunbar, a Professor of Evolutionary Psychology at the University of Liverpool, UK, argued that the relative volume of grey matter in the frontal lobe reflects the increasing level of intentionality to account for the communal religion which is the current stage of human evolution, which permits us to invoke a spiritual force that obliges, perhaps even forces us to behave in a certain way. It is hypothesised that the ability to attribute mental states, together with specific neural organization that makes such attribution possible, is a unique specialization of human species. However, there is evidence to suggest that many organisms routinely form neural states instantiating desire, intention and belief, it is possible that only a few species (and possibly on humans) have the ability to reflect upon these states (Povinelli and Preuss, 1995).

The phenomenal expansion of cognitive neuroscience and neuroimaging techniques since 1990s has resulted in dramatic growth of studies on cognition, attitude, moral and social judgement and religious experiences. Virtually a new field of neuro-theology has emerged as a result of these advances.

Some years ago VS Ramachandran predicted that the 'mirror neurons' will provide a unifying framework and help explain a host of mental abilities that have remained mysterious and inaccessible to experiments. Jeeves (2007) invoking the 'Theory of Mind' suggested that 'mind reading' refers to the activity of representing to oneself the specific mental state of others, their goals, their perceptions and their beliefs and their expectations. Whether they had in mind the subject of beliefs in the context discussed here is not clear.

It has been demonstrated that their goals, their emotions like happiness pleasure and joy were associated with increased activity in the medial prefrontal cortex (Broadmann's area 9), anterior and posterior temporal structures and thalamus. One wonders if bliss experienced by the mystics could also be activating these areas. Newberg et al. (2001) who studied Canadian Buddhist monks and Fransciscan Nuns observed that all mystics describe their peak meditative, spiritual and mystical moment as the absorption of self into something larger, while others described, "this moment as a tangible sense of closeness to God and a mingling with him".

This brings me to the consideration of neuro-physiological correlates of meditation a unique component of some belief systems. It must be recognized that in different traditions the practice of meditation is not identical yet practitioners of all systems, yogic, Zen, transcendental, Vipassana or some of the Christian sects, all report alteration of consciousness characterized by calmness, serenity, heightened awareness/peace and bliss during such practices. In certain, enhanced state of meditation it is claimed that the individual's identify merges with the universal self or God. It is surprising that the feelings aroused across different religions and cultures are so similar that one is forced to believe that there should be a common neuro-physiological basis for this experience. Beginning in 1960s, with the pioneering studies of EEG on Yogis by Anand and colleagues (1961) a large number of such studies have been carried out on a variety of practitioners of meditation. These were greatly stimulated by the popularization of "transcendental meditation" (TM) by Maharishi Mahesh Yogi. Just to quote a few are those by Wallace (1970), Wallace et al. (1971), Benson et al. (1982). Wallace, alongwith Herbert Benson carried out extensive physiological studies on a number of TM meditators which were published in several prestigious scientific journals like American Journal of Physiology, New England Journal of Medicine, Scientific American etc. They observed that, "All kinds of medical tests showed that TM relieved stress, you become clearer, you can do more, _ _ _ the mind doesn't wander". These mental states can markedly alter physiological functions like producing hypometabolic state, lowering the blood pressure, changing the skin resistance and altering the EEG to predominantly alpha. Lutz et al. (2004) carried out EEG studies on 8 Buddhist monks practicing Tibetan Nuingmapa and Kagyupa tradition meditation. Long term practitioners could self-induce sustained electro-encephalographic high amplitude gamma – band (25-42 Hz) oscillations, and phase – synchrony during meditation. The activity was particularly observed over lateral frontal-parietal electrodes. An essential aspect of these gamma oscillations is that their amplitude monotonically increased over the time of practice. These data were interpreted to suggest that mental training involved temporal integrative mechanisms and may induce short-term and long-term neural changes. These results are different from those mentioned above that found an increase in slow alpha and theta rhythms during meditation practices followed by different practitioners. The TM practitioners use voluntary concentrative meditation on an object such as on a mantra – a technique which represents a top-down control, while the Buddhist monks practice on objectless meditation. Further studies on larger number of practioners using different type of practices of meditation are required to resolve these observations.

Richard Davidson, a psychologist at the University of Wisconsin, Madison, USA and the coordinator of the Dharmsala Conference, in his interaction with Tibetan monks found, that certain neural processes in the brain are more coordinated in people with extensive training in meditation that may be linked to the heightened awareness reported by meditating monks. It is interesting to note that a short program in mindfulness meditation, as practiced by Buddhist monks, produced demonstrable effects on brain and immune function in positive ways. (Davidson 2003)

Newberg et al. (2001) found significant increase in rCBF in inferior, orbital frontal and dorsolateral prefrontal cortex, cingulate gyrus and thalamus during meditation. Brain mapping of event – related electrical activity showed that the frontal theta rhythm was active during meditation (Kubota et al. 2001). The areas which were active during hypnosis, sleep and even simple rest are different than those during meditation (Rao 2003). Experienced meditators who reported periods of blissful experiences were found to have characteristic changes in EEG (Caftan and Golocheikine, 2001).

In contrast to the above it has been reported that Zen monks in Japan during meditation could go right into alpha, relaxed but alert and the alpha amplitude got bigger and slower as meditation progressed. A control group sitting in the same position but not meditating stayed in beta. Yogis following Indian yogic tradition, who sat in lotus position, eyes closed, for upto two and a half hours also showed a strong alpha rhythm in EEG. They were in a state of deep relaxation physiologically with no evidence of drowsiness and sleep. In some yogis outside stimulus – noise, lights – didn't show any change in the alpha activity. It may be pointed out that one difference between the yogis and the Zen monks was in the reaction to an outside disturbance. The yogis simply shut it out and after the experiment they weren't aware that there had been any noises or lights. In contrast, the Zen monks registered each of a series of clicks and the alpha went right on.

"The Soul in the Brain : The Cerebral Basis of Language, Art and Belief: (Michael R. Trimble, 2007, The Johns Hopkins University Press.) Soul traditionally an esoteric and controversial concept is variously defined as "psyche, inspiration, and energy". If, one were to consider, "the soul" as the vital force that inspires, energises and stimulates us, then it may be possible to study its manifestations and effects in all human activities having those qualities.

The possibility that one could study the soul by associating inspirational human experience, religion, music, poetry, and literature, with the brain has been proposed. In his book Trimble, Emeritus Professor of Behavioural Neurology at University College of London, expounds the neurological correlates of such inspirational human experiences. Limbic system has been considered as the seat of emotions ever since Papez postulated this in 1937. More recent imaging studies have provided much needed support to this. It is now well established that other brain regions outside the limbic system play important role in various emotions including joy, happiness, etc. Trimble describes how the use of the language of poetry and metaphor produces heightened activity of the right hemisphere of the brain.

It has been shown that the practice of Yoga improves performance in various tasks for motor skills, visual perception, spatial memory and specific tasks for planning. Positron Emission Tomography (PET) study by Herzog et al. 1990-91, as well as fMRI studies by Khushu et al. 2000 demonstrated an increase in frontal blood flow during meditation. There is some scientific evidence of changes in the brain with Yoga. Some of the studies suggested changes at specific levels, for example in the pre-frontal cortex and the mesencephalic – diencephalic region (Telles, 2002). Jacob Abraham, a neurosurgeon in his book, "The Quest for the Spiritual Neuron" theoretically traced this function to the neurons in the prefrontal lobes of the brain.

"After years of scientific study, and careful consideration of our results, Gene and I further believe that we saw evidence of a neurological process that has evolved to allow us humans to transcend material existence and acknowledge and connect with a deeper, more spiritual part of ourselves perceived as an absolute universal reality that connects us to all that is. The transcendent state we call Absolute Unitary Being refers to states known by various names in different cultures – the Tao, Nirvana, the *Unio Mystica*, Brahman – atman – but which every persuasion describes in strikingly similar terms. It is a state of pure awareness, a clear and vivid consciousness of *no-thing*. Yet it is also a sudden, vivid consciousness of everything as an undifferentiated whole". While our neurological model offers a plausible explanation of how we experience the mystical state of pure awareness, it proves nothing about the ultimate nature of Absolute Unitary Being. Yet our work has convinced us that the mystics, at the very least, are not delusional or psychotic. They are certain beyond a shadow of doubt that their experience are real. Mystical reality holds, and the neurology does not contradict it, that beneath the mind's perception of thoughts, memories, emotions and objects, beyond the subjective awareness we think of as the self, there is a deeper self, a state of pure awareness that sees beyond the limits of subject and object, and rests in a universe where all things are one. (Newberg et al., 2001)

King et al. (1995) critically examined relations between religion or spirituality and any physical and mental health conditions covering more than 1200 studies and 400 reviews. A 60-80% relation between better health and religion or spirituality was found in both coorelational and longitudinal studies covering heart disease, hypertension, cerebrovascular disease, immunological dysfunction, cancer, pain and disability and health behaviours. Psychiatric topics covered included psychosis, depression, anxiety, suicide and personality problems. The benefits are threefold: aiding prevention, speeding recovery and fostering equanimity in the face of ill health. [King M et al. 1995]

Steve Blinkhorn (2000) the common view, informed by 400 years of philosophy merging into 150 years of neurology, is that, as a settled matter of fact, the brain is the organ of mind, with the only real puzzle being how the brain generates consciousness. Cognitive neuroscience now leaves little doubt that specific cognitive acts require the transient integration of numerous widely distributed, constantly interactivity areas of the brain. [Varela FJ et al., 2001]

"John Eccles in his book, What is Mind?" stated, "_ _ _ new concepts of brain and of evolution are infact leading to the supremacy of the spirit and also divine creation of us all. The more we know about the brain, the less materialistic we should be"

CONCLUSION

It could thus be said of belief or faith that philosophers have discussed it, priests have propounded it, poets have sung pions about it, voluminous scriptures from every religion attempt to inculate its supreme value for human emancipation from the very beginning of civilization. Now cognitive and neuroscientists have begun to explore its neurobiological basis. Recent advances in these fields alongwith neuroimaging are expected to provide answers if belief, like many other higher mental functions is hard wire and what regions of the brain are involved in its evolution.

REFERENCES

Abraham Jacob (2004). *The Quest for the Spiritual Neuron*. Dharmaram Publications.

Aftanas, L.I. and Golocheikine, S.A. (2001). Human anterior and frontal midline theta and lower alpha reflect emotionally positive state and internalised attention: high resolution EEG investigation of meditation. *Neurosc. Letters* 310, 57-60.

Alexander and Samuel: Space, Time and Deity. Quoted by Prof. Kireet Joshi in his Key Note Address at the National Seminar on Supramental Consciousness held at New Delhi on 27th March 2004

Anand, B.K., Chhina, G.S., and Singh, B. (1961). Some aspects of electroencephalographic studies in Yogis Electroencephalogr, *Clinical Neurophysiolgy.* 13,452-456.

Ayala Francisco, (2006). Book Review of Dennett's book. *New Scientist*, p 48.

Benson, H., Lehmann, J.W., Malhotram M.S., Goldman, R.E., Hopkins, J. and Epstein, M.D. (1982). Body temperature changes during the practice of g Tum-mo Yoga. *Nature* 295, 234-35.

Brooks, Allen and Eccles John. *What is the Mind?* Round Table discussion.

Cacioppo, L. (2005). In Report of a study by scientists at the University of Chicago: Times International.

Davidson, R.J. (2003). Alterations in brain and immune function produced by mindfulness meditation. *Psychosomatic Medicine*, 65, 564.

Dennett, Daniel. (2006). Breaking the Spell: *Religion as a natural phenomenon*, Penguin.

Dunbar Robin, (2006). We believe In Beyond Belief. *New Scientist*, pp 30-36

Dunbar, R. (2006). We believe: *New Scientist*, pp 30-33.

Hamer Dean, H. (2004). *The God Gene.*

Herzog, H., Lele, V.R., Kuwert, T. et al. (1990-91). Changed pattern of regional glucose metabolism during yoga meditative relaxation. *Neuropsychobiology* 23,182-187.

James, W. (1890). *The Principle of Psychology.*

Jeeves, M. (2007). Soul-searching and mind-reading issues raised by 21st century neuropsychology. In What is our Real Knowledge about the Human Beings (ed) Marcelo Sanchez Sorondo, Pontificia Academiae Scientiarum, Scripta Varia 109, *Vatican City* pp 36-41.

Khusu, S., Telles, S., Kumaran, S. et al. (2002). Frontal activation during meditation based on functional magnetic resonance imaging (fMRI). *Indian J. Physiol. Pharmacol. Supplement* (2000) Quoted by Telles.

Krauss, L. (2005). *Hiding in the Mirror.* Vicking.

Kubota, Y.W., Sato, M., Toichi, T., et al. (2001). Frontal midline theta rhythm is correlated with cardiac autonomic activities during the performance of an attention demanding meditation of procedure. *Cogn. Br. Res.* 11, 281-87.

Lutz, A., Greischar, L.L., Rawlings, N.B. et al. (2004). Long term meditators self-induce high amplitude gamma synchrony during mental practice. *PNAS* 101,16369-73.

Marcozzi and Vittorio, (1992). The brain and the psyche in the spiritualist scholastic view. In Brain Research and Mind-Body Problem: Epistemological and Metaphysical Issues: Ed. Giuseppe Del Re: *Pontificiae Academiae Scientiarum Scripta Varia* 79.

Marcus, J. and Plannery, K. (DOI: 10-1073/pnas.0408551102).

Motluk, A. (2006). Particle of faith. *New Scientist*, p 34

Newberg , A. D'Aquili, E. and Rause, V. (2001). Brain Science and the Biology of Belief: *Why God Won't GO Away*. Ballantine Publishing Group.

Newberg, A., Alive, A., Baime, M., et al. (2001). The measurement of regional cerebral blood flow during the complex cognitive task of meditation: a preliminary SPECT study. *Psychiat. Res.* 106,113-122.

Nicholas, W. (2009). The Faith Instinct: How Religion Evolved and Why It Endures?

Pande, G.C. Talk delivered at Indian Institute of Advanced Study, Shimla.

Povinelli, D.J. and Preuss, T.M. (1995). Theory of Mind, evolutionary history of a cognitive specialization. *TINS* 18,418.

Pribram Karl, H. and Bradey, R. (1998) "The Brain, the Me and the I. In M. Ferrari and R. Sternberg (eds) *Self Awareness: Its nature and Development*, New York: Guilford Publication.

Ramachandran, V.S. (2006) Quoted by Alison Motluk.

Ramachandran, V.S., Quoted by Malcom Jeeves

Rao, S.L. (2003). Neuropsychology of Consciousness. *In On Mind and Consciousness (eds)* C. Chakraborti, M.K. Mandal, R.B. Chatterjee, Indian Institute of Advanced Study, Shimla, pp 359-363.

Shirley T. (1990). Swami Vivekananda Yoga Anusandhana Samsthana, Bangalore.

Sorondo and Marcelo Sanchez. (2007). Philosophy, Science and Faith. In What is our Real Knowledge About the Human Being: Pontificia Academia Scientiarum, Scripta Varia 109, *Vatican City*, pp 127-160.

Sperry, R.W. (1995). A search for beliefs to live by consistent with science. In Cosmic Beginning and Human Emotion. Ed. C.L. Mathews and A.A. Varghese, *Open Court, Illinois,* chapter 18, pp 313-336.

Sri Aurobindo, (2001). *The Life Divine: Sri Aurobindo Ashram Trust,* Pondicherry, Sixth Edition.

Swami Harshanand, (2002). Science, religion and spirituality. In *Consciousness and Genetics ed.* S. Menon, A. Sinha, B.V. Sreekantan, National Institute of Advanced Study, Bangalore, pp. 200-207.

Swami Jitatmananda: Unity of Mind and Matter: Science Spirituality and Future Society: Address given at a Seminar held at New Delhi on 5th January 2003, organized by the Indian Institute of Advanced Study, Shimla.

Tandon, P.N. (1992). *Consciousness: Clinical and beyond.* Presidential Address, Proc. Indian National Science Academy 1993

Telles, S. (2002). Neural plasticity and yoga. In Sangeetha Menon, Anindya Sinha and B.V. Sreekantan (Eds.) *Consciousness and Genetics* (pp 275-282), National Institute of Advanced Studies, Bangalore, India.

Trimble Michael, R. (2007). The Soul in the Brain: *The Cerebral Basis of Language, Art and Belief.* The Johns Hopkins University Press.

Van Inwagen and Peter, (2006). Our deepest beliefs about ourselves. *Pontifical Academy,* p 110.

Varela, F., Lachan, J.P., Rodrigues, E. and Martinez, J. (2001). The brain web: phase synchronization and large scale integration. *Nat. Rev. Neu*rosci. 2,229-39.

Wallace, R.K. (1970). Physiological effects of Transcendental Meditation: *Science* 167,1751-54.

Wallace, R.K., Benson, H.M. and Wilson, A.F. (1971). A Wakeful hypometabolic physiological state. *Amer. J. Physiol.* 221,795-799.

Ward and Keith, (2004). Quoted in New Scientist.

Zubieta and Jon-Kar. (2005): *Nature Neuroscience* 25,7754.

In: Expanding Horizions of the Mind Science(s)   ISBN: 978-1-62808-705-5
Editors: P.N. Tandon, R.C. Tripathi and N. Srinivasan   ©2013 Nova Science Publishers, Inc.

Chapter 14

COMPARATIVE AND EVOLUTIONARY ASPECTS OF COGNITION

Soumya Iyengar
National Brain Research Centre, Manesar, India

ABSTRACT

Cognitive functions such as attention, learning, memory, language, perception, emotions, social interactions, judgement and empathy have long been thought to be the preserve of mammals, especially humans. These aspects of cognition have been attributed to the presence of an expanded neocortex (six-layered cortex) in mammalian brains, which is connected to the thalamus and basal ganglia. Different kinds of information which are processed by these cortical-thalamic-basal ganglia loops are ultimately conveyed to the motor system, through corticospinal tracts for the performance of behaviour. Recently however, a large body of literature has demonstrated that birds, an equally successful vertebrate species, also perform some of the 'higher' cognitive functions at par with mammals such as primates. The ability of some species of birds such as crows, parrots and songbirds to perform tasks such as tool use, food caching (which requires memorization), the ability to count objects, vocal mimicry and birdsong (akin to language) stems from the presence of the pallial component of their brains which is not organized into laminae but is homologous to the mammalian cortex. Further, birds also possess well-defined basal nuclei, thalamic nuclei and other brain regions interconnected by neural circuits which are homologous to the cortical-thalamic-basal ganglia loops present in mammalian brains. Since birds and primates exist in the same ecological niches, it is important to study how differently organized brains have evolved to perform various cognitive functions and face similar challenges to survive.

One of the definitions of the term 'contrapunto' is 'a contrasting but parallel element, item, or theme' as in a musical composition. Perhaps it is surprising to begin with this definition, but this chapter attempts to compare the structure of mammalian brains with another highly successful vertebrate species (birds) and also make the case that some of the higher cognitive functions are not exclusively performed by organisms which have long been considered 'higher' on the evolutionary scale. The term 'cognition' encompasses the different processes by which the brain represents and makes sense of the myriad forms of information it receives, to give a unified sense of the world. Some of the main aspects of cognition include attention, learning and memory, language and perception as well as motor responses. A number of studies have focused on emotions, judgement, empathy and social interactions in mammals, especially in primates. Humans undoubtedly excel in and outperform all other species in all aspects of cognition. Recent studies have however, demonstrated that a number of the so-called higher cognitive functions are performed with equal ease by different species of birds, whose mental faculties have been referred to in a derogatory manner as 'bird-brains'.

1. EVOLUTION OF THE CORTEX/PALLIUM AND ITS CONNECTIONS

The misconception that birds are inferior to mammals in terms of their cognitive abilities came about with the idea that they do not possess a laminated cortex – which has been suggested to endow humans and other primates with cognitive capabilities that other organisms lack. However, besides finding that certain species of birds are adept at performing certain tasks, recent evidence has also suggested that birds do indeed possess a cortex, albeit differently organized than the mammalian cortex (Reiner, 2005; Jarvis et al., 2004 for review).

Edinger presented a unified theory of brain evolution nearly a century ago and classified the mammalian cortex as paleocortex (oldest cortex, three layered), archicortex (archaic cortex with three to four layers) and neocortex (new or six-layered, Edinger, 1908, Ariens-Kapper, 1909). According to his theory, the basal ganglia formed a core around which were added the paleocortex, archicortex and finally the neocortex. The outer part of the telencephalon (the neocortex) was dramatically expanded in humans compared to smaller mammals and was correlated with higher cognitive capabilities in humans. Since large parts of the avian brain do not possess the classic six-layered pattern of the neocortex seen in mammals (Figure 1A), Edinger further posited that the avian brain consists mostly of basal ganglia which was divided into the hyperstriatum (hypertrophied striatum), paleostriatum (older striatum which was the precursor of the mammalian amygdala), neostriatum (new striatum which would give rise to the caudate and putamen in mammals) and archistriatum. The fact that birds only possessed a hypertrophied basal ganglia lead to the belief that they were only capable of reacting instinctively to their environment and not higher cognitive functions. Darwin's book "The Origin of Species" (1859) further fuelled the belief that evolution of different species (and their brains) was progressive – that is, from fish to amphibians, followed by reptiles, birds, 'lower' mammals, primates and culminating in humans.

By the end of the 20th century, a battery of newer techniques were developed to study brain structure such as immunohistochemistry, methods to detect gene products and the advent of neuroanatomical tracing techniques for visualizing connections between different divisions of the brain lead to the accumulation of vast amounts of connectional, morphological and genetic data. A number of comparative studies were performed using these newly developed methods to demonstrate that there were several homologies between avian and mammalian brains.

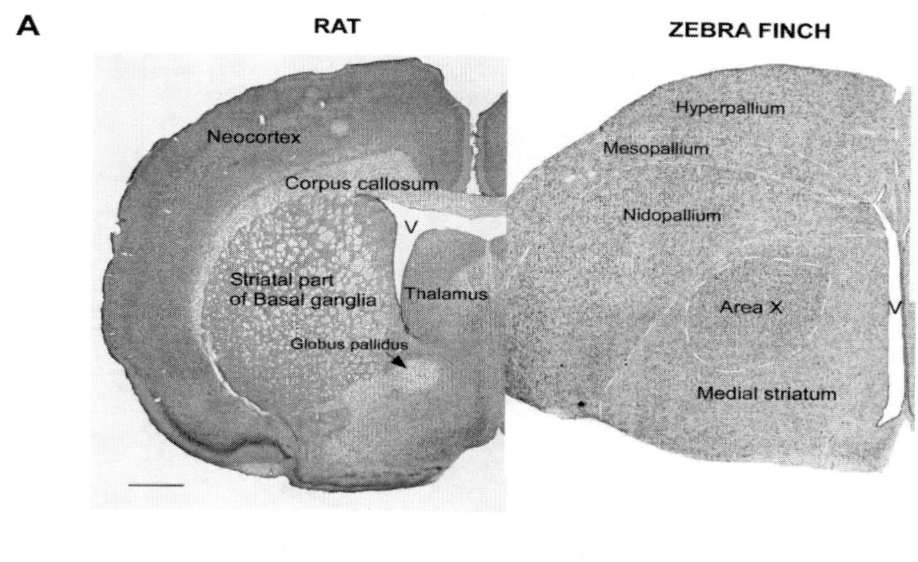

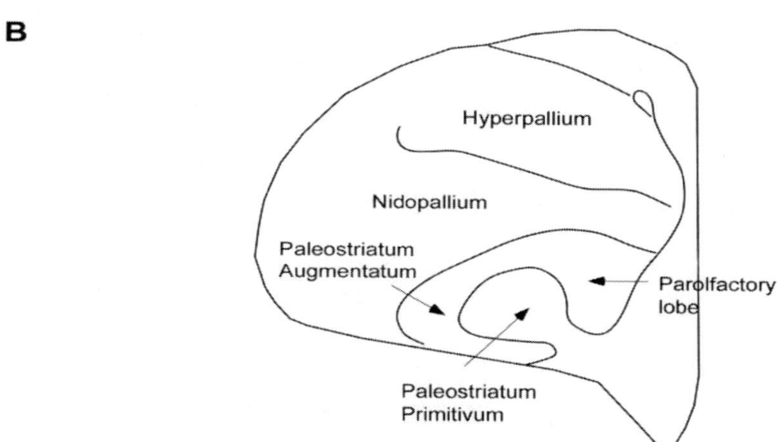

Figure 1. (A) Comparison between mammalian and avian brain. Coronal sections of the left hemisphere of a rat brain (left) and a zebra finch (songbird) brain (right). Dorsal is at the top and medial is towards the right. The neocortex is typically six-layered and is dorsal to the corpus callosum and the basal ganglia. The avian homologue of the cortex in birds consists of the hyperpallium, mesopallium and nidopallium at this level of the brain. Note that the organization of neurons in the avian brain is nuclear and not laminated. At this level, the basal ganglia consist of Area X and medial striatum, earlier called the parolofactory lobe. (B) A schematic of the brain of a non-singing bird showing the hyperpallium, mesopallium and nidopallium which surround the avian basal ganglia [composed of the parolfactory bulb, paleostriatum augmentatum (a counterpart of the mammalian striatum) and paleostriatum primitivum (a counterpart of the globus pallidus in mammals)).

The data on avian brain structure was also supported by a large body of behavioural research (see next section) which demonstrated that the avian brain performs many more cognitive functions than it was earlier given credit for.

Initial studies demonstrated that high levels of AChE (acetylcholinesterase, the enzyme that metabolizes the neurotransmitter acetylcholine) were present in the neostriatum of mammals but not in the cortex.

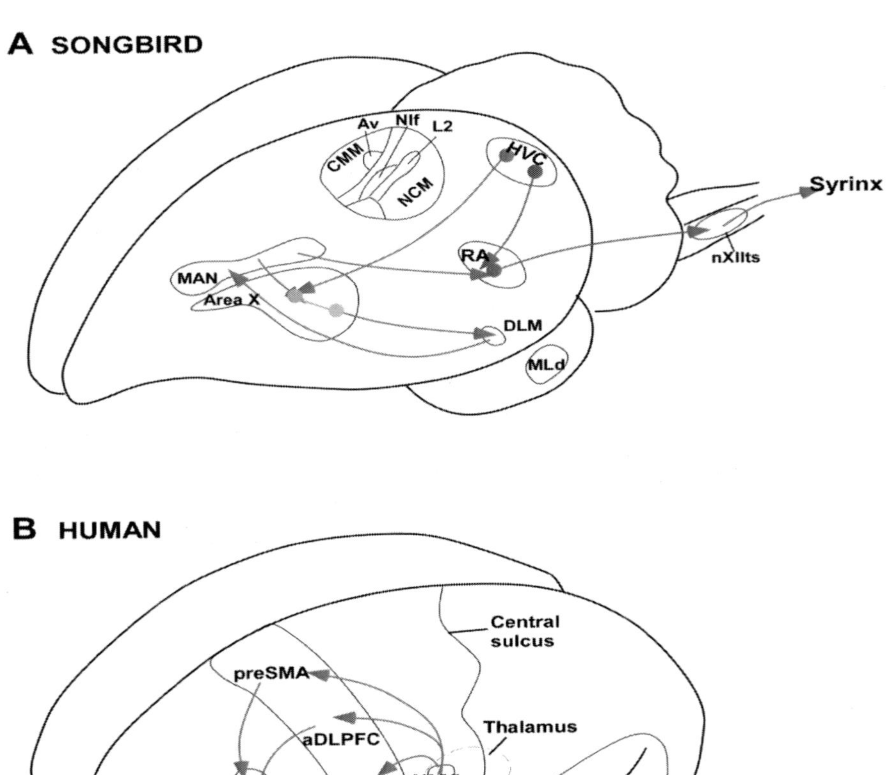

Figure 2. (A) Schematic figure of the neural circuitry in songbirds (adapted from Jarvis, 2004). Distinct sets of neurons in HVC (higher vocal centre, a nidopallial nucleus), project to RA (robust nucleus of the arcopallium) and Area X of the avian basal ganglia. The anterior pathway (shown in red) consists of projections from HVC to Area X, which consists of striatal (light blue) and pallidal (pink) neurons. The GABAergic pallidal neurons project to the thalamic nucleus DLM. DLM projects to the nidopallial nucleus LMAN which projects to Area X as well as RA. The posterior pathway (blue) consists of projections from HVC to RA which in turn innervates motor neurons in the tracheosyringeal part of the hypoglossal nucleus The nXIIts innervate muscles of the vocal organ, syrinx. (B) In humans, the anterior pathway (red) consists of projections from the premotor cortex to the striatum (light blue) which projects to the globus pallidus (pink). Pallidal neurons project to the ventral anterior and ventral lateral nucleus in the thalamus which then project back to the pre-supplementary motor cortex and the anterior dorsolateral prefrontal cortex. The posterior pathway (blue) consists of projections from the motor cortex representing the face, which projects to the nucleus ambiguus which lies in the brainstem and innervates muscles of the larynx. Abbreviations, aDLPC, anterior dorsolateral prefrontal cortex, pre-SMA, pre-supplementary motor cortex, FMC, face motor cortex, aC, anterior caudate, aP, anterior globus pallidus, Am, nucleus ambiguus, VA/VT, ventral anterior and ventral lateral nuclei of the thalamus.

In birds, it was found that the outer part of the brain (earlier thought to be part of a hypertrophied striatum) was not labelled with AChE. Instead, AChE expression was restricted to an inner core in the bird brain called paleostriatum augmentatum (PA) and lateral parolfactory lobe (LPO, Karten, 1969). Further, the mammalian neostriatum was rich in dopaminergic terminals originating from the substantia nigra pars compacta in the midbrain as well as neurons rich in substance P and enkephalin which project to the internal and external parts of the globus pallidus, respectively. As with the localization of AChE in the mammalian striatum, dopaminergic terminals from the midbrain were seen only in PA and LPO of birds (Jurio and Vogt, 1967, Karten and Dubbeldam, 1973) and they were found to possess substance P and enkephalin neurons which projected to an adjacent region called the paleostriatum primitivum (PP, Figure 1B, Graybiel, 1986, Smeets, 1992, Steiner and Gerfen, 1998, Jiao et al., 2000). Interestingly, the functions of these neurons may also be conserved in mammals as well as birds: substance P neurons may be important for planning, whereas enkephalin neurons may be important for inhibiting movements for the smooth performance of movements (Graybiel, 1997, Perkel and Farries, 2000). Connectional studies using neuroanatomical tracers have further shown that there are topographically organized cortico-thalamo-basal ganglia loops in the brains of mammals which are important for motor functions as well as higher cognitive processes such as speech (Alexander et al., 1986; Alexander and Crutcher, 1990; Parent and Hazrati, 1995). Interestingly, nuclei which were involved in learning and singing and were earlier designated 'striatal' in songbirds formed similar topographically organized loops with other nuclei in the avian basal ganglia and thalamus (Iyengar et al., 1999; Johnson et al., 1995; Luo et al., 2001; Nixdorg-Bergweiler et al., 1995; Vates and Nottebohm, 1995). Recent evidence from embryological studies has demonstrated that areas in the avian brain designated 'hyperstriatum' and 'neostriatum' express transcription factors such as EMX1, PAX6 and TBR1, which are only present in the mammalian pallium (cortex) and not in the striatum (Medina and Reiner, 2000; Puelles et al., 2000; Smith-Fernandez et al., 1998). Further, markers like BDNF (brain derived neurotrophic factor) and the glutamate receptor mGluR2 (Li et al., 2000; Wada et al., 2004) which are expressed mainly by the mammalian cortex are expressed in the avian hyperstriatum, neostriatum and archistriatum, also suggesting that these regions are in fact, cortical/pallial in origin.

A number of studies have shown that evolution is not linear or progressive as proposed by Darwin and Edinger. Similar traits or characters (such as similar behaviour patterns) can emerge independently and more than once during evolution. In fact, Carroll (1998) and Evans (2000, cf. Jarvis , 2004) have demonstrated that birds evolved *after* mammals and not before. Their studies have shown that the common ancestor of mammals and birds (stem amniotes) existed 320 million years ago (MYA). The ancestors of birds and reptiles (Sauropsids) and mammals (Therapsids) subsequently arose from the stem amniotes approximately 250 MYA in the Triassic period. Whereas modern-day mammals arose about 203 MYA, modern-day birds arose about 135 MYA. Karten's group (Karten, 1969; Karten and Shimizu, 1989; Nauta and Karten, 1970) have shown that the outer part of the avian brain (pallium) can constitute as much as 75% of the brain in different species of birds. They have also presented a nuclear-layered hypothesis which suggests that the common ancestor (the stem amniote) of birds, reptiles and mammals possessed a nuclear pallium which evolved into a laminar pallium in mammals. The six-layered neocortex was probably present in therapsids 250 million years ago, whereas the avian hyperpallium probably evolved more recently.

Since there were a number of discrepancies in earlier literature regarding the nomenclature of different brain regions in birds, a consortium of eminent neuroscientists (Jarvis, 2004) renamed these areas in 2004 with a view towards easier comparison between different brain regions in mammals and birds. Of particular relevance for the present essay is the fact that the subdivisions of the pallial domain in birds have been renamed Hyperpallium, Mesopallium, Nidopallium and Arcopallium, replacing hyperstriatum, neostriatum and archistriatum.

2. THE COGNITIVE ABILITIES OF BIRDS

Having established the fact that the avian brain possesses a fairly well-developed pallium, the next question to be answered is what is the avian pallium capable of, in terms of behaviour? Although it is not possible to elaborate on all aspects of avian behaviour, those seen in corvids (including species such as crows, jays, ravens, jackdaws) and learned vocalizations performed by passerines (perching birds), parrots and hummingbirds are the best studied and have been reviewed below.

2.1. Corvids

A well-known fable by Aesop relates the story of a thirsty crow (a corvid) which raises the level of water in a pitcher by placing stones in it. In a modern-day experiment, rooks (a related species of corvids) were faced with the complex task of reaching a worm floating on a small quantity of water in a container which they were unable to reach with their beaks (Bird and Emery, 2009). The experimenters showed that the rooks were able to raise the level of water by adding stones provided to them and even learned to add larger stones rather than smaller ones to get to their target sooner. Interestingly, the rooks also quickly learned that they would not be able to get a reward (the worm) if water was replaced with sawdust. Other examples of intelligence in corvids include tool use and manufacture by New Caledonian crows. In this species, tools of different kinds were manufactured to extract different kinds of insect larvae from holes in trees and under leaves (Chappell and Kacelnik, 2002; 2004). These behaviours were also tested in the lab, wherein crows were capable of choosing wire of the correct length and diameter and further, bending it to the correct angle to extract food from a narrow container (Weir et al., 2002).

It has recently been suggested that birds (in particular corvids) have a 'Theory of Mind', that is when an individual can think about another's mental state, earlier thought to be restricted to primates. A number of experiments have been performed on birds which store their food in caches during lean periods as in winter. Food caching is a formidable task considering some species such as Clark's nutcrackers can have up to 30,000 caches to be recovered over a period of 6 months (Balda and Kamil, 1992). It is important for food-storing birds to remember not only where the food was stored, but also whether it was perishable and when it was stored (since perishable food needs to be consumed before non-perishable food). Not only has it been shown that food-storing birds such as western scrub-jays will consume perishable food before they start on non-perishable stores (Bednekoff and Balda, 1996; Clayton and Dickinson, 1998; 1999; Clayton et al., 2001; Heinrich and Pepper, 1998), recent studies have shown that those birds who pilfer others' stores are more careful about protecting their own caches (Clayton et al., 2001; Emery and Clayton, 2004). The strategies that are employed (both in nature and in laboratory settings) are avoiding being seen by another bird, hiding food at a site that cannot be seen by an observer, re-caching stores multiple times and hiding inedible material in place of caches when observed (Bugnyar and Kotrschal, 2002; 2004; Dally et al., 2004; Dally et al., 2006a; Emery and Clayton, 2001; Emery et al., 2004). Interestingly, it has also been seen observed that food-storing birds will not re-cache their food if their mate is present at the time of caching (Dally et al., 2006b). These results suggest that birds can attribute their own experience as thieves to others and take measures to protect their food stores. As explained in the previous section, the higher cognitive abilities of birds likely stem from the avian homologue of the mammalian prefrontal cortex, that is, a part of the pallium called the caudolateral nidopallium (CDLN). Interestingly, the organization of this region is similar to that of the mammalian prefrontal cortex in that it is connected to secondary sensory areas of all modalities (Emery, 2006 for review) and receives information from multiple sources in the brain (Kroener and Gunturkun, 1999; Leutgeb et al., 1996; Metzger et al., 1998). Further, lesions in the CDLN or blocking D1 (dopamine Type 1) receptors present in this region lead to impairments in delayed alternation, reversal learning, working memory tasks and some kinds of visual discrimination tasks (Reiner, 1986; Mogensen and Divac, 1993; Gagliardo et al.,1996; Hartmann and Gunturkun, 1998; Diekamp et al., 2002a; Aldavert-Vera et al., 1999).

2.2. Songbirds and Parrots

Vocal learning is the 'ability to acquire vocalizations through imitation rather than instinct' (Jarvis, 2004). Vocal learners imitate the sounds produced by conspecifics and can use these sounds for cultural transmission. Whereas the most advanced form of vocal learning is seen in humans, the only other species of mammals that learn their vocalizations are bats and cetaceans (dolphins). Three distantly related species of birds (parrots, hummingbirds and songbirds) also possess neural circuits which are highly specialized to enable them to learn their vocalizations. Of the different species of birds, some can imitate and produce thousands of songs whereas others can only produce one highly stereotyped song. Jarvis (2004) has proposed that genetic and epigenetic constraints may have influenced the evolution of vocal learning and that the auditory pathways for vocal learning in birds and mammals may have been inherited from the stem amniote ancestors (320 MYA) of both orders. It is also possible that the vocal control system evolved independently in birds and mammals or that most birds and mammals can learn vocalizations to some degree, and this system has become highly specialized in certain species of songbirds and mammals. Finally, Jarvis (2004) suggests that possibly some degree of each of these three hypotheses could be applied to the evolution of vocal learning and the circuits underlying them.

The three groups of avian vocal learners have homologous nuclei arranged as anterior and posterior pathways. Interestingly, the human brain also has anterior and posterior neural circuits which are similar in organization to their avian counterparts and are specialized for vocalization. The posterior pathway in songbirds (and homologous projections in parrots) consists of a projection from a region called NIf (nucleus interfacialis) to the pallial nucleus HVC (higher vocal centre, McCasland, 1987). HVC projects to RA (Robust nucleus of the arcopallium, Nottebohm et al., 1976; 1982) which then projects to the nXIIts (12th nerve nucleus). The 12th nerve nucleus consists of a pool of motor neurons which innervate the musculature of the syrinx or the vocal organ in songbirds (Vicario, 1994, Figure 2A). In humans, a similar pathway exists between the motor cortex representing the face. The face motor cortex projects to nucleus ambiguus which controls laryngeal muscles as well as the 12th nerve nucleus which innervates tongue musculature (Kuypers 1958a, b; Jurgens, 1998; Zhang et al., 1995, Figure 2B). The nucleus RA is similar to Layer 5 of the motor cortex in mammals in that it projects to the nXIIts, as in the case of the pyramidal tract. RA-projecting neurons in HVC are similar to Layers 2 and 3 of the mammalian cortex since RA represents the mammalian Layer 5 (Jarvis, 2004).

The anterior forebrain pathway has been considered the avian homologue of the cortical − thalamic − basal ganglia loops in mammals (Johnson et al. 1995; Vates and Nottebohm, 1995; Iyengar et al., 1999; Luo et al., 2001, Figure 2A and B). In songbirds, a population of neurons in the pallial nucleus HVC distinct from those which project to RA send projections to Area X, a nucleus of the avian basal ganglia (Nottebohm et al., 1976; 1982). Area X is not identical to the mammalian basal ganglia since it contains a mixture of striatal and pallidal neurons (Reiner et al., 2004a, b; Carrillo and Doupe, 2004; Farries and Perkel, 2002; Farries et al., 2005). Pallidal neurons in Area X project to the thalamic nucleus DLM (nucleus dorsolateralis anterior, pars medialis, which in turn projects to the pallial nucleus LMAN (lateral magnocellular nucleus of the anterior nucleus, Carrillo and Doupe, 2004, Reiner et al., 2004b). Loops are formed within the song control system since LMAN projects to Area X as well as RA (see above). Jarvis (2004) has suggested that projection neurons of LMAN are similar to neurons of Layer 3 and upper Layer 5 of the mammalian premotor cortex which project to spiny neurons in Area X as well as RA (homologous to motor cortex). The pallidal DLM-projecting neurons in Area X are similar to those in the mammalian globus pallidus which project to the VL (ventral lateral) and VA (ventral anterior) thalamic nuclei, both of which project to Layer 3 of the premotor cortex (equivalent to LMAN, Jacobson et al., 1975, Alexander et al., 1986, Alexander and Crutcher, 1990).

Besides the vocal control pathways, auditory pathways are very similar in mammals, birds and even reptiles (Vates et al., 1996; Carr and Code, 2000; Webster et al., 1992). Hair cells in the ear send auditory input to sensory neurons in the cochlea, which then send their input to nuclei in the brainstem in different species. Brainstem lemniscal nuclei send their output to the midbrain [a region called MLd (nucleus mesencephalicus lateralis, pars dorsalis) in birds and the inferior colliculus in mammals]. The midbrain projects to auditory thalamic regions [Ov (nucleus ovoidalis) in birds and the medial geniculate nucleus in mammals]. Auditory information is finally routed to the Layer 4 of the primary auditory cortex in mammals and Field L2 in birds. Avian pallial areas L1 and L3 appear to be similar to Layers 2 and 3 of the primary auditory cortex in mammals (mammalian Layer 3 receives sensory input from Layer 4, Wild et al., 1993, Karten, 1991). Besides L1 and L3, the pallial areas NCM (caudal part of the medial nidopallium) and CMM (caudal part of the medial mesopallium) in birds may also be considered similar to Layers 2 and 3 of the primary auditory cortex since they form reciprocal connections and receive input from Field L2 (Jarvis, 2004).

Songbirds have long been used as models for research on vocal learning since lesions in the vocal control circuits produce deficits which are similar to those seen in humans. Lesions in HVC and RA produce muteness which is similar to the effects of a stroke in the human face motor cortex (Nottebohm et al., 1976; Simpson and Vicario, 1990; Jurgens et al., 1982). In contrast, lesions in the anterior pathway (anterior forebrain pathway) of birds cause deficits similar to those caused by lesions in the anterior part of the human premotor cortex, that is, prevention of vocal learning and inducing problems in learning the sequence of sounds. Whereas lesions in LMAN during the sensitive period for song learning produce premature stereotypy in vocalizations (Bottjer et al., 1984), lesions in the human insula and Broca's area also lead to poor imitation of speech, poor syntax and the construction of phonemes into words and words into sentences (Benson and Ardila, 1996). Lesions in Area X (avian basal ganglia) lead to the production of songs which do not become stereotyped (poor imitation, Scharff and Nottebohm, 1991) and those in the head of the caudate and putamen in humans produce verbal aphasias and amusias (Cummings, 1993; Leicester, 1980; Damasio et al., 1982; Alexander et al., 1987). Finally, lesions of VA/VL in humans and their avian counterpart (DLM) produce aphasias in both species (Halsema and Bottjer, 1991; Graff-Radford et al., 1985).

In addition to similarities between the mammalian cortex and avian pallium, other regions of the brain in both orders are markedly similar. One well-known example is the mesolimbic dopaminergic pathway which is important for the motivation to perform various goal-directed behaviours (Berridge and Robinson, 1998; Robinson et al., 2005). It consists of a dopaminergic projection from the VTA-SNc complex which innervates the nucleus accumbens in the ventral striatum in mammals (De Vries and Shippenberg, 2002; Tomkins and Sellers, 2001; Xi and Stein, 2002). The dopaminergic neurons are maintained in a state of tonic inhibition by GABAergic neurons in VTA-SNc. When endogenous opioids bind to μ-opioid receptors expressed on GABAergic neurons in VTA-SNc, the tonic inhibition that they maintain over dopaminergic VTA-SNc neurons is released, leading to the release of dopamine in nucleus accumbens and the modulation of motivated behaviours. The mesolimbic dopaminergic pathway is similarly organized in songbirds such as zebra finches (Lewis et al., 1981; Gale and Perkel, 2006; Khurshid et al., 2010) and the effects of opioid action (Khurshid et al., 2010) as well as the release of dopamine in the striatum (Sasaki et al., 2006) during song (a highly motivated behaviour) is also similar to that in mammals.

Songbirds have also lent themselves to studies on FOXP2, a gene linked to language disorders (Enard et al., 2002), which is known to be involved in Specific Language impairment. Interestingly, FoxP2 (the avian homologue of the FOXP2 gene) was detected in the pallial, striatal, and thalamic regions important for singing and song learning in songbirds such as zebra finches (Ferland et al., 2003; Haesler et al., 2004; Lai et al., 2003; Takahashi et al., 2003; Teramitsu et al., 2004). The role of FoxP2 in learned vocalizations has been demonstrated in zebra finches in an experiment which would have been difficult in other vocal learners. Haesler et al. (2007) used shRNAs to knock down the expression of FoxP2 in Area X of juvenile zebra finches which led to the significant impairment in the songs that they performed. This study also demonstrated that FoxP2 was required for normal learning since FoxP2 knockdown birds did not imitate their tutors precisely.

Besides songbirds, parrots have long been held in fascination for their abilities at vocal mimicry. Parrots are open-ended vocal learners (Cruikshank et al., 1993) and can learn to use human speech (Pepperberg, 1981; 1999) to label different objects as in the well-documented case of the African Grey parrot Alex. Pepperberg (1997a; 2004) found that parrots would learn well only when the training was referential, functional and rich. Alex was trained to vocally label a hundred objects with different colours, shapes and made from different materials and could produce 20,000 vocalizations (Pepperberg et al., 1991). He was capable of categorizing different objects and discriminating between them. He was also capable of combining labels for objects in novel ways and asking for or refusing specific objects (Pepperberg, 2007). Interestingly, Alex was found to practice his vocalizations when he was left to himself (Pepperberg et al., 1991). Further, Alex demonstrated object permanence [as do corvids (Pepperberg et al., 1997b)], that is, the ability to keep track of objects and individuals who are currently out of sight and counting, as well as the concept of personal pronouns ['I' versus 'you'(Pepperberg et al., 2000)]. Pepperberg (2006) has also demonstrated that Alex was able to quantify six or fewer items, with some understanding of the concept of zero or 'none' and his skills at addition were similar to non-human primates and young children.

CONCLUSION

There are over 9000 species of birds in diverse regions of the world. Some species of birds such as corvids, parrots and songbirds are similar to primates in that they occupy similar ecological niches, face similar challenges in foraging for food and live in large social groups. Whereas primates are equipped with the six-layered prefrontal cortex for cognitive functions such as working memory, executive control of behaviour, causal reasoning, flexibility and imagination to survive in their environment, the caudal nidopallium which is not laminated performs similar tasks in birds. A number of behavioural studies have shown that despite brain evolution being divergent in mammals and birds, evolution of brain function is convergent. Since there are many similarities in the functioning of the brain in mammals and birds, it would be important to study the evolution of intelligence and consciousness in both orders of vertebrates.

ACKNOWLEDGMENTS

Dr. Soumya Iyengar's laboratory is funded partly by NBRC and a grant from DBT (no. BT/PR6615/MED/14/857/2005).

REFERENCES

Aldavert-Vera, L., Costa-Miserachs, D., Divac, I. and Delius, J.D. (1999). Presumed 'prefrontal cortex' lesions in pigeons, effects on visual discrimination performance. *Behavioral. Brain Research, 102,* 165–170.

Alexander, G.E., Belong, M.R. and Strick P.L. (1986). Parallel organization of functionally segregated circuits linking basal ganglia and cortex. *Annual Reviews of Neuroscience, 9,* 357–381.

Alexander, G.E. and Crutcher, M.D. (1990). Functional architecture of basal ganglia circuits, neural substrates of parallel processing. *Trends in Neuroscience, 13,* 266–271.

Alexander, M.P., Naeser, M.A. and Palumbo, C.L. (1987). Correlations of subcortical CT lesion sites and aphasia profiles, *Brain, 110,* 961–991.

Ariens-Kappers, C.U. (1909). The phylogenesis of the paleo-cortex and archi-cortex compared with the evolution of the visual neo-cortex. *Archives of Neurology and Psychiatry 4*, 161–173.

Balda, R.P. and Kamil, A.C. (1992). Long-term spatial memory in Clark's nutcrackers, Nucifraga columbiana. *Animal Behaviour, 44*, 761–769.

Bednekoff, P.A. and Balda, R.P. (1996). Observational spatial memory in Clark's nutcrackers and Mexican jays. *Animal Behaviour, 52*, 833–39.

Benson, D.F. and Ardila, A. (1996). Aphasia, A Clinical Perspective. Oxford University Press, New York, 1996.

Berridge, K.C. and Robinson, T.E. (1998). What is the role of dopamine in reward, hedonic impact, reward learning, or incentive salience? *Brain Research Reviews, 28*, 309–69.

Bird, C.D. and Emery, N.J. (2009). Insightful problem solving and creative tool modification by captive nontool-using rooks. *Proceedings of the National Academy of Sciences of the United States of America, 106*, 10370–10375.

Bottjer, S.W., Meisner, E.A. and Arnold, A.P. (1984). Forebrain lesions disrupt dvevelopment but not maintenance of song in passerine birds. *Science, 224*, 901-903.

Bottjer, S.W. and Johnson, F. (1997). Circuits, hormones, and learning, vocal behavior in songbirds. *Journal of Neurobiology, 33*, 602–618.

Bugnyar, T. and Kotrschal, K. (2002). Observational learning and the raiding of food caches in ravens, Corvus corax, is it 'tactical' deception?, *Animal Behaviour, 64*, 185–195.

Bugnyar, T. and Kotrschal K (2004). Leading a conspecific away from food in ravens (Corvus corax)? *Animal Cognition, 7*(2), 69-76.

Carr, C.E., Code, R.A. (2000). The central auditory system of reptiles and birds. In, Dooling, R.J., Fay, R.R., and Popper, A.N., editors. *Comparative Hearing, Birds and reptiles. 13. Springer, New York.* p.197-248.

Carrillo, G.D. and Doupe, A.J. (2004). Is the songbird Area X striatal, pallidal, or both? An anatomical study. *Journal of Comparative Neurology, 473*, 415–437.

Carroll, R.L. (1988). Birds evolved after mammals. In *Vertebrate Paleontology and Evolution 1–13* (W. H. Freeman, New York, 1988).

Chappell, J. and Kacelnik, A. (2002). Tool selectivity in a nonprimate, the New Caledonian crow (Corvus moneduloides). *Animal Cognition, 5*, 71–78.

Chappell, J. and Kacelnik, A. (2004). Selection of tool diameter by New Caledonian crows Corvus moneduloides. *Animal Cognition, 7*, 121–127.

Clayton, N.S. and Dickinson, A. (1998). Episodic-like memory during cache recovery by scrub jays. *Nature, 395*, 272–274.

Clayton, N.S. and Dickinson, A. (1999). Scrub jays (Aphelocoma coerulescens) remember the relative time of caching as well as the location and content of their caches. *Journal of Comparative Psychology, 113*, 403–416.

Clayton, N.S., Griffiths, D.P., Emery, N.J. and Dickinson, A. (2001). Elements of episodic-like memory in animals. *Philosophical Transactions of the Royal Society B: Biological Sciences, 356*, 1483–1491.

Cruickshank, A.J., Gautier, J.P. and Chappuis, C. (1993). Vocal mimicry in wild African Grey parrots Psittacus erithacus. *Ibis, 135*, 293–299.

Cummings, J.L. (1993). Frontal-subcortical circuits and human behavior. *Archives of Neurology, 50*, 873–880.

Dally, J.M., Emery, N.J. and Clayton, N.S. (2004). Cache protection strategies by western scrub-jays (Aphelocoma californica), hiding food in the shade. *Philosophical Transactions of the Royal Society B: Biological Sciences, 271*, S387–S390.

Dally, J.M., Clayton, N.S. and Emery, N.J. (2006a). The behaviour and evolution of cache protection and pilferage. *Animal Behaviour, 72*, 13–23.

Dally, J.M., Emery, N.J. and Clayton, N.S. (2006b). Food-caching western scrub-jays keep track of who was watching when. *Science, 312,* 1662–1665.

Damasio, A.R., Damasio, H., Rizzo, M., Varney, N. and Gersh, F. (1982). Aphasia with nonhemorrhagic lesions in the basal ganglia and internal capsule. *Archives of Neurology, 39*, 15–24.

De Vries, T.J. and Shippenberg, T.S. (2002). Neural systems underlying opiate addiction. *Journal of Neuroscience, 22*, 3321–3325.

Diekamp, B., Kalt, T., Ruhm, A., Koch, M. and Gunturkun, O. (2000). Impairment in a discrimination reversal task after D1 receptor blockade in the pigeon 'prefrontal cortex'. *Behavioral Neuroscience 114*, 1145–1155.

Durand, S.E., Heaton, J.T., Amateau, S.K. and Brauth, S.E. (1997). Vocal control pathways through the anterior forebrain of a parrot (Melopsittacus undulatus). *Journal of Comparative Neurology, 377*, 179–206.

Edinger, L. (1908). The relations of comparative anatomy to comparative psychology, *Journal of Comparative Neurology and Psychology, 18*, 437–457.

Emery, N.J., and Clayton, N.S. (2001). Effects of experience and social context on prospective caching strategies by scrub jays. *Nature, 414*, 443–446.

Emery, N.J., and Clayton, N. (2004). The mentality of crows, convergent evolution of intelligence in corvids and apes. *Science, 306*, 1903–1907.

Emery, N.J., Dally, J.M., and Clayton, N.S. (2004). Western scrub-jays (Aphelocoma californica) use cognitive strategies to protect their caches from thieving conspecifics. *Animal Cognition, 7*, 37–43.

Emery, N.J. (2006). Cognitive ornithology, the evolution of avian intelligence, *Philosophical Transactions of the Royal Society B: Biological Sciences, 361*, 23–43

Enard, W., Przeworski, M., Fisher, S.E., Lai, C.S.L., Wiebe, V., Kitano, T., Monaco, A.P., and Pääbo, S. (2002). Molecular evolution of FOXP2, a gene involved in speech and language. *Nature, 418,* 869-872.

Evans, S.E. (2000), In *Evolutionary Developmental Biology of the Cerebral Cortex* (eds Bock, G. R. and Cardew, G.) 109–113. Chicester: John Wiley and Sons.

Farries, M.A., and Perkel, D.J. (2002). A telencephalic nucleus essential for song learning contains neurons with physiological characteristics of both striatum and globus pallidus. *Journal of Neuroscience, 22*, 3776–3787.

Farries, M.A., Ding, L., and Perkel, D.J. (2005). Evidence for "direct" and "indirect" pathways through the song system basal ganglia. *Journal of Comparative Neurology, 484*, 93–104.

Ferland, R.J., Cherry, T.J., Preware, P.O., Morrisey, E.E., and Walsh, C.A. (2003). Characterization of Foxp2 and Foxp1 mRNA and protein in the developing and mature brain. *Journal of Comparative Neurology, 460*, 266–279

Gagliardo, A., Bonadonna, F., and Divac, I. (1996). Behavioural effects of ablations of the presumed 'prefrontal cortex' or the corticoid in pigeons. *Behavioural Brain Research, 78*, 155–162.

Gale, S.D. and Perkel, D.J. (2006). Physiological properties of zebra finch ventral tegmental area and substantia nigra pars compacta neurons. *Journal Neurophysiology, 96*, 2295–306.

Graff-Radford, N.R., Damasio, H., Yamada, T., Eslinger, P.J. and Damasio, A.R. (1985). Nonhaemorrhagic thalamic infarction. Clinical, neuropsychological and electrophysiological findings in four anatomical groups defined by computerized tomography. *Brain, 108*, 485–516.

Graybiel, A.M. (1986). Neuropeptides in the basal ganglia. In: *Neuropeptides in Neurologic and Psychiatric Disease*, J.B. Martin and J.D. Barchas, eds., Raven Press, NY, 135-161

Graybiel, A.M. (1997). The basal ganglia and cognitive pattern generators. *Schizophrenia Bulletin, 23*, 459–469.

Haesler, S., Wada, K., Nshdejan, A., Morrisey, E.E., Lints, T., Jarvis, E.D., Scharff, C. (2004). FoxP2 expression in avian vocal learners andnon-learners. *Journal of Neuroscience 24(13)*, 3164-3175.

Haesler, S., Rochefort, C., Georgi, B., Licznerski, P., Osten, P. and Scharff, C. (2007). Incomplete and inaccurate vocal imitation after knockdown of FoxP2 in songbird basal ganglia nucleus Area X. *PLoS Biology, 5(12)* 2885-2897.

Halsema, K.A. and Bottjer, S.W. (1991). Lesioning afferent input to a forebrain nucleus disrupts vocal learning in zebra finches. *Society for Neuroscience Abstracts 1991, 17*, 1052.

Hartmann, B. and Gunturkun, O. (1998). Selective deficits in reversal learning after neostriatum caudolaterale lesions in pigeons, possible behavioral equivalencies to the mammalian. prefrontal system. *Behavioural Brain Research, 96*, 125–133.

Heinrich, B. and Pepper, J.W. (1998). Influence of competitors on caching behaviour in the common raven, Corvus corax. *Animal Behaviour, 56*, 1083–1090.

Iyengar, S., Viswanathan, S.S. and Bottjer, S.W. (1999). Development of topography within song control circuitry of zebra finches during the sensitive period for song learning. *Journal of Neuroscience, 19*, 6037–6057.

Jacobson, S. and Trojanowski, J.Q. (1975). Corticothalamic neurons and thalamocortical terminal fields, an investigation in rat using horseradish peroxidase and autoradiography. *Brain Research, 85*, 385–401.

Jarvis, E.D., Scharff, C., Grossman, M.R., Ramos, J.A., and Nottebohm, F. (1998). For whom the bird sings, context-dependent gene expression. *Neuron, 21*, 775–788.

Jarvis, E.D. (2004). Learned Birdsong and the Neurobiology of Human Language. *Annals of the New York Academy of Sciences, 1016*, 749–777.

Jiao, Y., Medina, L., Veenman, C.L., Toledo, C., Puelles, L. and Reiner, A. (2000). Identification of the anterior nucleus of the ansa lenticularis in birds as the homologue of the mammalian subthalamic nucleus. *Journal of Neuroscience, 20*, 6998–7010.

Johnson, F., Sablan, M.M., and Bottjer, S.W. (1995). Topographic organization of a forebrain pathway involved with vocal learning in zebra finches. *Journal of Comparative Neurology, 358*, 260–278.

Jurgens, U., Kirzinger, A. and Von Cramon, D. (1982). The effects of deep-reaching lesions in the cortical face area on phonation. A combined case report and experimental monkey study. *Cortex 1982, 18*, 125–139.

Jurgens, U. (1998). Neuronal control of mammalian vocalization, with special reference to the squirrel monkey. *Naturwissenschaften, 85*, 376–388.

Karten, H.J. (1969). The organization of the avian telencephalon and some speculations on the phylogeny of the amniote telencephalon, in, J. Pertras (Ed.), Comparative and Evolutionary Aspects of the Vertebrate Central Nervous System, *Annals of the New York Academy of Sciences, 167,* 164–179.

Karten, H.J. and Dubbeldam, J.L. (1973). The organization and projections of the paleostriatal complex in the pigeon (Columba livia), *Journal of Comparative Neurology, 148,* 61–90.

Karten, H.J. and Shimizu, T. (1989). The origins of neocortex, connections and lamination as distinct events in evolution, Journal of Cognitive Neuroscience, 1, 291–301.

Karten, H.J. (1991). Homology and evolutionary origins of the "neocortex". *Brain, Behavior and Evolution, 38,* 264–272.

Khurshid, N., Jayaprakash, N., Hameed, L.S., Mohanasundaram, S. and Iyengar, S. (2010). Opioid modulation of song in male zebra finches (Taenopygia guttata), *Behavioural Brain Research, 208,* 359–370.

Kroner, S. and Gunturkun, O. (1999). Afferent and efferent connections of the caudolateral neostriatum in the pigeon(Columba livia). *Journal of Comparative Neurology 407,* 228–260.

Kuypers, H.G.J.M. (1958). Corticobulbar connexions to the pons and lower brain-stem in man. *Brain, 81,* 364–388.

Kuypers, H.G.J.M. (1958). Some projections from the peri-central cortex to the pons and lower brain stem in monkey and chimpanzee. *Journal of Comparative Neurology, 100,* 221–255.

Lai, C.S., Gerrelli, D., Monaco, A.P., Fisher, S.E. and Copp, A.J. (2003). FOXP2 expression during brain development coincides with adult sites of pathology in a severe speech and language disorder. *Brain, 126,* 2455–2462.

Leutgeb, S., Husband, S., Riters, L.V., Shimizu, T. and Bingman, V.P. (1996). Telencephalic afferents to the caudolateral neostriatum of the pigeon, *Brain Research, 730,* 173–181.

Leicester, J. (1980). Central deafness and subcortical motor aphasia. *Brain Language, 10,* 224–242.

Lewis, J.W., Ryan, S.M., Arnold, A.P. and Butcher, L.L. (1981). Evidence for a catecholaminergic projection to area X in the zebra finch. *Journal of Comparative Neurology, 196,* 347–54.

Li, X.C. and Jarvis, E. (2001). Sensory- and motor-driven BDNF expression in a vocal communication system, *Society for Neuroscience Abstracts 2001, 538.8.*

Li, X.C., Jarvis, E.D., Alvarez-Borda, B., Lim, D.A., and Nottebohm, F. (2000). A relationship between behavior, neurotrophin expression, and new neuron survival. *Proceedings of the National Academy of Sciences of the United States of America, 97*(15), 8584-9.

Luo, M., Ding, L., and Perkel, D.J. (2001). An avian basal ganglia pathway essential for vocal learning forms a closed topographic loop. *Journal of Neuroscience, 21*(17), 6836–6845.

McCasland, J.S. (1987)., Neuronal control of bird song production. *Journal of Neuroscience, 7,* 23-29.

Medina, L., and Reiner, A. (2000). Do birds possess homologues of mammalian primary visual, somatosensory and motor cortices?, *Trends in Neurosciences, 23,* 1–12.

Metzger, R.R., Jiang, S. and Braun, K. (1998). Organization of the dorsocaudal neostriatal complex, a retrograde and anterograde tracing study in the domestic chick with special emphasis on pathways relevant to imprinting. *Journal of Comparative Neurology, 395,* 380–404.

Mogensen, J. and Divac, I. (1993). Behavioural effects of ablation of the pigeon-equivalent of the mammalian prefrontal cortex. *Behavioural Brain Research, 55,* 101–107.

Nauta, W.J.H. and Karten, H.J. (1970). A general profile of the vertebrate brain, with sidelights on the ancestry of cerebral cortex, in, F.O. Schmitt (Ed.), *The Neurosciences, Second Study Program,* The Rockefeller University Press, New York, pp. 7–26.

Nottebohm, F., Stokes, T.M. and Leonard, C.M. (1976). Central control of song in the canary, Serinus canaries, *Journal of Comparative Neurology, 165*(4), 457-86.

Nottebohm, F., Kelley, D.B. and Paton, J.A. (1982). Connections of vocal control nuclei in the canary telencephalon, *Journal of Comparative Neurology, 207,* 344-357.

Nixdorf-Bergweiler, B.E., Lips, M.B., Heinemann, U. (1995). Electrophysiological and morphological evidence for a new projection of LMAN neurones toward Area X, *NeuroReport, 6,* 1729 –1732.

Parent, A. and Hazrati, L.N. (1995). Functional anatomy of the basal ganglia. The cortico-basal ganglia thalamo-cortical loop. *Brain Research Reviews, 20,* 91–127.

Pepperberg, I.M. (2010). Vocal Learning in Grey parrots, A brief review of perception, production, and cross-species comparisons, *Brain and Language, doi, 10.1016/j.bandl.2009.11.002.*

Pepperberg, I.M. (2006). Grey Parrot (Psittacus erithacus) Numerical Abilities, addition and further experiments on a zero-like concept. *Journal of Comparative Neurology, 120*(1), 1-11.

Pepperberg, I.M., Brese, K.J. and Harris, B.J. (1991). Solitary sound play during acquisition of English vocalizations by an African Grey parrot (Psittacus erithacus), Possible parallels with children's monologue speech. *Applied Psycholinguistics, 12,* 151–177.

Pepperberg, I.M. (1981). Functional vocalizations by an African Grey parrot (Psittacus erithacus), *Zeitschrift für Tierpsychologie, 55,* 139–160.

Pepperberg, I.M. (1999). The Alex studies. Cambridge, MA, *Harvard University Press.*

Pepperberg, I.M. (1997a). Social influences on the acquisition of human-based codes in parrots and nonhuman primates. In C. T. Snowdon and M. Hausberger (Eds.), *Social influences on vocal development* (pp. 157–177). Cambridge, Cambridge University Press.

Pepperberg, I.M., Willner, M.R. and Gravitz, L.B. (1997). Development of Piagetian object permanence in a grey parrot (Psittacus erithacus). *Journal of Comparative Psychology, 111*(1), 63-75.

Pepperberg, I.M., Sandefer, R.M., Noel, D.A. and Ellsworth, C.P. (2000). Vocal learning in the Grey parrot (Psittacus erithacus), effects of species identity and number of trainers. *Journal of Comparative Psychology, 114*(4), 371-80.

Pepperberg, I.M. (2004). Human speech, Its learning and use by Grey parrots. In P. Marler and H. Slabbekoorn (Eds.), *Nature's music* (pp. 363–373). London, Elsevier.

Pepperberg, I.M. (2007). Grey parrots do not always ''parrot'', Roles of imitation and phonological awareness in the creation of new labels from existing vocalizations. *Language Science, 29,* 1–13.

Perkel, D. and Farries, M. (2000). Complementary 'bottom-up' and 'top-down' approaches to basal ganglia function. *Current Opinions in Neurobiology, 10,* 725–731.

Puelles, L., Kuwana, E., Puelles, E., Bulfone, A., Shimamura, K., Keleher, J., Smiga, S, and Rubenstein, J.L. (2000). Pallial and subpallial derivatives in the embryonic chick and mouse telencephalon, traced by the expression of the genes Dlx-2, Emx-1, Nkx-2.1, Pax-6, and Tbr-1, *Journal of Comparative Neurology, 424*, 409–438.

Reiner, A. (1986). Is prefrontal cortex found only in mammals?, *Trends in Neurosciences, 9*, 298–300.

Reiner, A., Perkel, D.J., Bruce, L.L., Butler, A.B., Csillag, A., Kuenzel, W., Medina, L., Paxinos, G., Shimizu, T., Striedter, G., Wild, M., Ball, G.F., Durand, S., Gunturkun, O., Lee, D.W., Mello, C.V., Powers, A., White, S.A., Hough, G., Kubikova, L., Smulders, T.V., Wada, K., Dugas-Ford, J., Husband, S., Yamamoto, K., Yu, J., Siang, C., and Jarvis, E.D. (2004a). Avian brain nomenclature forum revised nomenclature for avian telencephalon and some related brainstem nuclei. *Journal of Comparative Neurology, 473*, 377–414.

Reiner, A., Meade, C.A., Cuthberston, S.L., Laverghetta, A. and Bottjer, S.W. (2004b). An immunohistochemical and pathway tracing study of the striatopallidal organization of Area X in the zebra finch, *Journal of Comparative Neurology, 469*, 239–261.

Reiner, A., Medina, L. and Veenman, C.L. (1998). Structural and functional evolution of the basal ganglia in vertebrates, *Brain Research Reviews, 28*, 235–285.

Robinson, S., Sandstrom, S.M., Denenberg, V.H. and Palmiter, R.D. (2005). Distinguishing whether dopamine regulates liking, wanting, and/or learning about rewards, *Behavioral Neuroscience, 199*, 336–41.

Sasaki, A., Sotnikova, T.D., Gainetdinov, R.R. and Jarvis, E.D. (2006). Social context-dependent singing-regulated dopamine. *Journal of Neuroscience, 26*, 9010–9014.

Scharff, C. and Nottebohm, F. (1991). A comparative study of the behavioral deficits following lesions of variousparts of the zebra finch song system, implications for vocal learning. *Journal of Neuroscience, 11*, 2896–2913.

Simpson, H.B. and Vicario, D.S. (1990). Brain pathways for learned and unlearned vocalizations differ in zebra finches. *Journal of Neuroscience, 10*, 1541–1556.

Smeets, W.J. (1992). Comparative aspects of basal forebrain organization in vertebrates. *European Journal of Morphology, 30*, 23–36.

Smith-Fernandez, A., Pieau, C., Reperant, J., Boncinelli, E. and Wassef, M. (1998). Expression of the Emx-1 and Dlx-1 homeobox genes define three molecularly distinct domains in the telencephalon of mouse, chick, turtle and frog embryos, implications for the evolution of telencephalic subdivisions in amniotes. *Development, 125*, 2099–2111.

Steiner, H. and Gerfen, C.R. (1998). Role of dynorphin and enkephalin in the regulation of striatal output pathways and behavior. *Experimental Brain Research, 123*, 60–76.

Takahashi, K., Liu, F.C., Hirokawa, K. and Takahashi, H. (2003). Expression of Foxp2, a gene involved in speech and language, in the developing and adult striatum. *Journal of Neuroscience Research, 73*(1), 61-72.

Teramitsu, I., Kudo, L.C., London, S.E., Geschwind, D.H., and White, S.A. (2004). Parallel FoxP1 and FoxP2 expression in songbird and human brain predicts functional interaction. *Journal of Neuroscience, 24*, 3152–3163.

Tomkins, D.M. and Sellers, E.M. (2001). Addiction and the brain, the role of neurotransmitters in the cause and treatment of drug dependence. *Canadian Medical Association Journal, 164*, 817–21.

Vates, G.E. and Nottebohm, F. (1995). Feedback circuitry within a song-learning pathway. *Proceedings of the National Academy of Sciences of the United States of America, 92*, 5139–5143.

Vates, G.E., Broome, B.M., Mello, C.V. and Nottebohm, F. (1996). Auditory pathways of caudal telencephalon and their relation to the song system of adult male zebra finches. *Journal of Comparative Neurology, 366*, 613–642.

Vates, G.E., Vicario, D.S. and Nottebohm, F. (1997). Reafferent thalamo-"cortical" loops in the song system of oscine songbirds. *Journal of Comparative Neurology, 380*, 275–290.

Vicario, D.S. (1994). Motor mechanisms relevant to auditory-vocal interactions in songbirds. *Brain, Behavior and Evolution, 44*(4-5), 265-78.

Wada, K., Sakaguchi, H., Jarvis, E.D. and Hagiwara, M. (2004). Differential expression of glutamate receptors in avian neural pathways for learned vocalization, *Journal of Comparative Neurology, 476*, 44–64.

Webster, D.B., Popper, A.N. and Fay, R.R. (1992). editors. *The Mammalian Auditory Pathway, Neuroanatomy.* Springer-Verlag, New York, 1992.

Weir, A.A.S., Chappell, J. and Kacelnik, A. (2002). Shaping of hooks in New Caledonian crows. *Science, 297*, 981.

Wild, J.M., Karten, H.J. and Frost, B.J. (1993). Connections of the auditory forebrain in the pigeon (Columba livia*). Journal of Comparative Neurology, 337*, 32–62.

Xi, Z.X. and Stein, E.A. (2002). GABAergic mechanisms of opiate reinforcement. *Alcohol Alcohol (Oxford, Oxfordshire), 37*, 485–94.

Zhang, S.P., Bandler, R. and Davis, P.J. (1995). Brain stem integration of vocalization, role of the nucleus retroambigualis, *Journal of Neurophysiology, 74*, 2500–2512.

In: Expanding Horizions of the Mind Science(s)
Editors: P.N. Tandon, R.C. Tripathi and N. Srinivasan

ISBN: 978-1-62808-705-5
©2013 Nova Science Publishers, Inc.

Chapter 15

FACIAL EXPRESSIONS OF EMOTION: A NEURO-CULTURAL PERSPECTIVE

Manas K. Mandal[1], Amri Sabharwal[2] and Avinash Awasthi[1]

[1]Defence Institute of Psychological Research, Delhi
[2]University of Essex, Colchester, United Kingdom

ABSTRACT

There have been two predominant approaches to the study of facial expression of emotion – the evolutionary-biological approach and the socio-cultural approach. While the former, based on the Darwinian evolutionary perspective, suggests that facial emotions are expressed by an innate mechanism and are, therefore, universally expressed, the latter views emotional experience as a social phenomenon and maintains that people understand facial expressions of emotions based on their knowledge of societal norms. Researchers have presented evidence in favour of each of these theoretical perspectives, although neither of these perspectives has been able to conclusively explain the multitude of findings in the research on expressions of emotion. In the present paper, we discuss the findings of some studies on lateralisation of facial expressions of emotion and universality and culture specificity of emotion with a view to put forward an interactionist perspective on facial expressions of emotion. A neuro-cultural approach to examine the association between laterality and universality is suggested, which would simultaneously deal with the biological and social contexts of emotions.

1. INTRODUCTION

The face is considered to be the most important channel of communication of emotions. As Aristotle noted, "Everyone knows that grief involves a gloomy and joy a cheerful countenance There are characteristic facial expressions which are observed to accompany anger, fear, erotic excitement, and all the other passions" (Aristotle, nd/1913, as cited in Russell, 1994, p. 805, 808). Tomkins (1962), in fact, suggested that the face is the primary locus of emotional behaviours, and emphasised that the study of face is in other words the study of emotion. Not surprisingly then, research in the study of human emotions has predominantly focused on the face as the medium for expression of emotions. As Oatley and Jenkins (1992) observed, "by far the most extensive body of data in the field of human emotions is that on facial expressions of emotion" (p. 67).

The psychological theory of emotional expression perhaps began with Darwin's influential *The Expressions of Emotions in Man and Animals* (1872), in which he explained facial expressions of emotions on the basis of his theory of evolution. According to Darwin, the survival function of human behaviour is constituted by emotions, which initiate the fight or flight response in threatening situations. Subsequent laboratory investigations of this initial idea by psychologists yielded encouraging results. Following the Darwinian evolutionary perspective of emotions, later researchers like Izard (1979) and Ekman (1982) suggested that facial emotions are expressed by an innate mechanism.

Izard (1979) speculated that expressive (facial) behaviours are a source of subjective feeling. He noted that "there appears to be evolutionary-biologically based connection between expressive behaviour and particular feeling." (Izard 1994, p. 288). Izard supported this proposition with evidence from his research on young children and preverbal infants (Izard, Huebner, Risser, McGinnes, and Dougherty, 1980; Izard, Hembree, Dougherty, and Spizzirri, 1983; Abe and Izard, 1999). The findings of these studies suggested that newborn infants are capable of making facial expressions of basic emotions and that their facial expression patterns are morphologically similar to those of adults (Izard et al., 1995).

Ekman (1982) substantiated this viewpoint of Izard and further suggested that since emotions are the result of an innate facial affect programme, they are universally expressed, although these expressions may be modified by social learning or culture-constant display rules. By facial affect programme, Ekman meant the expression of primary emotions via a combination of facial muscular movement which are neurally connected. In support of this suggestion, Ekman conducted a series of experiments in which subjects were asked to voluntarily perform certain facial muscular actions. He found that this instructionally-produced facial muscle activity generated changes in the autonomic nervous system activity (Ekman et al., 1983). Ekman's theory, therefore, was based on a neuro-cultural perspective, and maintained that the expressions of emotion are triggered by an innate facial affect programme and the culture-specific display rules modify these emotion expressions to suit social relevance and context. The universal aspects of Ekman's theory are discussed later in the chapter.

Russell (1994), on the other hand, argued that facial expressions of emotion are not biologically triggered, although judged similarly. He observed that evidence of universality of emotion expression, as proposed by Ekman and Izard, is methodologically derived, and is based merely on the association between certain facial configurations and categorical labels of emotion (Russell, 1994). Russell further stated that universality in the recognition of facial expressions of emotion is found primarily because of the facial stimuli (which are posed and not spontaneous) and the characteristic of the (literate) observer. Reviewing the studies on expression and recognition of facial expressions of emotion, Mesquita, Frijda, and Scherer (1997, p. 259) concluded that "universality and cross-cultural specificity of emotion should not be treated as mutually exclusive (concepts)."

Cultural psychologists challenge the biologically oriented researches with the notion that 'humans are biologically programmed to become cultural beings' (Norasakkunkit, 2000). They view emotional experience as a social phenomenon and, therefore, explain emotional expressions in terms of the social construction. Cultural psychologists posit that emotions are expressed or experienced within a social network and that people understand the facial expressions of emotions based on their knowledge of societal rules and norms (e.g., Averill, 1992; Shweder, 1991).

The two predominant approaches to the study of emotion have, therefore, been the evolutionary-biological perspective and the socio-cultural perspective. While researchers have presented evidence in favour of each of these theoretical perspectives, they have generally argued upon which single theoretical perspective is capable of conclusively explaining the multitude of findings in the research on expressions of emotion.

Are emotions biologically triggered as proposed by Darwin and supported by the numerous neuroscientific studies of emotions? Or are emotions socio-culturally constructed as advocated by the cultural psychologists? In the present chapter, we will revisit some of the evidence that forms the basis of the evolutionary-biological perspective of emotion expression, which has led to the universality thesis. We will then revisit the studies that oppose the universality thesis and, instead, advocate for the culture-specific influences on expression and recognition of emotion. Finally, we will present the findings of some of the studies on lateralisation of facial expressions of emotion with a view to put forward an interactionist perspective to explain how the universality and culture-specific influences in facial expressions of emotion are to be examined in future studies. The impetus behind the utilisation of evidence from lateralisation studies was drawn primarily from some recent studies that have shown that facial expressions are asymmetrical in nature with the right side more social (culturally moderated) and the left side more innate (biologically triggered; see Mandal, Harizuka, Bhushan, and Mishra, 2001). It has been found that observers take note of these differential expressions in the human face while judging emotion (see Mandal, Asthana, Madan, and Pandey, 1992).

2. THE EVOLUTIONARY-BIOLOGICAL PERSPECTIVE

The evolutionary-biological perspective of emotions has stemmed from Darwin's (1872) evolutionary theory of emotions as mentioned earlier. This perspective suggests that expressions of emotion send signals that help regulate the social interaction and increase the likelihood of survival (see Westen, 1996). Knapp (1963) later on noted that "emotional phenomena were among the evolving attributes of man which had developed like man himself from antecedents in his animal forebears" (p. 5). Since then, the unique patterns of neural and physiological activity that accompany different emotions have been a central subject of research in the study of human emotion. After possibly the first systematic investigations by James (1984), Canon, and Bard (Canon 1927, 1931; Bard, 1928; Bard and Rioch, 1937), a wealth of studies has explored the physiological and neural correlates of emotions. While earlier researchers relied on behavioural experiments (e.g., Reuter-Lorenz and Davidson, 1981), lesion methods (e.g., Heilman, Scholes, and Watson, 1975), and electrophysiological recordings (e.g. Davidson and Schaffer, 1983) in laboratory animals and human volunteers, the more recent ones have additionally employed functional neuroimaging techniques such as Positron Emission Tomography (PET; e.g., Morris et al., 1996) and functional Magnetic Resonance Imaging (fMRI; e.g., Buchanan et al., 2000), which have aided the study of the neural basis of emotion with unprecedented precision.

2.1. Development of Expressions of Emotions

In order to further the idea of innateness of emotions, psychologists have examined infants to determine the age at which they begin to express facial expressions and recognise those of their parents or caregivers. Reviewing a number of studies, Oster and Ekman (1978) observed that the facial muscles are fully functional in neonates. Preverbal infants have been found capable of expressing elementary facial emotions in recognition studies (Barrera and Maurer, 1981; Spieker, 1982; Serrano, Iglesias, and Loeches, 1992), which are reliably recognised by adults (Izard et al., 1980). In 1983, Izard developed the *maximally discriminative facial movement coding system* (MAX) to examine the facial expressions of emotion in infants. He used this coding system to collect further evidence in support of his proposition of the innateness of facial expressions. Plutchik (1980) strengthened Izard's viewpoint by proposing that all organisms acquire at their earliest stage (i.e., infancy), (emotional) behaviours which may indicate their need for survival and nurturance. These behaviours have adaptational functions and have a root common to all species (see Mandal, 2004).

Further support for the biological basis of emotions comes from studies of the congenitally blind. Dumas (1932) found that congenitally blind people expresses spontaneous expressions adequately as normally sighted people express, although they are not able to express posed expressions adequately. Galati, Miceli, and Sini (2001) found no significant differences after analysing video of facial expressions of congenitally blind people and normal people by using the Maximally Discriminative Facial Movement Coding System (Max; Izard, 1983). More recently, Matsumoto and Willingham (2009) compared the facial expressions of congenitally and non-congenitally blind athletes of the Paralympic Games 2004 with normal athletes of 2004 Olympic Games and found no significant differences in the level of facial emotion configurations.

2.2. The Universality Thesis

The evolutionary-biological perspective of Izard (1971, 1994) was supported by Ekman (1984) who found that individuals across cultures display the same facial expressions when experiencing the same emotion until when the culture-specific display rules do not interfere. Ekman (1972) suggested that emotions are expressed in universally equal manner. He suggested that emotions are expressed through different combinations of facial muscles which are governed by a subset of neural network. Ekman followed the universality thesis of Darwin and explained emotions as the result of facial affect program that may be modulated by cultural display rules. Both Izard (1979) and Ekman (1982) advocated that there is an innate programme to express and recognise facial emotion expressions.

Universality refers to accurate recognition of facial expressions across cultures at better than chance levels (Ekman, et al., 1987). On the basis of some classic studies on emotion, it has been concluded that emotions are recognised cross-culturally in western-oriental populations (Ekman, Sorenson, and Friesen, 1969; Izard, 1971; Ekman, 1972; Sorenson, 1975; Ekman et al., 1987), and literate and preliterate populations (Ekman et al., 1969; Ekman and Friesen, 1971; Sorenson, 1975; Boucher and Carlson, 1980). To eliminate the exposure and familiarity factor, Ekman et al., (1987) studied the isolated and preliterate South Fore and Dani people of New Guinea. The subjects were given different situations (e.g. 'pretend your son has died') and were asked to express themselves. Photographs of these expressions were then shown to the western literate populations. High recognition accuracy was found across ten different cultures for all emotions except sadness.

However, this universality theory has been criticised by the cross-cultural studies of facial expressions of emotions, which suggest that facial expressions are not universal, but differ across cultures (Russel, 1994). Indeed, universality theorists (Ekman, 1972; Izard, 1971) emphasised on the similarities of facial expressions and recognition across cultures but did not address the differences that have been found in the accuracy levels across different cultures (Matsumoto and Assar, 1992).

Later addressing the issue of cultural differences, Ekman (1982) proposed a neuro-cultural theory, which suggests that the expression of emotion via face is the outcome of an elicitor that generates the innate facial affect programme. It is proposed by the universality theorists that elicitors may change from culture to culture (as a function of the social situation and prevalent norms) but the facial behaviour in response to that situation conveys the same meaning in all cultures. Ekman, therefore, proposed a set of primary emotions – happiness, sadness, fear, anger, surprise, and disgust – the expression and recognition of which he believed to be universal.

Cultural psychologists suggest that differences in the accuracy of recognition of emotion are a result of cultural differences. In their study of three to six months-old infants and their mothers, Malatesta and Haviland (1982) found a wide range of changes in facial expressions of infants, as well as similarities between the infants and their mothers in the expression type and the use of mouth or brow region. These findings thus suggested socialisation of emotion expressions during early infancy, since infants were found to follow the expressions of their mothers. Later in socialisation, these expressions are modulated by social/cultural norms and demands of the situation.

The culture-bound issues in facial emotion have been accommodated by the introduction of the concept of display rules. Ekman (1972) described display rules as socially and culturally learned norms, which specify when and how emotions can be expressed with respect to the social situation and the cultural demands. Thus, display rules guide emotion expressions to some extent according to social/cultural context (Ekman, 1982). In a recent study Matsumoto, Yoo, and Fontaine (2008) collected data from 32 countries regarding the cultural display norms. They found greater expressions towards the in-group than the out-group, and therefore, suggested cultural differences in expressivity endorsement between individualistic and collectivistic cultures.

3. THE CULTURAL PERSPECTIVE

Although the correlation between emotion and neuro-biological processes has been established beyond doubt, the social significance and origins of emotions cannot be overlooked. It is commonly believed that an individual's emotional response is often guided by their evaluation of their social situation. Emotions are considered responsive to social events and entities. Oatley and his colleagues (Oatley, Keltner, and Jenkins, 2006) posit that all emotions are social in nature because their evolution is a result of the need of individuals to deal with the complexities of human social life.

Individuals learn to express subjective feelings through the process of socialisation, in which they learn about self and others through different emotional behaviours and events (Markus and Kitayama, 1991; Mesquita and Albert, 2007). The socio-cultural environment influences one's expressions of emotion in a systematic manner right from birth. Socialisation differs from culture to culture since cultures differ in their characteristics and nature. These cultural differences, in turn, may influence an individual's expressions of emotion. Researchers believe that emotions are not solely the result of innate programs but are the result of different types of cultural variations across its components (e.g., Frijda, 1986; Mesquita and Frijda, 1992; Scherer, 1984). Kitayama and Markus (1994) suggested that emotions depend upon the cultural situations and they cannot be separated from cultural influences. The individualistic and collectivistic characteristics of culture have also been studied as variables influencing facial expressions and recognition of emotion.

Despite such theoretical positions, cross-cultural studies of facial expressions have revealed that facial expressions are recognised amongst different cultures with a high degree of agreement, although the differences in recognition accuracy have been substantial when the emotional lexicon (see Russell, 1994) used in a given culture and the emotion category (e.g., happy, sad, fear, etc.; see Mandal, Bryden, and Bulman-Fleming, 1996) are given due consideration. The debate between universality and culture-specificity has been addressed by numerous cross cultural psychologists. Russell and Fernandez-Dols (1997), for example, have suggested this controversy as 'minimal universality'.

3.1. Culture-Specificity

Cultural psychologists emphasise on the cultural aspect of emotion and suggest that emotions are not universal, but are culture-specific, for both recognition and expression of facial emotion. Individuals learn emotion expression and recognition through their interaction with the culture and other members of the culture. Emotions, therefore, are seen as the result of social programmes which are governed by social and cultural norms and rules. In a meta-analysis of studies on emotion, Russell (1994) concluded that facial emotion recognition accuracy indeed differs from one culture to another. This finding has been replicated in several other studies using various research designs (Luce, 1974; Kilbride and Yarczower, 1983; Matsumoto, 1989; Russell, 1994; Markham and Wang, 1996; Mandal et al., 1996; Biehl, Matsumoto, Ekman, Hearn, Heider, Kudoh, and Ton, 1997; Elfenbein and Ambady, 2002; Elfenbein and Ambady, 2003a; van Hemert, Poortinga, and van de Vijver., 2007). Certain emotions like happiness and sadness have been found to be recognised equally across cultures, while emotions like fear and anger are not (Mandal, Saha, and Palchoudhary, 1986; Russell, 1994).

Psychologists who support the culture-specificity of emotion recognition suggest that the universal affect programme of facial expressions may be characterised by the differences across cultures, and as the contact between cultures increases, familiarity becomes high (Elfenbein and Ambady, 2003b). Individuals are able to recognise familiar faces easily across large variations in image quality, although the ability to match unfamiliar faces is strikingly poor (Burton, Jenkins, Hancock, and White, 2005). Kitayama and Markus (1994) illustrated that emotions depend upon the dominant culture frame.

In-group advantage: Research has shown that individuals recognise facial expressions of emotion with greater accuracy (Elfenbein and Ambady, 2002; Beaupre and Hess, 2006; Thibault, Bourgeois, and Hess, 2006) and lesser response time (Elfenbein and Ambady, 2003b) when they are displayed by members of their own culture (in-group) as compared to members of other cultures (out-group). This social process is known popularly as in-group advantage, or less commonly as ethnic bias (e.g., Kilbride and Yarczower, 1983; Markham and Wang, 1996) or other-race-effect (e.g., Michel, Rossion, Han, Chung, and Caldara, 2006).

In support of their findings suggesting an in-group advantage in recognition of facial expressions of emotion, Elfenbein and Ambady (2003a) have proposed the dialect theory. According to this theory, there is a universal pattern of emotion recognition at better than chance levels, but because of cultural differences, the recognition of facial expressions varies across cultures. The dialect theory, thus, views cultural norms and patterns as responsible for the differences in recognition of emotions across cultures. Elfenbein and Ambady further suggested that facial expressions of emotion contain certain nonverbal accents that are identified differently by observers across cultures. These nonverbal accents are intensified during the facial expressions of emotion, which, in turn, facilitate the in-group advantage (Marsh, Elfenbein, and Ambady, 2003).

4. LATERALISATION OF EMOTION

During the last few years, attention has been paid to the association between laterality and universality of facial expressions of emotion (see review by Mandal and Ambady, 2004). The interest has been generated ever since neuroscientific studies have established the fact that expressions, as well as perception of facial expressions, of emotion are asymmetrical in nature. Researchers have found that the two sides of human face do not behave identically during expressions of emotion; the left of face expresses emotion more intensely (see Borod and Koff, 1990), while the right of face displays emotion more congruent to the social context (see Mandal et al., 2001). It has also been observed that human beings do not perceive the two sides of face simultaneously; the human face elicits a left visual-field advantage of the observer (see review by Leventhal and Tomarken, 1986; Mandal and Singh, 1990). Before we further discuss the association between laterality and universality, it is important to revisit studies that have established the construct of laterality in the domain of facial expressions of emotion.

One of the central issues in the neuroscientific study of emotion has been the observed asymmetry of emotion processing across the hemispheres. A considerable amount of research in emotion has been devoted to understanding the nature and significance of these asymmetries, as well as the interaction between the two hemispheres in processing different aspects of emotion. Two competing theories have been proposed in this regard (see Borod, 1992): (i) the Right-Hemisphere Hypothesis and (ii) the Valence Hypothesis. The former proclaims the superiority of the right hemisphere for processing all emotions regardless of the valence (Borod et al., 1998; Borod, Koff, Lorch, and Nicholas, 1988; Sackeim and Gur 1978; Schwartz, Davidson, and Maer, 1975), while the latter asserts that both hemispheres process emotions, but the right hemisphere (RH) is specialised for processing negative emotions and the left hemisphere (LH) for positive emotions (Adolphs, 2002; Gur, Skolnick, and Gur, 1994; Reuter-Lorenz and Davidson, 1981; Davidson, Mednick, Moss, Saron, and Schaffer, 1987). Through the years, researchers have presented results in support of each of these theories of lateralisation of emotion.

4.1. The Right-Hemisphere Hypothesis

Numerous studies with both brain-damaged and neurologically normal populations have provided evidence in support of the right-hemisphere hypothesis. In 1975, Heilman et al. observed severe impairment of the right hemisphere damaged (RHD) participants in a task that required them to match neutral sentences with differing emotional intonation with pictures of facial expressions of different emotions. Following this, several other clinical studies using a variety of tasks confirmed the finding that damage to the right hemisphere impairs the processing of emotion (Tucker, Watson, and Heilman 1977; Dekosky, Heilman, Bowers, and Valenstein 1980; Mandal, Tandon, and Asthana, 1991; Mandal et al., 1992; Adolphs, Damasio, Tranel, and Damasio, 1996), although some studies failed to find lateralisation of emotion (Mandal, Asthana, Tandon, and Asthana, 1992).

More recently, in a large sample of over 100 brain-damaged participants, Adolphs, Damasio, Tranel, Cooper, and Damasio (2000) found that RHD participants performed significantly worse than left hemisphere damaged (LHD) participants in recognising facial expressions of emotion. In addition, RHD patients have been found to show deficits in remembering emotional material (Wechsler, 1973), comprehending emotional tone of the voice (Heilman, Bowers, Speedie, and Coslett, 1984), and appreciating humorous stimuli (Bihrle, Brownell, Powelson, and Gardner, 1986). The right temporal and parietal cortices, in particular, have been reported as discrete sectors in the RH, lesions to which result in deficits in processing emotion (Bowers, Bauer, Coslett, and Heilman, 1985; Rapcsak, Comer, and Rubens, 1993; Gur et al., 1994; Ojemann, Ojemann, and Lettich, 1992). The performance of RHD patients in the recognition of emotion has been found to be impaired across modalities (Kucharska-Pietura, Phillips, Gernand, and David, 2003; Heilman et al., 1984).

Studies with neurologically intact subjects have reported similar results. Most of these studies use the split visual-field technique for the visual modality or the dichotic-listening technique for the auditory modality. In the split visual-field technique, emotional stimuli are presented to a specific hemisphere through exposure in the contralateral visual field for a very brief period of time (150 ms or less). Studies using this technique have found a left visual field advantage for processing emotional stimuli (Suberi and McKeever 1977; Buchtel, Campari, De Risio, and Rota 1978; Strauss and Moscovitch 1981). In the dichotic-listening technique, emotional stimuli are presented to specific hemispheres by simultaneously stimulating the two ears with different stimuli. Using this technique, researchers have shown a left ear advantage for emotional intonation recognition (Ley and Bryden, 1982; Bryden, Ley, and Sugarman, 1982; Safer and Leventhal, 1977; Saxby and Bryden, 1984; Hartje, Villmes, and Weniger, 1985; King and Kimura, 1972; Erhan, Borod, Tenke, and Bruder, 1998) as well as for nonverbal emotional vocalisations (Carmon and Nachshon, 1973; King and Kimura, 1972; Mahoney and Sainsbury, 1987). These findings have been corroborated by findings of electroencephalographic (Pihan, Altenmueller, and Ackermann, 1997) as well as functional neuroimaging studies (Buchanan et al., 2000; George et al., 1996; Kanwisher, McDermott, and Chun, 1997; Sergent, Ohta, and MacDonald, 1992)

While the majority of studies have focused on determining lateralisation of emotion in the recognition or identification of emotion, many researchers have also found right hemisphere superiority in the expression of emotion. There is considerable evidence that emotions are more expressed more intensely in the left hemiface, which has been interpreted as support for the right hemisphere hypothesis, given that the facial muscles are innervated by the contralateral hemisphere (Borod, Haywood, and Koff, 1997; Adolphs et al., 1996; see Skinner and Mullen, 1991). In expression studies typically, actors are required to produce facial expressions in posed or spontaneous conditions. Facial composites (chimeric faces) are created by splitting normal and mirror-reversed images of these faces down the vertical midline and recombining them to form left-left or right-right composites, and are rated by observers in terms of intensity of expression. Findings from these experiments suggest a left hemifacial bias for both posed and spontaneous expressions of emotion (Asthana and Mandal 1998; Borod et al., 1988; Mandal, Asthana, and Tandon, 1993).

3.1. The Valence Hypothesis

A considerable number of studies of lateralisation of emotion have found that negative emotional states are preferentially processed by the right hemisphere, whereas positive emotions are processed by the left hemisphere. Unilateral brain damage to the LH impairs the perception of positive emotions while comparable RH lesions impair perception of negative emotions (Mandal et al., 1991). A disproportionate number of patients who have suffered damage to the LH have been found to be depressed (eg Morris et al., 1996; Paradiso, Chemerinski, Yazici, Tartaro, and Robinson, 1999; Sackeim et al., 1982), while patients with RH damage have been found to show signs of inappropriate cheerfulness and mania (Starkstein et al., 1989).

Behavioural studies using tachistoscopic stimuli have reported left visual field bias for the perception of negative emotions and a right visual field bias for perceiving positive emotions, providing further support to the valence hypothesis of lateralisation of emotion (Adolphs, Jansari, and Tranel 2001; Davidson et al., 1987; Reuter-Lorenz and Davidson, 1981). Studies using chimeric faces and a free viewing administration reported that right-handed participants judge faces as happier when the smile is on the left side of the face, as compared to the smile on the right side of the face (Hoptman and Levy, 1988; Levy, Heller, Banich, and Burton, 1983; Luh, Ruechert, and Levy, 1991).

In their study of emotion experience, Leventhal and Tomarken (1986) reported that LH damage results in depression, pessimism, and crying, while RH damage results in joviality, indifference, and laughing. Similar findings have been reported by Sackeim et al. (1982).

Gainotti and colleagues (Gainotti, Caltagirone, and Zoccolotti, 1993) found that patients with lesions in the LH presented depressive catastrophic reactions more often as compared to patients with lesions to the RH, who presented reactions of euphoria or indifference. EEG studies have shown greater activation of the LH in response to positive emotional experiences, as compared to the increased activation of the RH in response to negative emotional experiences (Harman and Ray, 1997; Davidson and Schaffer, 1983). However, many studies have failed to demonstrate valence lateralisation (Collet and Duclaux, 1987; Gotlib, Ranganath, and Rosenfeld, 1998; Reid, Duke, and Allen, 1998).

Closely related to the valence hypothesis is the approach-withdrawal theory proposed by Davidson and colleagues (Davidson and Erwin, 1999). The approach-withdrawal theory integrates the dimension of motivation with the dimension of emotional valence. According to this theory, the approach system facilitates goal directed behaviour and generates related positive emotions, while the withdrawal system facilitates the withdrawal from aversive stimulation and generates related negative emotions. EEG studies (Davidson, 1988; Davidson, Ekman, Saron, Senulis, and Friesen, 1990; Davidson, Marshall, Tomarken, and Henriques, 2000), fMRI studies (Canli, Desmond, Zhao, Glover, and Gabrieli, 1998), and behavioural studies (Nicholls, Ellis, Clement, and Yoshino, 2004; Richardson, Bowers, Bauer, Heilman, and Leonard, 2000; Schiff and Bassel, 1996) have shown associations between the LH and positive or approach-related emotions, and between the RH and negative or withdrawal related emotions.

5. A Neuro-Cultural Perspective

Laterality studies have, therefore, abundantly suggested that the two sides of the human face are asymmetrical in the expression of emotion, and that the right hemisphere generates the trigger for the facial affect programme as advocated earlier by Ekman (1984). Interestingly, Darwin (1872), in his original theoretical work, hinted that the two sides of the human face may not be equally expressive. Subsequent laboratory research has taken the hint from Darwin's conceptualisation and generated extensive research in support of this proposition. Using a variety of techniques like symmetrical composite photographs of faces (Sackeim, Gur, and Saucy, 1978), the Facial Action Coding System that examines facial muscle activation (Hager and Ekman, 1985), slow motion replay of video taped faces (Borod, Caron, and Koff, 1981), and electromyography (Schwartz, Ahern, and Brown, 1979), psychologists have concluded that for facial expressions, the left hemiface (a contralateral function of the RH) is relatively more mobile than the right, and therefore, emotions expressed on the left hemiface are perceived as more intense. It has also been observed that the left-visual-field (a contralateral function of the RH) is superior in the recognition of facial expressions of emotion, although such findings have been found to be true more for negative emotions (see Borod, 1992). The evidence thus conclusively favours a right hemisphere hypothesis in triggering facial expression of emotion.

As early as 1933, Wolff suggested that the right side of the human face offers social or public expressions whereas the left side of the face reveals hidden, personalised feelings. The greater intensity and control of left hemifacial expressions has been attributed to the dominance of the right hemisphere for emotion processing (Borod, 1992, 1993). On the other hand, evidence indicates that the right side of the face is under greater conscious control than the left (Sackeim et al., 1978) since it is under the motor control of the more cognitively controlled left hemisphere (Rinn, 1984). It has been suggested that the left hemisphere exerts an inhibitory effect on the facial expression of emotion (Gainotti, 1983), thus moderating the right hemifacial expression by the public intent of social communication in terms of display norms (Ekman and Friesen, 1975).

We studied hemispheric differences to test for cultural differences in the expressions of emotion (Mandal et al., 2001) and in-group advantage for the same (Elfenbein, Mandal, Ambady, Harizuka, and Kumar, 2004). In the first study (Mandal et al., 2001), photographs of hemifacial composites (e.g., left-left, right-right) of individuals from three cultures (Japanese, Indian, and North American) expressing the six basic emotions were shown to observers for their judgement in terms of distinctiveness of the expression. The subjects did not show variation for their judgement across cultures, which indicated universality. However, culture-specificity was also observed with respect to the hemifacial asymmetry and valence of emotion expression. The Japanese showed a right hemifacial bias for positive emotions and a left hemifacial bias for negative emotions. Negative emotions were also the least noticeable in Japanese faces as compared to Indians and North Americans. The Indians and North Americans showed left hemifacial bias for all emotions. The second study (Elfenbein et al., 2004) examined American, Indian, and Japanese participants judging the facial expressions of photographs from all three cultures displaying basic emotions. This study was designed to separate 'absolute' cultural differences from 'relational' effects characterising the relationship between expressor and expressed. Using a response bias correction method, the findings of the study indicated that there was a greater in-group advantage for the judgement of the left hemifacial expression as compared to the right hemifacial expression. These studies, thus, clearly indicate an interaction between Expressor X Expressed X Emotional Valence.

We also investigated, in an earlier study, the human face at rest (Mandal et al., 1992). The asymmetrical nature of the resting face was examined with the help of facial composites in order to test the hypothesis that the left of the face will be emotionally more involved than both, the right of the face and the whole face. This was done to identify 'facial leakage' of the general affective state of an individual at resting state. Our findings supported the hypothesis and indicated that the left of face is more emotionally tuned in comparison to the right side or the whole face. The question that remained unanswered, however, was how the emotional content in the left hemiface eludes notice in our day-to-day interaction.

Research on hemispatial bias suggests that the hemiface which falls in the observer's left visual hemifield (a function of the right hemisphere) largely determines the judgment of facial emotion (Asthana and Mandal, 2001; Reuter-Lorenz and Davidson, 1981). It is possible that such hemispatial bias in the perception of facial emotion is learned in the course of human development (see Mandal et al., 1992), which allows a temporal priority of cognition in the understanding of emotional expressions, as advocated by Lazarus (1984).

The hemifacial differences in the expression of emotions, thus, present a unique opportunity to explore social/cultural differences in the expressions of emotion from a neuroscientific perspective. While the left or the "personalised" side of the face (controlled by the right hemisphere) may represent the neuroanatomical underpinnings of emotion, the right or the "public" side of the face represents the socially and culturally driven aspect of emotion. Future research incorporating cultural differences in the lateralisation paradigm will, therefore, present further insights into the dynamics of expressions of emotion.

In conclusion, research on facial expressions of emotion, has been at the centre of scientific pursuits and an area of extensive debate, and evidence suggests its biological as well as social roots. Unarguably, facial expressions serve the adaptive functions of conveying one's emotional reactions to the external world, and this process of adaptation takes place through evolution. Emotional reactions are indeed important for the survival of human beings and the evolution theory of human behaviour is not only relevant at the biological level but also at the social level. However, to bridge the gap between the biological and social evolutions, these two contexts need to be studied simultaneously. A neuro-cultural approach may, therefore, be adopted to examine the association between laterality and universality which deals simultaneously with the biological and social contexts of emotions.

REFERENCES

Abe, A. A. J., and Izard, C. E. (1999). The developmental functions of emotions: an analysis in terms of differential emotions theory. *Cognition and Emotion, 13*, 523-550.

Adolphs, R. (2002). Neural systems for recognizing emotion, *Current Opinion in Neurobiolology, 12*, 169-177.

Adolphs, R., Damasio, H., Tranel, D., and Damasio, A. R. (1996). Cortical Systems for the Recognition of Emotion in Facial Expressions. *The Journal of Neuroscience, 16*(23), 7678-7687.

Adolphs, R., Damasio, H., Tranel, D., Cooper, G., and Damasio, A. R. (2000). A role for somatosensory cortices in the visual recognition of emotion as revealed by 3-D lesion mapping. *Journal of Neuroscience, 20*, 2683-2690.

Adolphs, R., Jansari, A., and Tranel, D. (2001). Hemispheric Perception of Emotional Valence From Facial Expressions. *Neuropsychology, 15*(4), 516-54.

Aristotle. (1913). Physiognomonica. In W. D. Ross (Ed.) and T. Loveday and E. S. Forster (Trans.), *The works of Aristotle* (pp. 805-813). Oxford, England: Clarendon.

Asthana, H. S., and Mandal, M. K. (1998). Hemifacial asymmetry in emotion expression. *Behaviour Modification, 22*, 177-183.

Asthana, H. S., and Mandal, M. K. (2001). Visual-field bias in the judgment of facial expression of emotion. *The Journal of General Psychology, 128*, 21-29.

Averill, J. R. (1992). The structural bases of emotional behaviour: A methodological analysis: In M. S. Clark (Ed.), *Emotion*. Newbury Park, California: Sage.

Bard, P. (1928). A diencephalic mechanism for the expression of rage with special reference to the central nervous system. *American Journal of Physiology, 84*, 490-513.

Bard, P., and Rioch, D. M. (1937). A study of four cats deprived of neocortex and additional portions of the forebrain. *Bulletin of the John Hopkins Hospital, 60*, 73-153.

Barrera, M. E., and Maurer, D. (1981). The perception of facial expression by three months old. *Child Development, 52*, 203-206.

Beaupre, M. G., and Hess, U. (2006). An ingroup advantage for confidence in emotion recognition judgments: The moderating effect of familiarity with the expressions of outgroup members. *Personality and Social Psychology Bulletin, 32*(1), 16-26.

Biehl, M., Matsumoto, D., Ekman, P., Hearn, V., Heider, K., Kudoh, T., et al. (1997). Matsumoto and Ekman's Japanese and Caucasian Facial Expressions of Emotion

(JACFEE): Reliability Data and Cross-National Differences. *Journal of Nonverbal Behavior, 21*(1), 3-21.

Bihrle, A. M., Brownell, H. H., Powelson, J. A., and Gardner, H. (1986). Comprehension of humorous and nonhumorous materials by left and right brain-damaged patients. *Brain and Cognition, 5,* 399-411.

Borod, J. C. (1992). Interhemispheric and intrahemispheric control of emotion: A focus on unilateral brain damage. *Journal of Consulting and Clinical Psychology, 60,* 339-348.

Borod, J. C. (1993). Cerebral mechanisms underlying facial, prosodic and lexical emotional expression: A review of neuropsychological studies and methodological issues. *Neuropsychology, 7,* 445-463.

Borod, J. C., and Koff, E. (1990). Lateralization of facial emotional behaviour: A methodological perspective. *International Journal of Psychology, 25,* 155-177.

Borod, J. C., Cicero, B. A., Obler, L. K., Welkowitz, J., Erhan, H. M., Santschi, C., et al. (1998). Right hemisphere emotional perception: evidence across multiple channels. *Neuropsychology, 12,* 446-458.

Borod, J. C., Caron, H. S., and Koff. E. (1981). Asymmetry of facial expression related to handedness, footedness and eyedness: A quantitative study. *Cortex, 17,* 381-390.

Borod, J. C., Haywood, C. S., and Koff, E. (1997). Neuropsychological aspects of facial asymmetry during emotional expression: a review of the normal adult literature. *Neuropsychological Review, 7,* 41-60.

Borod, J. C., Koff, E., Lorch, M. P., and Nicholas, M. (1988). Emotional and non-emotional facial behaviour in patients with unilateral brain damage. *Journal of Neurology, Neurosurgery and Psychiatry, 51*(6), 826–832.

Boucher, J.D., and Carlson, G.E. (1980). Recognition of facial expression in three cultures. *Journal of Cross-Cultural Psychology, 11,* 263-280.

Bowers, D., Bauer, R. M., Coslett, H. B., and Heilman, K. M. (1985). Processing of faces by patients with unilateral hemisphere lesions. *Brain and Cognition, 4,* 258-272.

Bryden, M. P., Ley, R. G., and Sugarman, J. H. (1982). A left ear advantage for identifying the emotional quality of tonal sequences. *Neuropsychologia, 20,* 83-87.

Buchanan, T. W., Lutz, K., Mirzazade, S., Specht, K., Shah, N. J., Zilles, K., et al. (2000). Recognition of emotional prosody and verbal components of spoken language: an fMRI study. *Cognition Brain Research, 9,* 227-238.

Buchtel, H., Campari, F., De Risio, C., and Rota, R. (1978). Hemispheric differences in discriminative reaction time to facial expressions. *Italian Journal of Psychology, 5,* 159-169.

Burton, A. M., Jenkins, R., Hancock, P. J. B., and White, D. (2005). Robust representations for face recognition: The power of averages. *Cognitive Psychology, 51(3),* 256-284.

Canli, T., Desmond, J. E., Zhao, Z., Glover, G., and Gabrieli, J. D. E. (1998). Hemispheric asymmetry for emotional stimuli detected with fmri. *Neuroreport, 9,* 3233-3239.

Cannon, W. B. (1927). The James–Lange theory of emotions: a critical examination and an alternative theory. *American Journal of Psycholology, 39,* 106-124.

Cannon, W. B. (1931). Against the James–Lange and the thalamic theories of emotions. *Psychological Review, 38,* 281-295.

Carmon, A., and Nachshon, I. (1973). Ear asymmetry in perception of emotional non-verbal stimuli. *Acta Psychologica, 37,* 351-357.

Collet, L., and Duclaux, R. (1987). Hemispheric lateralization of emotions: absence of electrophysiological arguments. *Physiology and Behaviour, 40*(2), 215-220.

Darwin, C. (1872/1998). *The Expressions of Emotion in Man and Animals*. New York/Oxford: University Press.

Davidson, R. J. (1988). EEG measures of cerebral asymmetry: conceptual and methodological issues. *International Journal of Neuroscience, 39*(1-2), 71-89.

Davidson, R. J., and Irwin, W. (1999). The functional neuroanatomy of emotion and affective style. *Trends in Cognitive Sciences, 3*(1), 11-21.

Davidson, R. J., and Schaffer, C. E. (1983). Affect and disorders of affect: Behavioural and electrophysiological studies. In P. Flor-Henery and J. Gruzelier (Eds.), *Laterality and psychopathology* (pp. 249-268). Amsterdam: Elsevier Science.

Davidson, R. J., Ekman, P., Saron, C., Senulis, J., and Friesen, W. V. (1990). Approach/withdrawal and cerebral asymmetry: Emotional expression and brain physiology, I. *Journal of Personality and Social Psychology, 58*, 330-341.

Davidson, R. J., Marshall, J. R., Tomarken, A. J., and Henriques, J. B. (2000). While a phobic waits: regional brain electrical and autonomic activity in social phobics during anticipation of public speaking. *Biological Psychiatry, 47*(2), 85-95.

Davidson, R. J., Mednick, D., Moss, E., Saron, C., and Schaffer, C. E. (1987). Ratings of emotion in faces are influenced by the visual field to which stimuli are presented. *Brain and Cognition, 6*, 403-411.

Dekosky, S. T., Heilman, K. M., Bowers, D., and Valenstein, E. (1980). Recognition and discrimination of emotional faces and pictures. *Brain and Language, 9*, 206-214.

Dumas, E. (1932). La mimique des aveugles [Facial expression of the blind]. *Bulletin de l'Acaddmie de M~decine, 107*, 607-610.

Ekman, P. (1972). Universals and cultural differences in facial expression of emotion. In J. R. Cole (Ed.), *Nebraska Symposium on Motivation: Vol. 19* (pp. 207–283). Lincoln: University of Nebraska Press.

Ekman, P. (1982). *Emotion in the human face*. Cambridge: Cambridge University Press.

Ekman, P. (1984). Expression and the nature of emotion. In K. Scherer and P. Ekman (Eds.), *Approaches to emotion* (pp. 319-343). Hillsdale, N. J.: Lawrence Erlbaum.

Ekman, P., and Friesen, W. V. (1971). Constant across cultures in the face and emotion. *Journal of Personality and Social Psychology, 17*, 124-129.

Ekman, P., and Friesen, W. V. (1975). *Unmasking the face. A guide to recognizing emotions from facial clues*. Englewood Cliffs, New Jersey: Prentice-Hall.

Ekman, P., Friesen, W. V., O'Sullivan, M., Chan, A., Diacoyanni-Tarlatzis, I., Heider, K., et al. (1987). Universals and cultural differences in the judgments of facial expressions of emotion. *Journal of Personality and Social Psychology, 53*(4), 712-717.

Ekman, P., Levenson, R. W., and Friesen, W. V. (1983). Autonomic nervous system activity distinguishes among emotions. *Science, 221*, 1208-1210.

Ekman, P., Sorensnn, E. R., and Friesen, W. V. (1969). Pan-cultural elements in facial displays of emotions. *Science, 164*, 86–88.

Elfenbein, H. A., and Ambady, N. (2002). On the universality and culture specificity of emotion recognition: A meta-analysis. *Psychological Bulletin, 128*(2), 203-235.

Elfenbein, H. A., and Ambady, N. (2003a). Universal and cultural differences in emotion recognition. *Current Directions in Psychological Science, 12*(5), 159-164.

Elfenbein, H. A., and Ambady, N. (2003b). When familiarity breeds accuracy: Cultural exposure and facial emotion recognition. *Journal of Personality and Social Psychology, 85*(2), 276-290.

Elfenbein, H. A., Mandal, M. K., Ambady, N., Harizuka, S., and Kumar, S. (2004). Hemifacial differences in the in-group advantage in emotion recognition. *Cognition and Emotion, 18*(5), 613-629.

Erhan, H., Borod, J.C., Tenke, C.E., and Bruder, G.E. (1998). Identification of emotion in a dichotic listening task: event-related brain potential and behavioral findings. *Brain and Cognition, 37*(2), 286-307.

Frijda, N. H. (1986). *The emotions.* Cambridge: Cambridge University Press.

Gainotti, G. (1983). Laterality of affect: The emotional behaviour of right and left brain damaged patients. In M. S. Myslobodsky (Ed.), *Hemisyndromes: Psychobiology, Neurology, Psychiatry.* New York: Academic Press.

Gainotti, G., Caltagirone, C., and Zoccolotti, P. (1993). Left/right and cortical/subcortical dichotomies in the neuropsychological study of human emotions. *Cognition and Emotion, 7,* 71-93.

Galati, D., Miceli, R., and Sini, B. (2001). Judging and coding facial expression of emotions in congenitally blind children. *International Journal of Behavioral Development, 25*(3), 268-278.

George, M. S., Parekh, P. I., Rosinsky, N., Ketter, T., Kimbrell, T. A., Heilman, K. H., et al. (1996). Understanding emotional prosody activates right hemisphere regions. *Archives of Neurology, 53,* 665-670.

Gotlib, I. H., Ranganath, C., and Rosenfeld, J. P. (1998). Frontal EEG alpha asymmetry, depression, and cognitive functioning. *Cognition and Emotion, 12*(3), 449-478.

Gur, R. C., Skolnick, B. E., and Gur, R. E. (1994). Effects of emotional discrimination tasks on cerebral blood flow: regional activation and its relation to performance. *Brain and Cognition, 25,* 271-286.

Hager, J. C., and Ekman, P. (1985). The asymmetry of facial actions is inconsistent with models of hemispheric specialisation. *Psychophysiology, 22,* 307-318.

Harman, D. W., and Ray, W. J. (1997). Hemispheric activity during affective verbal stimuli: An EEG study. *Neuropsychologia, 15,* 457-460.

Hartje, W., Villmes, K., and Weniger, D. (1985). Is there parallel and independent hemispheric processing of intonational and phonetic components of dichotic speech stimuli? *Brain and Language, 24,* 83-99.

Heilman, K. M., Bowers, D., Speedie, L., and Coslett, H. B. (1984). Comprehension of affective and nonaffective prosody. *Neurology, 94,* 917-921.

Heilman, K. M., Scholes, R., and Watson, R. T. (1975). Auditory affective agnosia: Disturbed comprehension of affective speech. *Journal of Neurology, Neurosurgery and Psychiatry, 38,* 69-72.

Hoptman, M. J., and Levy, J. (1988). Perceptual asymmetries in left- and right-handers for cartoon and real faces. *Brain and Cognition, 8,* 178-188.

Izard, C. E. (1971). *The Face of Emotion.* New York: Appleton-Century-Crofts.

Izard, C. E. (1979). *Emotions in Personality and Psychopathology.* New York: Plenum Press.

Izard, C. E. (1983). *The Maximally Discriminative Facial Movement Coding System.* Newark, DE: Instructional Resources Center, University of Delaware.

Izard, C. E. (1994). Innate and Universal Facial Expressions: Evidence from Developmental and Cross-Cultural Research. *Psychological Bulletin, 115*, 288-99.

Izard, C. E., Fantauzzo, C. A., Castle, J. M., Haynes, O. M., Rayias, M. F., and Putnam, P. H. (1995). The ontogeny and significance of infants' facial expressions in the first nine months of life. *Developmental Psychology, 31*, 997-1013.

Izard, C. E., Hembree, E., Dougherty, L. M., and Spizzirri, C. C. (1983). Changes in facial expressions of 2 to 19 months old infants following acute pain. *Developmental Psychology, 19*, 418-426.

Izard, C. E., Huebner, R. R., Risser, D. McGinnes, G., and Dougherty, L. M. (1980). The young infants' ability to produce discrete emotion expressions. *Developmental Psychology, 16*, 13-140.

James, W. (1984). What is an emotion? *Mind, 4*, 188-204.

Kanwisher, N., McDermott, J., and Chun, M. M. (1997). The fusiform face area: a module in human extrastriate cortex specialized for face perception. *Journal of Neuroscience, 17*, 4302-4311.

Kilbride, J. E., and Yarczower, M. (1983). Ethnic bias in the recognition of facial expressions. *Journal of Nonverbal Behavior, 8*, 27–41.

King, F. L., and Kimura, D. (1972). Left-ear superiority in dichotic perception of vocal nonverbal sounds. *Canadian Journal of Psychology, 26*, 111–116.

Kitayama, S., and Markus, H. R. (1994). *Emotion and Culture: Empirical Studies of Mutual Influences.* Washington, D. C.: American Psychological Association.

Knapp, P. H. (1963). *Expressions of the Emotions in Man.* New York: International Universities Press.

Kucharska-Pietura, K., Phillips, M. L., Gernand, W., and David, A. S. (2003). Perception of emotions from faces and voices following unilateral brain damage. *Neuropsychologia, 41*, 1082-1090.

Lazarus, R. S. (1984). On the primacy of the cognition. *American Psychologist, 39*, 124-129.

Leventhal, H., and Tomarken, A. J. (1986). Emotion: Today's problems. *Annual Review of Psychology, 37*, 565-610.

Levy, J., Heller, W., Banich, M. T., and Burton, L. A. (1983). Asymmetry of perception in free viewing of chimeric faces. *Brain and Cognition, 2*, 404-419.

Ley, R. G., and Bryden, M. P. (1982). A dissociation of right and left hemispheric effects for recognizing emotional tone and verbal content. *Brain and Cognition, 1*(1), 3-9.

Luce, T. S. (1974). The Role of Experience in Inter-rater Recognition. *Personality and Social Psychology Bulletin, 1*, 39-41.

Luh, K. E., Ruechert, L. M., and Levy, J. (1991). Perceptual asymmetries for free viewing of several types of chimeric stimuli. *Brain and Cognition, 16*, 83-103.

Mahoney, A. M., and Sainsbury, R. S. (1987). Hemispheric asymmetry in the perception of emotional sounds. *Brain and Cognition, 6*, 216-233.

Malatesta, C. Z., and Haviland, J. M. (1982). Learning Display Rules: The Socialization of Emotion Expression in Infancy. *Child Development, 53*, 991-1003.

Mandal, M. K., and Singh, S. K. (1990). Lateral asymmetry in identification and expression of facial emotions. *Cognition and Emotion, 4*, 61-70.

Mandal, M. K. (2004). *Emotion.* Delhi: East-West Press.

Mandal, M. K., and Ambady, N. (2004). Laterality of facial expressions of emotion: Universal and culture-specific influences. *Behavioural Neurology, 15*, 23-34.

Mandal, M. K., Asthana, H. S., and Tandon, S. C. (1993). Judgement of facial expression of emotion in unilateral brain-damaged patients. *Archives of Clinical Neuropsychology, 8,* 171-183.

Mandal, M. K., Asthana, H. S., Madan, S., and Pandey, R. (1992). Hemifacial display of emotion in the resting face. *Behavioural Neurology, 5,* 169-171.

Mandal, M. K., Asthana, H. S., Tandon, S. C., and Asthana, S. (1992). Role of cerebral hemispheres and regions in processing hemifacial expression of emotions. Evidence from brain damage. *International Journal of Neuroscience, 63,* 187-195.

Mandal, M. K., Bryden, M. P., and Bulman-Fleming, M. B. (1996). Similarities and variations in facial expressions of emotion: Cross-cultural evidence. *International Journal of Psychology, 31,* 49–58.

Mandal, M. K., Harizuka, S., Bhushan, B., and Mishra, R. C. (2001). Cultural variation in hemifacial asymmetry of emotion expressions. *British Journal of Social Psychology, 40,* 385-398.

Mandal, M. K., Saha, G. B., and Palchoudhury, S. (1986). A cross-cultural study on facial affect. *Journal of Psychological Researches, 30,* 140–143.

Mandal, M. K., Tandon, S. C., and Asthana, H. S. (1991). Right brain damage impairs recognition of negative emotions. *Cortex, 27,* 247-254.

Markham, R., and Wang, L. (1996). Recognition of emotion by Chinese and Australian children. *Journal of Cross-Cultural Psychology, 27,* 616–643.

Markus, H. R., and Kitayama, S. (1991). Culture and Self Implications for Cognition, Emotion, and Motivation. *Psychological Review, 98,* 224-53.

Marsh A. A., Elfenbein H. A., and Ambady N. (2003). Nonverbal "accents": Cultural differences in facial expressions of emotion. *Psychological Science, 14*(4), 373-376.

Matsumoto, D. (1989). Cultural influences on the perception of emotion. *Journal of Cross-Cultural Psychology, 20,* 92–105.

Matsumoto, D., and Assar, M. (1992). The effects of language on judgments of universal facial expressions of emotion. *Journal of Nonverbal Behavior, 16,* 85-99.

Matsumoto, D., and Willingham, B. (2009). Spontaneous Facial Expressions of Emotion of Congenitally and congenitally Blind Individuals. *Journal of Personality and Social Psychology, 96*(1), 1-10.

Matsumoto, D., Yoo, S. H., and Fontaine, J. (2008). Mapping Expressive Differences around the World: The Relationship between Emotional Display Rules and Individualism versus Collectivism. *International Journal of Cross-Cultural Psychology, 39,* 55-74.

Mesquita, B., and Albert, D. (2007). The cultural regulation of emotions. In J. J. Gross (Ed.), *Handbook of emotion regulation* (pp. 486-503). New York: Guilford.

Mesquita, B., and Frijda, N. H. (1992). Cultural variations in emotions: A review. *Psychological Bulletin, 112,* 179-204.

Mesquita, B., Frijda, N. H., and Scherer, K. R. (1997). Culture and emotion. In J. W. Berry, P. R. Dasen, and T. S. Sarawasthi (Eds.), *Handbook of Cross-Cultural Psychology, Vol. 2* (254-297). Boston: Allyn and Bacon.

Michel, C., Rossion, B., Han, J., Chung, C-S., and Caldara, R. (2006). Holistic processing is finely tuned for faces of our own race. Psychological Science, *17*(7), 608-615.

Morris, J. S., Frith, C. D., Perrett, D. I., Rowland, D., Young, A. W., Calder, A. J., et al. (1996). A differential neural response in the human amygdala to fearful and happy facial expressions. *Nature, 383,* 812-815.

Nicholls, M. E., Ellis, B. E., Clement, J. G., and Yoshino, M. (2004). Detecting hemifacial asymmetries in emotional expression with three-dimensional computerized image analysis. *Proceedings of the Royal Society of London – Biological Sciences, 271*(1540), 663-668.

Norasakkunkit, V. (2000). *Culture, ethnicity and measures of emotional distress: The role of self-construal and self-enhancement.* Master's thesis. Boston: University of Massachussetts.

Oatley, K., and Jenkins, J. M. (1992). Human emotions: Function and dysfunction. *Annual Review of Psychology, 43,* 55-85.

Oatley, K., Keltner, D., and Jenkins, J. M. (2006). *Understanding emotions, Second edition.* Malden, MA, and Oxford, UK: Blackwell.

Ojemann, J. G., Ojemann, G. A., and Lettich, E. (1992). Neuronal activity related to faces and matching in human right nondominant temporal cortex. *Brain, 115,* 1-13.

Oster, H., and Ekman, P. (1978). Facial Behaviour in Child Development. *Minnesota Symposium on Child Psychology, 11,* 231-76.

Paradiso, S., Chemerinski, E., Yazici, K. M., Tartaro, A., and Robinson, R. G. (1999). Frontal lobe syndrome reassessed: comparison of patients with lateral or medial frontal brain damage. *Journal of Neurology, Neurosurgery, and Psychiatry, 67*(5), 664-667.

Pihan, H., Altenmueller, E., and Ackermann, H. (1997). The cortical processing of perceived emotion: a DC-potential study on affective speech prosody. *Neuroreport, 8,* 623-627.

Plutchik, R. (1980). *Emotion: A Psychobioevolutionary Synthesis.* New York: Academic Press.

Rapcsak, S. Z., Comer, J. F., and Rubens, A. B. (1993). Anomia for facial expressions: neuropsychological mechanisms and anatomical correlates. *Brain and Language, 45,* 233-252.

Reid, S. A., Duke, L. M., and Allen, J. J. B. (1998). Resting frontal electroencephalographic asymmetry in depression: inconsistencies suggest the need to identify mediating factors. *Psychophysiology, 35,* 389-404.

Reuter-Lorenz, P., and Davidson, R. (1981). Differential contributions of the two cerebral hemispheres to the perception of happy and sad faces. *Neuropsychologia, 19*(4), 609-613.

Richardson, C. K., Bowers, D., Bauer, R. M., Heilman, K. M., and Leonard, C. M. (2000). Digitizing the moving face during dynamic displays of emotion. *Neuropsychologia, 38,* 1028-1039.

Rinn, W. (1984). The neuropsychology of facial expressions: A review of the neurological and psychological mechanism for producing facial expressions. *Psychological Bulletin, 95,* 52-77.

Russell, J. A. (1994). Is there universal recognition of emotion from facial expression? A review of the cross-cultural studies. *Psychological Bulletin, 115,* 102–141.

Russell, J. A., and Fernandez-Dols, J. M. (1997). What does a facial expression mean? In Russell, J. A. and Fernandez-Dols, J. M. (Ed.), *The Psychology of Facial Expression.* Cambridge University Press.

Sackeim, H. A., and Gur, R. E. (1978). Emotions are expressed more intensely on the left side of the face. *Science, 202,* 434-436.

Sackeim, H. A., Greenberg, M. S., Weiman, A. L., Gur, R. C., Hungerbuhler, J. P., and Geschwind, N. (1982). Hemispheric asymmetry in the expression of positive and negative emotions: Neurologic evidence. *Archives of Neurology, 39,* 210-218.

Sackeim, M. A., Gur, R. C., and Saucy, M. C. (1978). Emotions are expressed more strongly on the left side of the face. *Science, 202*, 434-435.

Safer, M. A., and Leventhal, H. (1977). Ear differences in emotional tone of voice and verbal content. *Journal of Experimental Psychology: Human Perception and Performance, 3*, 75-82.

Saxby, L., and Bryden, M. P. (1984). Left ear superiority in children for processing auditory emotional material. *Developmental Psychology, 21*, 72-80.

Scherer, K. R. (1984). Emotions as a multicomponent process: A model and some cross cultural data. In P. R. Shaver (Ed.), *Review of personality and social psychology, Vol. 5* (pp. 37-63). Beverley Hills, CA: Sage.

Schiff, B. B., and Bassel, C. (1996). Effects of asymmetrical hemispheric activation on approach and withdrawal responses. *Neuropsychology, 10*(4), 557-564.

Schwartz, G. E., Ahern, G. L., and Brown, S. L. (1979). Lateralised facial muscles response to positive and negative emotional stimuli. *Psychophysiology, 16*, 561-571.

Schwartz, G. E., Davidson, R. J., and Maer, F. (1975). Right hemisphere lateralization for emotion in the human brain: Interactions with cognition. *Science, 190*, 286-288.

Sergent, J., Ohta, S., and MacDonald, B. (1992). Functional neuroanatomy of face and object processing. A positron emission tomography study. *Brain, 115*(1), 15-36.

Serrano, J. M., Iglesias, J., and Loeches, A. (1992). Recognition of facial expressions of anger, fear and surprise in 4 to 6 month-old infants. *Developmental Psychology, 25*, 411-425.

Shweder, R. A. (1991). *Thinking Through Cultures: Expeditions Indigenous Psychologies.* Cambridge: Harvard University Press.

Skinner, M., and Mullen, B. (1991). Facial asymmetry in emotional expression: A meta-analysis of research. *British Journal of Social Psychology, 30*, 113-124.

Sorenson, E. R. (1975). Culture and the expression of emotion. In T. R. Williams (Ed.), *Psychological anthropology* (pp. 361–372). Chicago: Aldine Publishing.

Spieker, S. J. (1982). Infant recognition of invariant categories of faces: Person identity and facial expression. *Dissertation Abstracts International, 42*, 12.

Starkstein, S. E., Robinson, R. G., Honig, M. A., Parikh, R. M., Joselyn, J., and Price, T. R. (1989). Mood changes after right-hemisphere lesions. *British Journal of Psychiatry, 155*, 79-85.

Strauss, E., and Moscovitch, M. (1981). Perception of facial expressions. *Brain and Language, 13*, 308-332.

Suberi, M., and McKeever, W. F. (1977). Differential right hemisphere memory storage of emotional and nonemotional faces. *Neuropsychologia, 15*, 757-768.

Thibault, P., Bourgeois, P. and Hess, U. (2006). The effect of group-identification on emotion recognition: The case of cats and basketball players. *Journal of Experimental Social Psychology, 42*(5), 676-683.

Tomkins, S. S. (1962). *Affect, Imagery, Consciousness, Vol. 1. The Positive Affects.* New York: Springer-Verlag.

Tucker, D. M., Watson, R. T., and Heilman, K. M. (1977). Affective discrimination and evocation in patients with right parietal disease. *Neurology, 27*, 947-950.

van Hemert, D. A., Poortinga, Y. H., and van de Vijver, F. J. R. (2007). Emotion and culture: A meta-analysis. *Cognition and Emotion, 21*(5), 913-943.

Wechsler, A. F. (1973). The effect of organic brain disease on recall of emotionally charged versus neutral narrative texts. *Neurology, 23*, 130-135.

Westen, D. (1996). *Psychology: Mind, Brain, and Culture*. New York: Wiley.

Wolff, W. (1933). The experimental study of forms of expression. *Character and Personality, 2*, 168-173.

In: Expanding Horizions of the Mind Science(s)   ISBN: 978-1-62808-705-5
Editors: P.N. Tandon, R.C. Tripathi and N. Srinivasan   ©2013 Nova Science Publishers, Inc.

Chapter 16

ECO-CULTURAL CONTEXTS AND COGNITIVE FUNCTIONING

R.C. Mishra

Department of Psychology, Banaras Hindu University, India

ABSTRACT

This chapter presents a brief state of current thinking and research on cognition in a historical context. The focus is on the work carried out by Indian psychologists, and how this work is linked with those reported from other parts of the world. The ecological perspectives and their impact on cognitive functioning are examined in some detail with reference to psychological differentiation theory. The key issues in the relationship of ecology, culture and cognition are discussed, and the way in which these issues are being addressed is described in the context of recent theoretical advancements in the field. Difficulties in adopting a value neutral approach to cognitive functioning are discussed, and a course of action for future research is suggested.

1. INTRODUCTION

Cognition refers to every process by which people know about the world around them and utilize this knowledge to solve problems. It includes such elementary processes as recognition, labeling, analysis, categorization, thinking and reasoning. Psychologists, educators, child development personnel, policy planners and many others have been interested in the study of these processes for a long time. Consequently, we have now a substantial body of knowledge about people's cognitive functioning, which has both theoretical and practical implications in several domains of human life.

Far a long time psychology as a scientific discipline has been tied to the apron strings of the natural sciences (Berry et al., 2003). Few decades ago the laboratory was regarded as the ideal place of work, and idealization of a behavioral phenomenon as the ultimate goal of psychological studies. The pursuit of this goal necessitated rigorous controls on extraneous variables. As a result, "environment" was considered as a "nuisance" in psychological studies, and every effort was made to control it as much as possible to understand various psychological phenomena in their purest forms. Failure to control environmental variables was regarded as an open invitation to potential sources of error, which not only contaminate the findings of the studies, but also brought disrespect for the researcher from the scientific community.

2. THE ECOLOGICAL PERSPECTIVE

An important departure from this dominant tradition of thinking and research in psychology during the last decades is the emergence of an "ecological perspective" in which knowledge of the characteristic features of the environment are considered essential for understanding human behavior. It is argued, and also empirically demonstrated, that the total setting in which individuals function has to be taken into consideration for comprehending their behavioral characteristics. As a result, psychologists today are looking toward field biology, ethology, and more importantly, toward ecology for developing wider principles of human behavior.

The application of ecological principles and methods in psychology began with Lewin (1951). He argued that the first step in understanding behavior of individuals or groups was to examine the opportunities and limitations of their environments. Wright (1967) and Barker (1968) also strongly advocated the idea of developing an ecological perspective in psychology. They drew attention to the limitations of the traditional psychological approach of bringing people into the laboratories and subjecting them to certain prearranged conditions or tasks in mechanical ways, as it is usually done in the physical sciences.

The ecological approach to the study of human behavior puts emphasis on the description of various events in a given environment. These may include such descriptions as how a mother cares for her child, how a child prepares to go to school or cattle grazing, what the child does in the school or forest, or how a family interacts at the meal. It also advocates for the preparation of "maps of the habitats", which may include the distinctive features of the physical and social environments that surround individuals and exert potential influences on their perception, feeling and behavior. These records may be effectively utilized in the comparison of environments and behavior of people from different social, economic, racial, ethnic and cultural contexts. In this way, it is possible to derive knowledge about inputs, which people receive from their everyday environment, and which may cumulatively shape their perceptions, feelings, behaviors and the overall pattern of life. In short-lived experimental situations, it is not possible to create events that are as frequent, complex or durable as events in everyday life.

The ecological movement in psychology was also greatly influenced by Brunswik (1957), who argued that the task of psychology was the analysis of interrelation between the environment and the behaving subject. These interrelations were regarded as "adaptive events". In his view, the ecological environment consists of all measurable characteristics of the objective surroundings of an organism. For cultural ecologists, this notion of ecology conveys the core meaning of the term "natural-cultural habitat". Making use of the term in this sense, Brunswik distinguished ecology from both the psychological environment (environment as perceived by the organism) and the immediate environment at the surface of the organism (the stimulus in experimental psychology). He regarded culture as part of the overall habitat to which the individual adapts.

This conception of ecology is slightly different from that of Barker, who considered psychological environment also as a part of ecology. Barker believed that behavior took place within ecological units, called "behavior settings". Environment and behavior were regarded as "mutually causally related systems", suggesting that the relationship and regulation of the two systems are not a one-way affair. Barker also made a distinction between physical and socio-cultural components of the environment. In this context, he proposed a promising hypothesis: "in relatively stable environments, people are the source of behavior variance", while "in varied and changing environments, the contribution of environmental inputs to the variance of behavior is enhanced" (Barker, 1965, p.13). Methodologically this proposition appears very sound, but in practice, it poses a great challenge by asking researchers to isolate the two components, which looks like an impossibility.

3. THE ECOLOGICAL MODELS

A number of models have been developed over the years to provide a description of the ecological variables, their interrelationships, and their probable linkages with behavioral variables. In this respect, two approaches seem to be evident. One is represented in the "eco-cultural" model of Berry, which has evolved through a series of studies devoted to understanding similarities and differences in cognition and social behaviors (Berry 1966, 1976; Berry et al. 1986; Mishra et al. 1996; Mishra and Berry 2008). In this approach, both ecological and cultural variables are treated together in understanding behavior. The basic argument is that as one crosses ecological boundary, one also crosses the cultural one. A major proposition of Berry's model is that *behavior is adaptive to culture, and culture is adaptive to ecology.*

The second approach is represented in the "ecological model" developed by Bronfenbrenner (1974). In this approach, we find greater emphasis on "experiences" embedded in ecological contexts in explaining behavioral characteristics of individuals than on "adaptations" that develop due to the demands of one's ecology. In describing the enduring environment of a child, Bronfenbrenner (1974) made a distinction between outer and inner ecology. In a later elaboration of this model, Bronfenbrenner (1979) conceived of ecological environment as a set of nested structures, each embedded in the next, like a set of Russian dolls. The different systems that form the ecological environment include the microsystem, the mesosystem, the exosystem and the macrosystem. Each system contains different set of factors, although in an overall treatment, they all seem to be highly interlinked.

Following Bronfenbrenner, Sinha (1977) developed an "ecological model" for understanding the context of child development in the Indian cultural setting. In this model, ecology is viewed as consisting of two concentric layers, called *supporting* and *surrounding* layers. Home, school and peer groups with physical facilities, nature of social interactions and prevalent activities in each constitute the supporting layer of ecology. This layer is embedded in a larger and more pervasive setting, called the surrounding layer of ecology.

Characteristics of the geographical environment, including the mode of economic pursuit and density of population, constitute the first layer of the surrounding ecology. The institutional setting of the child provided by caste, class and other factors with attendant role expectancies and interactive limitations constitute the second layer. General amenities available to individuals residing in those places (e.g., drinking water, power, municipal and civic amenities, means of entertainment, etc.) constitute the third layer of ecology. The general environment is made up of these surrounding layer factors, which combine and constantly interact with upper layer factors. These factors determine not only the economic pursuits of people and their way of life, but also their socialization processes and interpersonal relationships, which significantly influence cognitive and perceptual processes, motivation, style of coping with problems, and general personality development (Sinha 1982).

In a relatively recent work, Dasen (2003) has developed an integrated model by combining the elements of various ecological models, specifically those of Bronfenbrenner and Berry, but also of Super and Harkness (1986) model, called "ecological niche". This effort is certainly very ambitious. It brings out not only the complexity of the ecological and cultural systems, but also of human behavior, which receives inputs in different combinations from all of the elements represented in various systems. Apparently, the model looks a bit complex, and the predictive validity of the overall model has yet to be tested. However, in a study carried out in India, Indonesia, Nepal and Switzerland, Dasen and Mishra (in press) have collected evidence that supports the linkages of a number of ecological and cultural elements of the model with spatial language and cognitive development of children.

4. HOLISTIC APPROACHES TO ECOLOGY

Some five decades ago Murdock (1969) had presented a major review of the specific variables constituting ecology. These variables include exploitive pattern, settlement pattern and mean size of the group. Employing a sample of 320 societies from the Ethnographic Atlas, Murdock classified these societies as fully nomadic, semi-nomadic, semi-sedentary and fully sedentary. Rating of these societies in terms of population density revealed that different exploitive patterns were highly related to settlement patterns and the size of the community. For example, gathering and hunting societies were nomadic or semi-nomadic, whereas agricultural societies were semi-sedentary or sedentary. Gathering and hunting societies had low population concentration; agricultural societies had high population concentration; the pastoral and fishing communities were in between. Exploitive and settlement patterns were also highly related to food accumulation. Agricultural societies were high on food accumulation; hunting and fishing were low. Root crop and grain agriculturists were in between. These distributions were highly interrelated suggesting that they could be treated as important elements of ecology besides temperature and rainfall that were earlier documented in biological sciences.

Berry (1976) has indicated that although physical environmental variables of temperature and rainfall do usually permit human organisms to engage in certain economic pursuits, it should not be taken as deterministic of such pursuits. Societies may adopt other exploitive patterns if any possibility of the kind exists. Thus, even without considering temperature, rainfall or soil composition (the basic elements of physical ecology), the geographical setting, the exploitive and settlement patterns, including population density and food accumulation, may be taken as other constituents of human ecology to account for cultural and behavioral variability in human populations.

With regard to socio-cultural component of ecology, psychological research indicates that these variables are highly adaptive to physical ecological parameters. For example, "role diversity" and "social stratification" of groups in a society are two such variables. These determine the institutional setting of the groups in terms of their caste, class or religious status, and constitute an essential element in the scale of socio-cultural evolution. Nimkoff and Middleton (1960), Ember (1963) and Pelto (1968) have discussed various forms of stratification involving general social stratification, family structure, and political and economic authority. These factors have been found to be significantly related to child socialization practices (Berry, 1969).

Research carried out in different eco-cultural settings also suggests that the relationship of physical and socio-cultural components of ecology with behavior is modified through the influences of urbanization and education. These are generally referred to as "acculturation influences" (Berry, 1976; Berry et al., 1986; Mishra et al., 1996). These findings suggest that for an adequate understanding of cognitive behavior of a group of people, it is necessary to examine the role of eco-cultural variables in the context of their acculturation.

5. ECOLOGICAL AND CULTURAL VARIABLES IN COGNITIVE RESEARCH

In psychological research of the earlier decades, the ecological factors discussed above have received considerable attention in different combinations. Long time ago Porteus (1937) reported that some aspects of the environmental background were related to intellectual functions. Kardiner (1945) argued for a similar relationship between socio-cultural background and personality. Whiting and Child (1953) emphasized on the role of economic "maintenance system" in the origin of cultural and behavioral variations. Whiting's model (1973) in its present form recognizes the importance of ecological, cultural and behavioral elements. Ecological elements include the physical and the learning environment along with some economic and demographic features of the maintenance system. Cultural variables include the indigenous social structure and the diffusion from other cultures. Behavioral variables consist of both innate and acquired behaviors as well as the "projective expressive" systems characteristic of a group of people.

Segall, Campbell, and Herskovits (1966) initiated another tradition of research on ecological influences on behavior. They studied susceptibility to visual illusions in the context of natural and cultural environment related factors. In order to determine the extent of susceptibility to visual illusions, they analyzed the differential frequencies with which various geometric shapes were experienced by samples in their visual ecology. In this ambitious research scheme, the cultural products were allowed to become a part of the subject's physical environment, and there was a lack of concern for such cultural variables as socialization, social structure, social relations and language.

Dawson (1967a; 1967b; 1969) developed a "biosocial" framework for understanding ecological-cultural-behavioral relationships. In a later elaboration of this framework, Dawson (1973) mentioned that ecology forms the fundamental basis of the biosocial system. In this context, he particularly discussed the role of factors like physique, malnutrition, disease and hormonal changes on the one hand, and laterality and responses to Western influence on the other.

The general distinction between behavior that is a function of physical environment and behavior that is mediated by cultural influences was outlined by Berry (1976). He argued that cross-cultural/cross-ecology studies should consider independently two kinds of functional relationship between ecology and behavior. One may be called a *direct relationship*, in which the focus is on intracultural analysis of behavior in relation to the frequency with which different aspects of ecology are experienced by individuals (e.g., Segall et al. 1966 study). Another may be called an *indirect relationship*, in which culture is brought in as a mediating variable between ecology-behavior relationships.

The latter kind of relationships can be understood only through cross-cultural studies, not through intracultural studies, because ecological variation in this case is usually not available. Berry (1968) has shown how variation in illusion response is associated with cultural variables like education and socialization (besides ecology). Later work also points out the importance of combining ecological and cultural variables for a precise understanding of human cognitive functioning (Altarriba, 1993; Mishra et al., 1996; Mishra, 1997, 2001; Rogoff, 1981). Since it is difficult to determine empirically the extent of co-variation of ecological and cultural factors, most of the studies of cognition focus either on ecological or cultural factors in examining their influences. However, there are a few studies in which ecology and culture have been considered simultaneously for analyzing cognitive functioning.

6. FURTHER STUDIES

In the following pages, we will briefly discuss some research that has adopted an eco-cultural approach to the study of cognition. The focus is mainly on studies carried out in India, but there are also others, which either use samples from the Indian cultural context, or have been seminal to Indian studies.

In India Professor Durganand Sinha at Allahabad University initiated research on the effect of ecological and cultural factors on cognition. In early studies, he attempted a comparative analysis of the experiences available to children of nonscheduled castes, scheduled castes and scheduled tribes in their respective ecological settings, and linked them with perceptual and cognitive skills that developed among them (Sinha, 1977). Later on, the emphasis shifted to a thorough analysis of the overall circumstances in which children of different groups were raised with a view to understanding the effect of these circumstances on the development of cognitive styles (Sinha, 1979, Sinha and Bharat, 1985; Sinha and Shrestha, 1992). Theory of psychological differentiation (Witkin et al. 1962) and eco-cultural model (Berry, 1976) broadly guided these later studies.

The search for the relationship between ecology, culture and cognition has been governed by both empirical and theoretical questions. In research, the question pertains to the role of physical, social and cultural environmental experiences in the development of cognition. In theory the question relates to (1) the patterning or organization of eco-cultural experiences in different environmental settings, (2) the patterning or organization of performances on a variety of cognitive tasks, and (3) the relationship between eco-cultural experiences and cognitive performances (Berry et al., 2002).

Different theories of cognitive development (e.g., general intelligence, genetic epistemology, specific skills, and psychological differentiation) deal with these issues in their own peculiar ways (Mishra, 1997). For understanding cognition and its development in diverse eco-cultural settings, however, the theory of psychological differentiation has been utilized more than any other theory (Mishra, 1997; 2001).

The theory postulates that an individual's cognitive development proceeds from a lesser differentiated to a more differentiated pattern of functioning, which is manifested in the segregation of psychological activities. One of the indicators of greater segregation is the "articulated cognitive functioning". The process involves "perceptual analysis and disembedding" of certain items from a complex field. Because it attempts to analyze the extent to which an individual relies on external field in making judgments, the process is referred to as field independent (FI)-field dependent (FD) cognitive functioning. In perception, the FI refers to the ability to disembed a part from the total field. In cognition, it refers to the ability to analyze, restructure, reorganize or break up certain cognitive materials or problems for achieving their solutions. People differ in the extent to which they rely on internal or external referents. Because the ability is consistently reflected in cognitive, interpersonal and neuro-physiological domains, it has been referred to as "cognitive style" (Witkin and Goodenough, 1981).

In understanding the development of cognition in diverse eco-cultural settings, the cognitive style approach begins with an analysis of what people do cognitively in their respective ecological and cultural settings. Thus, it attempts to analyze the demands on people's life in different ecological contexts. Elements of these contexts are believed to be highly textured (Berry, 1983), and not easily amenable to isolation. Assuming that the elements of experience may be structured differently in different ecological settings, the approach looks for interrelationships (patterns) in cognitive performances. It also postulates that different patterns of cognitive abilities develop in different ecological and cultural settings depending on the nature of demands placed on the life of individuals.

Several components of cognitive style have been identified and refined during the course of research (Gruenfeld and Lin, 1984; Goodenough, 1986). Witkin and Goodenough (1981) have presented an excellent review of studies in this field, and Berry (1981; 19912004) has discussed the cross-cultural issues in comparative study of the development of cognitive style. A unique feature of the FD-FI construct is that it is treated as a bipolar dimension. The FI individuals generally do better in analytic domain, and FD in the social or interpersonal domain. Hence, no absolute value judgments about good or bad can be made at any one of the poles (Witkin and Goodenough, 1981). Individuals are also not considered as "fixed" into their usual places on the dimension; rather they can move from one pole to the other, depending on the circumstances in which they are placed.

Cross-cultural research on cognitive style has largely used the "eco-cultural framework" developed by Berry (1966; 1976;1987). This framework describes ecology and acculturation as two major sets of "input" variables. The culture of groups and behavior of individuals are considered as adaptive to the demands placed on them in their respective ecological and acculturation settings. Comprehensive review of research on cognitive style, using the eco-cultural framework, is available in Witkin and Berry (1975), Berry (1981; 1991), and Mishra (1997; 2000).

7. Major Research Issues

Research studies have raised a number of issues relating to cognitive style. These issues have emerged from the patterns of performance obtained on different tests used for the assessment of FD-FI cognitive style. In earlier studies, Rod and Frame Test (RFT), Body Adjustment Test (BAT), and Embedded Figures Test (EFT) were widely used. In later studies, a portable version of the RFT, called Portable Rod and Frame Test (PRFT), has been employed. Several forms of EFT have also been developed and used with children and adults in various cultural groups. The Children's Embedded Figures Test (CEFT), African Embedded Figures Test (AEFT), Indo-African Embedded Figures Test (IAEFT) and Story-Pictorial Embedded Figures Test (SPEFT) have been widely used by researchers. Tests measuring FD-FI style in tactual and auditory domains have also been given to samples.

A central issue in these studies relates to the consistency in test scores obtained in different cognitive domains. Gender difference in the level of field dependence-independence is another important issue. Individual and group differences in the level of field dependence-independence constitute a third major issue. The adaptability of cognitive style to ecological pressures and acculturative influences (e.g., urbanization, wage employment, education) constitutes the fourth major issue in research.

With respect to *self-consistency,* earlier studies largely explored into the perceptual domain, and generally found support for this. Evidence for less consistency appeared between perceptual and other domains. Studies seeking *sex difference* found its presence in agricultural samples, but not in hunting and gathering populations (Berry 1966; Mac Arthur 1967). Developmental studies revealed a shift from FD to FI from childhood to adulthood, with an evidence for the stability of the styles over time.

The search for *individual and group differences* in cognitive style has led to the study of the characteristics of family and child-rearing practices, including socialization pressure on "conformity", ecological engagements of people (hunting-gathering or agriculture), and acculturative experiences. Findings suggest that it is possible to predict the cognitive style of individuals and groups from the knowledge of their ecological and cultural characteristics. These findings have considerable practical implications for dealing with the problems of education, health, social change and development of individuals and groups (Mishra et al. 1996). Hence, the study of the conditions in which people develop different cognitive styles has been a more attractive issue of psychological inquiry than other issues that seem to fulfill mainly academic concerns.

8. Empirical Studies

Cross-cultural research on FD-FI cognitive styles during the last decades has continued along these lines. There is evidence for both the sporadic and programmatic kind of research.

The large-scale research programs have been pursued in Central Africa and northern India, using the eco-cultural framework of Berry (1976; 1987).

In an early study, Sinha (1979) worked with two subgroups of the Birhor Tribal culture. One of these lived a nomadic hunting-gathering life; the other one had made transition to a sedentary agricultural life. A long-standing agricultural group of the Oraon Tribal culture was also included. Boys and girls of 8-10 years, sampled from each of these groups, were administered the SPEFT (Sinha, 1978; 1984). This test was modeled on the CEFT by embedding local familiar stimuli (e.g., squirrels, snakes, and butterflies) in larger organized natural scenes (e.g., forests, and gardens) in order to ensure its cultural appropriateness. Findings revealed that children of the hunting-gathering group showed greater disembedding of stimuli in comparison to those of the agricultural group.

In another study, Sinha (1980) compared Tribal and non-Tribal samples, with a view to analyzing sex difference in performance of the SPEFT. The difference between boys and girls in the tribal sample was not significant. On the other hand, a clear gender difference (favoring boys) in the non-tribal sample was noted in the 4-5, 7-8 and 9-10 year groups.

The results of these studies were interpreted as supporting Berry's (1976) prediction that hunters and gatherers would be psychologically more differentiated than agriculturists, and that in complex and stratified societies, distinct sex roles are culturally prescribed, leading to differing psychological outcomes. In a later review of the findings of Indian studies on perceptual and cognitive skills, Sinha (1982) offered interpretations that were distinctly in favor of the eco-cultural model.

The effect of ecology of hills and plains on the development of cognitive style has been demonstrated in another study with 7-10 year children of the Brahmin and Gurung cultural groups of Nepal (Sinha and Shrestha, 1992). Both schooled and unschooled children were administered the SPEFT. Findings revealed that schooled children generally scored higher than the unschooled. On the other hand, hill Brahmin children scored higher than hill Gurung, but plains Brahmin children scored lower than the plains Gurung. The overall findings indicated that ecology reinforced the process of differentiation by being associated with certain cultural practices of the groups.

Shrestha and Mishra (1996) worked with children (7-10 years) of the Brhamin and Gurung cultural groups of Nepal. They found a significant gender difference in the level of psychological differentiation in the Brahmin sample, but not in the Gurung sample. On the other hand, Brahmin children appeared to be psychologically more differentiated than Gurung children in hill than in the plain ecology. Differing cultural practices of Brahmins and Gurungs in the ecology of hills and plains were held responsible for the obtained inconsistency in gender difference.

In these studies, however, the socialization practices and social structure of the groups were not assessed directly. This problem was addressed by Sinha and Bharat (1985) in a study carried out with children reared in monogamous, polyandrous and polygynandrous families found in the cis-Himalayan region of India. The monogamous families comprised one husband, one wife, and their children. The polyandrous families comprised one wife with two or more husbands (often brothers) and their joint children. In polygynandrous families, there were two or more husbands (again often brothers) sharing two or more wives, and their common children. The SPEFT and Block Designs Test (BDT) were administered to children of 7-9 and 13-15 years. Family experiences (e.g., mother and father involvement), disciplinary practices used with children, and mother-father dominance were observed, assessed through interviews, and rated for different parameters of child socialization. While some of the socialization variables showed variations across family types, they did not significantly influence the level of psychological differentiation of children of the concerned groups.

Mishra and Singh (2008) examined the role of specific socialization variables in the level of field dependence-independence in a study in which 8-10 year boys and girls negotiating life in a rural ecological setting of Varanasi were tested. Children were given a House Building Task (Mishra et al., 1996) in the presence of their parents. Parent-child interaction during the task performance was assessed. The socialization variables included parental utterances (positive, negative, and task-specific), parental help, and looking to parents and researcher (indicators of help-seeking behavior of children). The findings revealed no significant difference in the pattern of parental interaction with boys and girls on any of the measures indicated above. On the other hand, these variables played significant role in distinguishing FD and FI children: less parental utterances, less positive utterances, more task-specific utterances, less parental help, and less help-seeking behavior of children were associated with higher scores on the SPEFT.

G. Sinha (1988) examined the role of acculturation context (another component of Berry's model) in the emergence of cognitive style among children of the Santhal Tribal culture. The effects of schooling, and exposure to urbanization and industrialization on SPEFT scores of children were found to be in the predicted direction. The findings indicated a stronger influence of industrialization than schooling and urbanization on the SPEFT performance. A less pervasive effect of schooling was attributed to some qualitative (less stimulating) features of schools that Santhal children had attended.

A major test of psychological differentiation theory and eco-cultural model has been attempted in Central Africa (Berry et al., 1986). A comprehensive assessment of ecological, cultural and acculturation contexts was attempted. Male and female children and adults of the Biaka (Pygmy hunters and gatherers), the Bangandu (agriculturists with some hunting and gathering), and the Gbanu (full-fledged agriculturist) cultural groups were studied. The cognitive tests included the African Embedded Figures Test, Portable Rod and Frame Test, Body Adjustment Test, Sophistication of Body Proportioning Scale, Block Designs Training Test, Auditory Sequential Tones Test, Auditory Embedded Tones Test, and Tactile Embedded Figures Test. These tests measured cognitive differentiation in visual, auditory and tactual domains. Evidence for differentiation in the social domain was collected by the assessment of looking, telling and sitting behaviors. Socialization emphases of the groups were studied through parent and neighbor interviews, child ratings, and observation of parent-child interaction on a given task. Acculturation was measured at both the subjective and objective levels.

The results broadly supported the notion of cognitive style. Evidence for socialization emphases on independence and self-reliance in the Gbanu also corresponded to the general predictions of the eco-cultural model. On the other hand, the prediction of test performance from socialization measures was not as strong as expected. The difference with respect to the experiences of acculturation was substantiated by ethnographic accounts of the groups as well as standardized test- and contact-acculturation measures, and there was clear evidence for their influence on test performance. The overall findings revealed a significant effect of ecological variables on performance even after statistically controlling for the effect of acculturation.

Taking into consideration some of the limitations of this study, Mishra et al. (1996) carried out a large-scale research in the State of Bihar (now Jharkhand) in India. By focusing on ecological and cultural characteristics of groups they selected children and adults of the Birhor (nomadic hunters-gatherers group), Asur (recent settlers pursuing a mixed economy of hunting gathering and agriculture), and Oraon (long-standing agriculturists) Tribal cultural groups. In each group, sampling variations were obtained with respect to a number of objective and subjective measures of contact-acculturation. The test-acculturation of individuals was also measured. Socialization emphases (pressure towards compliance or assertion) in the groups were assessed through a combination of observation, interview and testing. SPEFT, TEFT, and Kohs Block Designs Test were used as the measures of cognitive style.

Findings again provided evidence for the existence of cognitive style; the above mentioned measures (SPEFT, BDT, Kohs Blocks) loaded on a common factor. With regard to the effect of ecology and acculturation on cognitive performance of adults and children, results were in the predicted direction. The hunting-gathering samples scored significantly higher than the agricultural sample almost on all tests. On the other hand, the effect of acculturation was significant mainly for the Oraon cultural group, suggesting an interaction between ecological background and acculturation.

To explain this culture-bound effect of acculturation the authors used a "threshold hypothesis". It was argued that acculturation could change the psychological characteristics of individuals only if it could penetrate into their life beyond a threshold point; below that level, the effects might remain superficial and incapable of bringing about changes in their psychological make-up. Test-acculturation appeared to be an important predictor of people's test performance; however, it could not displace the effect of the long-standing eco-cultural adaptation of groups. The findings revealed that socialization emphasis reported by parents or children were not strong predictors of children's cognitive style. On the other hand, variables like helping and feedback (extracted from factor analysis of the House Building Task data) predicted children's cognitive style in the expected direction. K. Mishra (1996) obtained support for these results in another study with children of hunting-gathering, agricultural and wage-earning samples of the Tharu culture inhabiting the foothill region of the Himalayas.

R.C. Mishra (1996a) adopted another strategy for studying the influence of ecology on cognitive development among unschooled children of the Birjia cultural group in Bihar. Distances traveled away from home, either in the forest or within the village, and self-directed activities of children were assessed to develop ecological parameters. Children were administered the SPEFT and the Indo-African EFT as tests of cognitive style. The findings revealed that, in general, children moving into the forests traveled longer distances and engaged in more self-directed activities than those moving in the village surroundings. Children moving away in the forest scored significantly higher on both the measures than village children. These findings were explained in terms of greater self-exploration opportunities and high differentiation demands placed on children in the ecology of the forest than in the village ecology.

9. CURRENT DEVELOPMENTS IN RESEARCH

The studies mentioned in the preceding pages indicate that research on cognitive style in relation to ecological and cultural variables has addressed similar issues during the last three decades. While earlier studies provide clear results, findings reported from diverse cultural settings in recent years make the picture complex. Consequently, some theoretical advancement has surfaced in this field. These developments involve conceptualization of cultural level variables, psychological level phenomena, and the relationship between them.

At the cultural level, the main concern lies with respect to specifying cultural experiences that are important for understanding psychological differentiation. In most of the studies, Berry's (1976) eco-cultural framework has been used as a guide to look for psychologically relevant variables. As indicated earlier, in this framework, the ecological (economic activity, settlement pattern, societal size) and cultural (role diversity, socio-cultural stratification, socialization) variables have been combined into a single index, called "eco-cultural index". The dimensions involved in this index are those of "societal size" and "social conformity" (Berry, 1987). Boldt (1978) has argued that these dimensions should be treated independently, because they vary considerably as a function of the subsistence strategies of the groups. Societal size and role diversity tend to increase linearly from hunter-gatherer (low), through agriculturist (medium) to urban-industrial samples (high). On the other hand, social conformity and role obligations tend to be low in hunting-gathering and urban-industrial samples, but high in agricultural samples (Boldt and Roberts 1979; Gamble and Ginsberg 1981). Relatively high field-independence reported for Western samples can be explained in terms of the importance of "role obligation" for samples during acculturation. It may be argued that decrease in role obligation and norm imposition within a group would increase the probability of "self-nonself segregation", and produce cultural group differences in cognitive style.

This possibility has been examined in two major studies. Mishra (1996b, see also Mishra and Berry, 2008) tested samples from four subsistence level strategy groups (hunting-gathering, dry agriculture, irrigation agriculture, wage earning). These were drawn from the Tribal regions of Bihar in India. Measures of societal size, social conformity, social-connectedness, individual-connectedness and socialization patterns were used for the assessment of cultural dimensions. SPEFT, Hidden Words Test, Locating Objects Test, Syllogistic Reasoning Test, Unknown Words Test, Visual Closure Test and Object Enumeration Test were used for the assessment of cognitive dimensions of differentiation (called intraunit distinctiveness) and contextualization (called extraunit connectedness). The relationship between the two cultural and cognitive dimensions was also examined.

With respect to cultural dimensions, the findings revealed that societal size showed a progressive increase from the hunting-gathering to wage earning through agricultural samples, whereas social conformity was low in the hunting-gathering and wage earning samples, but high in the agricultural samples. Social as well as individual connectedness was found to be highly placed in agricultural than the hunting gathering and wage earning samples. The cultural variables (societal size, social conformity, social and individual connectedness) formed a cohesive cluster, indicating that the variables were highly interrelated.

With respect to cognitive dimensions, the analyses revealed that the level of differentiation in hunting-gathering and wage earning samples was higher than that of the agricultural samples. On the other hand, the level of contextualization in the hunting-gathering sample was lower than that of the agricultural and wage earning samples respectively. Differences between boys and girls in all test performances were less evident in the hunting-gathering than other samples. There was also evidence for "task specificity" with respect to gender effect on performance across different subsistence level groups. Integrative analyses (MRA) revealed that "years of school education" was the strongest predictor of children's performance on various measures. Subsistence economy and urban contact also made significant contribution to a number of measures.

In another study, Berry et al. (2000) examined variations along contextualization-decontextualization and differentiation-integration cognitive dimensions, using measures of problem solving, perceptual judgment, disembedding, linguistic comprehension, visual integration and syllogistic reasoning. Seven samples that varied in ecological adaptation were studied. These represented gathering (India), hunting (Canada, Ghana), agricultural (China, Ghana, India) and industrial-based (Canada) populations. Results provide moderate support for the hypothesized linear decrease in contextualization and the curvilinear variation in differentiation across this ecological range.

As far as the role of socialization in cognitive style is concerned, the evidence is not very consistent to make any strong claim in this respect. On the one hand, data obtained with various methods do not show coherence. On the other hand, data obtained with interview method pose challenges of reliability and predictive validity. While there is some consistent evidence about the role of socialization in cognitive style across cultures, within-culture studies do not reveal much of consistency in findings. In order to claim socialization as a factor in cognitive test performance of groups across cultures, its influential role within culture has to be essentially established. Thus, within-culture studies are seriously required before making strong theoretical assertions about the postulated role of socialization in the development of psychological differentiation. There is also need to examine other dimensions of socialization besides the usually implicated "compliance-assertion" dimension made available to us by ethnographic literature.

Application of psychological differentiation theory in different ecological and cultural settings has provided considerable evidence for differences in the level of differentiation among individuals and groups at various age levels. These differences may be taken as indicating different levels (high or low) of cognitive development of individuals or groups. In this case, individuals or groups, who obtain lower scores on the tests of perceptual or cognitive differentiation, can be easily marked as showing "developmental deficit" or "developmental lag", a concern seriously shared by Cole and Scribner (1977).

A cognitive style interpretation, however, postulates that different individuals (or different groups) react differently to a cognitive problem (task, test, experiment, etc.), but in some systematic ways. This differential reaction is believed not to be due to the differential level of their cognitive capacity or competence. A number of individual and group level factors, such as age, gender, previous experience, socialization, etc., may be responsible for their particular way of responding to the task or test. The choice seemingly occurs at the unconscious level; and is often found to be linked to people's habits, customs, or preferred values. These factors constitute part of one's "culture". As indicated earlier, an important aspect of cognitive styles is that there is no judgmental aspect to this choice: it is not inherently "better" or "more advanced" to respond to stimuli or situations in one way than another. Differences in cognition can be clearly interpreted in this way if one prefers to adopt a value neutral approach to cognitive development.

The nature of cognitive functioning in different cultural groups still continues to be an issue of debate in the field. Nisbett (2003) holds "that there are indeed dramatic differences in the nature of Asian and European thought processes" (p. xviii), while others seem to take a moderate stand in this respect. For example, Norenzayan et al. (2007) hold the view that "Although both systems of thought are in principle cognitively available to all normal adult humans, cultural experiences may encourage reliance of one system at the expense of another, giving rise to systematic cultural differences", and that "these differences in cognitive orientations are believed to be rooted in the different social worlds of East Asians and Westerners today" (p. 578). This line of interpretation goes very well with the one coached in terms of cognitive styles.

Dasen et al. (1979; also Berry et al., 1982) and Mishra (1998) have argued for a value free interpretation of cognitive styles (and of cognitive development in general) in the context of cross-cultural studies. To be FI is not inherently better than being FD, although this is what apparently seems to be valued in many situations. Compared to FI, someone who is more FD has more empathy, more social skills, can easily live with a group, and appreciate things in a more holistic fashion. These characteristics are very similar to Nisbett's description of Asian mode of thinking.

The problem of keeping FD-FI cognitive style value free can be situated in the way it is measured. Some of the tests currently used for the assessment of FD-FI style show a strong association with ability, or even with general intelligence (Dornyei, 2005; Sternberg and Grigorenko, 2001; Zhang and Sternberg, 2006). The early tests devised by Witkin to measure psychological differentiation are quite specific. For example, in the RFT, the subjects are presented with a line surrounded by a tilted square, and are asked to adjust the line so that it looks vertical. The FD people, who are influenced by the square, tilt the line in the same direction as the square, while the FI people set the line independently of the square, according to their body-based perception of verticality. In such cases, it is rather easy to accept that there are two ways to react to the same situation, and that one is not inherently better than the other.

But in tests like the EFT, where small elements have to be located in a complex picture, FI subjects locate the elements more easily than the FD. In some tests (e.g., SPEFT) they can find more number of hidden objects in less time, providing evidence for an efficient performance. The positive correlation noted between some standard psychometric tests used in the assessment of intelligence (e.g., the Kohs Block Designs Test or Raven's Progressive Matrices Test) with RFT and EFT (see for example Berry, 1976; Mishra et al., 1996) present other evidences that may implicate an ability component in cognitive style.

Although cognitive ability vs. cognitive style debate still seems to continue in the field, data obtained from cross-cultural studies with the application of cognitive tests, however, are generally not suggestive of an ability-based "developmental deficit" interpretation. Since the development of interpersonal skills among less differentiated (i.e., FD) people has not been thoroughly studied, there is still the problem of claiming this component to be more developed in the FD than FI individuals. Demands of the tasks used for the study of interpersonal skills (e.g., of human face recognition) in many of these studies seem to favor the FI people (e.g., Arya, 2009). Hence, the bipolarity of the dimension (that FI are high on analytical skills and FD on interpersonal) is less strongly established. This needs serious research, especially in the Indian cultural context, which is characterized mainly by its emphasis on enduring relationships of individuals and groups.

CONCLUSION

The theory of psychological differentiation makes great promise to the study of cognition and cognitive development across cultures. It demonstrates different patterning of cultural elements, cognitive performances as well as of the central processes that connect them. It reveals that the effect of culture on cognition cannot be explored simply by observing the performance of certain cultural groups on some cognitive tasks. What we need to do is to analyze the cultural life of groups, behavioral competencies that are valued in a given culture, and the manner in which these are nurtured among individuals in that culture. A distinction of cognitive process, ability, competence and performance has also to be maintained in order to reveal the complexity of their relationships.

As we have mentioned in the beginning, theories differ enormously with respect to the patterning of the elements of culture, cognition and the processes through which these are connected. This diversity does not permit an easy conclusion with respect to the relationship between culture and cognition. Psychological differentiation theory maintains that cognitive processes are universal, but cognitive competencies develop in different ways according to the demands of one's ecology and culture. A valid inference about these competencies can be drawn only by situating assessment into the broader ecological settings and cultural life of individuals or groups. Recent developments with respect to the conceptualization of these dimensions are encouraging. We may hope that researchers will pay more attention to these elements in future studies. This is particularly important in the context of social and cultural changes, which are taking place globally, and are generating new demands for adaptation to the changed ecological circumstances.

REFERENCES

Altarriba, J. (Ed.) (1993). *Cognition and Culture: A Cross-cultural Approach to Cognitive Psychology*. Amsterdam: Elsevier Science.

Arya, Y. (2009). Field dependent-field independent cognitive style, learning and memory of children. Unpublished manuscript, Department of Psychology, Banaras Hindu University.

Barker, R.G. (1965). Explorations in ecological psychology. *American Psychologist, 20,* 1-14.

Barker, R.G. (1968). *Ecological psychology: Concepts and methods for studying the environment of human behavior*. California: Stanford University Press.

Berry, J.W. (1966). Temne and Eskimo perceptual skills. *International Journal of Psychology, 1,* 207-229.

Berry, J. (1968). Ecology, perceptual development and the Muller Lyer illusion. *British Journal of Psychology, 59,* 205-210.

Berry, J.W. (1969). Ecology and socialization as factors in figural assimilation and the resolution of binocular rivalry. *International Journal of Psychology, 4*, 271-280.

Berry, J.W. (1976). *Human ecology and cognitive style: Comparative studies in cultural and psychological adaptation.* New York: Sage/Halsted.

Berry, J.W. (1981). Developmental issues in the comparative study of psychological differentiation. In R.H. Munroe, R.L. Munroe, and B.B. Whiting (Eds.), *Handbook of Cross-cultural Human Development* (pp. 475-498), New York: Garland.

Berry, J.W. (1983). Textured contexts: Systems and situations in cross-cultural psychology. In S.H. Irvine and J.W. Berry (Eds.), *Human Assessment and Cultural Factors* (pp. 117-126), New York: Plenum.

Berry, J.W. (1987). The comparative study of cognitive abilities. In S.H. Irvine and S. Newstead (Eds.), *Intelligence and Cognition: Contemporary Frames of Reference* (pp. 393-420). Dordrecht: Nijhoff.

Berry, J.W. (1991). Cultural variation in field dependence-independence. In S. Wapner and J. Demick (Eds.), *Field Dependence-Independence: Cognitive Style Across the Life Span* (pp. 289-308). Hillsdale, NJ: Lawrence Erlbaum.

Berry, J.W. (2004). An ecocultural perspective on the development of competence. In R. Sternberg and E.L.Grigorenko (Eds.), *Culture and competence: Contexts of life success* (pp. 3-22). Washington, DC: American Psychological association.

Berry, J.W., Bennett, J.A., Denny, J.P., Mishra, R.C., and Turner, N. (2000). Ecology, culture and cognitive processing. Unpublished manuscript.

Berry, J.W., Dasen, P.R., and Witkin, H.A. (1983). Developmental theories in cross-cultural perspective. In L. Adler (Ed.), *Cross-cultural Research at Issue* (pp. 13-21). New York: Academic Press.

Berry, J.W., Mishra, R.C., and Tripathi, R.C. (2003). *Psychology in Human and Social Development.* New Delhi: Sage.

Berry, J.W., Poorting, Y.H., Segall, M.H., and Dasen, P.R. (2002). *Cross-cultural Psychology: Research and Applications* (2nd edition). New York: Cambridge University Press.

Berry, J.W., van de Koppel, J.M.H., Senechal, C., Annis, R.C., Bahuchet, S., Cavalli-Sforza, L.L., and Witkin, H.A. (1986). *On the Edge of the Forest: Cultural Adaptation and Cognitive Development in Central Africa.* Lisse: Swets and Zeitlinger.

Boldt, E.D. (1978). Structural tightness and cross-cultural research. *Journal of Cross-Cultural Psychology, 7*, 21-36.

Boldt, E.D., and Roberts, L.W. (1979). Structural tightness and social conformity. *Journal of Cross-Cultural Psychology, 10*, 221-230.

Bronfebrenner, U. (1974). Development research, public policy and ecology of childhood. *Child development, 45*, 1-5.

Bronfenbrenner, U. (1979). *The ecology of human development.* London: Harvard University Press.

Brunswik, E. (1957). Scope and aspects of the cognitive problem. In A Gruber (Ed.), *Cognition: The Colorado Symposium* (pp. 5-31). Cambridge, Mass.: Harvard University Press.

Cole, M., and Scribner, S. (1977). Developmental theories applied to cross-cultural cognitive research. In: Issues in Cross-Cultural Research. *Annals of the New York Academy of Sciences, 285*, 366-373.

Dasen, P. R. (2003). Theoretical frameworks in cross-cultural developmental psychology: An attempt at integration. In T. S. Saraswathi (Ed.), *Cross-cultural Perspectives in Human Development* (pp. 128-165). New Delhi/Thousand Oaks, CA: Sage.

Dasen, P.R., Berry, J.W., and Witkin, H.A. (1979). The use of developmnental theories cross-culturally. In L. Eckensberger, W. Lonner, and Y.H. Poortinga (Eds.), *Cross-cultural Contributions to Psychology* (pp. 69-82). Lisee: Swets and Zeitlinger.

Dasen, P. R., and Mishra, R.C. (2010). *Development of Geocentric Spatial Language and Cognition*. Oxford: Cambridge University Press.

Dawson, J.L.M. (1967a). Cultural and psychological influences upon spatial perceptual processes in West Africa, Part 1. *International Journal of Psychology, 2*, 115-125.

Dawson, J.L.M. (1967b). Cultural and psychological influences upon spatial perceptual processes in West Africa, Part 2. *International Journal of Psychology, 2*, 171-185.

Dawson, J.L.M. (1969). Theoretical and research bases of biological psychology. *University of Hong Kong Gazzette, 16*, 1-10.

Dawson, J.L.M. (1973). Effects of ecology and subjective culture on individual traditional-modern attitude change, achievement motivation and potential for economic development in the Japanese and Eskimo societies. *International Journal of Psychology, 8*, 215-225.

Dornyei, Z. (2005). *The Psychology of the Language Learner*. Mahwah, NJ: Lawrence Erlbaum.

Ember, M. (1963). The relationship between economic and political development in non-industrialized societies. *Ethnology, 2*, 228-148.

Gamble, T., and Ginsberg, P. (1981). Differentiation: Cognitive and social evolution. *Journal of Cross-Cultural Psychology, 12*, 445-459.

Goodenough, D.R. (1986). History of field dependence construct. In M. Bertini, C. Pizzamiglio, and S. Wapner (Eds.), *Field Dependence in Psychological Theory, Research and Application* (pp. 5-14). Hillsdale, NJ: Erlbaum.

Gruenfeld, L.W., and Lin, T. (1984). Social behavior in field independents and dependents in an organic group. *Human Relations, 37*, 721-741.

Kardiner, A. (1945). *Psychological Frontiers of Society*. New York: Columbia University Press.

Lewin, K. (1951). *Field Theory in Social Psychology*. New York: Harper.

Mac Arthur, R.S. (1967). Sex difference in field dependence for the Eskimo: Replication of Berry's finding. *International Journal of Psychology, 2*, 139-140.

Mishra, K. (1996). Cognitive style of Tharu children in relation to daily life activities and experience of schooling. Unpublished doctoral thesis, Banaras Hindu University.

Mishra, R.C. (1996a). Perceptual differentiation in relation to children's daily life activities. *Social Science International, 12*, 1-11.

Mishra, R.C. (1996b). *Cognitive Processes, Cultural Adaptations, and Education of Children of Some Tribal Groups*. New Delhi: NCERT Project Report.

Mishra, R.C. (1997). Cognition and cognitive development. In J.W. Berry, P.R. Dasen, and T.S. Saraswathi (Eds.), *Handbook of Cross-cultural Psychology*, Vol. 2 (pp. 143-175). Boston: Allyn and Bacon.

Mishra, R.C. (1998). A cross-cultural perspective on cognitive development theories used in Indian settings. *Psychology and Developing Societies, 10*, 21-33.

Mishra, R.C. (2000). Perceptual, learning, and memory processes. In J. Pandey (Ed.), *Psychology in India revisited: Developments in the discipline, Vol. 1* (94-150). New Delhi: Sage.

Mishra, R.C. (2001). Cognition across cultures. In D. Matsumoto (Ed.), *The Handbook of Culture and Psychology* (pp. 119-136). New York: Oxford University Press.

Mishra, R.C., and Berry, J.W. (2008). Cultural adaptation and cognitive processes of tribal children in Chotanagpur. In N. Srinivasan, A.K. Gupta, and J. Pandey (Eds.). *Advances in Cognitive Science Vol. 1* (pp. 289-301). New Delhi: Sage.

Mishra, R.C., and Singh, D.V. (2008). Psychological differentiation in relation to some socialization variables: A study with rural children. *Psychology and Developing Societies, 20,* 241-256.

Mishra, R.C., Sinha, D., and Berry, J.W. (1996). *Ecology, Acculturation and Psychological Adaptation: A Study of Adivasis in Bihar.* New Delhi: Sage.

Murdock, G.P. (1969). Correlations of exploitive patterns. In D. Damas (ed.), *Ecological Essays* (pp. 129-149). Ottawa: National Museum of Canada.

Nimkoff, M.F., and Middleton, R. (1960). Types of family and types of economy. *American Journal of Sociology, 66,* 215-225.

Nisbett, R. E. (2003). *The Geography of Thought. How Asians and Westerners Think Differently, and Why.* New York: Free Press.

Norenzayan, A., Choi, I., and Peng, K. (2007). Perception and cognition. In S. Kitayama and D. J. Cohen (Eds.), *Handbook of Cultural Psychology* (pp. 569-594). New York: Guilford Press.

Pelto, P. (1968). The difference between tight-loose societies. *Transaction (April),* 37-40.

Porteus, S.D. (1937). *Intelligence and Environment.* New York: Mac Millan.

Rogoff, B. (1981). Schooling and the development of cognitive skills. In H.C. Triandis and A. Heron (Eds.), *Handbook of Cross-cultural Psychology, Vol. 4* (pp. 233-294). Boston: Allyn and Bacon.

Segall, M.H., Campbell, D.T., and Herskovits, M.J. (1966). *The Influence of Culture and Visual Perception.* Indianapolis: Bobbs-Merrill.

Shrestha, A.B., and Mishra, R.C. (1996). Sex differences in cognitive style of Brahmin and Gurung children from the hills and plains of Nepal. In D.P.S. Bhawuk, D. Sinha, and J. Pandey (Eds.), *Asian Contributions to Cross-cultural Psychology* (pp. 165-174). New Delhi: Sage.

Sinha, D. (1977). Some social disadvantages and development of certain perceptual skills. *Indian Journal of Psychology, 52,* 115-132.

Sinha, D. (1978). Story-pictorial E.F.T.: A culturally appropriate test of perceptual disembedding. *Indian Journal of Psychology, 53,* 160-171.

Sinha, D. (1979). Perceptual style among nomadic and transitional agriculturalist Birhors. In L. Eckensberger, W. J. Lonner, and Y.H. Poortinga (Eds.), *Cross-cultural Contributions to Psychology* (pp. 83-93). Lisse: Swets and Zeitlinger.

Sinha, D. (1980). Sex differences in psychological differentiation among different cultural groups. *International Journal of Behavioral Development, 3,* 455-466.

Sinha, D. (1982). Socio-cultural factors and the development of perceptual and cognitive skills. *Review of Child Development Research, 6,* 441-472.

Sinha, D. (1984). *Manual for Story-Pictorial E.F.T and Indo-African E.F.T.* Varanasi: Rupa Psychological Corporation.

Sinha, D., and Bharat, S. (1985). Three types of family structure and psychological differentiation: A study among the Jausar-Bawar society. *International Journal of Psychology, 20*, 693-708.

Sinha, D., and Shrestha, A.B. (1992). Eco-cultural factors in cognitive style among children from hills and plains of Nepal. *International Journal of Psychology, 27*, 49-59.

Sinha, G. (1988). Exposure to industrial and urban environments and formal schooling as factors in psychological differentiation. *International Journal of Psychology, 23*, 707-719.

Sternberg, R. J., and Grigorenko, E. L. (2001). A capsule history of theory and research on styles. In R. J. Sternberg and L.F. Zhang (Eds.), *Perspectives on Thinking, Learning, and Cognitive Styles* (pp. 1-22), Mahwah, NJ: Lawrence Erlbaum.

Super, C., and Harkness, S. (1986). The developmental niche: A conceptualization at the interface of child and culture. *International Journal of Behavioral Development, 9*, 545-570.

Whiting, J.W.M. (1973). A model for psycho-cultural research. *American Anthropological Research,* Annual Report, 1-14.

Whiting, J.W.M., and Child, I. (1953). *Child Training and Personality*. New Haven: Yale University Press.

Witkin, H.A., and Berry, J.W. (1975). Psychological differentiation in cross-cultural perspective. *Journal of Cross-Cultural Psychology, 6*, 4-87.

Witkin, H.A., Dyk, R.B., Faterson, H.F., Goodenough, D.R., and Karp, S. (1962). *Psychological differentiation*. New York: Wiley.

Witkin, H.A., and Goodenough, D.R. (1981). *Cognitive style: Essence and origins*. New York: International University Press.

Wright, H. F. (1967). *Recording and analyzing child behavior*. New York: Harper and Row.

Zhang, L.-F., and Sternberg, R. J. (2006). *The nature of intellectual styles*. Mahwah, NJ: Lawrence Erlbaum.

SECTION 3. MENTAL AND NEURAL DISORDERS

In: Expanding Horizions of the Mind Science(s)   ISBN: 978-1-62808-705-5
Editors: P.N. Tandon, R.C. Tripathi and N. Srinivasan   ©2013 Nova Science Publishers, Inc.

Chapter 17

LANGUAGE ACQUISITION, LEARNING AND LEARNING DISABILITIES IN THE INDIAN CONTEXT

Prathibha Karanth
Com DEALL Trust, Bangalore, India

ABSTRACT

The paper explores our current understanding of language acquisition/language learning and its role in learning and learning disabilities. These issues will be discussed from the perspective of the multilingual context of India, with its particular languages and writing systems and related issues such as the official three language policy and the choice of the medium of instruction. An effort will be made to relate these aspects of learning to our language policies and decisions on contentious issues such as medium of instruction, the number of languages and writing systems that a child is required to learn, the stages and the manner in which they are introduced to the typically developing child and its implications for the child with learning difficulties. Finally it is argued that our policies on language learning and learning as well as those that are concerned with the learning disabled should be based on scientific premises derived from carefully done research studies of these issues in our multilingual context.

1. INTRODUCTION

The acquisition of language in children has, since the latter half of the twentieth century held the unabated interest of scientists. These include not only those scientists who are interested in language or children but a bewildering array of professionals from a wide range of professional backgrounds and training. No small part of this interest is due to the fact that the understanding of the phenomenon of language acquisition by the human child is seen as a potential source for arriving at answers to the many unanswered questions that permeate both cognitive and neuro sciences.

The paper explores our current understanding of the many aspects of language acquisition-language learning and its role in learning and learning disabilities with a special focus on our socio linguistic context. The learning of languages in schools – how many and when, and which of these will be the chosen one for the medium of instruction still remains a contentious issue in India. At the same time while the role of language in learning is well acknowledged very little attention is paid to what exactly this role is, how it plays out during the school years and in subsequent years. How the methods of language teaching at different stages of schooling need to reflect these subtleties, is largely ignored. Equally importantly, one needs to understand how all of these are affected by the multilingual context of India and whether and how our educational programs take these into account. Finally, how do these impact children with learning difficulties and how best could we accommodate their interests in the context of these complex issues. But, first an attempt at defining these terms as used in this paper.

1.1. Language Acquisition

Scientific literature on the acquisition of language by children has seen a phenomenal increase over the last few decades, thanks to the explosive interaction between many key disciplines such as behaviorist psychology, linguistics and child development, to name a few. This led to the subsequent emergence of new disciplines such as psycho, neuro and clinical linguistics, as well as a re-emergence of interest in cognitive neuropsychology with a special emphasis on the relation between language and cognition. What this has unraveled is the extraordinary complexities that underpin the child's acquisition of language and its use over the human lifespan. To understand these key issues one needs to distinguish and appreciate the thin divide that exists between language acquisition and language learning.

The term 'acquisition' as applied to language, as in 'children's acquisition of language' or 'the acquisition of language by children' or 'child language acquisition' would denote learning that takes place as a normal part of development in a typically developing child, without conscious effort on his part or those in his environment. Languages may be acquired during the entire life span. Nevertheless, it is also established that early childhood extending up to 12 years of age is the most conducive for acquiring languages – the 'critical age' for language acquisition.

A related phenomenon that is now acknowledged is that language acquisition, even in the first/dominant language, is not complete by school entry age. While the young child becomes proficient in the basics of phonology and semantics as well as the less complex syntactic structures at school entry age, the higher aspects of language and language use such as those used in narrative text, logical reasoning and humor, are now known to be acquired during the school years, up to adolescence and early adulthood – the high school and college years.

1.2. Language Learning

Language learning as opposed to language acquisition refers to a more conscious process of learning a language, as in school or when one enrolls in a language course. By and large this is a more conscious effortful process and generally applies to the learning not of the mother tongue but to the second, third or subsequent language/s that a person learns. Unlike as in the acquisition of the mother tongue or first language, the individual already has a linguistic system and this native language influences the subsequently learnt languages depending on the age and the manner in which the subsequent languages are learnt. Language acquisition is more likely to be an early childhood phenomenon and language learning more of a phenomenon in later childhood and or adulthood.

Language learning can take place in childhood as in school or even in the later years when an individual chooses to learn one or more additional languages. However, while in the latter it is often a matter of one's conscious choice, in the former it is often necessitated and influenced by factors such as schooling, medium of instruction (MOI) one's sociolinguistic environment and other such factors.

1.3. Language Teaching

Though language teaching is necessarily involved in language learning, there is a further distinction to be made in terms of the teaching methods that are adopted in the teaching of languages as a subject matter in school. Language teaching may involve building on the existing language skills of a native speaker as in grammar classes or in the teaching of a second, third or fourth language. Again, it may take place as a subject at school or in a specialized language teaching school. Many methods have been used to teach languages, some of which are widely used at present.

The major approaches to language teaching have been the structural approach which is the traditional approach of teaching the grammar of the language as against the functional approach to language which views language as a medium of social function. The interactive view which has been dominant in the recent past, views language as the medium of social and other communicative purposes such as information gathering, negotiation and interaction, with set patterns. Specific teaching methods and techniques are selected within these broad approaches.

1.4. Language in Learning

The connection between language skills and classroom learning is reasonably well understood. Appropriate linguistic skills at different stages of schooling are a must for learning to take place. While in the early school years the language environment is predominantly spoken, situation dependent and non abstract; as the child goes through the higher grades his linguistic skills play an ever increasing role in his academic learning. S/he is expected to master an increasing language facility for form and function as the language environment becomes informational, explicit and abstract. The relationship between linguistic skills and learning are now seen to be reciprocal with each enhancing the other, through the school years and beyond.

1.5. Learning Disabilities (LD)

Our understanding and definitions of 'learning disabilities' have undergone several changes since they were first identified in the 1930's. There is as yet no universally accepted definition of 'learning disability'. In general parlance however a child who has an unaccounted failure in academic success not explained by any mitigating biological or environmental factor, is considered to be learning disabled.

1.6. Language Learning Disabilities

There is increasing consensus that many children with learning disabilities have clearly identifiable language learning difficulties. Many children with reading problems have spoken language problems. It is also conceded that several otherwise typically developing children, have a specific delay in language development in the absence of any other developmental difficulty, which may or may not resolve by school age and many of the latter will end up having learning difficulties. These children are referred to as children with "language-learning disability' or 'language-based learning disability'.

2. LANGUAGE ACQUISITION, LEARNING, TEACHING AND DISABILITIES: WHAT WE KNOW ABOUT IT

A brief description of what we currently know about each of these aspects is presented in this section. It is important to note that the current literature on language and learning are largely based on the primarily monolingual populations of the west. Children from these countries acquire their mother tongue and subsequently learn at school in their mother tongue. The acquisition of literacy skills in these populations therefore is essentially one of learning to apply the secondary system of reading and writing to the primary system of understanding and speaking their mother tongue. In addition, given the historical reason of much of this work being restricted to the western hemisphere until late, the work on early literacy is also largely restricted to the acquisition of the alphabetic writing systems prevalent in the west.

2.1. Language Acquisition

Acquisition of language is the process though which infants pick up linguistic skills with seemingly little adult input or training. Children acquire their mother tongue or the dominant language spoken in their immediate environment with very little direct assistance or teaching from adults. This is true of all children irrespective of race, class or creed. How this occurs has been the topic of much scientific interest in the last century. While around the middle of the twentieth century the theoretical explanations of language acquisition in children centered around the 'nurture versus nature' debate, between Skinnerian behaviorism (Skinner, 1957) and Chomskyan innatism (Chomsky, 1959); since then there have been many other theoretical postulations such as the 'Social Interactionist theory (Bates, Elman, Johnson,

Karmiloff –Smith, Parisi and Plunkett, 1998) or the Emergentist theories (Macwhinney,1999) that are midway between the two.

At the two ends of this spectrum of theories, the behaviorists' postulation that language is simply learnt through operant conditioning and the use of reinforcers is diametrically opposed by the Chomskyan position that language is acquired by the child because of a certain innate biological mechanism that he is born with. The biological mechanism proposed by Chomsky and his followers was of a very specific nature - a unique mechanism that serves the acquisition of language alone, termed the 'language acquisition device' or 'language faculty', to which was attributed the set of principles that were universal across all languages, which in turn were modified for each particular language by a set of parameters. Other innatists postulate a more general cognitive mechanism that serves all learning including language. More recent models of language acquisition theories take more moderate positions than those of Skinner or Chomsky, in terms of the relative role of biology and learning in language acquisition. Theories such as the Social Interactionist theory of language (Bates, et. Al 1998) for instance, take the position that language is learnt by using mechanisms that are a part of cognitive learning which the child is born with rather than a separate dedicated language mechanism. It further emphasizes the role of social context and the functional use of language in child language acquisition.

More recent versions of the two basic and opposing viewpoints have also been put forth. The 'Relational Frame Theory' (Hayes, Barnes-Holmes and Roche, 2001), for instance while being based on Skinner's behaviorist theory of language acquisition, postulates a type of operant conditioning termed 'derived relational responding' which emphasizes the importance of thoughts and feelings in the manipulation of variables in the context and a system of inherent reinforcers. Empirical support for the predictions made on the basis of this theory is being documented. Macwhinney's competition model on the other hand views language acquisition as an emergentist phenomenon that results from competition between lexical items, phonological form, and syntactic patterns. Empirical studies based on the 'Competition Model' have shown that learning of language forms is based on the accurate recording of many exposures to words and patterns in different contexts - an ongoing cognitive simulation based on linguistic abstractions grounded on perceptual realities. The predictions made by the model are supported by work in the related disciplines of cognitive neuroscience.

Despite the voluminous work on the topic and the heated theoretical debates, there is as yet no single theoretical account of how children acquire language in all of its complexities. The universal capacity of children to acquire language and the universalities across all human languages is however generally believed to be linked to and depend on human biology, the broad consensus being that language emerges through an interaction between genetic inheritance and the linguistic environment. The child's ability to acquire language is seen to be dependent on neural systems that are genetically influenced to do so.

The discovery of mutations in the FOXP2 gene along with under activation of speech areas in fMRI studies of silent verb generation and word repetition tasks, in several members of a large family who exhibited specific speech-language disorders such as 'Specific Language Impairment' and 'Developmental Verbal Dyspraxia' with little or no cognitive handicaps, added impetus to the search for the genetic basis and neural expression of speech and language skills in humans (Vargha Kadem, Ghadian, Copp and Mishkin, 2005). While the initial euphoria at the 'discovery of the language gene' has since abated, it nevertheless has opened up exciting new arenas in the researchthe biological basis of language acquisition.

Among the recent theories of language acquisition that have received considerable empirical support are those that argue that language acquisition takes place much like all other learning, on the basis of statistical probability and computations. Empirical evidence for these 'Connectionist Models' are built through artificial neural networks that are fed naturalistic input from children acquiring language, which then produce utterances that can be compared to children's utterances. The connectionist models for cortical processing of language have converged on related probabilistic descriptions. Further these have been simulated across languages. It is expected that such targeted interdisciplinary study of language, drawing from engineering and computer science, linguistics, biology, psychology, neuroscience, and cognitive science, can result in tangible breakthroughs in our understanding of the biology of language (Blumstein, Carew, Kanwisher, and Sejnowski, T. 2010).

The developmental steps of language acquisition in children are universal. Substantial mastery of the distinct aspects of language such as phonology, syntax and semantics take place during particular phases of language acquisition by the child in early childhood, followed by mastery of the more complex social and instructional use of language in early adolescence and adulthood. However there is no clear cut division in the acquisition of these different aspects of language, often the acquisition of one aspect bootstraps another. At the same time all these aspects are embedded in each 'Speech Act'. Importantly the child masters most of these aspects of language by being an active participant of his language environment.

2.2. Language Learning

Language is best mastered when acquired, not learnt or taught. Nevertheless languages are taught and learnt. The term language learning is generally used to refer to the learning of a second language, though it could also refer to an individual learning the more complex formal aspects of his native language. The requirements for learning of languages are varied. In multilingual situations like ours the child is often forced to learn a language other than his mother tongue at school entry, by virtue of the medium of instruction being a language other than his mother tongue. The demands and the pressures of language learning are therefore considerably different across situations.

As in the theories of language acquisition, so too in the theories of second language acquisition three principal views are held – the structural view with its emphasis on the acquisition of the second language through the structural aspects of language and a mapping of these onto those of the first language, the functional view with a focus on the function of language in the manipulation of the environment and the interactive view that emphasizes the social functions of language. Several methods of second language teaching such as the direct method, the oral method, the audio lingual method, the immersion method and the communicative language teaching method have been evolved over the years within each of these approaches. Since the 1980s the latter has been more dominant with many of the second language acquisition models being developed within the communicative approach. These models try to replicate the conditions of communicative function of first language acquisition in language teaching, since it is acknowledged that acquisition through normal exposure as happens in early childhood could be the best way to master a language. The latter approach emphasizes the functions that language serves and provides appropriate linguistic input for various communicative functions. Specific methods within these broad approaches emphasize and combine the form versus functional dichotomy to different degrees with conflicting claims of their efficacy in enabling language learning.

2.3. Language Teaching

The teaching of language is generally addressed in language classes in schools and language teaching schools. The methods of language teaching are said to have originated from the study and teaching of Latin in the 17th century. Historically there has been a clear dichotomy with two major and conflicting approaches to language teaching - the empirical and theoretical. The empiricists promoted verbatim repetition and imitation followed by drills for language teaching. On the theoretical side were those, who advocated language teaching methods that were similar to the traditional model of language learning and teaching which is grammar based and designed within the grammatical or grammar-translation approach. The lessons are organized on the basis of linguistic or grammatical forms – introduction of lexical items or words, their phonological and morphological forms, and an emphasis on the ways in which these forms may be combined to form grammatical sentences or syntactic patterns, complementing the work of Chomsky and the proponents of his school of thought. However, just as the theories of language acquisition have so far failed to account for the process of language acquisition in children adequately, so have the language teaching programs largely failed to succeed.

Much of the intellectual debate on early language teaching in schools has also focused on the manner of introducing the child to reading and writing. Two diametrically opposite methods – the 'phonic method' which prioritizes the teaching of the sound patterns of a language vis a vis the letters of that language versus the 'whole word method' which prioritizes the teaching of word meaning as a whole; have been prevalent in teaching early reading over the last 4-5 decades. These educational methods have been based on the teaching of opaque alphabetic writing systems such as English, to proficient speakers of that language and in all likelihood influenced by the particular complexities of their writing systems. The question of their relative efficacy remains unresolved despite half a century of heated debate (Snow 2000). Yet, they have been universally adopted by educators across the world, including those in India, without consideration of their appropriateness to other languages and scripts.

In the recent past, with the increasing acknowledgement of the role of metalinguistic skills in learning to read, particularly the role of phonological awareness in early reading of the alphabetic scripts, there has been a shift to introducing metaphonological tasks that facilitate phoneme awareness, such as syllable segmentation, in teaching children to read. Once again, there is insufficient awareness that the empirical findings on increased phonological awareness and success in reading have been largely based on children learning to read opaque alphabetic writing systems like English, which may not be as crucial to learning to read other types of writing systems such as the Indian writing systems.

Welcome changes that are now seen in the teaching of languages are a greater sensitivity to the changes in the nuances of the use of language as a medium in information seeking and gathering in the school years and the importance of metalinguistic awareness and skills.

Metalingusitics as defined by Chomsky is the subject's knowledge of the characteristics and functioning of language or, from a more functionalist perspective, of its structure, its functioning and its usage. The student's awareness of language and the components of language is an essential aspect of language development, particularly in the later school years. Metalinguistic skills that are contingent on metalinguistic awareness as related to reading skills, and academic success are also being seen as linked to cognitive development, beginning in childhood and continuing through school years (Karanth, Kudva, andVijayan, 1995). More flexible approaches in teaching methods that sensitize the learner to metalinguistic skills are now used to enable the child to master these subtleties, as he progresses in school.

2.4. Language in Learning

The advent of the discipline of psycholinguistics in the latter half of the twentieth century with its combination of methods from linguistics and psychology and a focus on empirical studies of children through the language acquisition period, had a tremendous impact on our understanding of the nuances of the process of acquisition of language as seen in typically developing children. With the refinement in methodology and tools that this necessitated came a greater understanding of the subtler aspects of language acquisition in children right through adolescence and even more importantly the role of language in learning.

The changes that occur in the linguistic skills of a child as he progresses through school and its importance for academic learning are increasingly being understood. As children move to higher grades, from elementary to secondary school, the language of the classroom gradually becomes quite distinct from that used in everyday life. Classroom discourse in the higher classes is increasingly characterized by lexical density and complex syntactic constructions. These changes in language skills and use have been described as the 'communication hierarchy' (van Kleeck 1994), in which the child moves from a dependence on the non-linguistic cues for communication in the early years to the linguistic in the early school years and to the metalinguistic in the later school years. That is, initially the child depends largely on the contextual non-linguistic factors for effective communication. When he enters school there is an increasing emphasis on the linguistic aspects such as grammar. During the later school years and in adulthood, there is a need for increased awareness and use of metalinguistic skills in order to become a competent communicator. From the preschool to the high school years the 'communication hierarchy' takes the child from a dependence on the contextual non-linguistic to the linguistic and finally to the abstract metalinguistic. The latter are now seen as providing a link for moving children from social to increasingly instructional uses of language. These aspects of communication are particularly relevant to the more formal aspects of communication such as those present in the written medium. A child who has not mastered these complex forms of language in his medium of instruction (MOI) would necessarily have difficulties in classroom learning. During the later school years there appears to be a reciprocal relationship between academic learning and language learning, each contributing to the other. Language becomes the object of thought and discussion in reading and writing and the reading and writing of prose leads to metalinguistic reflection and further enhances literacy (Olson and Torrance, 1991).

2.5. Learning Disabilities

Three significant changes can be identified in the area of learning disability in the recent past. First, the definition of Learning Disability (LD) has shifted from the traditional approach, arrived at through a negation of all possible identifiable biological and environmental factors that might contribute to the learning difficulties of a typically developing child, to a newer dimensional approach of individual differences. Second, LD is no longer seen as an early childhood disorder but as a disorder that changes but persists over the life span. Thirdly there is a recognition that the communication deficits seen in children with LD are not restricted to those related to reading and writing alone, but also encompass the more basic communicative functions of speaking and listening too (for more details see Karanth, 2003a, 2008 and Thapa 2008). There is in particular a focus on phonological awareness and phonological processing difficulties leading to theories of LD as a linguistic / metalinguistic problem stemming from empirical evidence of phonological processing difficulty in children with LD, stimulating language based research and theories of LD.

The secondary consequences of LD such as poor motivation, self esteem, self efficacy and metacognition have also received increased attention. The snowballing effect of LD on the child's overall development and metacognition in particular, are of increasing concern. All of this is to be seen in the context of the increasing importance of and the rising demands for literacy, as well as the consequences of a lack of, or inadequate literacy skills, in modern day society.

2.6. Language Learning Disabilities

There is now a renewed interest in the connection between language learning and learning disability. It has been documented that many children diagnosed with learning disability in school have had either a personal or familial history of speech disorders such as delayed speech-language acquisition, articulation and fluency disorders. It is also documented that several children have a specific delay in language development which may or may not resolve by school age and many of the latter will end up having learning difficulties. There is a robust body of scientific literature on the language-learning disability connection that has been built over the last couple of decades.

There is increasing consensus that many children with learning disabilities have clearly identifiable language learning difficulties often related to specific language problems (see Karanth, 2003b, 2008). Empirical studies on language acquisition in children with learning disabilities resulting from the availability of fine toothed language assessment procedures and tools, that were available for the first time after the emergence of the discipline of psycholinguistics in the latter half of the 20th century, have thrown up a plethora of language difficulties in children with language learning disability. Many children with learning disorders identified with problems with age-appropriate reading, spelling, and/or writing, have in addition, spoken language problems such as learning new vocabulary, understanding questions and following directions that are heard and/or read. They are also poor at expressing ideas clearly, with excessive use of fillers. Their higher language skills are characterized by poor understanding and retaining of the details of a story's plot or a classroom lecture, they have difficulties in verbal analogical reasoning, their narrative discourse processing and production are often characterized by the use of the oral style and the understanding and production of written expository discourse structures such as comparison, contrast and problem solving are reported to be particularly difficult for them. The comprehension and production of non-literal language, such as metaphor, idioms, similes and irony, which are important not only in social situations but also increasingly necessary for academic success as children progress in school are also reported to be particularly poor. Their executive functioning skills or the ability to plan, organize, and attend to details such as in planning and organizing writing or keeping track of assignments and school materials, are also reported to be inadequate.

Cross linguistic work is seen to be of increasing importance in understanding some of the complex issues in both child language acquisition and the disorders of language acquisition in children (Slobin, 1985, 1992). Much of the cross linguistic work until recently however has been restricted to the different European languages with some amount of literature on Japanese – English bilinguals. There is however, now a robust and growing research interest in the languages of the Asia Pacific regions. That the many bi/multilinguals in these regions speak, read and write in English is an added bonus as it provides a point of comparison with the existing scientific body of literature on the subject.

3. LANGUAGE, LEARNING AND LEARNING DISABILITIES IN THE MULTILINGUAL CONTEXT OF INDIA

In comparison, to the western countries the linguistic realities in countries like India are much, much more complex. Despite multilingual speakers outnumbering monolingual speakers in the world, the existing knowledge on languages, their acquisition, learning, teaching and disorders come from monolingual populations and how these apply to multilingual contexts such as those prevalent in India needs to be examined.

Attitudes towards bi/multilingualism which was once dismissed as being detrimental to overall language proficiency and mastery is now being looked at as being of potential value and there has been a surge of scientific interest in the phenomenon. Empirically established possibilities that bilinguals may in fact exhibit more cognitive elasticity (Collier 1992, Ramirez, 1992), is increasingly negating the perception propagated in the 1960's that bilingualism was at the cost of proficiency in either (Hakuta, 1990).

India is reported to have had 1652 different spoken languages and dialects in 1961(Census of India, 1961). According to Census of India of 2001, 29 languages are spoken by more than a million native speakers, 122 by more than 10,000. The languages of India belong to four language families but have significantly influenced each other over a period of 3000 years. In addition there has been the considerable influence of two contact languages – Persian and English. Consequently there is an extraordinarily high level of linguistic diversity in the country. The reach of the audio visual media and the increasing mobility of current day society have complicated the issue further. The multilingual situation here warrants an understanding of the ground realities. Some attempts at the same on tribal populations have been reported recently (Mohanty, Panda, Phillipson and Skutnabb-Kangas, 2009). Similar studies on nontribal, rural and urban populations are needed urgently.

3.1. Language Acquisition in Indian Children

While the majority of children in India like elsewhere grow up acquiring their mother tongue, considerable numbers of Indian children acquire more than one language simultaneously in their early childhood, at times even within the early home environment with the two parents addressing the child in two different languages. It is likely that truly monolingual populations in India are more likely to be found only in rural and tribal India and even in these places, exposure to other languages would be present to some extent through mass media.

In the urban areas on the other hand growing up as a multilingual is more the norm than the exception. Even within urban areas a variety of combinations in terms of language acquisition and use may be prevalent, with wide differences between cities. In two recently conducted studies, for instance it was found that while the majority of children growing up in Mysore city reported their mother tongue as being the dominant language (Shanbal and Prema 2007), in a similar study on children from the metros of Bangalore and Delhi, 10 out of 31 children studied did not report their mother tongue to be their most dominant language (Dash 2009). These 10 reported the language of the state to be their dominant language. While one of these children, ranging in age from 4- 9 years, spoke only one language the remaining 30 spoke from 2 to 4 languages. A majority reported English, their second language, as having been acquired through formal exposure by means of reading and writing in school, with the other languages having been acquired through the oral modality in informal situations including family, neighborhood, TV and friends. Such high levels of linguistic diversity are reported even within rural blocks (Jhingran 2009). Patterns of language acquisition and use are indeed complex, varied and difficult to capture in India.

The complex phenomenon of language learning in India is barely studied or understood. The few studies on bilingualism/bilingual language acquisition in India are largely under the influence of existing western models of bilingual language acquisition such as the sequential, simultaneous and coordinate types, which are seldom applicable on the ground in India. It is important to recognize that bilingualism, nay multilingualism that is seen in India can be quite different from that of the west. It is not for instance uncommon to find a child from a family in Bangalore where two different languages are spoken at home, owing to the parents origin in different parts of India and the presence of grandparents at home, a third language – the language of the state being spoken by the neighbors and the maid at home, with expectations of a fourth language, most often English in this case, being learnt in school. Despite these many complexities several of us are reasonably competent bi and trilinguals. How this competence is acquired, what facilitates it and how competent are we really in each of these languages, need to be studied. There is therefore a very great and urgent need to understand the conditions here and develop models of multilingualism based on our socio linguistic context.

3.2. Language Learning in India

The relatively large number of languages that are spoken in India in comparison with the limited number of languages that are offered as medium of instruction (MOI) in our schools necessarily result in a large number of our children having to study in a language other than their mother tongue. In addition economic motivations dictate that the MOI of choice is English, irrespective of the child's exposure to and competence in the same. In such instances, learning of the language that is the MOI, through formal instruction at school appears to follow patterns and effects that are similar to what has been reported in the west, with inadequate linguistic skills in the second (and at times third or fourth) language to justify its use as a medium of instruction.

3.3. Language Teaching in India

Language teaching in schools in India continues to be largely influenced by the traditional grammarian approach with an inordinate emphasis on the teaching of grammar and the neglect of higher language skills. The practice of teaching reading and writing when the child does not understand or speak the language is widely prevalent. Language teaching/learning at school seldom takes in to account facts about language acquisition. That our language teaching methodology pays no heed to any of these established scientific facts is amply exemplified by what happens in our classrooms. Most often the emphasis in schools is on language taught through writing in the medium of instruction and begins at learning the alphabet, even when the child so often has no understanding of the spoken form of the language and or the basic expressive skills in that language. Speaking is a skill that is not given adequate attention in the traditional classroom.

Ideally writing should be introduced for the spoken system that the child is most proficient in and not for an alien language which he has no clue about. This is particularly poignant in a multilingual context like ours where many a young child is exposed to his medium of instruction, for the first time on entry to school. Those who come from a language background that is different from their medium of instruction would obviously face a total disconnect. It is noteworthy that policy makers in similar situations elsewhere recommend that, bilingual children with a mother tongue other than English when enrolled in schools with English as their MOI should be taught to read in their native language/dialect while acquiring proficiency in spoken English and then extend their reading skills to English. Where reading instruction in the native language is not possible they recommend 'the instructional priority should be to develop the children's proficiency in English and the postponement of formal reading instruction is appropriate until an adequate level of *proficiency in spoken English* has been achieved' (Snow, 2000).

In addition, language teaching methods in our schools need to be informed by the nature of the language and particularly the nature of the writing system. We have for long unthinkingly applied the methods and techniques of teaching reading and writing that were developed for the alphabetic scripts of the west to the teaching of reading and writing of our writing systems. We have failed to acknowledge the substantial differences in the structures of the scripts and added to the existing confusion (see Karanth 2003c for more details).

3.4. Language in Learning in India

The education system in India has not as yet come up with an adequate response to the complex linguistic diversity in the country. Our language in education policies provide some standardized solutions which have not proved very effective. Attempts at addressing these issues have been perfunctory with some short modules on spoken language introduced as add on's rather than of a substantive nature. We need to identify and understand the factors that influence naturally occurring bi/trilingual acquisition in our children and develop strategies for multilingual education based on empirical findings. It has been documented that children in immersion programs, including dual immersion programs where academic content is delivered through the medium of the immersion language for part of the school day, and through English the rest of the school day, demonstrate much higher levels of proficiency (Genesee, 1987). Similar strategies that respond to our conditions need to be evolved and implemented by us. Appropriate early grades teaching strategies particularly language teaching methods that focus on oral work, conversation and meaning and flexible use of language need to be introduced. Pre school exposure based on models that have a strong empirical base in our sociolinguistic context could be an ideal precursor of multilingual education strategy.

Dash (2009) reported that among the multilingual Indian children that she studied, the majority of the languages were acquired by the child in the oral modality in informal environments. A second important finding is that she listed parental proficiency as being an important contributor to the child's mastery of the language, with parental proficiency being rated higher than frequency of usage of the language in the environment. Family factors emerged as being very important and these include family motivation and attitudes. Findings such as these need to be tested on larger number of subjects who live in diverse geographical locations in India, and when established our language policies and teaching should be influenced by them. There has been some recent interest in the specific issue of education in multilingual contexts such as ours, highlighting the nature of our multilingualism and its impact (Mohanty, Panda, Phillipson and Skutnabb-Kangas 2009). However, much of the work here represents the tribal populations in India. As the studies described above would suggest, similar but differently compounding factors are in operation both in urban and nontribal rural areas and all of them need to be studied and understood.

The importance of higher language skills such as discourse, narration and text processing are also not adequately acknowledged in the increasing complexities of communication and linguistic skills that are required as the student moves through school. Deficits in these advanced skills of language also result in problems in reading and writing as well as critical, logical and scientific reasoning, all of which are essential for academic success during and beyond the school years. The linguistic/communicative skills that a child is required to learn in high school years are more subtle and difficulties in these are much harder to identify. Deficits in these skills significantly interfere with comprehension and the thinking process (see Karanth 2009).

As elsewhere, the connection between language skills and classroom learning is reasonably well understood or at least acknowledged, but neglected in practice in India. As pointed out by Jhingran language not only serves as a medium of communication but is inextricably linked with learning and thinking and "In this age where everyone talks of Education for ALL (EFA) it is surprising that this serious learning issue has received scant attention" in India (2009, p255).

On the positive side, metalinguistic awareness is now reported to be enhanced in bilingualism. In fact the term is said to have been first used to demonstrate the shift of linguistic intelligence across languages (Cazden, 1974). Bilingual skills such as code switching and translation are necessarily built on metalinguistic skills. This heightened metalinguistic awareness in bilinguals is said to have positive effects on language ability, symbolic development and literacy skills.

Multidimensional models such as those advocated by Nag (2006) which include dimensions such as the 'discontinuity between home and school and the scope of the literacy culture in the child's environment' and whether 'literacy acquisition is in a non-dominant language and the nature of the orthography' taken along with the inherent positive contributions of our bi/multilingualism need to be developed and issues of language in learning dealt within these perspectives.

3.5. Learning Disabilities in India

In such a complex milieu what is the fate of the child with language learning disability or a learning disability? Thanks to the impact of a popular movie, India is just waking up to the existence of a large number of children who are apparently growing like all other children of their age but in reality have subtle delays and difficulties in learning language and in learning, both of which have a tremendous negative impact on their schooling and eventual academic outcome. These children labeled as learning disabled and developmentally language disordered have largely gone unheeded hitherto, with a few exceptions of a small number who have been acknowledged and helped, largely thanks to the concern and attention of their educated families and a few enlightened educators.

Given the neglect that LD has faced in India, compounded by the complexity of the definition and identification procedure, we do not as yet have clear cut figures for the incidence and prevalence of language learning and learning disorders in India. However an extrapolation of available statistics from the western world would suggest a prevalence figure of over 18 million of school age children (Source - Learning Disabilities NIMH, USA: National Health Interview Survey, 1999, NCHS, CDC). It is imperative that we recognize the magnitude of this problem which is found in both sexes and affects children of all socioeconomic classes, creed, race or religion. It is time that their needs be taken in to account in framing our education policies and implementation.

Our multilingual context gives rise to many challenges in the identification and management of children with learning disabilities. At times it could be difficult for the teacher to tell whether a given child has a learning disability or whether his difficulties are because of the bi/multilingual background and lack of proficiency in the MOI. Our solutions, limited as they have been, have also been trigger happy, instant recipes and short sighted in nature. For instance, the one achievement that is claimed by LD activists in India is the exemption for the child with LD, from having to learn more than one language at school that is now granted by some state governments in India. While this may provide some immediate relief to the overburdened LD child at school, the other related aspects vis a vis the choice of the medium of instruction, in terms of its appropriateness for the specific nature of the child's learning difficulties, the environmental supports available for the same, and its eventual impact on the child's overall well being are seldom considered.

3.6. Language Learning Disabilities in India

What follows elucidates the particular complexities that a child with language learning or learning difficulty could face in a complex multilingual society such as ours; what supports, if any are provided in our academic contexts and what could be and needs to be done by careful study and understanding of the realities on ground and a careful application of our understanding of the nature of our multilingualism, our languages and their scripts.

It is, by now, well established that even minor lags in language acquisition could result in learning disabilities. Similar findings of considerable language gaps that have gone unnoticed by caregivers and teachers alike have been established among children with learning disabilities in India (Karanth 2008). This lack of sensitivity to the role of language is due in part to the common perception (misperception) that much of language acquisition is complete by school entry age and that in a child with learning difficulties but no 'apparent' delay in language acquisition at school entry time, the two are unlikely to be connected or causally linked. However we now know that the process of acquiring mastery over one's language is complex and long drawn with a two way interaction between language learning and learning, particularly through the school years.

CONCLUSION

In conclusion, it is clear that there is an urgent need for a rethink on our policies on language in education and our approaches and methods for language teaching. Our policies on language learning and learning as well as those that are concerned with the learning disabled should be based on scientific premises derived from carefully done research studies of these issues in our multilingual context. Our current official language policy does not consider the ground realities. Importantly, no effort is made to base these decisions on what is known from scientific studies of developmental language acquisition.

Both official language policy and parental choice of MOI should be informed by science and our sociolinguistic ground realities. Resolution of contentious issues such as MOI should be child centered and based on what is known from the psycholinguistic studies of child language acquisition rather than driven by political agendas. We urgently require carefully done studies of the factors that facilitate bi/multilingual language acquisition in our environment and the cognitive development that bi/multilingual acquisition could facilitate. The long term outcomes of our multilingual capacities and competences need to be carefully looked into and choice of MOI has to be guided by these long term implications and outcomes.

Language learning in children should be facilitated by a replication of the model environment for bilingualism to occur. This would include adequate opportunities to hear the two (or more) languages spoken in their environment by family and peers, as well as other sources such as the media. Frequent and sustained exposure to the spoken form of the languages in a situation in which active use of these languages by the child is necessary for socio communicative reasons is more likely to enable them to acquire the MOI, than the current form of language teaching at school.

Equally importantly we need to understand bi and multilingualism as exists in our country rather than base our policies and educational formulas on western research where the bilingual situation is quite different from ours. A related aspect is the teaching methods that are again based on western data and science on reading, and does not take into consideration the nature of our languages and their scripts.

Out of the box thinking, such as provision of incentives (such as awards, job quotas or preference) for those who excel in the local languages, over and above the basic competence in the mainstream language or English, that is required for the job; should be considered to preserve our fast diminishing wealth of languages, rather than forcing local languages as MOI through legislation and decrying the expansion of English for economic reasons.

Finally there is also a need to empathize with the plight of the child with learning disability and/or language learning disability in our multilingual context. Additional research on understanding the interaction between our specific languages and writing systems could in particular lead to better informed choice of MOI for the child with LD. The setting up of good quality vernacular schools and the exploration of options for learning that do not depend excessively on language would provide alternate solutions for supporting children with LD and LLD.

REFERENCES

Bates, E., Elman, J., Johnson, M., Karmiloff-Smith, A., Parisi, D. and Plunkett, K. (1998). Innateness and emergentism. *A companion to cognitive science* (pp. 590-601), Oxford: Basil Blackwell.

Blumstein, S. Carew, T., Kanwisher, N and Sejnowski, T. (2010). A Report on Grand Challenges of Mind and Brain, National Science Foundation, USA. www.nsf.gov/pubs/reports/grand_chall.pdf - 2006-08-04.

Cazden , C.B. (1974). Play with language and metalinguistic awareness: one dimension of language experience. *International Journal of Early Childhood, 6,* 12-24.

Chomsky, N. (1959). A Review of B. F. Skinner's Verbal Behavior. *Language, 35,* 26-58.

Collier, V.P. (1992). A synthesis of studies examining long-term language-minority student data on academic achievement. *Bilingual Research Journal, 16,* 187-212.

Dash, T. (2009). Assessing the influence of language environment on bi/multilingual language acquisition. Bangalore University: Unpublished Master's Dissertation.

Genesee, F. (1987). *Learning through Two Languages: Studies of Immersion and Bilingual Education.* Cambridge, Mass: Newbury House

Hakuta, K. (1990). Bilingualism and bilingual education: A research perspective. Occasional Papers in Bilingual Education. Washington, DC: Delta Systems and the Center for Applied Linguistics.

Hayes, S.C., Barnes-Holmes, D. and Roche, B. 2001. *Relational Frame Theory: A Post-Skinnerian Account of Human Language and Cognition.* Plenum Press.

Jhingran, D. (2009). Hundreds of home languages in the country and many in most classrooms – coping with diversity in primary education in India. In A. Mohanty, M. Panda, R. Phillipson and T. Skutnabb-Kangas (Ed.), *Multilingual Education for Social Justice* (pp. 250-267), Hyderabad: Orient Blackswan.

Karanth P., Kudva, A., and Vijayan, A. (1995). "Literacy and Linguistic Awareness". In B. de Gelder and J. Morais (Ed), *Speech and Reading: A Comparative Approach*, Erlbaum: London.

Karanth, P. (2003a). Introduction. In P. Karanth and J. Rozario (Eds.), *Learning Disability in India: Willing the Mind to Learn* (pp. 17-29) New Delhi: Sage.

Karanth, P. (2003b). Language and Learning Disabilities or Language Learning Disabilities". In P. Karanth and J. Rozario (Eds.), *Learning Disability in India: Willing the Mind to Learn* (pp. 127-137), New Delhi: Sage

Karanth, P. (2003c). *A Cross-Linguistic Study of Acquired Reading Disorders: Implications for Reading Models, Disorders, Acquisition and Teaching.* New York: Kluwer Academic

Karanth, P. (2008). Learning Disability and Language Learning. In K. Thapa, G. M. van der Aalsvoort and J. Pandey (Eds.), *Perspectives on Learning Disabilities in India: Current Practices and Prospects* (pp. 80-96), New Delhi: Sage

Karanth, P. (2009). *Children with Communication Disorders.* Hyderabad: Orient Blackswan.

MacWhinney, B. (1999). *The Emergence of Language.* Lawrence Erlbaum Associates.

Mohanty, A., Panda, M., Phillipson, R. and Skutnabb-Kangas, T. (2009). Multilingual Education for Social Justice (pp. 278-294), Hyderabad: Orient Blackswan.

Nag, S. (2006). Literacy for All: Chipping away at the ceiling. NORRAG NEWS, No. 37, Special Issue on Educating and training out of poverty, UK.

Ramirez, J.D. (1992). Executive summary of the Final Report: Longitudinal study of structured English immersion *Bilingual Research Journal, 16,* 1-62.

Olson, D. R., and Torrance, N. (1991). *Literacy and Orality.* Cambridge University Press.

Shanbal, J. and Prema, K.S. (2007). 'Languages of School Going Children – A sample survey in Mysore. *Language in India, 7,* 1-14.

Skinner, B.F. (1957). *Verbal Behavior.* Acton, Massachusetts: Copley Publishing Group.

Slobin, D.I. (Ed) (1985). *The crosslinguistic study of language acquisition Vol 1and2.* Hillsdale, NJ: Lawrence Erlbaum Associates

Slobin, D.I. (1992). *The crosslinguistic study of language acquisition Vol 3,* Hillsdale, NJ: Lawrence Erlbaum Associates.

Snow, C.E., Burns, M.S., and Griffin, P. (2000). Preventing Reading Difficulties in Young Children. US National Council Report. Washington DC: National Academy Press.

Thapa, K. (2008). Learning Disabilities: Issues and Concerns. In K. Thapa, G. M. van der Aalsvoort and J. Pandey (Eds). *Perspectives on Learning Disabilities in India: Current Practices and Prospects* (pp. 23-47), New Delhi: Sage.

van Kleeck, A. (1994). Metalinguistic Development. In G. P. Wallach and K. G. Butler (Eds), *Language learning disabilities in school-age children and adolescents* (pp 53-88), New York: Macmillan.

Vargha-Khadem, F., Gadian, D.G., Copp, A., and Mishkin, M. (2005). FOXP2 and the neuroanatomy of speech and language. *Nature Reviews Neuroscience, 6,* 131–137.

In: Expanding Horizions of the Mind Science(s) ISBN: 978-1-62808-705-5
Editors: P.N. Tandon, R.C. Tripathi and N. Srinivasan   ©2013 Nova Science Publishers, Inc.

Chapter 18

DYSLEXIA ACROSS ORTHOGRAPHIES – INSIGHTS FROM NEUROIMAGING

N. C. Singh[1], A. Chakravarty[1] and P. Padakannaya[2]
[1]National Brain Research Centre, Manesar, India
[2]Department of Psychology, University of Mysore, Mysore, India

ABSTRACT

Developmental dyslexia is now recognised as a neurobiological disorder with a genetic basis. It is characterized by a core deficit in reading, despite normal intelligence and conventional teaching methods. Recent neuroimaging findings have shown anatomical differences in the manifestation of dyslexia. This article reviews the functional anatomy of reading in different orthographies and subsequent differences in the manifestation of dyslexia. We also discuss implications for children in India, who acquire reading skills in two or more different scripts and argue for the need to extend dyslexic research to a bilingual population.

1. INTRODUCTION

The term, "Dyslexia" – was first coined by Rudolf Berlin (Berlin, 1887) to describe a boy, severely impaired in reading and writing but with typical intellectual and physical abilities. Subsequently, a disorder or difficulty in reading fluently and accurately, despite normal intelligence, adequate instruction, and socio-cultural opportunity and in the absence of any physical brain tumour or sensory deficit in vision and hearing (DSM IV) has been termed as developmental dyslexia. Dyslexia is found across cultures and languages and affects 5-10% of school children in India (Rumsey, 2000).

Unlike speech that is acquired through normal social exposure, reading is a relatively recent skill in human civilization and requires explicit instruction. Reading involves the decoding of a writing system and since all writing systems represent spoken language, the principles governing the decoding of written scripts suggests a common cognitive and neuro-anatomical network for reading across language (Perfetti et al., 2003).

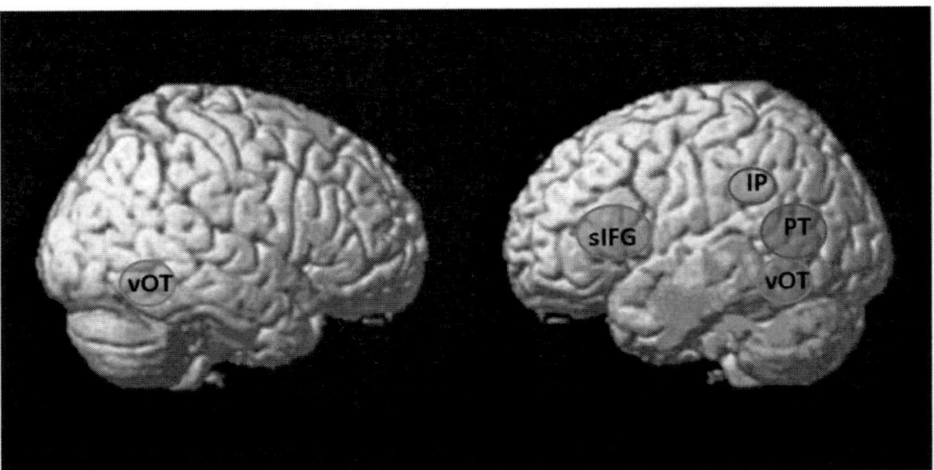

Figure 1. Brain areas involved in reading alphabetic scripts.

Reading requires two processes, one involving letter-sound mapping also called orthography and a second, wherein entire text is direct accessed (semantic storage) (Coltheart 1985). The first process requires the ability to decode, in that each printed letter of a word is first converted to its corresponding sound before accessing its meaning. Studies from alphabetic scripts like English have shown that to acquire basic reading skills one must learn to map alphabetic characters (graphemes) to the basic sound segments in speech (phonemes) that they represent. Alphabetic words thus are predominantly read out by assembling fine-grained phonemic units, i.e., by assembled phonology (Coltheart et al., 1993). Acquisition of speech is a prerequisite before one learns to read. However speech acquisition is a natural process whereas learning to read requires extensive instruction. This is because spoken language appears seamless to the listener, with no clues to its segmental nature. Thus, the word "bat" comprises of the sounds "b", "aa" and "t" but one hears the holistic word "bat" and not as three separate sounds. In order to learn how to read, however, one must muster multiple skills including developing an awareness that spoken language can be segmented into smaller elements (i.e., phonemes), identifying letters, learning the rules of how print maps onto sound, recognizing whole words not only accurately but also rapidly (automatically), acquiring a vocabulary, and extracting meaning from the printed word(s). One of the critical steps in this process is recording the letters (orthography) onto their sounds (phonology).

In case of words that are irregular, like "yacht" or "champagne" where the spellings do not follow the rule, the second process is employed wherein the reader looks up the printed word in the mental lexicon (referred to as addressed phonology), reads the word aloud and is followed by sound decomposition. Functional magnetic resonance imaging (fMRI), a non-invasive high-resolution anatomical technique has shown that these two processes are accomplished via two distinct routes in the brain. Reading regular words wherein, each letter (grapheme) can be mapped to a sound (phoneme) has been shown to activate areas along the dorsal route namely the angular gyrus, inferior parietal lobule, supramarginal gyrus and superior temporal gyrus. For irregular words, (examples 'bough' and 'cough'), wherein a mapping of combination of letters to sound is necessary, reading is believed to be accomplished by areas along the ventral route primarily the inferior and middle temporal gyri (Jobard et al., 2003; Price, 2000).

Individuals with dyslexia primarily demonstrate deficits in decoding (sound-letter mapping). Behavioural studies have shown that dyslexic readers are unable to perform satisfactorily on tasks involving word segmentation (Liberman, 1973). Equally poor performance is seen on tasks of categorical perception (/ba/ vs /da/) (Godfrey et al., 1981) as well as phonology awareness tasks like single-letter matching (does T sound like V?) and non-word rhyming (leat vs jete), short-term memory and spelling tests, (Snowling, 2000, Shaywitz et al., 2002). In school-going children in particular, this causes serious issues in classroom performance. However, sound-letter mapping or orthography varies across languages. For languages like Italian or Hindi, it is consistent whereas for languages like English it is inconsistent. This suggests that the incidence of dyslexia should differ across languages, which is indeed the case (Paulesu et al. 2000). In the following paragraphs we review these differences in orthography across languages and discuss the neural pathways for reading in different orthographies. We show how reading networks are different for alphabetic and logographic scripts, which in turn lead to differences in the cortical areas of the brain affected in dyslexia. In section 2 we discuss dyslexia in alphabetic scripts and in section 3 we discuss dyslexia in logographic scripts. We discuss the implications of these findings for Indian scripts in section 4. We conclude by discussing the need for urgently studying dyslexia in bilinguals.

2. DYSLEXIA IN ALPHABETIC SCRIPTS

As described earlier, mapping the phonological units of spoken words onto the grapheme units of written texts is believed to be necessary for fluent reading. For an alphabetic script like English, Italian or Spanish this involves mapping each letter (grapheme) or letter combinations to a distinct sound (phoneme). However for a language like English, there are 1120 ways of representing 40 sounds (phonemes). The orthography of English is deep or inconsistent in that a single sound may be represented by multiple different letters or letter combinations and vice-versa (the simplest example is the letter /u/ in 'but' and in 'put'). On the other hand in a language like Italian 33 graphemes are able to represent the 25 phonemes of the language and has a transparent or consistent orthography. One would therefore expect the incidence of dyslexia to be less in Italian as compared to English, which is indeed the case. A behavioral study conducted in Italy and the United States in … showed that the prevalence of dyslexia in Italian was half that of the United States. Functional imaging (fMRI) studies in skilled readers, have shown that the reading network in alphabetic languages like English and Italian, consists of occipito-temporal, temporo-parietal and inferior frontal regions (Shaywitz et al., 1998, Paulesu et al., 2001) (Figure 1).IP – Inferior Parietal Lobule ; PT is parieto-temporal area vOT – ventral region ; Occipital areas ; sIFG – superior region of Inferior Frontal Gyrus.

Studies on individuals exhibiting dyslexia in English, Italian and French, all alphabetic scripts, have shown functional disruption in the dorsal neural systems (Paulesu et al., 2001). A study by Shaywitz et al. (1998) wherein cortical activation were examined in dyslexic and nondyslexic children while performing a series of phonological tasks showed under activation in several components of posterior cortical system namely posterior superior temporal gyrus (Wernicke's area), BA 39 (angular gyrus) and BA 17 (striate cortex) in dyslexics as compared to normal readers. However the study also observed a pattern of over activation in the anterior regions – inferior frontal gyrus and BA 46/47 (anterior-frontal-brain region) of dyslexics. They concluded that dyslexic readers demonstrated functional disruption in the traditional language and visual system and certain areas of association cortex (angular gyrus) necessary for cross modal integration. Brain activation patterns in dyslexics in higher order phonological processing tasks corresponded to the behavioral findings of phonological order deficits. Similar findings were reported by several other studies (Booth et al., 2007).

Hoeft et al., (2006) addressed the hyper- and hypo-activation in the anterior and posterior brain areas respectively in dyslexic population. They measured brain activation pattern of dyslexic adolescents against age-matched normal readers and against reading matched but not age-matched control readers. They reported hypo-activation in left parietal and fusiform regions in dyslexics as against the control groups in a visual word rhyme judgment task and hyper activation in left inferior and middle frontal gyrus, caudate and thalamus as against age-matched control group but not against reading matched control group. They concluded that the pattern of activation shifts developmentally from frontal (IFG) to parieto-temporal region in beginner readers to fluent readers. So the hyperactivation in IFG is not atypical to dyslexia per se. However the hypoactivation in the parieto-temporal region in phonological tasks indicates a functional deficit that is characteristic to dyslexia. The group also performed voxel based morphometry to show structure-function correlation and observed reduced gray matter volume in dyslexic population as compared to the control groups.

However Cao et al. (2006) reported contrasting results in activation pattern of left inferior frontal gyrus in phonological analysis tasks. They designed a judgmental task paradigm with words that varied on orthography to phonology discrepancy. Their fMRI study showed no difference in pattern of activation between dyslexics and nondyslexics. However, children with dyslexia showed less activation than the controls in left inferior frontal gyrus (BA 45/44/47/9), left inferior parietal lobule (BA 40), left inferior temporal gyrus/fusiform gyrus (BA20/37) and left middle temporal gyrus (BA 21) for the more difficult conflicting trials. They concluded that the inferior frontal gyrus may act as an executive system that controls the selection, access, and retrieval of information by the modulation or reactivation of representations in posterior brain regions (Poldrack et al., 1999, Wagner, Desmond, Demb, Glover, and Gabrieli, 1997;) and the decreased activation observed in dyslexic children reflects their deficiency in orthography to phonology mapping skill. Thus a common finding from functional imaging studies on dyslexics reading in alphabetic scripts suggests reduced activation in the temporo-parietal area.

In the first fMRI study on functional connectivity in phoneme mapping task Richards et al. (2008) found that there was no significant difference in fMRI connectivity between dyslexic and control children in the middle frontal gyrus, occipital lobe and cerebellum. However there was significantly greater connectivity from left inferior frontal gyrus to right inferior frontal gyrus in dyslexic children. Further, following intervention there was no significant difference between dyslexic and control children in activation of this particular loop. It was concluded that dyslexic children differed from control children in their temporal coordination of brain regions during phonological task. However instructional intervention designed to improve time sensitive procedure may overcome deficit in the temporal coordination. Thus the role of left inferior frontal gyrus in phonological analysis task in dyslexics and whether it plays a role in neural basis of developmental dyslexia is still not clear.

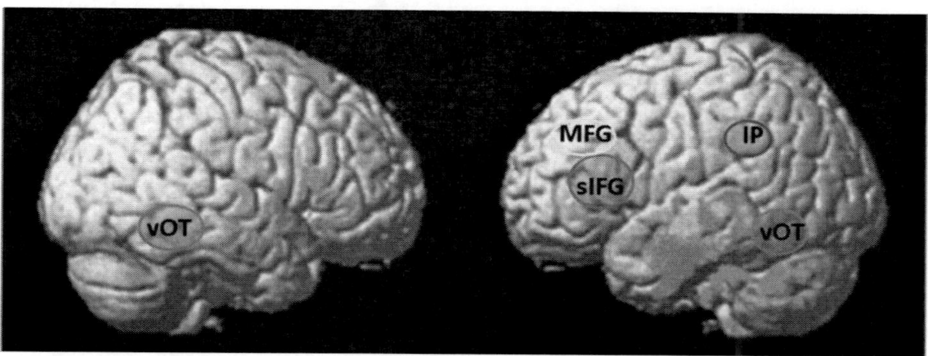

Figure 2. Reading network for Chinese.

Silani et al. (2005) looked at the brain activation pattern of dyslexic adults during reading task using PET scan and correlated it with structural morphometry using VBM as a technique across three cultures English, French and Italian. The scripts of these three languages differ on regularity. They observed reduced activation in left inferior parietal in the dyslexic population across all three cultures which was significantly correlated with reduced gray matter density due to atropy in that region. Thus the brain differences were replicated across dyslexic samples with the conclusion that the neurological disorder underlying dyslexia is same across culture.

3. DYSLEXIA IN CHINESE

However in languages like Chinese, though individuals exhibit phonological deficits at the behavioural level (Perfetti et al., 2005) fMRI studies have shown that Chinese dyslexics exhibit reduced activation in left middle frontal regions and not posterior parts of left temporo-parietal areas seen in alphabetic languages (Siok et al., 2004). This has been attributed to the fact that reading in Chinese, which is logographic, requires that graphic forms be mapped to syllables, which is markedly different from reading in an alphabetic system (English, for example) in which graphic units are mapped to phonemes. Hence, reading in Chinese depends more on memorization and visuo-spatial processing, which is accomplished by the left middle frontal gyrus whereas reading in alphabetic languages like English and Italian depends more on letter to sound conversion carried out by the temporo-parietal regions. Skilled readers of Chinese activate left middle and inferior frontal gyrus (Tan et al., 2000) as shown in Figure 2. IP – Inferior Parietal Lobule; vOT – ventral region Occipital areas, sIFG – superior region of Inferior Frontal Gyrus; MFG – Middle Frontal Gyrus.

In Chinese, reading verbal memory is very important as it is required for rote learning; however, behavioural manifestation of this deficit may not be distinguishable from phonological processing deficit. Thus when Siok et al. (2008) studied Chinese dyslexic children in phonological analysis tasks using fMRI and correlated the brain activation pattern with structural morphometry using VBM they found that regional gray matter volume was significantly less in left middle frontal gyrus in dyslexic Chinese readers than control children. The left middle frontal gyrus is believed to be involved in the allocation and coordination of cognitive resources in working memory and hence its relevance for Chinese script reading. We thus find that at the behavioral level individuals with dyslexia exhibit deficits in phonological processing – however the cortical areas involved depend on orthography and script and as a result are manifested in different cortical areas, temporo-parietal in alphabetic scripts and middle frontal gyrus in logographic Chinese.

4. DYSLEXIA IN ALPHASYLLABARIES

The situation in India is quite complex and varied. There are groups of children who speak one language at home and go to school to learn through a regional language (e.g., tribal children). There are another group of children who speak one language and go to learn and get educated through English from the very beginning (e.g., urban lower middle class children). There are children who go through literacy training in two or three languages simultaneously from the very beginning or at different points of time during their primary education. Along with initiating studies of dyslexia in bilinguals, it is equally important to understand the manifestation of dylexia in Indian writing systems which are alphasyllabaries. The Indian writing systems are unique in several ways. They are all derived from a single source, *brahmi*, and share features (see Padakannaya and Mohanty, 2004 for details). All of them, including Devanagari used to write Hindi, are semisyllabic or alphasyllabic and transparent with regard to *akshara* – sound mapping. This is in contrast to English, taught all over India, which is written alphabetically in Roman and is opaque in nature. Indian scripts differ from the Roman derived English script in terms of grain size, transparency as well as orthographic layout (Padakannaya and Mohanty, 2004; Patel, 2004; Prakash and Joshi, 1995; Sproat and Padakannaya, 2008; Vaid and Gupta, 2002).

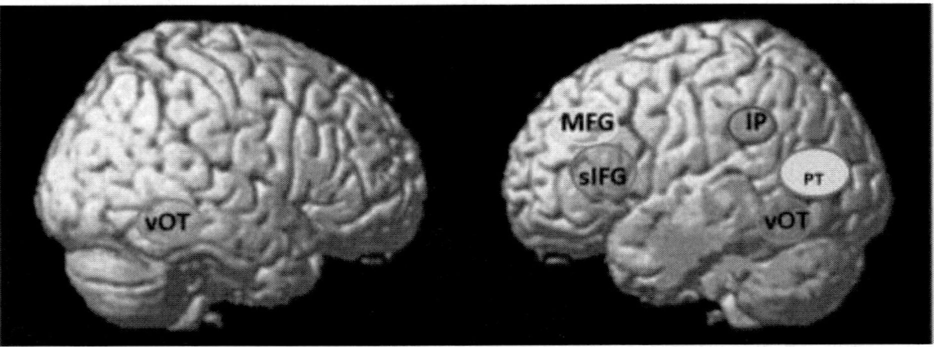

Figure 3. Reading network for Devnagari.

One Indian script for which some studies have been recently initiated is Devnagari. The Devanagari system originated as the script used to write down Sanskrit. It consists of 48 basic letters and additional diacritical signs that together represent every sound of the Sanskrit language (Bright, 1996). Devanagari has syllabic as well as alphabetic properties. It differs from other alphabetic scripts in that each consonant in Devanagari has an inherent associated vowel. Further it differs from purely syllabic writing systems in that it does not employ unique symbols for distinct syllables (Vaig & Gupta, 2002). The basic phonological unit that corresponds to a grapheme in Hindi is an akshara that is always a syllable. For an alphabetic language like English this is a phoneme. These aksharas in written form contain one or more consonant(s) and vowel that are often symbolically attached to each other by matras. These matras often precede and follow the core consonant and have an impact on visual manipulation of phonological units (Vaid & Gupta, 2002). A single akshara can consist of three or four symbolic attachments of different sounds on a single base representation, making this a complex reading system. Research on reading in Indo-Aryan languages (Karanth, 2002; Vaid & Gupta, 2002) has indicated the relevance of script specific features in processing written languages like Devanagari as opposed to English or French. Studies on segmentation in Devanagari by native Hindi speakers have shown how alphasyllabic nature of the script influences one's performance (Vaid & Gupta, 2002). Since Devanagari presents a mixture of syllabic and alphabetic scripts it is not clear how dyslexia would manifest in Devnagari? Few studies on phrase reading in Devnagari that have been initated suggest a reading network that involves both the temporo-parietal and the middle frontal gyrus besides the occipital areas. IP – Inferior Parietal Lobule; vOT – ventral region Occipital areas. sIFG – superior region of Inferior Frontal Gyrus; MFG – Middle Frontal Gyrus, PT- Parieto-temporal.

Neuroimaging studies on reading in Indian languages are still in infancy. No fMRI study on dyslexia has yet been reported. There is a great need for such studies in Indian languages for several reasons. First, Indian orthographic system is very unique. Second, it does not fit exactly into the general taxonomy of orthographies normally followed. Third, imaging studies on reading Indian scripts would provide a new platform for unravelling the plasticity of human brain as well as for enhancing our insight on brain-orthography relationship.

5. READING IN TWO LANGUAGES

More than 65% of children in the world today are bilinguals and a great majority of those who go to schools among them are also biliterates. Learning to read two or more distinct orthographies with very different scripts is a remarkable accomplishment by many bilinguals as in the acquisition of reading skills in distinct scripts such as Chinese and English or English and Devanagari. There are general principles applicable to any kind of writing system (Perfetti, 2003) suggesting a common neuro-anatomical network for reading across languages and also orthography specific pathways suggesting separate processing networks for different orthographies (Meschyan and Hernandez, 2006; Tan et al., 2003). Bolger, Perfetti, and Schneider (2005), suggested that different writing systems may engage the same gross cortical regions, but activation within those regions differs across writing systems. Their metaanalysis of over 150 studies on visual word recognition revealed various regions associated with orthography, phonology and semantics of a writing system. Orthography associated activations included bilateral occipital/posterior fusiform gyrus (BA 18, 37/19), left mid-fusiform and posterior inferior temporal gyrus (BA37). Areas associated with phonological processing were superior temporal sulcus/inferior parietal lobe (BA 22/40/39), inferior frontal sulcus/insula/premotor cortex (BA 45/6/9) whereas activations related to semantic processing included anterior fusiform/inferior temporal gyrus, middle temporal gyrus, anterior inferior frontal gyrus (BA 37/21, 44). Bilingual readers in alphabetic scripts also showed similar patterns of cortical activations (Frenck-Mestre et al., 2005). A neuroimaging study by Dietz, Jones, Gareau, Zeffir and Eden (2005) suggested that the left medial fusiform gyrus activation is more prominent in reading Chinese whereas for English, activation in the left lateral fusiform gyrus is stronger though cortical activation in left temporo-occipital circuits that mediate phonological processing is present in reading both alphabetic and logographic scripts. However studies on bilinguals reading distinct scripts such as English and Chinese showed that the orthography learnt earlier affects reading in a later learned language (Liu, Dunlap, Fiez, and Perfetti, 2007; Perfetti, Liu, Fiez, Nelson, and Bolger, 2007; Tan et al., 2003).

Meschyan and Hernandez (2006) studied the influence of orthographic transparency among Spanish-English bilinguals readers. They reported greater activation in left superior temporal gyrus (BA 22) while reading a transparent script- a region implicated in assembled phonological strategy of reading (Paulesu et al., 2000) – and for English, which has an opaque orthography, reported greater activity in visual processing and word recoding regions such as left occipito-parietal border and right inferior parietal lobe (BA 40). Kumar et al. (2009, in press), in an fMRI study on reading phrases in Hindi and English found activation in bilateral temporal pole (BA 38) and the caudate nucleus in the right hemisphere for Hindi. Involvement of right hemispheres region was ascribed to spatially complex and highly non-linear characters of Devanagari. Studies on dyslexia would also help us understanding the neuroanatomical basis of a developmental disorder that affects a significant number of school going children leading to failures, dropouts, delinquency and often antisocial behaviours. More than all these, neurobiological understanding of normal and dyslexic reading would help us in devising effective pedagogy and appropriate interventional methods to help children with reading problems.

ACKNOWLEDGMENT

The authors would like to thank Sarika Cherodath for help with the figures and discussions.

REFERENCES

Bolger, D. J., Perfetti, C. A., and Schneider, W. (2005). Cross-cultural effect on the brain revisited: Universal structures plus writing system variation. *Human Brain Mapping, 25,* 92-104.

Booth, J. R., Burman, D. D., Meyer, J. R., Gitelman, D. R., Parrish, T. B. and Mesulam, M. M. (2004). Development of brain mechanisms for processing orthographic and phonologic representations. *Journal of Cognitive Neuroscience, 16,* 1234-1249.

Bright, W. (1996). The devanagari script. In The World's Writing Systems (Eds.) P. Daniels and W. Bright (pp. 384–390), NY: Oxford University Press.

Castles, A., and Coltheart, M. (1993). Varieties of developmental dyslexia. *Cognition, 47,* 149–180.

Cao, F., Bitan, T., Chou, T., Burman, D.D., and Booth, J.R. (2006). Deficient orthographic and phonological representations in children with dyslexia revealed by brain activation patterns. *Journal of Child Psychology and Psychiatry, 47,* 1041–1050.

Coltheart, M. (1985). Cognitive neuropsychology and the study of reading. In M. I. Posner and O. S. M. Marin (Eds.), *Attention and performance IX* (pp. 3–37). Hillsdale, NJ: Erlbaum.

Das, T., Singh, L., and Singh, N. C. (2008). Rhythmic structures of Hindi and English: New insights from a computational analysis. *Progress in Brain Research, 168,* 207-214.

Dietz, N. A., Jones, K. M., Gareau, L., Zeffiro, T., Eden, G. (2005). Phonological processing involves left posterior fusiform cortex. *Human Brain Mapping, 26,* 81-93.

Gottardo, A., Yan, B., Siegel, L. S., and Wade, W. L. (2001). Factors related to English reading performance in children with Chinese as a first language: More evidence of a cross-language transfer of phonological processing. *Journal of Educational Psychology, 93,* 530-542.

Haller, S., Radue, E. W., Erb, M., Grodd, W., and Kircher, T. (2005). Overt sentence production in event-related fMRI. *Neuropsychologia, 43,* 807-814.

Hoeft, F., et al. (2006) Neural basis of dyslexia: a comparison between dyslexic children and non-dyslexic children equated for reading ability. *Journal of Neuroscience, 26,* 10700-10708.

Horwitz, B., Rumsey, J. M., and Donohue, B. C. (1998) Functional connectivity of the angular gyrus in normal reading and dyslexia. *Proceedings of the National Academy of Sciences USA, 95,* 8939–8944.

Jobard, G., Crivello, F., and Tozourio-Mazoyer. (2003). Evaluation of dual route theory of reading: a metaanalysis of 35 neuroimaging studies. *Neuroimage, 20,* 693-711.

Karanth, P. (2002). The search for deep dyslexia in syllabic writing systems. *Journal of Neurolinguistics, 15,* 143–155.

Kim, K. H., Relkin, N. R., Lee, K. M., and Hirsch, J. (1997) Distinct cortical areas associated with native and second languages. *Nature, 388,* 171-174.

Liberman, I. Y. (1973) Segmentation of the Spoken Word and Reading Acquisition. *Bulletin of the Orton Society, XXIII,* 65-77.

Liu, Y., and Perfetti, C.A., (2003). The time course of brain activity in reading English and Chinese: An ERP study of Chinese bilinguals. *Human Brain Mapping, 18,* 167-175.

Liu, Y., Dunlap, S., Fiez, J., and Perfetti, C. (2007). Evidence for neural accomodation to a writing system following learning. *Human Brain Mapping, 28,* 1223-1234.

Meschyan, G., and Hernandez, A. E. (2006). Impact of language proficiency and orthographic transparency on bilingual word reading: An fMRI investigation. *Neuroimage, 29,* 1135-1140.

Nelson, J., Liu, Y., Fiez, J., and Perfetti, C. A. (2005). Learning to read Chinese as a second language recruits Chinese-specific visual word-form areas. Paper presented at the Society for the Scientific Study of Reading Conference, Toronto.

Padakannaya, P., and Mohanty, A. K. (2004). Indian orthography and teaching how to read: A Psycholinguistic framework. *Psychological Studies, 49,* 262-271.

Patel, P. G. (2004). *Exploring Reading Acquisition and Dyslexia in India,* New Delhi: Sage.

Paulesu, E., McCrory, E., Fazio, F., Menoncello, L., Brunswick, N., Cappa, S. F., Cotelli, M., Cossu, G., Corte, F., Lorusso, M., Pesenti, S., Gallagher, A., Perani, D., Price, C., Frith, C., D., and Frith, U. (2000). A cultural effect on brain function. *Nature Neuroscience, 3*, 91-96.

Paulesu, E., Demonet, J. F., Fazio, F., McCrory, E., Chanoine, V., Brunswick, N., Cappa, S. F., Cossu, G., Habib, M., Frith, C. D. (2001). Dyslexia: cultural diversity and biological unity. *Science, 291*, 2165-2167.

Perani, D., Paulesu, E., Galles, N. S., Dupoux, E., Dehaene, S., and Bettinardi, V. (1998). The bilingual brain. Proficiency and age of acquisition of the second language. *Brain, 121*, 1841-1852.

Perfetti, C. A. (2003). The universal grammar of reading. *Scientific Studies of Reading, 8*, 3-24.

Perfetti, C. A., Liu, Y., Fiez, J., Nelson, J., and Bolger, D. J. (2007). Reading in two writing systems: Accomodation and assimilation of the brain's reading network. *Bilingualism: Language and Cognition, 10*, 131-146.

Prakash, P., and Joshi, M. (1995). Orthography and Reading in Kannada : A Dravidian Language. In I.Taylor and D.Olson (Eds.), *Scripts and Reading: Reading and Learning to Read World's Scripts,* London: Kluwer Academic Publishers.

Price, C. J. (2000) The anatomy of language: contributions from functional neuroimaging. *J. of Anatomy, 197*, 335–359.

Richards, T. L. and Berninger, V. W. (2008) Abnormal fMRI Connectivity in Children with Dyslexia During a Phoneme Task: Before But Not After Treatment, *Journal of Neurolinguistics, 21*, 294-304.

Rumsey, J. M., Nace, K., Donohue, B., Wise, D., Maisog, J. M., Andreason, P. (1997). A positron emission tomographic study of impaired word recognition and phonological processing in dyslexic men. *Archives of Neurology*, 54, 562-573.

Shaywitz, S. E. (1998). Dyslexia. *New England Journal of Medicine, 338*, 307-312.

Silani, G., Frith, U., Demonet, J. F., Fazio, F., Perani, D., Price, C., Frith, C. D., Paulesu, E. (2005). Brain abnormalities underlying altered activation in dyslexia: a voxel based morphometry study. *Brain, 128*, 2453-2461.

Snowling M.J. (2000). Language and literacy skills: who is at risk and why? In Bishop, D.V.M. and Leonard, L.B. (Eds). Speech and language impairments in children: Causes, characteristics, intervention and outcome (pp. 245-260), Hove, UK: Psychology Press.

Sproat, R., and Padakannaya, P. (2008). Script Indices. In N. Srinivasan, A. K. Gupta and J. Pandey (Eds.), *Advances in Cognitive Science Vol. 1* (pp. 62-70), New Delhi: Sage.

Tan, L. H., Liu, H-L., Perfetti, C. A., Spinks, J. A., Fox, P. T. and Gao, J-H. (2001) The neural system underlying Chinese logograph reading. *Neuroimage, 13*, 836–846.

Tan, L. H., Spinks, J. A., Feng, C-M., Siok, W. T., Perfetti, C. A., Xiong, J., Fox, P. T., and Gao, J-H. (2003). Neural systems of second language reading are shaped by native language. *Human Brain Mapping, 18*, 158-166.

Tan, L. H., Liard, A. R., Li, K. and Fox, P. T. (2005). Neuroanatomical correlates of phonological processing of Chinese characters and alphabetic words: A meta-analysis. *Human Brain Mapping, 25,* 83-91.

Vaid, J., and Gupta, A. (2002). Exploring word recognition in a semialphabetic script: the case of Devanagari. *Brain and Language, 81,* 679–690.

Vaid, J., and Hull, R. (2002). Re-envisioning the bilingual brain using functional neuroimaging: Methodological and interpretive issues. In F. Fabbro (Ed.), *Advances in the neurolinguistics of bilingualism*, Udine (Italy) Forum.

In: Expanding Horizions of the Mind Science(s) ISBN: 978-1-62808-705-5
Editors: P.N. Tandon, R.C. Tripathi and N. Srinivasan ©2013 Nova Science Publishers, Inc.

Chapter 19

DEMENTIAS

Manjari Tripathi and Deepti Vibha
All India Institute of Medical Sciences, New Delhi

ABSTRACT

Dementia is a common disorder in the elderly, involving as many as 10% of those over 65 years of age. There are no known curative or preventive measures for most types of dementia. Diet and lifestyle could influence risk, and studies suggest that midlife history of disorders that affect the vascular system, such as hypertension, type 2 diabetes, and obesity, increase the risk for dementia including Alzheimer's disease (AD) which is the most common dementia. Reversible causes enumerated above may be found in up to 20% of cases. Having classified the patient into the type of dementia, it should always be kept open to revision in terms of diagnosis, associated reversible factors, addition and modification of treatment and counselling of care giver.

1. INTRODUCTION

Dementia is defined as an acquired deterioration in cognitive abilities that impairs the successful performance of activities of daily living. Those conscious mental tasks that we perform every waking second of every day, from remembering our name to performing complicated mathematical calculations. Memory is one of the most essential cognitive functions, and it is often the first and most crucial one that dementia impairs. Dementia also affects problem-solving ability, decision making, judgment, our ability to orient ourselves in space, and our ability to put together simple sentences and understand and communicate with words. It is also often associated with personality change. Dementia is a permanent, progressive disease that affects mostly the elderly. People who suffer from dementia eventually are unable to take care of themselves and require round-the-clock care (Mayeus and Sano, 1999). Dementia is a common disorder in the elderly, involving as many as 10% of those over 65 years of age (Knopman et al., 2001).

The specific patterns of cognitive and other associated symptoms with pathological substrates characterize dementias into various types namely- Alzheimer's disease (AD), Frontotemporal Dementia (FTD), Dementia with Lewy Body (DLB) to name a few. This chapter would cover causes of dementia, approach to dementia diagnosis and each major types of dementia under the headings of brief description of their clinical features, pathophysiology and treatment.

2. Causes of Dementia: (Tripathi and Vibha, 2008)

2.1. Common Causes of Dementia

- Alzheimer's disease
- Vascular dementia
- Multi-infarct
- Diffuse white matter disease (Binswanger's)
- Alcoholism
- Parkinson's disease
- Drug/medication intoxication

2.2. Less Common Causes of Dementia

Vitamin Deficiencies
- Thiamine (B1): Wernicke's encephalopathy
- B12 (pernicious anemia)
- Nicotinic acid (pellagra)

Endocrine and other Organ Failure
- Hypothyroidism
- Adrenal insufficiency and Cushing's syndrome
- Hypo- and hyperparathyroidism
- Renal failure
- Liver failure
- Pulmonary failure

Chronic Infections
- HIV
- Neurosyphilis
- Papovavirus (progressive multifocal leukoencephalopathy)
- Prion (Creutzfeldt-Jakob and Gerstmann-Sträussler-Scheinker diseases)
- Tuberculosis, fungal, and protozoal
- Whipple's disease

Head Trauma and Diffuse Brain Damage
- Dementia pugilistica
- Chronic subdural hematoma
- Post anoxia
- Post encephalitis
- Normal-pressure hydrocephalus

Neoplastic
- Primary brain tumor
- Metastatic brain tumor
- Paraneoplastic limbic encephalitis

Toxic Disorders
- Drug, medication, and narcotic poisoning
- Heavy metal intoxication
- Dialysis dementia (aluminum)

Psychiatric
- Depression (pseudodementia)
- Schizophrenia
- Conversion reaction

Degenerative Disorders
- Huntington's disease
- Pick's disease
- Dementia with Lewy bodies
- Progressive supranuclear palsy (Steel-Richardson syndrome)
- Multisystem degeneration (Shy-Drager syndrome)
- Hereditary ataxias (some forms)
- Motor neuron disease [amyotrophic lateral sclerosis (ALS); some forms]
- Frontotemporal dementia
- Cortical basal degeneration
- Multiple sclerosis
- Adult Down's syndrome with Alzheimer's
- ALS–Parkinson's–Dementia complex of Guam

Miscellaneous
- Sarcoidosis
- Vasculitis

CADASIL Etc
- Acute intermittent porphyria
- Recurrent nonconvulsive seizures

Additional Conditions in Children or Adolescents

Hallervorden-Spatz disease
Subacute sclerosing panencephalitis
Metabolic disorders (e.g., Wilson's and Leigh's diseases, leukodystrophies, lipid storage diseases, mitochondrial mutations)

2.3. Causes of potentially reversible dementia

Neurosurgical conditions	Neuroinfections and inflammation	Metabolic conditions	Others
Subdural hematoma	Meningitis (tubercular, fungal or malignant)	Hypo or hyperthyroidism	Depression
Normal Pressure Hydrocephalus		Hashimoto's encephalitis	Epilepsy
Intracranial tumours		Hypo or hyperparathyroidism	Drugs and toxins
Intracranial empyema or abscess	Encephalitis (limbic, viral)	Pituitary insufficiency	Alcohol abuse
	Cerebral vasulitis	Hypercalcemia	Sleep apnea
	Neurosyphilis	Cushing's disease	
	Lyme's disease	Addison's disease	
	Whipple's disease	Hypoglycaemia	
	Sarcoidosis	Vitamin deficiencies(B1, B6, B12 and folate)	
		Chronic liver failure	
		Chronic respiratory failure	
		Chronic renal failure	
		Wilson's disease	

3. APPROACH TO DEMENTIA DIAGNOSIS

The essential features of dementia are an acquired and persistent compromise in multiple cognitive domains that is severe enough to interfere with everyday functioning according to the DSM IV criteria (Table 1) (DSM IV, 1994). Reversible causes enumerated above may be found in upto 20% of cases (Hejl, Hogh, and Waldemar, 2002). A comprehensive history should determine the initial manifestations, mode of onset and course over time. *While this criterion helps to confirm dementia and also excludes treatable causes, it should be emphasized that treatable causes may frequently accompany degenerative dementias*; which may at times unmask the disease. However, even in a diagnosed case of primary degenerative dementia, *any unexplained deterioration should prompt the clinician to look for a reversible cause or even revise the diagnosis.*

Having kept a clinical possibility of dementia, the domains affected are assisted both by history and examination (Table 2). While MMSE (Folstein, Folstein, and McHugh, 1975) stands the test of time as a screening tool for dementia, detail lobar function assessment would delineate the cerebral areas dominantly affected. While cognitive batteries like Addenbrooke's (Mioshi et al., 2006) are also popular in a clinical setting, detailed cognitive batteries like ADAS-Cog (Alzheimer's disease Assessment Scale–cognitive subscale),

ADCS-ADL (Modified Alzheimer 's disease Cooperative Study activities of daily living inventory for severe Alzheimer's disease), CGIC-MCI (Cooperative Study Clinician's Global Impression of Change for MCI), ADFACS (Alzheimer's disease Functional Assessment and Change Scale), CDR (Clinical Dementia Rating), CIBIC-Plus (Clinician Interview Based Impression of Change incorporating caregiver information scale), CIBIS - The Clinician's Interview-Based Impression of Severity are used mainly for research purposes.

The cognitive assessment is supplemented by neuropsychiatric evaluation. The laboratory investigations help more in establishing or ruling out treatable causes while neuroimaging in the form of MRI, SPECT and PET should be done with specific indications. A brain imaging study on clinical history and examination alone have shown imperfect precision, although specificity and sensitivity may be approximately 90% (Alexander et al., 1995; Martin et al., 1987). A non-contrast CT or MRI scan may be done under most circumstances at the time of the initial dementia assessment to identify pathology such as brain neoplasms or subdural hematomas. A third condition, normal pressure hydrocephalus, which might be detected by CT or MR and might be responsive to treatment, is very rare (Vanneste, 1992). The AAN guidelines state that (Knopman et al., 2001):

1. Linear or volumetric MR or CT measurement strategies for the diagnosis of AD and are not recommended for routine use at this time.
2. For patients with suspected dementia, SPECT cannot be recommended for routine use in either initial or differential diagnosis as it has not demonstrated superiority to clinical criteria.
3. PET imaging is not recommended for routine use in the diagnostic evaluation of dementia at this time.

Table 1. DSM IV criteria for dementia

The development of multiple cognitive deficits such as manifested by both, Impaired memory, long or short-term, can't learn new information or can't recall information previously learned and is distinguished by:
One (or more) of the following cognitive disturbances: 3   Aphasia (language disturbance). 4   Apraxia (impaired ability to carry out motor activities despite intact motor function). 5   Agnosia (failure to recognize or identify objects despite intact sensory function). 6   Disturbance in executive functioning (i.e., planning, organizing, sequencing, abstracting) The cognitive deficits above each cause significant impairment in social or occupational functioning and represent a significant decline from a previous level of functioning.
The decline in mental functioning begins gradually and worsens steadily
The cognitive deficits above are not due to any of the following: Other central nervous system conditions that cause progressive deficits in memory and cognition (e.g., cerebrovascular disease, Parkinson's disease, Huntington's disease, subdural hematoma, normal-pressure hydrocephalus, brain tumor). Systemic conditions that are known to cause dementia (e.g., hypothyroidism, vitamin B-12 or folic acid deficiency, niacin deficiency, hypercalcemia, neurosyphilis, HIV infection). Substance-induced conditions.
They aren't better explained by another Axis I disorder such as a Depressive Disorder or Schizophrenia

4. DEMENTIA TYPES

The following dementias would be discussed:

- 4.1. Alzheimer's disease
- 4.2. Frontotemporal dementia
- 4.3. Vascular dementia
- 4.4. Parkinsonian dementias

4.1. Alzheimer's Disease (AD)

4.1.1. Introduction

Alzheimer's disease is the most common form of dementia in the elderly. The disorder was originally described in 1907 by the German psychiatrist and neuropathologist Alois Alzheimer. His notes were rediscovered in 1996 and published, giving details of his findings (Maurer, Ihl, Dierks, and Frolich, 1997). Kraeplin subsequently named the condition Alzheimer's disease. An estimated 4.5 million Americans have AD and, as the elderly population continues to grow, the prevalence could increase by threefold to 13.2 million by 2050 (Hebert et al., 2003). The percentage of persons with Alzheimer's disease increases by a factor of two with approximately every five years of age, meaning that 1 percent of 60-year-olds and about 30 percent of 85-year-olds have the disease (Jorm, 1991). AD is the disease process that ultimately results in Alzheimer's dementia. Alzheimer's dementia has a characteristic cognitive pattern. Early in a patient's course, AD may cause memory loss of insufficient severity to warrant the designation of dementia. Other patients with AD may follow an atypical course with progressive aphasia or progressive apraxia rather than a typical Alzheimer's dementia. Most of the time, however, AD causes Alzheimer's dementia.

4.1.2. Clinical Features

AD is a progressive disorder of recent episodic memory, language, visuospatial function and executive function associated with high frequency of neurobehavioral abnormalities at some point in the course. AD has been considered the paradigm of a cortical dementia syndrome. The hallmarks of cortical dementia include not only memory loss, which is common to many dementia syndromes, but also elements of aphasia, apraxia, and agnosia. Further, in general, there is an absence of subcortical features such as parkinsonism. Some patients with pathologically proven AD, however, also exhibit features of mild parkinsonism, emphasizing that diagnostic boundaries are relative, not absolute. National Institute of Neurological and Communicative Disorders and Stroke and the Alzheimer's Disease and Related Disorders Association (NINCDS-ADRDA) Work Group's diagnostic criteria for AD (Table 3) is commonly used for diagnosis and has a clinical diagnostic accuracy of about 90% (Knopman et al., 2001).

Definite AD is defined by tissue confirmation. Probable AD is defined by the clinical picture of dementia in the absence of certain atypical features such as focal neurological abnormalities and for which another cause cannot be found. Possible AD is defined by the clinical picture of dementia, but with either atypical features in the course of the illness or with a second potentially contributory disease not believed to be the primary cause. In the absence of histological confirmation, therefore, the clinical picture of dementia is the most important diagnostic feature.

Table 2. Clinic-anatomical correlates of memory disorders

Anatomic Site of Damage	Memory Finding	Other Neurological and Medical Findings
Frontal lobe	Lateralized deficits in working memory. Right spatial defects, left verbal defects, impaired recall with spared recognition	Personality change, Perseveration, Chorea, dystonia, Bradykinesia, tremor, rigidity
Basal forebrain	Domain-independent declarative memory deficits	
Ventromedial cortex	Frontal lobe-type declarative memory deficits	Upper visual field defects
Hippocampus and parahippocampal cortex	Bilateral lesions yield global amnesia, unilateral lesions show lateralization of deficits--left: verbal deficits; right: spatial deficits	Myoclonus, Depressed level of consciousness, Cortical blindness, Autonomations
Fornix	Global amnesia	
Mammillary bodies	Declarative memory deficits	Confabulation, ataxia, nystagmus, signs of alcohol withdrawal
Dorsal and medial dorsal nucleus thalamus	Declarative memory deficits	Confabulation
Anterior thalamus	Declarative memory deficits	
Lateral temporal cortex	Deficits in autobiographical memory	

The development of multiple cognitive deficits such as manifested by both, Impaired memory, long or short-term, can't learn new information or can't recall information previously learned and is distinguished by:
One (or more) of the following cognitive disturbances: Aphasia (language disturbance). Apraxia (impaired ability to carry out motor activities despite intact motor function). Agnosia (failure to recognize or identify objects despite intact sensory function). Disturbance in executive functioning (i.e., planning, organizing, sequencing, abstracting) The cognitive deficits above each cause significant impairment in social or occupational functioning and represent a significant decline from a previous level of functioning.
The decline in mental functioning begins gradually and worsens steadily
The cognitive deficits above are not due to any of the following: Other central nervous system conditions that cause progressive deficits in memory and cognition (e.g., cerebrovascular disease, Parkinson's disease, Huntington's disease, subdural hematoma, normal-pressure hydrocephalus, brain tumor). Systemic conditions that are known to cause dementia (e.g., hypothyroidism, vitamin B-12 or folic acid deficiency, niacin deficiency, hypercalcemia, neurosyphilis, HIV infection). Substance-induced conditions.
They aren't better explained by another Axis I disorder such as a Depressive Disorder or Schizophrenia

Table 3. NINCDS-ADRDA diagnostic criteria for AD (Mckhann et al., 1984)

I. Clinical Diagnosis of Probable Alzheimer's Disease
1. Dementia established by clinical examination and mental status testing and confirmed by neuropsychological testing
2. Deficits in at least two cognitive domains
3. Progressive cognitive decline, including memory
4. Normal level of consciousness
5. Onset between ages 40 and 90 (most common aiter 65) years
6. No other possible medical or neurological explanation

II. Probable Alzheimer's Disease Diagnosis Supported by
1. Progressive aphasia, apraxia, and agnosia
2. Impaired activities of daily living
3. Family history of similar disorder
4. Brain atrophy on CT/MRI, especially if progressive
5. Normal CSF, EEG (or nonspecifically abnormal)

III. Other Clinical Features Consistent with Probable Alzheimer's Disease
1. Plateau in course
2. Associated symptoms: depression; insomnia; incontinence; illusions; hallucinations, catastrophic verbal, emotional, or physical outbursts; sexual disorders; weight loss; during more advanced stages increased muscle tone, myoclonus, and abnormal gait
3. Seizures in advanced disease
4. CT normal for age

IV. Features that Make Alzheimer's Disease Uncertain or Unlikely
1. Acute onset
2. Focal sensorimotor signs
3. Seizures or gait disorder early in course

V. Clinical Diagnosis of Possible Alzheimer's Disease
1. Dementia with atypical onset or course in the absence of another medical/neuropsychiatric explanation
2. Dementia with another disease not felt otherwise to be the cause of dementia
3. For research purposes, a progressive focal cognitive deficit

VI. Definite Alzheimer's Disease
1. Meets clinical criteria for probable Alzheimer's disease
2. Tissue confirmation (autopsy or brain biopsy)

VII. Research Classification of Alzheimer's Disease Should Specify
1. Familial
2. Early onset (before age 65)
3. Down's syndrome (trisomy 21)
4. Coexistent other neurodegenerative disease (e.g., Parkinson's disease)

The oldest memories appear to be the best preserved, with proportionately greater forgetting as the retrograde interval shortens (Beatty et al., 1988). In contrast, procedural memory appears to be relatively spared. Alzheimer's patients are able to learn simple skills as easily as normal controls and better than patients with subcortical patterns of dementia or patients with various types of sensorimotor deficits (Heindel et al., 1989). Memory loss is a cardinal feature of Alzheimer's dementia but it is not the only abnormality in most patients. Aphasia, apraxia, and agnosia are the other categories of cognitive impairment that typically occur in cortical dementia syndromes and particularly in AD. In Alzheimer's dementia, these aspects do not dominate the clinical picture. Rather, these aspects of cognitive pathology occur in discrete apportionments that define the dementia syndrome as typical Alzheimer's dementia. In mild to moderate stages of dementia, anomia is prominent and readily detectable with neuropsychological testing using a variety of naming tests (Christopher and Goetz, ***). Patients are, however, fluent and may have relatively good comprehension, so clinical detection is not always easy.

Anosognosia, the failure to recognize illness, is a cardinal feature of Alzheimer's dementia and is typically present, even in mild stages of the disease. It is a useful sign to distinguish progressive amnesic syndromes from Alzheimer's dementia. A typical patient with AD is brought or sent for evaluation rather than coming of his or her own accord. He or she may deny significant memory problems and will actively try to explain away the observations of concerned family members and friends, even to the point of becoming hostile and accusative. This is one of the most difficult aspects of the disease because such patients often should not be driving or managing their own finances but will do so anyway, sometimes to their detriment.

In moderate to advanced stages, patients exhibit other so-called agnosic types of disturbances. Arguably, the difficulty such patients have in recognizing familiar people reflects prosopagnosia, save that they are not reliably benefited by voice recognition either. Difficulty finding their way could be compared with atopographagnosia, another visual agnosic syndrome, but again this is not occurring in the pure form given the wealth of other cognitive problems accompanying it. Rarely, patients with histologically documented AD present with a disabling, primarily visual syndrome termed asimultanagnosia, in which there is an inability to view all parts of a complex visual scene in a single coherent time-space frame. Such patients fail to see a target object that is right in front of them, especially if it is surrounded by potentially distracting stray objects (Hof, Bouras, Constantinidis, and Morrison, 1989). Strictly speaking, this is not an agnosic disorder, but another type of complex visual disturbance that resembles agnosia and can occur in this setting. It is rarely the presenting feature. More commonly, it is a mild accompanying feature.

Psychiatric symptoms include both affective and psychotic disturbances. Depression interferes with a person's functional status and an accurate cognitive assessment, erroneously leading to the diagnosis of dementia; such is the nature of pseudodementia, which has been estimated to account for 4 to 5 percent of dementia cases (Clarfield, 1988). More commonly, depression may complicate the course of AD and other dementing illnesses. Antidepressant therapy should be strongly considered in such patients. Psychotic symptoms most commonly involve paranoid delusions and, less commonly, hallucinations. AD should be in the differential diagnosis of an elderly patient presenting de novo with an organic delusional syndrome or hallucinosis. Common paranoid themes involve infidelity and stealing. Hallucinations are generally complex, involving people (familiar or unfamiliar), animals, or both. The combination of anosognosia, paranoid delusions, and mild cognitive decline is potentially dangerous and very difficult to manage, especially if such individuals live alone, yet a significant minority of Alzheimer's patients have exactly that combination.

Late and preterminal stage AD ensues after roughly a 5- to 10-year course but is highly variable and can be shorter or longer. In very advanced stages, patients are no longer ambulatory, become mute and incontinent, and are totally dependent on their caregiver or nursing attendant. Eventually dysphagia signals the terminal phase, and unless a feeding tube is placed, patients eventually die from inanition or aspiration, or both. Few patients live to these stages, with death resulting from intercurrent illnesses in most.

Mild Cognitive Impairment (MCI): The most studied form of MCI is that of isolated recent memory loss, or amnestic MCI.

Diagnostic criteria include-

- A subjective or objective impairment of recent memory
- Relatively preserved cognitive functions in other domains
- Presence of normally performed activities of daily living

These patients have increased risk of developing AD at the rates of 12-15% per year as compared to age matched controls who have rates of 1-2% per year (Winbald et al., 2004). The multiple cognitive domain form or non- amnestic MCI, although more prevalent, has a less clear data on its nature of pathology. Of MCI cases, 60% or more are already associated with sufficient neuropathological changes to qualify for the diagnosis of AD (Bennet et al., 2005; Davis et al., 1999; DeKosky et al., 2002; Tiraboschi, 2000).

4.1.3. Pathology

AD is characterized by generalized cerebral cortical atrophy with widespread cortical neuritic (or senile) plaques (NP) and neurofibrillary tangles (NFT). Other typical pathological findings include neuropil threads, granulo- vacuolar degeneration, lipochrome accumulation, and Hirano bodies. NPs are composed of an amyloid core surrounded by dystophic neurites, whereas paired helical filaments are the major constituent in NFTs, neuropil threads, and dystrophic neurites. None of these histological findings individually are entirely specific for AD, because NPs can be seen in clinically nondemented patients (Crystal et al., 1988) and NFTs can occur in other neurodegenerative and prion disorders. Although recently debated, current neuropathological criteria for the diagnosis of AD requires that an adequate number of NPs be present within a specified age range in a clinically demented patient (Mirra et al., 1991).

Synapse loss appears to be the most important correlate of dementia severity (Terry et al., 1991) but the number of NPs and NFTs and the density of beta-amyloid load have each been associated with dementia severity. Mesial temporal structures, particularly the hippocampal formation, are involved early in AD, and this accounts for the amnestic syndrome in these patients. AD produces a lamina-specific pattern of damage to the entorhinal cortex that disrupts cortical input to the hippocampal formation from association and limbic cortices, and disrupts hippocampal outflow from the cornu Ammonis (CA) sectors and subiculum to the association cortices, diencephalon, basal forebrain, and amygdala. Hence, AD effectively disconnects the hippocampus from its major input and output pathways (Hyman, van Hoesen, Damasio, and Barnes, 1984).

4.1.4. Genetics of AD

A family history of AD is a major risk factor. Familial AD (FAD) can be Autosomal Dominant (AD) which has early onset and late onset FAD which does not have AD pattern but increased frequency in families. Three genes are responsible for more than 90% of these early onset familial cases: the presenilin- 1 gene on chromosome 14, the presenilin-2 gene on chromosome 1, and the amyloid precursor protein gene on chromosome 21.

The importance of an individual's apolipoprotein E (ApoE) gene status on chromosome 19 has received significant attention as an important genetic susceptibility risk factor for the development of the more typical, or "sporadic" AD. ApoE, a protein involved in cholesterol transport and probably neuronal repair, alters risk for AD but does not in itself cause the disease. This gene has three possible alleles: e2, e3, and e4. The e3 allele is the most common and the e2 allele the least common in the general population. The e4 allele increases the risk of developing AD in a dose-dependent manner, and the e2 allele appears to decrease the risk (Yaari and Corey-Bloom, 2007).

4.1.5. Neuroimaging

Several neuroimaging markers have also been evaluated for differentiating AD from normal aging and other dementias. Hippocampal and entorhinal cortical atrophy on MRI has been investigated, but quantification of hippocampal volumes requires a high degree of neuroanatomic expertise and is time-consuming (DeCarli, 2001). A pattern of bilateral temporoparietal hypometabolism on PET has been shown to have high specificity for AD and may be useful in confirming its diagnosis (Hoffman et al., 2000). However, PET imaging may not be widely available for clinical use and is expensive. SPECT, although more widely available and less expensive, appears to have a lower sensitivity and specificity than PET imaging for AD (Bergman et al., 1997). Further studies with pathological validation are needed.

4.2. Frontotemporal Dementia (FTD)

4.2.1. Introduction

Arnold Pick originally introduced the concept of a lobar atrophy with relevant progressive focal symptoms in his description of a patient with progressive aphasic dementia, who at autopsy had, in addition to generalized cortical atrophy, focally accentuated atrophy of the left temporal lobe.

The specific histological findings were later appreciated by Alzheimer in 1911, who observed distinctive intraneuronal inclusions, which were termed "Pick bodies." Swollen, ballooned, and poorly staining (chromatolytic) neurons were subsequently called "Pick cells." Neary divided the clinical syndromes into "frontotemporal dementia," "progressive nonfluent aphasia," and "semantic dementia."

4.2.2. Pathology

Over the following decades, most investigators have included three pathological varieties of Pick's disease: lobar atrophy with swollen chromatolytic neurons (SCN) and Pick bodies (Pick's disease Type A), lobar atrophy with SCN but not Pick bodies (Pick's disease Type B), and lobar atrophy without SCN or Pick bodies (Pick's disease Type C, or nonspecific degeneration) (Tissot, Constantinidis, and Richard, 1985). Whether the findings in Pick's disease Types B and C are indeed variants of Pick's disease or different disease processes is debated. The cardinal pathological changes in Pick's disease Type A are lobar atrophy with corresponding neuronal loss that is most apparent in the first three cortical layers.

Table 4. NINDS-AIREN criteria for Vascular Dementia

Criteria for clinical diagnosis of probable vascular dementia include all the following: • Dementia is defined by cognitive decline from a previously higher level of functioning and manifested by impairment of memory and of two or more cognitive domains • Evidence of cerebrovascular disease by focal signs on neurological examination consistent with stroke • Evidence of relevant cerebrovascular disease by brain imaging including multiple large vessel infarcts or a single strategically placed infarct as well as multiple basal ganglia and white matter lacunae or extensive periventricular white matter lesions or combinations thereof
Clinical features consistent with the diagnosis of probable vascular dementia include the following: • Early presence of gait disturbance (small step gait or marche a petits pas, or magnetic gait) • History of unsteadiness and frequent, unprovoked falls • Early urinary frequency, urgency, and other urinary symptoms not explained by urologic disease • Pseudobulbar palsy • Personality and mood changes, abulia, depression, emotional incontinence, or other subcortical deficits, including psychomotor retardation and abnormal executive function

The maximally involved cortical gyri are very thin, and they are often described as walnut or knife-edge in appearance. When the temporal lobe atrophy is focally accentuated, the posterior one-third of the superior temporal gyrus appears relatively preserved. Marked caudate and hippocampal atrophy are also typical findings. Variable degrees of co-existing subcortical gliosis and degeneration of other subcortical nuclei can occur as well, although the basal forebrain tends to be spared. Pick bodies are argentophilic, eosinophilic, rounded cytoplasmic masses that appear densely black on silver stains. They are composed of 10- to 20-nm straight filaments that share antigenic properties with neurofilaments. Pick bodies tend to predominate in the most severely atrophic cortical regions and in the hippocampi. Pick cells also immunostain positively to neurofilaments, but they are not regarded diagnostic of Pick's disease because similar-appearing cells have been observed in several other disorders (Mizukami and Kosaka, 1989).

4.2.3. Clinical Features and FTD Syndromes

The criteria for the syndromes in the Neary classification (Neary et al., 1998) are based on the patient's symptoms but also turn out to be anatomically related. Nonfluent primary progressive aphasia is related to the left frontal lobe, semantic dementia to the left temporal lobe, and frontal dementia to the frontal lobes. As a subgroup of semantic dementia, the criteria recognize that when patients have the right temporal lobe affected out of proportion to other areas, they present with prosopagnosia. The criteria do not include the fluent aphasias that do not meet semantic dementia criteria (Graff-Radford and Woodruff, 2007). In 2001, McKhann and colleagues (McKhann et al., 2001) have proposed simpler guidelines that would facilitate recognition of the various FTD syndromes by general physicians. The six proposed criteria are:

- Early and progressive change in personality or language;
- Impairment in social or occupational functioning;
- A gradual and progressive course;
- Exclusion of other causes;
- Presence of deficits in the absence of delirium; and
- Exclusion of psychiatric causes such as depression.

Frontal Dementia (FTD)

The most striking presentation of FTD is that typified by behavioural changes. This is often manifested by lack of insight, impulsive or inappropriate behaviour, change in libido, mental inflexibility, and blunting of appropriate emotional responses. Family members often initially suspect mental illness, and these patients may initially present for psychiatric evaluation.

Primary Progressive Aphasia

Patients with this syndrome present with language difficulty with a combination of some or all of the following: word-finding difficulties, abnormal speech patterns, decreased comprehension, and impaired spelling. To be diagnosed as having primary progressive aphasia, a patient's language difficulty should remain the only feature for at least 2 years before more generalized deficits develop. Other features include dysarthria, ideomotor apraxia, dyscalculia, constructional deficits, and impaired executive function (Graff-Radford and woodruff, 2007).

Frontotemporal Dementia with Parkinsonism

The best known of the familial FTD syndromes is FTDP-17, a clinical syndrome linked to chromosome 17, with autosomal-dominant inheritance. About 200 families and 39 mutations have been reported. The major manifestations in affected individuals include behavioural changes (disinhibition, apathy, defective judgment, compulsive behaviour, hyper religiosity, psychosis, alcoholism, verbal and physical aggressiveness, hyperorality and hyperphagia, neglect of personal hygiene), dementia, and parkinsonism.

FTD with Motor Neuron Disease

The nosological distinction among frontotemporal lobe dementia (FTD), FTD with motor neuron disease (MND), and classic MND has been debated, with some arguing that FTD with MND is a distinct disorder and others suggesting each represents a phenotype along a continuum of the same disease (Strong et al., 2003). The motor findings in these patients may precede, coincide, or follow the development of cognitive and behavioural changes.

Corticobasal Degeneration

It is a clinical syndrome characterized by asymmetric rigidity, apraxia, and alien limb phenomena. Corticobasal degeneration is also a pathological diagnosis; in more than half the cases meeting the pathological criteria of CBD, the patients do not have asymmetric rigidity and apraxia (Josephs et al., 2006). A recent clinicopathologic study found that the majority of cases of progressive nonfluent aphasia had underlying CBD or PSP as the primary pathology (Hachinski et al., 1975).

4.3. Vascular Dementia

4.3.1. Introduction

Vascular dementia is a clinical syndrome of acquired intellectual and functional impairment resulting from the effects of cerebrovascular disease. Vascular dementia can be considered a term used to describe a particular constellation of cognitive and functional impairment and is seen as a subset of the larger syndrome of vascular cognitive impairment, which includes all forms of cognitive impairment associated with cerebrovascular disease including mixed dementias where both cerebrovascular and Alzheimer's disease(AD) processes may co-occur.

4.3.2. Clinical Features and VaD Syndromes

Hachinski et al. (1975) first introduced the term multi-infarct dementia and developed the Ischemia Scale that rated clinical features. Vascular dementia represents 4 to 22% of all dementias in the Western world, making it the second most common form of dementia after AD (Chui, 1989). Despite a sensitivity and specificity of 70 to 80% in separating vascular dementia from AD (Chui, 1989), the Ischemia Scale was less able to reliably diagnose the co-occurrence of the two conditions in the same individual and may actually have overdiagnosed vascular dementia in patients later found to have AD. The final criteria, proposed from a consensus conference sponsored by the National Institute of Neurological Disorders and Stroke and Association Internationale pour la Neurosciences (NINDS-AIREN) (Roman et al., 1993) (Table 4), has been used the most in clinical trials as the criteria for vascular dementia.

Multi Infarct Dementia

Multi-infarct dementia is attributable to multiple infarcts with occlusion, usually in large vessels that cause infarcts in the cortical and subcortical areas. The resulting clinical picture is characterized by variable impairment across several areas of intellectual function. Patients may demonstrate ''patchiness'' of deficits with sparing of some intellectual functions in the setting of severe impairments in others. With recurrent stroke, most areas of intellectual functions become disturbed and the dementia syndrome emerges. Importantly, patients with multi-infarct dementia often have lateralized sensorimotor changes, such as hemiparesis, hemisensory loss, visual field disturbance, and pathological reflex asymmetries that accompany cognitive deficits supporting evidence of a large vessel stroke Graff-Radford and Woodruff, 2007).

Strategic Infarct Dementia

In strategic single infarct dementia, ischemic damage is focal, often small, and involves functionally important cortical or subcortical sites. The thalamus, frontal white matter, basal ganglia, and/or angular gyrus may be involved (Konno et al., 1997).

Table 5. Classification of Parkinsonian Dementia Syndromes

Etiology/syndrome	Distinguishing features
Degenerative (Sporadic)	
Parkinson's disease with dementia (PDD)	Initial PD with later onset of dementia
Dementia with Lewy Body (DLB)	Recurrent visual hallucinations, fluctuating cognition with parkinsonism
Progressive Supranuclear Palsy (PSP)	Balance and bulbar dysfunction with downgaze palsy
Corticobasal Degeneration (CBD)	Asymmetric limb signs (apraxia, myoclonus),
Degenerative (Familial)	
Huntington's disease	AD, chorea, personality changes
Neuroacanthocytosis	AD, acanthocytic RBCs, HD mimic
Macahdo Joseph Disease	AD, ataxia, cerebellar atrophy
Familial frontotemporal dementia (FTDP 17)	AD, Chr 17 linked
Secondary Parkinsonism Syndromes	
Drug induced	Neuroleptic, antiemetics
Vascular parkinsonism	Multiple subcortical infarcts
Normal Pressure Hydrocephalus	Gait disturbance, incontinence, subcortical dementia
Whipple's disease	CSF pleocytosis withGI symptoms
Dementia pugilistica	Repetitive head trauma
Inherited metabolic syndromes	
Wilson's disease	AR, early onset, impaired copper clearance
Neuro degeneration with brain iron accumulation (NBIA)	Variable age onset with subcortical iron deposit

Table 6. Management of dementia

For Improving Symptoms	Behavioral management	For Slowing Progression	Reducing vascular risks	Non pharmacological
Cholinesterase inhibitors Donepezil Rivastigmine Galantamine Huperzine A Metrifonate NMDA recetor blocker Memantine	Psychotropic drugs for behavior Medications for sleep disturbances	Anti-oxidants (vitamin E) Anti-inflammatory drugs Hormones (estrogen) Alpha Lipoic Acid Ginkgo Biloba Lecithin Neuroprotective agents Piracetam	Statins, homocysteine reduction Hydergine	TENS Family support

Small Vessel Disease with Dementia

In small vessel disease with dementia, occlusion of small vessels most prominently causes lesions of subcortical structures such as the basal ganglia, thalamus, internal capsule, and subhemispheric white matter. This state is characterized by the insidious onset of symptoms, often without evident stroke. Common symptoms include psychomotor slowing, memory impairment, changes in speech, and behavioral alterations such as depression and apathy (Cummings, 1990). Two types of patterns may be present:

- Lacunar state (etat lacunaire) refers to a pattern of infarction in which multiple small lesions in the basal ganglia and internal capsule produce dementia. The deficits gradually accumulate to produce a state of dementia combined with multifocal motor, reflex and sensory disturbances, rigidity, spasticity, pseudobulbar palsy, limb weakness, exaggerated muscle stretch reflexes, and extensor plantar responses.
- Ischemic injury of the frontal hemispheric white matter produces a chronic progressive state called subcortical arteriosclerotic leukoencephalopathy or Binswanger's disease (Caplan, 1995). Like lacunar infarction, the signs and symptoms of Binswanger's disease reflect pathology localized to subcortical structures, and as a result of the multiple and cumulative occlusion of deep penetrating arterioles supplying the white matter, a more severe pattern of subcortical dementia can emerge, with abulia, incontinence, and limb rigidity.

4.3.3. Neuroimaging and Pathology

One of the most interesting and challenging aspects of imaging in vascular dementia is how to interpret white matter hyperintensities. These hyperintense lesions on T2-weighted images are seen on many routine MRI scans. The term leukoariaosis, generally taken to mean lucency of white matter as imaged by CT, has also been used interchangeably with the term white matter hyperintensities (Pantoni and Garcia, 1995). Pathologically, white matter hyperintensities most often represent areas of demyelination and enlargement of perivascular spaces.

The more extensive the white matter hyperintensities, the more likely the patient was to have cerebrovascular risk factors and to experience cognitive decline (Longstreth et al., 2000). Other studies have investigated the relationship of white matter hyperintensities in individuals with varying ranges of cognitive ability and suggested that white matter hyperintensities were associated with cognitive impairment, including memory loss (DeGroot et al., 2000; Wu et al., 2002), and that white matter hyperintensities as seen on MRI in nondemented individuals were associated with cardiovascular risk factors and lower scores on cognitive tests of psychomotor speed but not on tests of memory (Au, 2006).

4.4. Parkinsonian Dementias

4.4.1. Introduction

There are several forms of primary degenerative parkinsonism, including idiopathic (or sporadic) PD, sporadic PD with superimposed pathological features of AD, familial PD (or parkinsonian syndromes), and Parkinson's-ALS- dementia complex of Guam. Pathological substrates differ among these conditions, as do the frequency of cognitive disturbances. Although PSP was the original model for subcortical dementia, this model is applicable to PD, which is much more common than PSP. The concept was intended to distinguish subcortical dementia from the cortical dementia of AD.

4.4.2. Types and Clinical Features

Parkinsonian dementia syndromes encompasses a broad range of etiological disorders that share a heterogeneous combination of cognitive and motor features (Table 5).

5. TREATMENT

The treatment of dementia encompasses definitive and symptomatic treatment and management of risk factors (Table 6).

In MCI (MMSE > 23): There is no evidence to support the use of donepezil for patients with MCI. The putative benefits are minor, short lived and associated with significant side effects (Birks and Flicker, 2006).

In Alzheimer's disease: The Cochrane review published in 2006 studied 15 trials and found statistically significant improvement for both 5 and 10 mg/day of donepezil at 24 and 52 weeks compared with placebo on the ADAS-Cog scale. However the debate on whether donepezil is effective continues despite the evidence of efficacy from the clinical studies because the treatment effects are small and are not always apparent in practice (Birks and Harvey, 2006). There is evidence that Rivastigmine is beneficial for people with Alzheimer's disease, in being associated with improvements in cognitive function and activities of daily living (Birks, Grimley, Iakovidou, and Tsolaki, 2000).

In Parkinson's disease with dementia: Rivastigmine appears to improve cognition and activities of daily living in patients with PDD. This results in clinically meaningful benefit in about 15% of cases (Maidment, Fox, and Boustani, 2006).

In Vascular Cognitive decline: Deficient cholinergic neurotransmission, a characteristic of Alzheimer's disease, has been postulated to contribute to the cognitive impairment of vascular disease of the brain. Evidence from the available studies support the benefit of donepezil in improving cognition function, clinical global impression and activities of daily living in patients with probable or possible mild to moderate vascular cognitive impairment after 6 months treatment (Malouf and Birks, 2004). From existing trial data there is some evidence of benefit of Rivastigmine in vascular cognitive impairment. However, this conclusion is based on studies which had small numbers of patients, which sought to compare Rivastigmine to treatments other than placebo or which used data extrapolated post hoc from large studies involving patients with Alzheimer's disease and vascular risk factors of unclear significance (Craig and Birks, 2004).

Estimated cases of dementia in various states of India (based on census 2001)

State/UT	Total population	% ≥60 yrs	Estimated people with dementia
Jammu and Kashmir	10143700	6.7	12913
Himachal Pradesh	6077900	6	10393
Punjab	24358999	9	41654
Ghandigarh	900635	5	856
Uttaranchal	8485349	7.7	12414
Haryana	21144564	7.5	30131

(Continued)

State/UT	Total population	% $\geq$60 yrs	Estimated people with dementia
Delhi	13850507	5.2	13684
Rajasthan	56507188	6.7	71934
Uttar Pradesh	166197921	7	221043
Bihar	82998509	6.6	104080
Sikkim	540851	5.4	555
Arunachal Pradesh	1097968	4.5	939
Nagaland	1990036	4.5	1701
Manipur	5166788	6.7	2758
Mizoram	888573	5.5	929
Tripura	3199203	7.3	4437
Meghalaya	2318822	4.6	2027
Assam	26655528	5.9	29881
West Bengal	80176197	7.1	108158
Jharkhand	26945829	5.9	30206
Orissa	36804660	8.3	58041
Chhattisgarh	20833803	7.2	28501
Madhya Pradesh	60348023	7.1	81409
Gujarat	50671017	6.9	66430
Daman and Diu	158204	5.1	153
Dadra Nagar Haveli	22490	4	17
Maharashtra	96878627	8.7	160140
Andhra Pradesh	76210007	7.6	110047
Karnataka	52850562	7.7	77320
Goa	1347668	8.3	2125
Lakshadweep	60650	6.1	70
Kerala	31841374	10.5	63523
Tamil Nadu	62405679	8.8	104342
Pondicherry	974345	8.3	1536
Andaman and Nicobar Islands	356152	4.9	332
Total	1454679		

Memantine has a small beneficial effect at six months in moderate to severe AD. In patients with mild to moderate dementia, the small beneficial effect on cognition was not clinically detectable in those with vascular dementia and was detectable in those with AD. Memantine is well tolerated (McShane, Areosa, and Minakaran, 2006).

The following table gives the present treatment recommendation for various types of dementias (Table 7).

Table 7. Treatment of dementia

Dementia type	1ˢᵗ line treatment	Alternative treatments
Alzheimer's disease	Donepezil, Rivastigmine, Galantamine, and Memantine	Symptomatic treatment for behavior
MCI	Donepezil, Rivastigmine	
Vascular Dementia	Cholinesterase inhibitors and Memantine produce small benefit	
Parkinson's Disease with Dementia	Donepezil, Rivastigmine	Levodopa, Symptomatic treatment for behavior
Dementia with Lewy Body	Donepezil, Rivastigmine	Levodopa, Symptomatic treatment for behavior
Fronto temporal dementia		Symptomatic treatment for behavior

6. DEMENTIA IN INDIA

Well-designed epidemiological research is relatively lacking in low and middle income countries where two-thirds of the world's estimated 24 million people with dementia live. The 10/66 Dementia Research Group (10/66) was founded ten years and refers to a population-based research into dementia that had been directed towards the two-thirds or more of people with dementia living in developing countries (10/66 Dementia research group, 2000). It is estimated that there are, in total, nearly 1.5 million people living with dementia in India today. These numbers are expected to increase dramatically in the years ahead due to the demographic transition (Dias and Patel, 2009).

There are no accurate estimates for the treatment gap for dementia in India, but it is estimated that this gap exceeds 90% in most parts of the country, with the exception of urban areas and the two southern states of Kerala and Tamil Nadu.

CONCLUSION

Dementia constitutes a common disease with significant morbidity and dependence in population more than 65 years of age and although the degenerative dementias have presently no 'cure' there are promising treatments. It is important that a good subset of dementias is completely reversible and treatable and should also be thoroughly looked into as an associated cause of deterioration in primary degenerative dementias.

REFERENCES

Alexander, E.M., Wagner, E.H., Buchner, D.M., Cain, K.C., and Larson, E. B. (1995). Do surgical brain lesions present as isolated dementia? A population-based study. *Journal of American Geriatrics Society, 43*, 138–143.

Au, R., Massaro, J.M., Wolf, P.A., Young, M.E., Beiser, A., Seshadri, S., D'Agostino, R.B., and DeCarli, C. (2006). Association of white matter hyperintensity volume is associated with decreased cognitive functioning: the Framingham Heart Study. *Archives of Neurology, 63,* 246–250.

Beatty, W.W., Salmon, D.P., Butters. N., Heindel, W.C., and Granholm, E.L. (1988). Retrograde amnesia in patients with Alzheimer's or Huntington's disease. *Neurobiology of Aging, 9,* 181-186.

Bennett, D.A., Schneider, J.A., Bienias, J.L., Evans, D.A., and Wilson, R.S. (2005). Mild cognitive impairment is related to Alzheimer disease pathology and cerebral infarctions. *Neurology, 64,* 834-841.

Bergman, H., Chertkow, H., Wolfson, C., Stern, J., Rush, C., Whitehead, V., and Dixon, R. (1997). HM-PAO (CERETEC) SPECT brain scanning in the diagnosis of Alzheimer's disease. *Journal of American Geriatrics Society, 45,* 15–20.

Birks, J., and Flicker, L. (2006). Donepezil for mild cognitive impairment. Cochrane Database of Systematic Reviews 2006, Issue 3. Art.No.CD006104. DOI: 0.1002/14651858.CD006104.

Birks, J., and Harvey, R.J. (2006). Donepezil for dementia due to Alzheimer's disease. Cochrane Database of Systematic Reviews 2006, Issue 1. Art. No.: CD001190. DOI: 0.1002/14651858.CD001190.pub2.

Birks, J., Grimley Evans, J., Iakovidou, V., and Tsolaki, M. (2000). Rivastigmine for Alzheimer's disease. Cochrane Database of Systematic Reviews 2000, Issue 4. Art. No.: CD001191. DOI: 10.1002/14651858.CD001191.

Boeve, B.F., Maraganore, D.M., Parisi, J.E., Ahlskog, J.E., Graff-Radford, N., Caselli, R.J., Dickson, D.W., Kokmen, E., and Petersen, R.C. (1999). Pathologic heterogeneity in clinically diagnosed corticobasal degeneration. *Neurology, 53,* 795–800.

Caplan, L.R. (1995). Binswanger's disease—revisited. *Neurology, 45,* 626–633.

Chui, H.C. (1989). Dementia: a review emphasizing clinicopathologic correlation and brain-behavior relationships. *Archives of Neurology, 46,* 806–814.

Chui, H.C., Victoroff, J.I., Margolin, D., Jagust, W., Shankle, R., and Katzman, R. (1992). Criteria for the diagnosis of ischemic vascular dementia proposed by the State of California Alzheimer's Disease Diagnostic and Treatment Centers. *Neurology, 42,* 473–480.

Clarfield, A.M. (1988). The reversible dementias: Do they reverse? *Annals of Internal Medicine, 109,* 476-486.

Craig, D., and Birks, J. (2004). Rivastigmine for vascular cognitive impairment. Cochrane Database of Systematic Reviews 2004, Issue 2. Art. No.: CD004744. DOI: 10.1002/14651858.CD004744.pub2.

Crystal, H., Dickson, D., Fuld, P., Masur, D., Scott, R., Mehler, M., Masdeu, J., Kawas, C., Aronson, M., and Wolfson, L. (1988). Clinico-pathologic studies in dementia: Nondemented subjects with pathologically confirmed Alzheimer's disease. *Neurology, 38,* 1682-1687.

Cummings, J.L. (1990). Subcortical Dementia. 3rd ed. New York: Oxford University Press.

Diagnostic and Statistical Manual of Mental Disorders: DSM IV, 4th edition, 1994, American Psychiatric Association, Washington DC.

Davis, K.L., Mohs, R.C., Marin, D., Purohit, D.P., Perl, D.P., Lantz, M., Austin, G., and Haroutunian, V. (1999). Cholinergic markers in elderly patients with early signs of Alzheimer disease. *Journal of the American Medical Association, 281*, 1401-1406.

DeCarli, C. (2001). The role of neuroimaging in dementia. *Clinics in Geriatrics Medicine, 17*, 255–279.

DeKosky, S,T., Ikonomovic, M.D., Styren, S.D., Beckett, L., Wisniewski, S., Bennett, D.A., Cochran, E.J., Kordower, J.H., and Mufson, E.J. (2002). Upregulation of choline acetyltransferase activity in hippocampus and frontal cortex of elderly subjects with mild cognitive impairment. *Annals of Neurology, 51*, 145-155.

De Groot, J.C., de Leeuw, F.E., Oudkerk, M., van Gijn, J., Hofman, A., Jolles, J., and Breteler, M.M. (2000). Cerebral white matter lesions and cognitive function: the Rotterdam Scan Study. *Annals of Neurology, 47*, 145–151.

Dias, A., and Patel, V. (2009). Closing the treatment gap for dementia in India. Indian Journal of Psychiatry, 51, 93-97.

Folstein, M.F., Folstein, S.E. and McHugh, P.R. (1975). Mini-Mental State: A practical method for grading the cognitive state of patients for the clinician. *Journal of Psychiatric Research, 12*, 189–198.

Goetz, C. G. (1999). *Textbook of Clinical Neurology*. W.B. Saunders company. A Division of Harcourt Brace and Company.

Graff-Radford, N. R., and Woodruff, B.K. (2007). Frontotemporal Dementia. *Seminars in Neurology, 27*, 48–57.

Hachinski, V.C., Iliff, L.D., Zilhka, E., Du Boulay, G.H., McAllister, V.L., Marshall, J., Russell, R.W., and Symon, L. (1975). Cerebral blood flow in dementia. *Archives of Neurology, 32*, 632–637.

Hebert, L.E., Scherr, P.A., Bienias, J.L., Bennett, D.A., and Evans, D.A. (2003). Alzheimer disease in the US population: prevalence estimates using the 2000 census. *Archives of Neurology, 60*, 1119–1122.

Heindel, W.C., Salmon, D.P., Shults, C.W., Walicke, P.A., and Butters, N. (1989). Neuropsychological evidence for multiple memory systems: A comparison of Alzheimer's, Huntington's, and Parkinson's disease patients. *Journal of Neuroscience, 9*, 582-587.

Hof, P.R., Bouras, C., Constantinidis, J., and Morrison, J.H. (1989). Balint's syndrome in Alzheimer's disease: Specific disruption of the occipito-parietal visual pathway. *Brain Research, 493*, 368-375.

Hejl, A., and Hogh, P. (2002). Waldemar G. Potentially reversible conditions in 1000 consecutive memory clinic patients. *Journal of Neurology, Neurosurgery and Psychiatry, 73*, 390-394.

Hoffman, J.M., Welsh-Bohmer, K.A., Hanson, M., Crain, B., Hulette, C., Earl, N., and Coleman RE. (2000). FDG PET imaging in patients with pathologically verified dementia. *Journal of Nuclear Medicine, 41*, 1920–1928.

Hyman, B.T., Van Hoesen, G.W., Damasio, A.R., and Barnes, C.L. (1984). Alzheimer's disease: Cell-specific pathology isolates the hippocampal formation. *Science, 225*, 1168-1170.

Josephs, K.A., Petersen, R.C., Knopman, D.S., Boeve, B.F., Whitwell, J.L., Duffy, J.R., Parisi, J.E., and Dickson, D.W. (2006). Clinicopathologic analysis of frontotemporal and corticobasal degenerations and PSP. *Neurology, 66*, 41–48.

Jorm, A.F. (1991). Cross-national comparisons of the occurrence of Alzheimer's and vascular dementias. *European Archives of Psychiatry and Clinical Neuroscience, 240*, 218-222.

Knopman, D. S., DeKosky, S. T., Cummings, J. L., Chui, H., Corey–Bloom, J., Relkin, N., Small, G. W., Miller, B. and Stevens, J.C. (2001). Practice parameter: Diagnosis of dementia. (an evidence-based review). *Report of the Quality Standards Subcommittee of the American Academy of Neurology, 56*, 1143-1153.

Knopman, D. S., DeKosky, S. T., Cummings, J. L., Chui, H., Corey–Bloom, J., Relkin, N., Small, G. W., Miller, B. and Stevens, J. C. (2001). Practice parameter: Diagnosis of dementia (an evidence-based review): Report of the Quality Standards Subcommittee of the American Academy of Neurology. *Neurology, 56*, 1143-1153.

Konno, S., Meyer, J.S., Terayama, Y., Margishvili, G.M., and Mortel, K.F. (1997). Classification, diagnosis and treatment of vascular dementia. *Drugs Aging, 11*, 361–373.

Longstreth, W.T. Jr., Arnold, A.M., Manolio, T.A., Burke G.L., Bryan, N., Jungreis, C.A., Enright, P.L., O'Leary, D., and Fried, L. (2000). Clinical correlates of white matter findings on cranial magnetic resonance imaging of 3301 elderly people: The Cardiovascular Health Study. *Neuroepidemiology, 19*, 30–34.

Maidment, I., Fox, C., and Boustani, M. (2006). Cholinesterase inhibitors for Parkinson's disease dementia. Cochrane Database of Systematic Reviews 2006, Issue 1. Art. No.: CD004747. DOI: 00.1002/14651858.CD004747.pub2.

Malouf, R., and Birks, J. (2004). Donepezil for vascular cognitive impairment. Cochrane Database of Systematic Reviews 2004, Issue 1. Art. No.: CD004395. DOI: 10.1002/14651858.CD004395.pub2.

Martin, D.C., Miller, J., Kapoor, W. Karpf, M., and Boller, F. (1987). Clinical prediction rules for computed tomographic scanning in senile dementia. *Archives of Internal Medicine, 147*, 77–80.

Maurer, K., Ihl, R., Dierks, T., and Frolich, L. (1997). Clinical efficacy of Ginkgo biloba special extract EGb 761 in dementia of Alzheimer's type. *Journal of Psychiatric Research, 31*, 645-655.

Mayeux, R., and Sano, M. (1999). Treatment of Alzheimer's disease. *New England Journal of Medicine, 341*, 1670-1679.

McKhann, G.M., Albert, M.S., Grossman, M., Miller, B., Dickson, D., Trojanowski, J.Q. (2001). Clinical and pathological diagnosis of frontotemporal dementia: report of the Work Group on Frontotemporal Dementia and Pick's Disease. *Archives of Neurology, 58*, 1803–1809.

Mckhann, G., Drachman, D., Folstein, M., Katzman, R., Price, D., and Stadlan, E.M. (1984). Clinical diagnosis of Alzheimer's disease: Report of the NINCDS-ADRDA Work Group under the auspices of Department of Health and Human Services Task Force of Alzheimer's Disease. *Neurology, 34*, 939-944.

McShane, R., Areosa Sastre A., and Minakaran, N. (2006). Memantine for dementia. *Cochrane Database of Systematic Reviews 2006, Issue 2.* Art. No.: CD003154. DOI: 10.1002/14651858.CD003154.pub5.

Mioshi, E., Dawson, K., Mitchell, J., Arnold, R., and Hodges, J.R. (2006). The Addenbrooke's Cognitive Examination Revised (ACE-R): a brief cognitive test battery for dementia screening. *International Journal of Geriatrics and Psychiatry, 21*, 1078-1085.

Mirra, S., Heyman, A., McKeel, D., Sumi, S.M., Crain, B.J., Brownlee, L.M., Vogel, F.S., Hughes, J.P., van Belle, G., and Berg, L. (1991). The consortium to establish a registry for Alzheimer's disease (CERAD). Part II. Standardization of the neuropathological assessment of Alzheimer's disease. *Neurology, 41*, 470-486.

Mizukami, K., and Kosaka, K. (1989). Neuropathological study on the nucleus basalis of Meynert in Pick's disease. *Acta Neuropathologica, 78*, 52-56.

Neary, D., Snowden, J.S., Gustafson, L., Passant, U., Stuss, D., Black, S., Freedman, M., Kertesz, A., Robert, P.H., Albert, M., Boone, K., Miller, B.L., Cummings, J., and Benson, D.F. (1998). Frontotemporal lobar degeneration: a consensus on clinical diagnostic criteria. *Neurology, 51*, 1546–1554.

Pantoni, L., and Garcia, J. (1995). The significance of cerebral white matter abnormalities 100 years after Binswanger's report. *Stroke, 26*, 1293–1301.

Roman, G.C., Tatemichi, T.K., Erkinjuntti, T., et al. (1993). Vascular dementia: diagnostic criteria for research criteria for research studies. Report of the NINCDS-AIREN International Work Group. *Neurology, 43*, 250–260.

Roy, Y., and Jody, C-. (2007). Alzheimer's Disease. *Semin Neurol, 27*, 32–41.

Strong, M.J., Lomen-Hoerth, C., Caselli, R.J., Bigio, E., Yang, W. (2003). Cognitive impairment, frontotemporal dementia, and the motor neuron diseases. *Annals of Neurology, 54 (suppl 5)*, S20–S23.

Terry, R., Masliah, E., Salmon, D., Butters, N., DeTeresa, R., Hill, R., Hansen, L.A., and Katzman, R. (1991). Physical basis of cognitive alterations in Alzheimer's disease: Synapse loss is the major correlate of cognitive impairment. *Annals of Neurology, 30*, 572-580.

The 10/66 Dementia Research Group. (2000). Dementia in developing countries. A preliminary consensus statement from the 10/66 Dementia research group. *International Journal of Geriatric Psychiatry, 15*, 14-20.

Tiraboschi, P., Hansen, L.A., Alford, M., Masliah, E., Thal, .LJ., and Corey-Bloom. J. (2000). The decline in synapses and cholinergic activity is asynchronous in Alzheimer's disease. *Neurology, 55*, 1278-1283.

Tissot, R., Constantinidis, J., and Richard, J. (1985). Pick's disease. In P. Vinken, G. Bruyn, and H. Klawans (Eds.) *Handbook of Clinical Neurology: Volume 46* (pp. 233-246), Amsterdam: Elsevier.

Tripathi, M., and Vibha, D. (2008). Evidence based treatment of dementia. Reviews in Neurology, CME Program of IAN, Page 1-29.

Vanneste, J., Augustijn, P., Dirven, C., Tan, W.F., and Goedhart, Z.D. (1992). Shunting normal-pressure hydrocephalus: do the benefits outweigh the risks? A multicenter study and literature review. *Neurology, 42*, 54–59.

Winbald, B., Palmer, K., Kivipelto, M. et al. (2004). Mild cognitive impairment- beyond controversies, towards a consensus: Report of International Working Group on Mild Cognitive Impairment. *Journal of Internal Medicine, 256*, 240-246.

Wu, C.C., Mungas, D., Petkov, C.I., Eberling, J.L., Zrelak, P.A., Buonocore, M.H., Brunberg, J.A., Haan, M.N., and Jagust, W.J. (2002). Brain structure cognition in a community sample of elderly Latinos. *Neurology, 59*, 383–391.

In: Expanding Horizions of the Mind Science(s)   ISBN: 978-1-62808-705-5
Editors: P.N. Tandon, R.C. Tripathi and N. Srinivasan   ©2013 Nova Science Publishers, Inc.

Chapter 20

AUTISM: UNDERSTANDING THE ENIGMA

Komilla Thapa

Department of Psychology, University of Allahabad, Allahabad, India

ABSTRACT

This chapter deals with the puzzling developmental disorder of autism. In recent years, many new perspectives have emerged which have challenged older notions and myths about autism. Starting with a definition of autism and autism spectrum disorders, the demographic characteristics of the disorders are discussed. A brief historical background is presented followed by an explication of the core characteristics (social and communicative impairments and repetitive behaviours and interests) of autism. Associated deficits are also discussed. The various explanations and theories of autism including both biological and psychological accounts are briefly discussed.

LIST OF ABBREVIATIONS

AD	Autistic Disorder
ASDs	Autism Spectrum Disorders
CNS	Central Nervous System
EF	Executive Functions
IQ	Intelligence Quotient
PDDs	Pervasive Developmental Disorders
SAS	Supervisory Attentional System
ToM	Theory of Mind

1. INTRODUCTION

She was at pains to keep her own life simple, she said, and to make everything very clear and explicit. She had built up a vast library of experiences over the years, she went on. They were like a library of videotapes, which she could play in her mind and inspect at any time – 'videos' of how people behaved in different circumstances. She would play these over and over again and learn, by degrees, to correlate what she saw, so that she could then predict how people in similar circumstances might act.

   She said that she could understand 'simple, strong, universal' emotions but was stumped by more complex emotions and the games people play. 'Much of the time,' she said,' I feel like an anthropologist on Mars.'

(Oliver Sacks, An Anthropologist on Mars: Seven Paradoxical Tales, 1995, p. 248).

The person referred to in the above excerpt from Sacks' eponymously titled book, is none other than Temple Grandin, one of the most remarkable of all autistic people: she holds a Ph.D. in animal science, teaches at Colorado State University and runs her own business. Narratives like those of Temple Grandin are important in the discourse on autism. They tell us that autism is not the unmitigated disaster it was previously thought to be. There is hope and we learn from an "understanding of other minds" (Baron-Cohen, Tager-Flusberg and Cohen, 2000).

In this chapter, efforts will be made to present more recent as well as varied perspectives on understanding and explaining the puzzling spectrum of autism. It is true that children and their families still face significant problems when coping with the devastating effects of childhood autism. Many children suffering from autism especially in countries like India, receive little or no empirically supported treatment and the average age of diagnosis hovers around five years of age (or even later), though some experts believe that early intervention should begin much sooner (Strock, 2004).

Unfortunately, clinical research and practice in the area of autism has been plagued by many myths, stereotypes and misconceptions, many of which have now been debunked. One prominent example concerns the "refrigerator" mother, who because of her chilly disposition, was not responsive to the child's emotional needs and thus was involved in the pathogenesis of autism. Thus Bettelheim (1967) stated "the precipitating factor in infantile autism is the parent's wish that his child should not exist" (p.125). This misunderstanding continued for decades until specific neurological and biological imbalances were posited as a cause of autism (Edelson and Rimland, 2003). A generation of parents was traumatized by the experience of being blamed for their child's condition.

Autism continues to remain the most captivating and mysterious of all childhood disorders. The disorder affects every aspect of the child's interaction with his or her world, involves many parts of the brain and undermines the very traits that make us human – our social responsiveness, ability to communicate, and feelings for other people. Autism is a spectrum disorder, which means that its symptoms and characteristics are expressed in many different combinations and degrees of severity. Not only do children with autism vary widely in their cognitive, language and social abilities, they also display many features not specific to autism, most commonly, mental retardation and epilepsy. Thus two children with a diagnosis of autism can be vastly different from each other (Mash and Wolfe, 1999).

This chapter will focus on the definition, its prevalence, and demographic characteristics. A brief history of the disorder is followed by an outlining the core and associated characteristics and impairments found in individuals with autism. The final section will include an overview of the multiple perspectives which have been used to explain and make sense of this disorder.

2. AUTISTIC DISORDER

2.1. Definition

Autistic Disorder (AD), also known as *childhood autism, infantile autism*, and *early infantile autism* is by far the best known of the pervasive developmental disorders (PDDs). These include AD, Asperger's disorder, childhood disintegrative disorder, Rett's disorder, and pervasive developmental disorder not otherwise specified/atypical autism (DSM-IV-TR). With the exception of Rett's disorder, and perhaps childhood disintegrative disorder, the highly overlapping symptoms of these conditions suggest that they may share common biological foundations. They are frequently referred to as autism spectrum disorders (ASDs), implying that the differences between disorders are better conceptualized along one or more continuous dimensions, as opposed to distinct categories with sharp discontinuities (Schultz and Anderson, 2004).

AD is a neurodevelopmental condition defined by persistent, severe social dysfunction and early communication failure. It is also characterised by idiosyncratic preoccupations and restricted interests, repetitive behaviours and motor stereotypies, unusual sensory sensitivities and resistance to change (DSM-IV-TR). Abnormalities in functioning in each of these areas must be present by age 3. Approximately 70 percent of individuals with autism function at the mentally retarded level, and mental retardation is the most common comorbid diagnosis (Volkmar, Klin and Schultz, 2005).

2.2. Prevalence

Although AD was once thought to be a rare disorder, recent studies place its prevalence rate in the range of 1.5 to 2.0 per 1000; for the more broadly defined ASDs, the prevalence is about 5 per 1000 (Chakrabarti and Fombonne, 2001). In more recent epidemiological studies, the median prevalence rate is 8.7 in 10,000. However, rates tend to be higher in studies with smaller samples and also tend to be higher more recently. If one excludes smaller and older studies, the median rate is slightly higher – approximately 9.5 in 10,000 (Volkmar, Klin and Schultz, 2005).

2.3. Sex Ratio

Studies based on both clinical and epidemiological samples have suggested higher incidence of autism in boys than in girls, with ratios averaging around 3.5 or 4.0 to 1 (Wing and Gould, 1979). This ratio varies, however, as a function of intellectual functioning. Some studies have reported ratios of up to 6.0 or higher to 1 in individuals with autism without mental retardation, whereas ratios within the moderately to severely retarded range have been reported to be as low as 1.5 to 1. It is still not clear why females are underrepresented in the nonretarded range. One possibility is that males have a lower threshold for brain dysfunction than females, or conversely, that more severe brain damage would be required to cause autism in a girl. According to this hypothesis, when the person with autism is a girl, she is more likely to be severely cognitively impaired.

2.4. Social Class

Although a few early studies supported Kanner's impression of an association between autism and upper socio-economic status, most epidemiological studies published in the 1980s and 1990s have failed to reveal such an association. In addition to the bias for more educated and successful parents to seek referral, it seems likely that families from disadvantaged backgrounds are still underrepresented. Autism clearly is seen in all social classes and in all countries.

3. HISTORY

While autism was described almost simultaneously by Leo Kanner and Hans Asperger in the 1940s, Kanner seemed to see it as a more formidable disorder, whereas Asperger felt that it might have certain positive or compensating features – a "particular originality of thought and experience, which may well lead to exceptional achievements in later life" (Asperger, 1944). It is clear even in these first accounts that there is a wide range of phenomena and symptoms in autism – and many more can be added to those that Kanner and Asperger listed. In 1943, Kanner first described 11 cases of what he termed *autistic disturbances of affective contact*. In these cases, there was a congenital "inability to relate" to people in usual ways. Kanner also noted unusual responses to the environment, which could include both stereotyped motor mannerisms and resistance to change and insistence on sameness, as well as unusual aspects of the child's communication skills, such as pronoun reversal and tendency to echo language. In addition, these children may have 'soft' neurological signs – a whole range of non-functional, repetitive or automatic movements (often referred to as "stim" behaviours, Hillman, Snyder and Neubrander, 2007), such as spasms, tics, rocking, spinning, finger play, or flapping of the hands; problems of coordination and balance; peculiar difficulties, sometimes, in initiating movements, akin to what is seen in Parkinsonism. There may also be, very prominently, a large range of abnormal (and often 'paradoxical') sensory responses, with some sensations being heightened and even intolerable, others (which may include pain perception) being diminished or apparently absent. There may be, if language develops, odd and complex language disorders – a tendency to verbosity, empty chatter, cliché-ridden and formulaic speech, indicating a semantic-pragmatic deficit. In contrast, Asperger-type children are often of normal and sometimes superior intelligence and generally have fewer neurological problems.

Kanner also spoke about marked scatter in skills with occasional "splinter skills". However, studies suggest that fewer than 5% of all individuals have such extreme splinter or savant skills, as evidenced in the stereotype of Dustin Hoffman's character in the film *Rainman* (Obler and Fein, 1988).

Kanner was careful to provide a developmental context for his observations. He emphasized the centrality of deficits in social relatedness, as well as unusual behaviours in the definition of the condition. During the 1960s, there was much confusion about the nature of autism and its etiology. In the early 1960s, a growing body of evidence began to accumulate to suggest that the condition resulted from a neuropathological process. Difficulties in consensual definitions and confusion regarding the similarities and differences between autism and childhood schizophrenia were complications. By the 1970s, a considerable body of evidence began to accumulate suggesting the neurobiological basis of the disorder. This included high rates of seizures as children were followed over time, the persistence of unusual "primitive" reflexes, and other neurological signs. A landmark in classification occurred in 1978 when Michael Rutter proposed a definition of autism based on the criteria of social delay and deviance, communication problems, unusual behaviours (insistence on sameness) with an onset before age 30 months. This definition and the growing body of work were influential, in that DSM-III recognized a new class of disorders- the PDDs.

The definition of autism in DSM-III was based largely on Rutter's synthesis of Kanner's original description and subsequent research. In DSM-III-R, the concern about the lack of developmental emphasis was addressed and the age of onset was specified as before or after 36 months. In DSM-IV-TR, autism is defined on the basis of behavioural features and age of onset. Behavioural difficulties must include some feature of social disturbance, communicative disturbance, and restricted interests or repetitive behaviours. The definitions of autism in both ICD-10 and DSM-IV-TR are conceptually identical.

4. CORE CHARACTERISTICS OF AUTISM

Considerable debate still exists about the core features of autism, despite more than 50 years of research by professionals from many disciplines. As autism can affect children in vastly different ways, it is a difficult problem to study. Findings from studies of small samples of children who differ widely in their levels of language and intellect are hard to interpret. Contributing to the difficulty is the interrelatedness during the first few years of life of the child's social, language, cognitive and emotional development; a disturbance in one area is likely to affect all other areas as well (Klinger and Dawson, 1996). These include social and communication impairments and repetitive behaviours and interests.

4.1. Social Impairments

Children with autism experience profound difficulties in relating to other people. From a very young age, they show deficits in the skills that are crucial for early social development – imitating others, orienting to social stimuli, sharing a focus of attention with others, understanding other people's emotions, and engaging in make-believe play. As they grow older, children with autism initiate few social behaviours and seem unresponsive to other people's feelings. Social expressiveness and sensitivity to others' social cues are limited, and little sharing of experiences or emotions takes place. Their lack of understanding of people as social agents may lead to their treating people as objects or to actions directed at the body parts of people (Carr and Kemp, 1989; Phillips, Gomez, Baron-Cohen, Laa, and Riviere, 1995).

Children with autism display impairments in all aspects of *joint social attention* which is the ability to coordinate one's focus of attention on another person and an object of mutual interest. Joint social attention which normally develops by 12 to 15 months of age, involves getting on the same wavelength with another person by directing that person's attention to objects or people by pointing, showing, and looking, and by communicating shared interest. Children with autism show little desire to share interest and attention with another person for the sheer pleasure of doing so.

Children with autism process social information in unusual ways. At a young age, they may have greater difficulty with imitation of body actions versus toy actions (Smith and Bryson, 1994; Stone, Ousley and Littleford, 1997), or in orienting to social versus non-social stimuli (Dawson, 1996). In processing information about the human face, they may focus on parts of the face, such as the mouth or nose, rather than its overall shape (Boucher and Lewis, 1992).

Although it was once believed that children with autism failed to form a social bond with their parents, research has proved this wrong. Most children with autism *do* respond differently to their care-givers than to unfamiliar adults, directing more social behaviour and seeking to be closer to them (Sigman and Mundy, 1989). Thus children with autism do not show a global deficit in their ability to form an attachment. Rather, the deficit seems to be in their ability to understand and respond to social information (Rogers, Ozonoff, and Maslin-Cole, 1993).

Children with autism also show problems in processing emotional information contained in body language, gestures, facial expressions, or the voice. In contrast to control children of the same mental age, children with autism may sort pictures of people according to the hat these people are wearing rather than their emotional expressions (Weeks and Hobson, 1987). In addition, the bodily expressions of emotions of these children are very different from those of normal children, often characterized by limited spontaneous use of expressive gestures, and bizarre, rigid or mechanical facial expressions (Loveland et al., 1994).

4.2. Communication Impairments

Children with autism display serious abnormalities in communication and language that appear early in life and persist over time. This is best illustrated in a vignette given by Donnellan (1988). A young man with autism was repeatedly asked by his mother to behave in a way which "looked normal". One day she asked for the definition of normal and her son's reply was "it's the second button from the left on the washing machine".

Before children learn to talk, they have at their disposal a rich array of facial expressions, vocalizations and gestures to communicate their needs, interests and feelings to others. One of the first signs of language impairment in children with autism is their inconsistency in using these early preverbal communications. They may use *protoimperative* gestures to express their needs but fail to use *protodeclarative* gestures to direct visual attention of other people to objects of shared interest.

As many as 50% of children with autism do not develop any useful language. They rely on primitive forms of communication and may use instrumental gestures to get someone else do something for them. However, they fail to use expressive gestures (Frith, 1989).

Children with autism who develop language do so before the age of 5 years. They show a *qualitatively deviant* form of communication. Most noticeable is their lack of social chatter – their failure to use language for interpersonal communication. Parents and teachers of children with autism describe their communications as nonsensical, silly, incoherent and irrelevant, having little meaningful connection with the situation in which they occur. Other qualitative language impairments include pronoun reversal and echolalia. Children with autism also display profound impairments in pragmatics, or the appropriate use of language in social and communicative contexts. Thus these children have problems in understanding nonliteral statements or adjusting their language to fit the situation (Dawson, 1996).

Children with autism rarely use language to share information with others or to ask for new information. This reflects a lack of recognition that each individual possesses unique information and ideas, and that language is a critical tool for understanding what others are thinking. It has been suggested that the common element underlying all the communication deficits in autism is a general failure to understand that language can be used to inform and influence other people (Tager-Flusberg, 1996).

Children with autism often fail to establish joint attention to sounds with the person they are speaking to, and as a result may speak too loudly or too softly, with very little intonational inflection. They fail to do so, because they lack the concept of the other person as an interested listener.

4.3. Repetitive Behaviours and Interests

Children with autism often display narrow patterns of interests, repetitive behaviours such as lining up objects, or stereotyped body movements such as rocking. They seem driven to engage in and maintain these behaviours. Stereotyped and repetitive behaviours occur at times when children are not explicitly directed to engage in some other activity, suggesting a possible deficit in their ability to generate alternative actions. Other stereotyped behaviours occur in novel, unpredictable or demanding situations, and may serve to provide the child with a sense of control over the environment and a way of coping with changes that are not understood.

Self-stimulatory behaviours are non-functional, repetitive movements which are referred to as "stim" behaviours or while a child is performing them as "stimming". A classic repetitive behaviour is hand flapping, toe walking, flicking fingers, head banging, walking in circles, and jumping repeatedly. Self-stimulation may involve one or more of the senses and common forms of self stimulation include visual, auditory, tactile, vestibular, taste and smell.

Several theories have been advanced to explain self-stimulatory behaviour. One theory is that these children crave stimulation and "stim" behaviours excite their nervous system. Another theory is that the environment may be too stimulating for them and they engage in repetitive self-stimulation as a way of blocking out unwanted stimulation.

5. ASSOCIATED CHARACTERISTICS OF AUTISM

5.1. Intellectual Deficits and Strengths

About 80% of children with autism have mental retardation, with approximately 60% having Intelligence Quotients (IQs) less than 50, and 20% having IQs between 50 and 70. Also the performance of children with autism on IQ tests is strikingly uneven or "scattered" across different subtests. One of the most common patterns found is a relatively low score on the verbal Comprehension subtest, a relatively high score on the nonverbal Digit Span (short-term memory for strings of numbers) and Block Design (arranging blocks to form a specific pattern) of the Wechsler Intelligence scales (Happe, 1994).

Despite their many intellectual deficits, a small but significant number of children with autism develop splinter skills or islets of ability. Their special talents may be in spelling, reading, arithmetic, music or drawing. As many as 25% of children with autism with an IQ greater than 35, display a special cognitive skill that is above average for the general population and well above their own general level of intellect (Goode, Rutter, and Howlin, 1994). About 5% of children with autism develop an isolated and often remarkable talent far in excess of that found in normal children of the same age (Obler and Fein, 1988). These children, referred to as *autistic savants*, display supernormal abilities in such areas as calculation, memory, jigsaw puzzles, music or drawing. At the same these children would be limited in activities of daily living and basic communications.

One explanation offered is that autistic savants have a tendency to segment information into parts which leads to exceptional performance in certain domains (Pring, Hermelin, and Heavey, 1995). Another is that children think in images rather than abstract ideas, which allows them to remember material in a camera or recorder like fashion (Hurlbert, Happe and Frith, 1994). As Temple Grandin (2006) stated in her forthright manner "I think in pictures. I translate both spoken and written words into full colour movies, complete with sound, which run like a VCR tape in my head" (p.1).

5.2. Sensory and Perceptual Impairments

Sensory abnormalities and deficits are common in children with autism and include over-sensitivities and under-sensitivities to certain stimuli, over-selective and impaired shifting of attention to sensory input, and impairments in mixing across sensory modalities.

Early theories of autism examined both sensory dominance (focus on certain types of sensory input over others) and stimulus over-selectivity (focus on one feature of an object or event while ignoring other equally important features).

5.3. Cognitive Deficits

Two types of dysfunctions that have been proposed to underlie autism are specific cognitive impairments in processing social and emotional information and general cognitive impairments in information processing, planning and attention.

5.3.1. Specific Deficits

Studies have investigated how children with autism process social, emotional, and personal information, such as emotional expressions, facial cues, and internal mental states. These studies have significant impairments in social sensitivities (Sigman, 1995) and an absence of spontaneous pretend play (Frith, 1993). This led to the hypothesis that these children would display impairments in their understanding of mental states – other people's and their own – such as beliefs and desires. This is referred to as the *theory of mind (ToM)* hypothesis, an influential theory which has received considerable research attention (Baron-Cohen, 1989). By the age of 4, most children can figure out what others might know, think, and believe, something that even older individuals with autism have great difficulty doing (Baron-Cohen, 1995). In addition, children with autism have difficulty in forming higher-order representations about invisible mental states (Dawson, 1996). ToM will be discussed in greater detail in Section 4.

5.3.2. General Deficits

It has also been suggested that children with autism display a general deficit in higher-order planning and regulatory behaviours. These executive functions are mediated by the frontal regions of the brain. In support of this general deficit, studies have shown that children and adults with autism have difficulties planning and organizing; changing to a new cognitive set; disengaging from salient stimuli; processing information in novel, unpredictable environments; and generalizing previously learned information to new situations (Bryson, Landry, and Wainwright, 1997; Klinger and Dawson, 1995; Ozonoff, 1994; Ozonoff, Pennington, and Rogers, 1991).

Another general cognitive deficit hypothesized to underlie autism is a weak drive for central coherence (Frith, 1993) which refers to a tendency of humans to interpret stimuli in a global way. It has been proposed that individuals with autism have a weak tendency for central coherence and tend to process information in bits and pieces rather than looking at the big picture (Frith and Happe, 1994). This hypothesis will also be discussed in the following section.

6. CAUSAL EXPLANATIONS AND THEORIES OF AUTISM

6.1. Field of Autism Research

Autism has defied all simple explanation. Perhaps, as pointed out by Happe (2000), in response to such a complex disorder, theories tend to attempt a simplification, leaving aside those features and symptoms for which they cannot account. In a hard-hitting review paper, Waterhouse (2008) has bemoaned the lack of any synthesis in the proliferating theories of autism. She has pointed out that that the heterogeneity of brain deficits, impaired behaviours and genetic variants in autism has challenged researchers and theorists and despite 45 years of research, no standard causal synthesis has emerged.

Raising the issue of whether theories of autism are progressive, Waterhouse cites the work of Lakatos (1970) who argued that a progressive research programme effectively explains anomalous data, generates new hypotheses, synthesizes findings additively and confirms and expands the hard core of fundamental assumptions. In this context, the four hardcore fundamental assumptions of the field of autism are as follows:

1. That a developmental disorder called autism does exist as defined by DSM-IV–TR (2000) and ICD-10 (1992).
2. That there are variants of this core disorder that can be included, along with autism itself, in a larger entity called ASDs.
3. That there is wide variability in diagnostic traits and associated impairments in individuals with autism and other forms of ASD and
4. That the majority of autism cases result from gene effects.(Waterhouse, 2008, p.277)

In a brief theory review, she demonstrates that autism theorists employ several strategies to account for the wide variability in symptoms. These include excluding mental retardation as unrelated to autism, settling for an explanation of a subset of symptoms, attempting to explain all symptoms in one large-capacity theory and splitting autism into phenotypic subgroups.

Loveland (2001) has classified current approaches to explaining autism into two categories: top-down explanations which deal with cognitive or other mental activity (e.g. impairment of the ToM module) and bottom-up explanations which deal with impaired brain structures and their effects on behavioural functioning. Recent studies have also attempted to link these two approaches by connecting laboratory tasks that measure cognitive or affective functioning with neuropsychological or neurobiological measures of brain integrity. She herself espouses the view that an ecological theory of autism offers a different epistemological stance which can draw together the diverse and scattered pieces of this puzzling developmental disorder. Based on Gibson's (1979) concept of "affordances" (opportunities for action and perception that the environment offers to an individual), she viewed people with autism as failing to act upon the same affordances as other people. Seen within the ecological framework, autism is not a static condition existing within the person but a developmental process that can be understood as taking place through the interaction of person and the environment.

Thus it is clear that no single theory is likely to explain all the impairments associated with autism. The existence of multiple causes is a more realistic alternative to account for the many different forms of autism, ranging from mild to severe. Although the causes of autism are still not known, our understanding of possible mechanisms has increased dramatically over the past two decades. It is now generally accepted that autism is a biologically based neurodevelopmental disorder. Accordingly the focus will first be on the biological bases of autism and factors such as early development, genetic influences, and neuropsychological and neurobiological findings will be considered. In the next section the cognitive and psychological explanations will be briefly presented.

6.2. Biological Bases of Autism

As children with autism were followed, various factors suggested a biological basis of the condition. The current consensus is that autism is a behavioural syndrome caused by one or more factors acting on the central nervous system (CNS). Efforts are now under way to delineate more precisely the neuropathological mechanisms of autism.

6.2.1. Genetic Factors

Studies of twins indicated high levels of concordance, especially for monozygotic same-sex twin pairs, with a reduced level of concordance for fraternal or dizygotic same-sex twin pairs. There was a finding that suggested high rates of cognitive difficulties in the unaffected monozygotic twin were associated with perinatal complications in the autistic co-twin, suggesting a perinatal insult related to autism in the face of some inherited liability for the disorder (Volkmar, Klin, Schultz, 2005).

Family studies have demonstrated a rate of recurrence in families in families of 2 to 7 percent of autism cases among siblings. This is, however, a 50- to 200-fold increase in the rate of autism relative to the general population. It remains unclear whether what is inherited is a specific predisposition to autism or a more general predisposition to developmental difficulties.

Efforts are under way to identify potential genetic mechanisms in autism, and promising leads have been identified through linkage analyses on several chromosomes, including regions on chromosomes 7, 2, 4, 15, and 19. It now appears likely that an estimated four or five genes are involved. Genetic subtypes in autism may exist, and the disorder has been linked to other genetic disorders such as fragile X syndrome and untreated phenylketonuria (Folstein and Rutter, 1987). The increased prevalence of mental retardation and specific deficits in the siblings of children with autism suggests a polygenic disorder that increases the risk for cognitive impairments including autism (Bailey et al., 1996).

6.2.2. Perinatal Factors

Several studies have shown increased rates of pre-, peri-, and neonatal complications in children with autism. Risk factors such as the mother's age, prematurity, viral infection or exposure and a lack of vigour after birth have been identified in about 25% of children with autism. Pregnancy and birth complications are not the primary cause of autism, but they do suggest that fetal or neonatal development has been compromised in some general way (Nelson, 1991; Piven et al., 1993).

6.2.3. Neuropsychological Impairments

Neuropsychological impairments in autism occur in many domains, including intelligence, attention, memory, language and executive functions. The widespread nature of these deficits suggests the involvement of multiple regions of the brain at both cortical and subcortical levels (Happe and Frith, 1996). The fact that there are both affected and unaffected areas of functioning within each domain, suggests that within each affected area some brain functions remain intact (Dawson, 1996).

Other neuropsychological explanations have linked social-affective deficits to impairments of the medial temporal lobe (Bachevalier and Merjanian, 1994; Dawson, Melttzoff, Osterling and Rinaldi, 1998); deficits in visuospatial attention to the cerebellum and parietal lobe (Townsend, Courchesene, and Egaas, 1996); deficits in executive functioning to impairment in the frontal lobe (Ozonoff, Pennington, and Rogers, 1991) and abnormalities of the hippocampus, amygdale and sensory cortex with social-affective, cognitive and attentional deficits (Waterhouse, Fein, and Modahl, 1996).

6.2.4. Postmortem and Neuroimaging Studies

The deficits that occur with autism suggest that the syndrome affects a functionally diverse and widely distributed set of neural systems. At the same time, however, the affected systems must be discrete, because autism spares many perceptual and cognitive systems.

Postmortem studies of a small number of individuals with autism have revealed a range of abnormalities, including a significant decrease in the number of Purkinje cells and granule cells in the cerebellum (Courchesene et al., 1995). The precise nature of these abnormalities, including a lack of gliosis indicative of scarring, suggests a prenatal origin.

Postmortem studies also implicate the limbic system in the pathophysiology of autism. There is consistent evidence for decreased neuronal size, decreased dendritic arborisation, and increased neuronal packaging density of neurons in the amygdala, hippocampus, septum, anterior cingulated, and mammillary bodies. These abnormalities are suggestive of a curtailment of normal development. Also the orbital and medial prefrontal cortices have dense reciprocal connections, with the amygdala providing the architecture for a system that can regulate social-cognitive processes. A parallel set of amygdale-cortical circuitry in the temporal lobes focus on social-perceptual processes. One hypothesis is that autism is largely caused by abnormalities in both of these amygdale-cortical loops.

The best-replicated functional neuroimaging finding concerns underactivation of a region in the fusiform gyrus on the ventral surface of the temporal lobe during face perception tasks (Schultz et al., 2000). Because of the specificity of this area for faces, it is known as the fusiform face area. Approximately six studies by different groups have shown that older children, adolescents, and adults with autism have reduced levels of responsivity to the human face in the fusiform face area, especially in the right hemisphere. These data are consistent with an extensive psychology literature documenting performance deficits in face and facial expression recognition in autism.

Older human studies using lower-resolution neuroimaging techniques reported general hypoactivation of the frontal lobes. More recent data suggests that subregions of the prefrontal cortices with strong connectivity to limbic areas are critical for social cognition-that is, thinking about other's thoughts, feelings, and intentions. Deficits in such ToM abilities are common in autism. Preliminary functional imaging evidence in ASDs suggests altered functional representation in prefrontal cortices during ToM tasks.

One of the more intriguing findings to emerge is that overall brain size appears to be increased in autism (by 2 to 10 percent). This could be a marker for a disturbance in the fine structure of the brain that actually causes autistic symptoms. Increased brain size might come at the expense of interconnectivity between specialized neural systems, giving rise to a more fragmentary processing structure. In fact, some evidence suggests that the corpus callosum is reduced in size in autism. Less neural integration would be consistent with one influential theory that attributes autistic symptoms to a lack of "central coherence", a cognitive processing style that makes integration of parts into wholes problematic.

Other studies have consistently identified structural abnormalities in the cerebellum and the medial temporal lobe and related limbic system structures (Courchesne, Chisum, and Townsend, 1994). Specific areas of the cerebellum have been found to be significantly smaller than normal in a majority of individuals with autism, a condition known as cerebellar hypoplasia (Courchesne et al., 1995). Ramachandran and Oberman (2006) have presented the Broken Mirrors theory of autism. They posit that there is a connection between autism and a newly discovered class of nerve cells in the brain called mirror cells. As these neurons appear to be involved in processes such as empathy and the perceptions of other's intentions, they hypothesized that a dysfunction of the mirror neuron system could result in some of the symptoms of autism. Over the past decade, several studies have provided evidence for this theory.

6.2.5. Neurochemistry

Beginning in 1961, a number of studies have reported that approximately one-third of children have increased peripheral levels of the neurotransmitter serotonin. The significance of this finding is unclear, as it is not specific to autism and the relation of peripheral levels to central levels of serotonin is unclear.

Another line of work has focused on other transmitters such as dopamine. A hypodopaminergic functioning of the brain might explain the overactivity and stereotyped movements in autism. Studies of cerebrospinal fluid of dopamine metabolites have been inconsistent.

6.3. Psychological Theories and Explanations

There has been a pendulum swing in the focus of psychological accounts of autism, from social to non-social facets of the disorder (Happe, 2000). While the 1980s saw a vital move towards explanation of the social handicap in autism, the 1990s have seen a reconsideration of the non-social features. A number of domain-general explanations have been proposed. The challenge to these theories is not to explain too much (Frith, 1989). That is any theory must allow people with autism the many things that they are good at. A domain-general theory, such as that people with autism are unable to shift attention rapidly is in danger of predicting too much — a blanket deficit across all tasks. Autism, however, is not a blanket deficit, but a fine landscape of peaks and troughs — parents, teachers, and clinicians are all unsurprised when a child with autism shows remarkable ability in a specific area out of line with his/her usual level of functioning.

As pointed out by Baron-Cohen (2000) abnormalities in understanding other minds is not the only cognitive feature of ASDs. Two other prominent phenomena include weak central coherence (Happe, 2000) and executive dysfunction (Russell, 1997). Thus this section will focus on ToM, weak central coherence and executive dysfunction and how they explain the puzzle of autism.

6.3.1. Theory of Mind (ToM) and Autism

By far, the most influential explanation of autism concerns ToM. ToM is one of the quintessential qualities that make us human (Whiten, 1993). ToM, means being able to infer the full range of mental states (beliefs, desires, intentions, imagination, emotions, etc.) that cause action. In brief, it refers to the ability to reflect on the contents of one's own and other's minds.

ToM is a 'theory' insofar as the "mind" is not "directly observable." The presumption that others have a mind is termed a "theory of mind" because each human can only prove the existence of his or her own mind through introspection, and one has no direct access to others' minds. ../../../../Users/Dr.Komila Thapa/autism1.htm - cite_note-5. Being able to attribute mental states to others and understanding them as causes of behavior implies, in part, that one must be able to conceive of the mind as a "generator of representations". If a person does not have a complete ToM it may be a sign of cognitive or developmental impairment.

The ToM impairment describes a difficulty someone would have with perspective taking. This is also sometimes referred to as *mind*. This means that individuals with a ToM impairment would have a hard time seeing things from any other perspective than their own. Individuals who experience a ToM deficit have difficulty determining the intentions of others, lack understanding of how their behavior affects others, and have a difficult time with social reciprocity. In 1985, Baron-Cohen, Leslie and Frith published research which suggested that children with autism do not employ a ToM, and suggested that these children have particular difficulties with tasks requiring the child to understand another person's beliefs. These difficulties persist when children are matched for verbal skills and have been taken as a key feature of autism.

Many individuals classified as having autism have severe difficulty assigning mental states to others, and they seem to lack ToM capabilities (Baron-Cohen, 1991). Researchers who study the relationship between autism and ToM attempt to explain the connection in a variety of ways. One account assumes that ToM plays a role in the attribution of mental states to others and in childhood pretend play. According to Leslie (1991) ToM is the capacity to mentally represent thoughts, beliefs, and desires, regardless of whether or not the circumstances involved are real. This might explain why individuals with autism show extreme deficits in both ToM and pretend play. However, Hobson (1995) proposes a social-affective justification which suggests that in persons with autism, deficits in ToM result from a distortion in understanding and responding to emotions. He suggests that typically developing human beings, unlike individuals with autism, are born with a set of skills (such as social referencing ability) which will later enable them to comprehend and react to other people's feelings. Other scholars emphasize that autism involves a specific developmental delay, so that children with the impairment vary in their deficiencies, because they experience difficulty in different stages of growth. Very early setbacks can alter proper advancement of joint-attention behaviors, which may lead to a failure to form a full theory of mind

It has been speculated that ToM exists on a continuum as opposed to the traditional view of a concrete presence or absence. While some research has suggested that some autistic populations are unable to attribute mental states to others, recent evidence points to the possibility of coping mechanisms that facilitate a spectrum of mindful behaviour (Dapretto et al., 2006).

6.3.2. Weak Central Coherence and Autism

Frith (1989) proposed that assets and deficits in autism can be explained in terms of a specific imbalance in integration of information at different levels. A characteristic of normal information-processing is the tendency to draw together diverse information to construct higher-level meaning in context; 'central coherence' in Frith's words. Central coherence is also demonstrated in the ease with which we recognize the contextually-appropriate sense of the many ambiguous words used in everyday speech (e.g. son-sun, meet-meat, sew-so, pear-pair).

Frith suggested that this feature of human information processing is disturbed in autism, and that weak central coherence could explain very parsimoniously the assets and deficits of autism. Frith predicted that individuals with autism would be relatively good at tasks where attention to local information (i.e. piece-meal processing) is advantageous, but poor at tasks requiring the recognition of global meaning or integration of stimuli in context. Shah and Frith (1993) found that people with autism were unusually good on the Block Design test of the Wechsler scales, and that this facility had to do with segmentation abilities.

Studies have explored weak central coherence in low-level perceptual processes, more complex visuospatial constructional tasks and at higher levels involving the semantic system. Many challenges remain to the central coherence explanation, not least to specify the mechanism for central coherence.

6.3.3. Executive Dysfunction and Autism

Executive function (EF) is a label for those processes in the control of behaviour, like planning, co-ordinating, and controlling sequences of actions, that are disrupted upon frontal lobe injury, and which have been documented by the work of Luria (1966). An important realization is that executive control is not required for all action but only for certain types of problems (Perner and Lang, 2000). Salient in this context is the model of Norman and Shallice (1980/1986) who distinguished between two levels of control. At the level of *contention scheduling* the control of action schemas takes place by mutual inhibition and activation. However, there are certain tasks for which a higher level of control is required, which is provided by the supervisory attentional system (SAS). SAS tasks include planning/decision making, trouble shooting, novel/ill-learned action sequences, dangerous or technically difficult actions, and overcoming strong, habitual response tendencies or temptation (e.g. Stroop).

Studies have shown that children with autism have problems with ToM tasks as well as impairment on tasks measuring executive functioning. Pennington and Ozonoff (1996) reviewed 14 studies that assessed children with autism on a variety of EF tasks and found significant impairment, with an average effect size of .98 (i.e. a mean of practically one standard deviation below that of the intelligence matched control group. There is also evidence that most children with a known ToM deficit are also severely impaired on EF (Ozonoff, et al., 1991).

CONCLUDING REMARKS

From the above review it is clear that the variability of autism and ASDs has posed a great challenge to researchers and theorists, not to mention parents and practitioners, always on the lookout for an empirically-supported, more promising intervention. The increasing prevalence of autism and ASDs has put pressure on the field for more productive and predictive theories. An explanatory model of autism will ultimately need to integrate behavioural, cognitive, and biological findings. Interventions based of such a model would then help to reduce the pain and uncertainty that surround efforts to help children with autism and their families.

ACKNOWLEDGMENTS

In the preparation of this chapter the help given by Shalini Choudhary is gratefully acknowledged.

REFERENCES

American Psychiatric Association (2000). *Diagnostic and statistical manual for mental disorders*, 4th edition, Text revision. Washington, DC: American Psychiatric Association.

Asperger, H. (1944). Autistic psychopathy in childhood. In U. Frith (Ed., 1991). *Autism and Asperger syndrome*. Cambridge: Cambridge University Press.

Bachevalier, J., and Merjanian, P. (1994). The contribution of medial temporal lobe structures in infantile autism: A neurobehavioural study in primates. In M.L. Bauman and T.L. Kemper (Eds.) *The neurobiology of autism* (pp. 146-169). Baltimore: John Hopkins.

Bailey, A., Phillips, W., and Rutter, M. (1996). Autism: Towards an integration of clinical, genetic, neuropsychological, and neurobiological perspectives. *Journal of Child Psychology and Psychiatry, 37*, 89-126.

Baron-Cohen, S. (1989). The autistic child's theory of mind: A case of specific developmental delay. *Journal of Child Psychology and Psychiatry, 30*, 285-297.

Baron-Cohen, S. (1991). Precursors to a theory of mind: Understanding attention in others. In A. Whiten (Ed.) *Natural theories of mind: Evolution, development and simulation of everyday mindreading* (pp. 233-251). Cambridge, MA: Basil Blackwell.

Baron-Cohen, S. (1995). *Mindblindness: An essay on autism and theory of mind*. Cambridge, MA: MIT Press.

Baron-Cohen, S. (2000). Theory of mind and autism: A fifteen year review. In S. Baron-Cohen, H. Tager-Flusberg, D.J. Cohen (Eds.) *Understanding other minds: Perspectives from developmental cognitive neuroscience*. (pp. 3-20). Oxford: Oxford University Press.

Baron-Cohen, S., Tager-Flusberg, H., Cohen, D.J. (Eds., 2000). *Understanding other minds: Perspectives from developmental cognitive neuroscience*. Oxford: Oxford University Press.

Baron-Cohen, S., Leslie, A.M., and Frith, U. (1985). Does the autistic child have a theory of mind? *Cognition, 21*, 37-46.

Bettelheim, B. (1967). *The empty fortress: Infantile autism and the birth of the self*. New York: Free Press.

Boucher, J., and Lewis, V. (1992). Unfamiliar face recognition in relatively able autistic children. *Journal of Child Psychology and Psychiatry, 33*, 843-849.

Bryson, S.E., Landry, R., and Wainwright, J. (1997). A componential view of executive dysfunction in autism: Review of recent evidence. In J.A. Burack and J.T. Enns (eds.), *Attention, development, and psychopathology* (pp. 232-259). New York: Guilford Press.

Carr, E.G., and Kemp, D.C. (1989). Functional equivalence of autistic leading and communicative pointing: Analysis and treatment. *Journal of Autism and Developmental disorders, 19*, 561-578.

Chakrabarti, S., and Fombonne, E. (2001). Pervasive developmental disorders in preschool children. *Journal of the American Medical Association, 285*, 3141-3142.

Courchesene, E., Townsend, J.P., and Chase, C. (1995). Neurodevelopmental principles guide research on developmental psychopathologies. In D. Cicchetti and D. Cohen (eds.) *Developmental psychopathology: Vol. 1. Theories and methods* (pp.195-226). New York: Wiley.

Courchesne, E., Chisum, H., and Townsend, J. (1994). Neural activity-dependent brain changes in development: Implications for psychopathology. *Development and Psychopathology, 6,* 697-722.

Dapretto, M., et al. (2006). Understanding emotion in others: Mirror neuron dysfunction in children with autism spectrum disorder. *Nature Neuroscience, 9,* 28-30.

Dawson, G. (1996). Neuropsychology of autism: A report on the state-of-the science. *Journal of Autism and Developmental Disorders, 2,* 179-181.

Dawson, G., Meltzoff, A.N., Osterling, J., and Rinaldi, J. (1998). Neuropsychological correlates of early symptoms of autism. *Child Development, 69,* 1276-1285.

Donnellan, A.M.(1988, February). Our old ways just aren't working. Dialect. (Newsletter of the Saskatchewan Association for the Mentally Retarded).

Edelson, S.M., and Rimland, B. (Eds., 2003). *Treating autism: Parent stories of hope and success.* San Francisco, CA: Autism Research Institute.

Folstein, S.E., and Rutter, M. (1987). Autism: Familial aggregation and genetic implications. In E. Schopler and G. Mesibov (Eds.) *Neurobiological issues in autism* (pp. 83 -105). New York: Plenum Press.

Frith, U. (1989). Autism: Explaining the enigma. Oxford: Basil Blackwell.

Frith, U. (1993, June). Autism. *Scientific American,* 108-114.

Frith, U., and Happe, F. (1994). Autism: Beyond "theory of mind". *Cognition, 50,* 115-132.

Gibson, J.J. (1979). *The ecological approach to visual perception.* Boston: Houghton-Mifflin.

Goode, S., Rutter, M., and Howlin, P. (1994). A twenty-year follow-up of children with autism. Paper presented at the 13th biennial meeting of the ISSBD, Amsterdam, the Netherlands.

Grandin, T. (2006). *Thinking in pictures: My life with autism* (Expanded edition). New York: Vintage.

Happe, F. (2000). Parts and wholes, meanings and minds: Central coherence and its relation to theory of mind. In S. Baron-Cohen, H. Tager-Flusberg, D.J. Cohen (Eds.,) *Understanding other minds: Perspectives from developmental cognitive neuroscience* (pp. 203-221). Oxford: Oxford University Press

Happe, F.G.E. and Frith, U. (1996). The neuropsychology of autism. *Brain, 119,* 1377-1400.

Happe, F.G.E. (1994).Wechsler IQ profile and theory of mind in autism: A research note. *Journal of Child Psychology and Psychiatry, 35,* 1461-1471.

Hillman, J., Snyder, S., and Neubrander, J.A. (2007). *Childhood autism: A clinician's guide to early diagnosis and integrated treatment.* Hove, East Sussex: Routledge.

Hobson, R.P. (1995). *Autism and the development of mind.* Hillsdale, NJ: Lawrence Erlbaum.

Hurlbert, R., Happe, F., and Frith, U. (1994). Sampling the form of inner experience in 3 adults with Asperger syndrome. *Psychological Medicine, 24,* 385-395.Klinger, L.G., and Dawson, G. (1995). A fresh look at categorization abilities in persons with autism. In E. Schopler and G.B. Mesibov (Eds.) *Learning and cognition in autism* (pp. 119-136). New York: Plenum.

Klinger, L.G., and Dawson, G. (1996). Autistic disorder. In E.J. Mash and R.A. Barkeley (Eds.) *Child psychopathology* (pp. 311-339). New York: Guilford Press.

Lakatos, I. (1970).Falsification and the methodology of scientific research programmes. In I. Lakatos and A. Musgrave (Eds.) Criticism and the growth of knowledge (pp.91-195). Reprinted in J. Worall and G. Curie (eds.) *Imre Lakatos: Philosophical papers, Vol 1:*

The methodology of scientific research programmes (pp.8-101). New York: Cambridge University Press, 1978.

Leslie, A.M. (1991). Theory of mind impairment in autism. In A. Whiten (Ed.) *Natural theories of mind: Evolution, development and simulation of everyday mindreading.* Cambridge, MA: Basil Blackwell.

Loveland, K.A. (2001). Toward an ecological theory of autism. In J.A. Burack, T. Charman, N. Yirmiya, and P.R. Zelazo (Eds.) *The development of autism: Perspectives from theory and research* (pp.15-33). Mahwah, NJ: Lawrence Erlbaum.

Loveland, K.A., Tunali-Kotoski, B., Pearson, D.A., Brelsford, K.A., Ortegon, J., and Chen, R. (1994). Imitation and expression of facial affect in autism. *Development and Psychopathology, 6*, 433-444.

Luria, A. (1966). *Higher cortical functions in man.* New York: Basic Books.

Mash, E.J., and Wolfe, D.A. (1999). *Abnormal child psychology.* Belmont, CA: Brooks/Cole, Wadsworth.

Nelson, K.B. (1991). Prenatal and perinatal factors in the etiology of autism. *Pediatrics, 87,* 761-.66.

Norman, D.A., and Shallice, T. (1980). Attention to action: Willed and automatic control of behaviour. Centre for Human Information Processing Technical Report No. 99. Reprinted in revised form in R.J. Davidson, G.E. Schwartz and D. Shapiro (Eds. 1986). *Consciousness and self-regulation*, Vol. 4 (pp.1-18).New York: Plenum.

Obler, L.K., and Fein, D. (Eds., 1988). *The exceptional brain: Neuropsychology of talent and special abilities.* New York: Guilford.

Ozonoff, S. (1994). Executive functions in autism. In E. Scopler and G.B. Mesibov (Eds.) *Learning and cognition in autism* (pp. 199-219), New York: Plenum.

Ozonoff, S., Pennington, B.F. and Rogers, S.J. (1991). Executive function deficits in high-functioning autistic individuals: Relationship to theory of mind. *Journal of Child Psychology and Psychiatry, 32*, 1081-1105.

Pennington, B.F., and Ozonoff, S. (1996). Executive functions and developmental psychopathology. *Journal of Child Psychology and Psychiatry, 37,* 51-87.

Perner, J., and Lang, B. (2000). Theory of mind and executive function: Is there a developmental relationship. In S. Baron-Cohen, H. Tager-Flusberg, D.J. Cohen (Eds.,) *Understanding other minds: Perspectives from developmental cognitive neuroscience.* Oxford: Oxford University Press.

Phillips, W., Gomez, J.C., Baron-Cohen, S., Laa, V., and Riviere, A. (1995).Treating people as objects, agents, or "subjects": How young children with or without autism make requests. *Journal of Child Psychology and Psychiatry, 36*, 1383-1398.

Piven, J., Simon, J., Chase, G.A., Wzorek, M., Landa, R., Gayle, J., and Folstein, S. (1993). The etiology of autism: Pre-, peri-, and neonatal factors. *Journal of the American Academy of Child and Adolescent Psychiatry, 32*, 1256-1263.

Pring, L., Hermelin, B., and Heavey, L. (1995). Savants, segments, art and autism. *Journal of Child Psychology and Psychiatry, 36, 1065-1076.

Ramachandran, V.S., and Oberman, L.M. (2006, September). Broken mirrors: A theory of autism. *Scientific American*, 62-69.

Rogers, S.J., Ozonoff, S., and Maslin-Cole, C. (1993). Developmental aspects of attachment behaviour in young children with pervasive developmental disorders. *Journal of American Academy of Child and Adolescent Psychiatry, 32*, 1274-1282.

Russell, J. (Ed.) *Autism as an executive disorder*. Oxford: Oxford University Press.

Rutter, M. (1978). Diagnosis and definitions of childhood autism. *Journal of Autism and Developmental Disorders, 8*, 139-161.

Sacks, O. (1995). *An anthropologist on Mars: Seven paradoxical tales*. New York: Picador.

Schultz, R.T., and Anderson, G.M. (2004). The neurobiology of autism and the pervasive developmental disorders. In D.S. Charney and Nestler, E.J. (Eds., 2004). *Neurobiology of mental illness*, 2nd edition (pp.954-967). Oxford: Oxford University Press.

Schultz, R.T., Gauthier, I., Klin, A. Et al. (2000) Abnormal ventral temporal cortical activity during face discrimination among individuals with autism and Asperger syndrome. *Archives of General Psychiatry, 57*, 331-340.

Shah, A., and Frith, U. (1993). Why do autistic individuals show superior performance on the Block Design task? *Journal of Child Psychology and Psychiatry, 34*, 1351-1364.

Sigman, M. (1995). Behavioural research in childhood autism. In M. Lenzenweger and J. Haugaard (Eds.,) *Frontiers of developmental psychopathology* (pp.190-206). New York: Springer- Verlag.

Sigman, M., and Mundy, P. (1989). Social attachments in autistic children. *Journal of American Academy of Child and Adolescent Psychiatry, 28*, 74-81.

Smith J.M., and Bryson, S.E. (1994). Imitation and action in autism: A critical review. *Psychological Bulletin, 116*, 259-273.

Stone, W.L., Ousley, O.Y., and Littleford, C.D. (1997). Motor imitation in young children with autism: What's the object? *Journal of Abnormal Child Psychology, 25*, 475-485.

Strock, M. (2004). *Autism spectrum disorders*. (NIH Publication No.NIH-04-5511). Bethesda, MD: National Institute of Health.

Tager-Flusberg, H. (1996). Brief report: Current theory and research on language and communication in autism. *Journal of Autism and Developmental Disorders, 26*, 169-172.

Townsend, J., Courchesene, E., and Egaas, B. (1996). Slowed orienting of covert visual-spatial attention in autism: Specific deficits associated with cerebellar and parietal abnormality. *Development and Psychopathology, 8*, 563-584.

Volkmar, F.R., Klin, A., and Schultz, R.T. (2005). Pervasive developmental disorders. In B.J. Sadock and V.A. Sadock (Eds.) *Kaplan and Sadock's comprehensive textbook of psychiatry*, 8th edition, Vol. 2 (pp.3164-3182). Philadelphia: Lippincott Williams and Wilkins.

Waterhouse, L. (2008). Autism overflows: Increasing prevalence and proliferating theories. *Neuropsychological Review, 18*, 273-286.

Waterhouse, L., Fein, D., and Modahl, C. (1996). Neurofunctional mechanisms in autism, *Psychological Review, 103*, 457-489.

Weeks, S. J., and Hobson, R.P. (1987). The salience of facial expression for autistic children. *Journal of Child Psychology and Psychiatry, 28*, 137-152.

Whiten, A. (1993). Evolving a theory of mind: The nature of nonverbal mentalism in other primates. In S. Baron-Cohen, H. Tager-Flusberg, D.J. Cohen (Eds.,) *Understanding other minds: Perspectives from developmental cognitive neuroscience*. Oxford: Oxford University Press.

Wing, L., and Gould, J. (1979). Severe impairments of social interaction and abnormalities in children. *Journal of Autism and Childhood Schizophrenia, 9*, 11-29.

World Health Organization (1992). ICD-10 classification of mental and behavioural disorders: Clinical descriptions and diagnostic guidelines. Geneva: W.H.O.

In: Expanding Horizions of the Mind Science(s)   ISBN: 978-1-62808-705-5
Editors: P.N. Tandon, R.C. Tripathi and N. Srinivasan   ©2013 Nova Science Publishers, Inc.

Chapter 21

CELLULAR AND MOLECULAR BASIS OF NEUROCOGNITIVE DEFICITS IN HIV/AIDS

Mamata Mishra and Pankaj Seth

Molecular and Cellular Neuroscience, National Brain Research Centre,
Nainwal Mode, Manesar Gurgaon, India

ABSTRACT

Approximately 40% of human immunodeficiency virus 1 (HIV-1) infected individuals demonstrate neurologic disorders of varying degree which encompass HIV-associated dementia, mild neurocognitive disorder and asymptomatic neurocognitive impairment. Infection of human central nervous system with HIV-1 can cause a range of clinical disorders, cognitive impairment being the most common among them. HIV-1-associated dementia (HAD) or AIDS dementia complex (ADC) is characterized by neurodegeneration resulting in progressive decline of cognitive and motor function. Impairment of cognitive and memory functions in HIV/AIDS has been related to three brain areas that are affected in HIV infection, namely prefrontal cortex - the executive hub, hippocampus - the memory hub, and the basal ganglion region. HIV/AIDS associated dementia develops in patients with advanced HIV-1 infection and severe immunosuppression. Despite the successful introduction of combinatorial anti-retroviral therapy (cART), HIV-related neurological disorders signify a substantial personal, economic, and societal burden on populations around the world. The advent of cART has modulated HAD to its milder form, the HIV-associated neurocognitive disorders (HAND). HAND is now getting common in HIV patients or AIDS survivors, however the underlying reasons for progressive decline of executive functions and neurocognitive performance remains ambiguous. It is believed that HIV-1 and proteins encoded by it affect functions of glial and neuronal cells at various levels. Detailed investigations into the molecular and cellular events occurring during the course of HIV-1 neuropathogenesis and regarding the prevalence of cognitive impairment in post cART era are urgently warranted. This chapter provides details of HIV-1, its viral proteins, effects of HIV-1 clades/subtypes on human brain, and their relation in modulation of neuronal function and neuronal damage which culminates into cognitive and memory impairments. The chapter also discusses some recent advances in neuroAIDS field that has evolved tremendously in last decade in terms of our understanding of HIV-1 neuropathogenesis.

Keywords: Human immunodeficiency virus, HIV dementia, HIV associated neurocognitive disorders, neuropsychology, neuropathogenesis, transactivating protein Tat

1. INTRODUCTION

Cognitive impairments define a broad group of severe mental and behavioral disorders etiologically traceable to brain disease, injury, or exposure to certain neurotoxins. The word cognition (Latin word *cognoscere* means to know, to conceptualize or to recognize) is used to refer to the mental functions, mental processes/thoughts and states of intelligent entities of human brain, a highly autonomous machine. In particular, the field focuses toward the study of specific mental processes such as comprehension, inference, decision-making, planning and learning.

The advanced stages of HIV-1 infection cause severe immunosuppression that may increase the risk of neuropsychological abnormalities. Following HIV-1 invasion of the parenchyma, the virus and its viral proteins damage neural cells that lead to a wide range of neuropsycological deficits, which are collectively termed as cognitive impairment. Cognitive impairment related to HIV-1 has been defined on the basis of a cognitive battery for which cutoff scores for each test are derived from normative values or from HIV-1 seronegative controls (Stern et al., 1996). A more precise consideration of cognitive impairment symptoms includes dementia, delirium, learning disability, mental retardation and delusions.

Cognitive impairment is one of the key neuroAIDS syndrome associated with HIV-1 infection. Although several components of human immunodeficiency virus are capable of causing cognitive impairment, the prime mediator is not known. In spite of several decades of extensive research, a complete understanding of the pathogenic mechanisms underlying HAD, that range from neurotoxicity to impaired neurogenesis, remains inexplicable. Long term HIV infection deteriorates the human brain function, particularly by damaging or impairing the glial and neuronal cell functions. Recently, perturbations due to HIV-1 to the glial-neural interplay has gathered attention for its possible role in cognitive dysfunction in HIV/AIDS patients and hence is being actively pursued by several investigators in the field, including the authors.

2. AREAS OF BRAIN INVOLVED IN HIV MEDIATED COGNITIVE IMPAIRMENTS

From breathing, blinking to memorize facts for a test, all actions are controlled by the central nervous system (CNS). Thomas Willis who is referred to as the father of neurology, was the first to suggest that not only the brain is the locus of the mind, but that different parts of the brain give rise to specific cognitive function. The major areas affected by HIV-1 infection are prefrontal cortex, hippocampus and basal ganglia. The deterioration of cognitive function and the reduction in volume of certain brain structures including the basal ganglia and caudate nucleus are highly correlated to clinical observations on HIV-1 infected individuals. Cortical atrophy associated with HIV infection may be caused by neuronal loss and demyelination and the degree of atrophy is correlated to the degree of cognitive motor dysfunction (Aylward et al., 1993; Hall et al., 1996). In the basal ganglia, particularly the dopaminergic systems are a major target of HIV infection, which cause psychomotor slowing, apathy, bradykinesia and altered posture and gait similar to those observed in advanced Parkinson's disease (Berger and Arendt, 2000). Post-mortem studies on human brains suggest there is a considerable loss of dopamine following CNS infection with HIV-1 (Kumar et al., 2009). Furthermore, a recent study suggests selective vulnerability of the dopamine system in developing brains of HIV-1 Tat transgenic rat (Webb et al., 2010). Combination of antiviral treatment has improved cognition, learning and memory. This suggests that the virus or its components is producing acute and reversible impairments which are a likely mechanism mediated through hippocampus (Miralles et al., 1998). In pediatric neuroAIDS cases, the neurodevelopmental progression is affected in cognitive areas of brain and affected children lag in their developmental milestones (Bruck et al., 2001; Abubakar et al., 2008; Van Rie et al., 2008).

3. RISK FACTORS FOR COGNITIVE IMPAIRMENT

Risk factors for early neuropsychological abnormalities in HIV/AIDS patients are controversial. In addition to HIV-1 infection and severe immunosuppression, several other factors may increase the risk of early prevalence of cognitive impairment in AIDS patients. Clinical studies based on neuropsychological evaluation and a psychiatric assessment implicate certain factors like age, sex, education, risk behaviors, HIV-1 stage, lymphocyte count, and antiretroviral therapy have been sporadically associated with a higher occurrence of dementia and cognitive impairment in HIV-1–infected patients (De Ronchi et al., 2002). Further investigations into this area of research are warranted before risk factors can be defined conclusively.

4. GLOBAL PICTURE OF HIV/AIDS AND HIV ASSOCIATED NEUROCOGNITIVE DISORDERS (HAND)

The central nervous system is vulnerable to infection by retroviruses of various species, particularly members of the lentivirus family. Human nervous system infection with human immunodeficiency virus type-1 can cause a range of clinical disorders that are debilitating due to irreversible damage to neuronal cells. Despite the success of antiretroviral therapy in recent years, globally around 33.2 million people are affected with HIV-1 (UNAIDS/WHO, 2008). Patients infected with HIV-1 invariably experience frequent and devastating neurological complications which include decreased memory, inability to concentrate, apathy and psychomotor retardation. Such conditions often develop in advanced stages of HIV-1 disease and are collectively referred to as HIV-associated dementia (HAD) (Goodkin et al., 2000; Boisse et al., 2008) or AIDS dementia complex (ADC) (Ho et al., 1989).

The incidence of the most severe form of HIV-associated dementia has decreased since the successful introduction of combination antiretroviral therapy (cART). In fact a subtle form of neurological deficit is now prevalent among the AIDS patients and is termed as HIV-associated neurocognitive disorders (HAND) (Ances and Clifford, 2008). The HAND often leads to sub-cortical dementia consisting of a triad of cognitive, behavioral, and motor dysfunction. In the post highly active antiretroviral therapy (HAART) era, the HAD is said to be evolving into a neurological disease that affects a wider population of HIV/AIDS individuals with reduced degree of the disease severity. Despite the success of HAART, till date it has been very difficult to eradicate HIV-1 because of its latent infection in sub-population of cells in the CNS (Alexaki, 2008). HAART decreases the efficiency of virus production as well as its effectiveness for causing the disease, but it may increase another facet of complications in CNS such as amyloid-beta accumulation (Pulliam, 2009) and surprising degree of microglial activation (Bell et al., 2006).

Since the beginning of HIV research, it was quite clear that HIV-1 and proteins encoded by it affect cells of the immune system particularly T-cells, however their role in HIV neuropathogenesis is still not clearly understood. Furthermore, while evidence for presence of HIV in brain is available, the production of HIV-1 in the CNS has not been detected in brain tissue of HIV-1 positive subjects between the time of seroconversion and symptomatic AIDS (e.g. pre-AIDS stages) (Kramer-Hammerle et al., 2005). The possibility of involvement of human brain cells in HIV-1 infection gathered attention following clinical observations of cognitive and motor deficits in some HIV/AIDS patients. Most of the knowledge on HIV-1 neuropathogenesis is recent and scanty hence it is essential to investigate the virological and cellular details that are important in the neuropathogenesis of HIV-1.

HIV-1 has been classified into at least 10 different subtypes, or clades, with distinct worldwide distributions and patterns of disease transmission and pathogenesis (Robertson et al., 2000). Different HIV-1 strains/clades (A-J) have been classified based upon the genetic diversity/phylogenetic relationship. There is a specific geographic distribution pattern for HIV-1 subtypes throughout the globe, except in sub-Saharan Africa where most subtypes have been reported (Hemelaar et al., 2006). According to recent studies the most prevalent HIV-1 genetic forms are subtypes A, B, and C. Currently, HIV-1 subtype C has spread disproportionately as compared to other subtypes or groups, the cause of which is not well understood (Tebit et al., 2007). Subtypes A, B, D and G account for 12%, 10%, 3% and 6% HIV-1 infections respectively (Hemelaar et al., 2006). Subtype C viruses predominate in southern Africa and India. Subtype B on the other hand is the main genetic form in Western and Central Europe, America, and Australia. It is also common in several countries of Southeast Asia and Northern Africa.

In western world, HIV-1 clade B is responsible for nearly all HIV-1 infections. HIV-1 clade C is responsible for more than half of new HIV-1 infections worldwide (Geretti, 2006) and more than 90% of HIV-1 infections in India (Siddappa et al., 2004). Most of our current knowledge on HIV-1 induced CNS complications is based on studies with HIV-1B. Limited information is available on how HIV-1C affects brain functioning in HIV-1 infected individuals. India has experienced a rapid and extensive spread of HIV-1 with an estimated 2.5 million people infected by this deadly virus, out of which one million are women (National AIDS Control Organization, NACO, 2007) (Nath, 2009). The burden of HIV-1C related pathology is quite extensive due to the enhanced risk of occurrence of opportunistic infections in this region. Unfortunately the related neuropathology is not well assessed.

Considering the large number of HIV-1 carriers in India, it is reasonable to expect that India may have a high incidence of HIV related neurological complications; interestingly that has not been the case. On the contrary, several reports suggest that India has an exceptionally low incidence of HAD cases (Sinha et al., 2004; Shankar et al., 2005; Gharu et al., 2009; Kandathil et al., 2009). Similarly, clinical studies from sub-Saharan Africa suggest clade/subtype specific differences in neurocognitive impairment (Liner et al., 2007; Sacktor et al., 2007; Sacktor et al., 2009). This has opened an interesting area of research which is being pursued by many investigators, including us at National Brain Research Centre.

HIV-1 associated dementia is quite common in western countries with prevalence up to 30-40% where clade B prevails. However, the incidence of HAD in India is believed to be quite low (1-3%) compared to western countries which may be due to a different clade HIV-1C, though there are no experimental studies to prove that. It is surprising that although HIV-1 clade C affects more than half the total HIV-1 infected individuals worldwide, there have been limited studies to understand the biology of HIV-1 clade C, particularly in relation to human CNS or neurological disorders.

The HIV-1 enters the CNS at a very early stage of infection, but it remains controlled apparently until the onset of significant immune compromise (Bell et al., 2006). Current theories regarding cellular and molecular mechanisms that lead to neuropathogenesis and host-viral interplay that may increase risk of developing HAND, lack clarity. A complete understanding of the role of viral proteins that result in neurological complications following HIV-1 infection of the CNS remains elusive till date (Jayadev and Garden, 2009).

5. HUMAN IMMUNODEFICIENCY VIRUS-1 AND BRAIN CELLS

The central nervous system consists of neurons as well as non-neuronal cells like astrocytes, oligodendrocytes macrophages and microglia. HIV-1 selectively affects different cell types in the brain. Soon after systemic infection, HIV-1 invades the CNS and establishes a life-long persistence due to the latent nature of the viral DNA and relative inadequacy of anti-HIV-1 drugs in CNS, because most of such drugs do not effectively cross the blood-brain barrier (Alexaki, et al., 2008). Entry of HIV-1 into neuroectodermal cells is independent of the CD4 receptor, and a number of different cell surface molecules have been implicated as alternate receptors of HIV-1 (Kramer-Hammerle, et al., 2005). The level of susceptibility of brain cells to HIV-1 infection differs dramatically (Gonzalez-Scarano and Martin-Garcia, 2005). The cell types that have been implicated in viral replication and pathogenesis are the perivascular macrophages (PM) and microglial cells where active replication of virus occurs (Wiley, et al., 1986; Fischer-Smith and Rappaport, 2005). HIV-1-infection of T-cells/monocytes occurs via viral engagement of the CD4 surface receptor and chemokines co-receptors, CXCR4/CCR5 and CCR3 (Deng, et al., 1996; Doranz, et al., 1996). HIV-1 traffics into brain through monocytes, as monocytes have privilege to enter the brain parenchyma. Perivascular macrophages are the primary cell type for productive infection of HIV within the CNS (Williams, et al., 2001). The microglial cells are also infected by HIV (Strizki, et al., 1996) and are in fact the most permissive cells to HIV-1 infection in the CNS. Following infection of brain cells, multi-nucleated giant cells (MNGCs) are formed by fusion of several infected cells that express HIV envelope glycoprotein on them; these also act as site for virus production. The presence of MNGCs is considered to be the hallmark of HAD (Sharer, et al., 1985).

Interestingly, other brain cells such as neurons, astrocytes and oligodendrocytes are not productively infected with HIV. Among the glial cells, astrocytes are the most abundant neuroglial cell type in the brain and they play important of roles in normal maintenance of CNS function (Seth and Koul, 2008). Dysfunction of glial cells, particularly astrocytes result in severe damage to neurons that largely account for the neurological complications observed in HAD. Viral proteins like Tat, Nef, and Rev are also expressed in astrocytes and brain mononuclear phagocytes. They induce inflammatory responses that lead to neuronal damage/cell death (Ranki, et al., 1995; Kaul and Lipton, 2006). Extracellular factors from other HIV-1 infected cells like monocytes, microglia and lymphocytes can trigger persistently infected astrocytes to synthesize HIV-1 proteins (Tornatore, et al., 1991). Similarly, oligodendrocytes are affected by the viral glycoprotein gp120 that alters calcium homeostasis leading to apoptosis (Codazzi, et al., 1995). A study of brain tissues from pediatric AIDS patients obtained before the era of highly active antiretroviral therapy suggest that HIV-1 is present in nestin positive cells, which supports the notion that neural stem cells/progenitors are also involved and perhaps permissive to HIV-1 infection (Schwartz, et al., 2007). Furthermore, in some *in vitro* culture models of neural cells, it has been reported that HIV-1 can persist in neural progenitor populations for a long time, and are capable of releasing the virus in different amounts depending on the extracellular environment and causing variable cellular changes in response to HIV-1 persistence (Rothenaigner, et al., 2007).

5.1. HIV-1 Neuropathogenesis, Direct and Indirect Damage to Human Neuron

So far terms like HAD or HAND have been used to describe the most severe or the more common milder form of neurological syndromes associated with HIV-1 infected individuals (Li, Li et al., 2009). The motor, cognitive, and behavioral impairments accompanying HIV-1 infection may be present in 30% of adult patients and up to 50% in pediatric AIDS patients. Neurons are very rarely infected by HIV-1 although neuronal deficits and neurotoxic events are very common in HAD (Kaul, et al., 2001; Kaul and Lipton, 2006; Kaul, 2008). There are two predominant hypothesis for the underlying mechanism of HIV-1 neuropathogenesis; 1) DIRECT PATHWAY – neuropathogenesis occurring by viral protein shed by virus i.e glycoprotein gp120 and the constitutive proteins like transactivating protein Tat or 2) INDIRECT PATHWAY – neuropathogenesis by endogenous components such as cytokines, chemokines, glutamate, quinolinate and NO released by non-neuronal-infected cells (Jones and Power, 2006). As mentioned earlier, the non-neuronal cells of the brain, like glial cells, macrophages, lymphocytes and microglia can be infected by HIV-1. Infected glial cells can activate uninfected macrophages or microglia to release potential neurotoxic substances, including quinolinic and arachidonic acids, Platelet-activating Factor (PAF), nitric oxide and superoxide ions, free radicals, tumor necrosis factor alpha (TNF-alpha) (Kaul, et al., 2001; Kaul and Lipton, 2006; Rumbaugh and Nath, 2006). In this context these substances have been collectively referred to as virotoxins (Nath and Geiger, 1998) as they are capable of inducing neuronal injury, dendritic and synaptic damage, and eventually cell death via apoptosis (Kaul and Lipton, 1999; Bezzi, et al., 2001; Kaul, et al., 2001; Mollace, et al.,

2001). Some direct neurotoxic effects of HIV-1 proteins like gp120 and Tat have also been reported (Kaiser, et al., 1990; Sabatier, et al., 1991; Buzy, et al., 1992; Lipton, 1994; Magnuson, et al., 1995; Nath, et al., 2000; Pocernich, et al., 2005; Wallace 2006; Mishra, et al., 2008).

The neurological complications in HIV-1 patients have been correlated to the amount of virion trafficking to the brain that is mostly dependent on secretion of the chemoattractive cytokine, monocyte chemoattractant protein-1 (MCP-1)/CC Chemokine Ligand-2 CCL2 from the glial cells (Eugenin, et al., 2006). Levels of MCP-1/CCL-2 correlate with viral load in cerebrospinal fluid (CSF) and severity of dementia in HIV-1 infected individuals (Conant, et al., 1998; Kelder, et al., 1998). As mentioned earlier, the direct viral-induced damage to brain cells has been attributed to two viral proteins namely the envelope glycoprotein, gp120, and the non-structural transactivating protein, Tat. Both these viral proteins are toxic to neuron because of their ability to act as excitotoxins. They can evoke release of endotoxins or proinflammatory cytokines (Neri, et al., 2007) that are directly implicated in development of HAD.

The activation of astrocytes leads to astrocytosis and to increased permeability of the blood brain barrier (BBB). Modulation in astrocytic functions results in increased release of Ca2+ and glutamate, and also decreased glutamate uptake by astrocytes. This results in the increased concentration of extra-cellular glutamate and other neurotoxins, ultimately resulting in neuronal death (Lipton, 1991; Haughey and Mattson, 2002).

Unlike other glial cells of CNS, the astrocytes harbour HIV-1 in a restrictive fashion. The mechanism of viral entry into astrocytes is unclear because they neither have detectable levels of CD4 nor the main HIV-1 co-receptors on their surface (Petito and Cash, 1992). Endothelial cells also lack the receptors commonly used by HIV-1. Since neurons are not infected by HIV-1 but die promiscuously in HAD patients, it indicates that indirect mechanisms are involved in the neuropathogenesis of HIV-1. In fact, a large number of viral as well as host cellular factors have been associated with HIV-1 induced neurodegeneration in the absence of direct neuronal infection with HIV-1. Interestingly, there is no known correlation between the number of HIV-1 infected macrophages or microglia and severity of symptoms in HAD.

A wide range of other neurological disorders are associated with HIV-1 which consist of (i) Primary Viral syndromes like AIDS dementia complex (ADC), HIV associated neurocognitive disorders (HAND), Mild cognitive motor disorder (MCMD), vascular myopathy, aseptic meningitis (ii) Opportunistic infections such as cerebral toxoplasmosis, cryptococcal meningitis, progressive multifocal leucoencephalities, Cytomegalovirus encephalitis (iii) Primary central nervous lymphoma and others which may include peripheral neuropathy, myopathy Guillain Barre syndrome, cerebrovascular complications etc.

5.2. Neurocognitive Impairment in Pre- and Post-HAART Era

Before combinatorial antiretroviral therapy was introduced, patients experienced the most severe form of HIV associated dementia that occurred in the advanced stage of HIV disease (McArthur, et al., 1993). Following introduction of cART/ HAART in the mid-1990s the severity of HAD reduced considerably and improvement in neurocognitive performances were reported in HAD patients (Cohen, et al., 2001; Sacktor, et al., 2006). Till date, introduction of cART is a major advancement in attenuating HIV infection in HIV/AIDS population. HAART / cART significantly delay the progression of HIV infection and help such patients to live longer. In recent years as patient live longer prevalence of several AIDS-defining illnesses have increased. Fortunately with this improvement in treatment, a less severe form of neurocognitive impairment is more prevalent than clear dementia and the clinical picture of neuroAIDS is said to be evolving (McArthur 2004; Power, et al., 2009). Over a period of years, it has been observed that HAART fails to provide complete protection from the development of HAD (Liner, et al., 2008; Brew, et al., 2009). It could be due to the limited penetration of many antiretroviral drugs into the CNS and in the long term, HAART poses a potential toxicity risk that may impair neurocognitive performances (Letendre, et al., 2008; Liner, et al., 2008). Before the HAART era, neuroinflammation was common in HIV patients and was generally termed as HIV encephalitis (HIVE). HIVE is characterized by activated microglia, macrophage infiltration, formation of multinucleated giant cells and pronounced astrocytosis. Neuroinflammation usually increases with the progression of disease culminating in HIV associated dementia or AIDS dementia complex (ADC), characterized with neuronal damage or loss. Other characteristic features include reduced synaptic and dendritic density, which are in fact the best neuropathological correlates in pre-HAART era (Glass, et al., 1995; Masliah, et al., 1997). Although it is expected that neuroinflammation would reduce by improved treatment, however some autopsy cases of HIV-1 related death suggest the opposite effect after introduction of HAART (Anthony and Bell, 2008) and hence needs further investigations on those lines.

5.3. Role of HIV-1 Proteins in Neurocognitive Disorders

HIV-1 proteins including envelop surface unit gp120 and the accessory proteins Tat, Nef and Vpr, have been demonstrated to directly induce neuronal cell death and cause neurotoxicity. HIV-1 neuropathogenesis is the complex interplay between viral proteins, pro-inflammatory and anti-inflammatory cytokines and several cellular factors. In this chapter, we restrict ourselves to the HIV-1 transactivating protein, a neurotoxic protein released by the virus that affects the neighbouring uninfected cells and has been studied in much more detail than other viral proteins. A brief account of these proteins and their actions on brain cells is provided below.

5.4. Tat (Transactivator of Transcription)

Tat is the primary transactivator protein for HIV and is released from infected cells and can be found in the serum and brain tissue of patients of HIV encephalitis. Tat appears to be a prominent contributor to neuronal damage in HIV infection. Tat induces neurotoxicity by stimulating numerous inflammatory cytokines and chemokines from monocytes/macrophages and astrocytes (Philippon et al., 1994; Nath et al., 1999; Kutsch et al., 2000). TNF-alpha and SDF-1 mediate neurotoxicity (New et al., 1998; Shi, et al., 1998) and the potent monocyte chemoattractant, MCP-1/CCL-2, is elevated in the CSF and brains of HIV-1-infected patients with dementia (Madani et al., 1998). Tat causes oxidative damage to neurons when it is released from glial cells (Aksenov et al., 2001). Interestingly the neurotoxic effect of Tat has been reported to be HIV-1 clade specific as differential neurotoxicity was observed in Tat derived from HIV-1 clade B and C (Li et al., 2008; Mishra et al., 2008; Rao et al., 2008). HIV-1 clade specific effects of Tat on neuronal damage correlate well with clinical observations of lower prevalence of HIV-1 associated neurological deficits in AIDS patients in India where clade C predominates.

5.5. Gp 120

The HIV-1 gp120 protein forms the surface protein of the HIV-1 virion and is reported in the brains of HIV-1 dementia patients, mainly in the perivascular regions (Jones et al., 2000). Neurotoxic effects are mediated largely by indirect effects from microglia, macrophages, and astrocytes, but also by direct effects on neurons (Lipton et al., 1991; Lipton 1993). Both in *in vitro* and *in vivo* studies, gp120 produces injury and apoptosis in primary rodent and human neurons (Muller et al., 1992; Aggoun-Zouaoui et al., 1996; Corasaniti, et al., 2001; Kaul and Lipton, 2006). Almost a decade ago, it was shown that gp120-induced neuronal injury occurs as a consequence of direct interaction with neurons via chemokine receptors and their cognate G protein-signaling systems (Meucci, et al., 1998), as well as, indirectly, via release of macrophage toxic factors.

5.6. Vpr (Viral Protein R)

The viral protein R (Vpr) is a small basic protein (14 kDa) of 96 amino acids that is well conserved in HIV-1 (Tristem, Marshall et al., 1992). Increased levels of Vpr have been detected in the cerebrospinal fluid of AIDS patients with neurological disorders which indicates the extracellular presence of Vpr, and possibly a role for this HIV-1 protein in the neuropathogenesis of HAD (Levy, et al., 1994; Pomerantz, 2004).

5.7. VPU

The protein vpu is an integral membrane protein, which is mainly present intracellularly. The main function of vpu is enhancement of virion-release from HIV-1-infected cells (Trono, 1995).

5.8. Nef (Negative Factor)

The viral negative protein of 27 kD is required by HIV-1 for proper budding from infected cells. Although Nef expression is abundant in brains of HIV-1-demented patients (Tornatore, et al., 1994; Ranki, et al., 1995), unlike Tat, Nef is not excreted; it is found only intracellularly, mostly in the nucleus of infected cells and also in the cytoplasm. The presence of extracellular nef in the brain of HIV-1-infected patients is uncommon. However, one study showed that nef possesses membrane fusion properties, which might allow the protein to pass through the cellular membrane (Curtain, et al., 1994).

5.9. Rev

Rev is a small regulatory protein of HIV-1 that is essential for virus replication. It controls the pattern of viral gene expression by promoting the transition from the early phase of infection, during which small regulatory proteins are expressed, to the late stage, when larger structural proteins are synthesized and assembled into viral particles. Like Tat and Vpr, Rev is a transactivator within the HIV genome. Like Nef, it is only available to the extracellular milieu upon rupture of infected cells, but may also mediate neuronal damage through indirect intracellular effects (Freed and Martin 2001).

5.10. Tat as a Factor for Neuropathogenesis and Cognitive Dysfunction

The presence of HIV-1 Tat protein in brains of HIVE patients has been demonstrated by several groups (Del Valle, et al., 2000; Hudson, et al., 2000; Giunta, et al., 2009). The HIV-1 Tat is released by HIV-1 infected cells and is taken up by neighboring uninfected cells (Ensoli, Buonaguro et al., 1993). Released Tat binds to heparan sulfate proteoglycans (HSPG) via its basic region that has heparin-binding properties and storage of Tat in extracellular matrix occurs in this binding form (Chang, et al., 1997). With help of low-density lipoprotein receptor-related protein (LRP), HIV-1 Tat protein binds to neurons and efficiently get internalized (Liu, et al., 2000).

Tat induced neuronal apoptosis is mediated via involvement of NMDA receptor, Post-synaptic density protein-95 (PSD-95) and neuronal nitric oxide synthase (nNOS) (Eugenin, et al., 2007). Tat can directly interact with NMDA receptor to cause neuronal apoptosis (Li, Huang et al., 2008). Furthermore, the neuronal function may be disrupted by HIV-1 Tat through interaction with CXCR4 (Xiao, et al., 2000). Along with the ability to induce production of chemokines and chemokine receptors, Tat by itself acts as a chemoattractant, especially for monocytes (Albini, et al., 1998). Interestingly, chemoattractive properties of HIV-1 Tat are clade specific. HIV-1 Tat B is a strong chemoattractant where as the clade C Tat has impaired chemoattractant nature. This decrease in the chemoattractant property of Tat has been attributed to the natural mutation at the 31 position of the Tat protein (Ranga, et al., 2004). Tat has significant sequence homology with several chemokine motifs such as CCF motif, SYXR motif and CXC/CC motif which also suggests its chemoattractive properties (Lusti-Narasimhan, et al., 1995).

5.11. Effect of Tat on Blood Brain Barrier

Blood brain barrier (BBB) consists of the cell layers of endothelial cells on the luminal surface and the astrocyte foot processes on the abluminal surface. The endothelial cells are connected together by tight junctions. In neuroAIDS patients, when Tat is released from infected perivascular macrophages it can disrupt the BBB (Toborek, et al., 2003). Different regions of brain like hypothalamus, occipital cortex and hippocampus have the highest level of uptake of Tat and are significantly correlated in affecting brain function (Banks, et al., 2005). Functional integrity of the tight junctions of brain endothelium is modulated by HIV-1 Tat (Annunziata, 2003). Tat disrupts the tight junction protein zonula occludens-1 (ZO-1) continuity (Pu, et al., 2005) and decreases expression of claudin-5 (Andras, et al., 2005). Extracellular HIV-1 Tat up-regulates matrix metalloproteinase-9 (MMPs) which impairs the integrity of blood brain barrier leading to enhanced infiltration of monocyte into the CNS, that may subsequently result in enhanced trafficking of HIV-1 into the brain (Ju, et al., 2009).

5.12. Human Fetal Brain Cells as a Model for NeuroAIDS Studies

Lack of animal models for investigating HIV-1 infections is a serious limitation for the advancement of neuroAIDS research. Most of the animal models have not been successful as HIV-1 is a human specific virus. Simian Immunodeficiency Virus (SIV) model can provide an excellent non-human primate model for studying HIV as it is closely related to HIV on a molecular basis (Lackner and Veazey, 2007), however the SIV model has major disadvantages like (i) rhesus macaques are costly and require accredited primate facilities, (ii) Vpx gene is unique to SIV where as Vpu is present in HIV, (iii) AIDS generally develops within six to twelve months of infection with SIV, where as HIV takes several years to develop human AIDS (Shacklett, 2008). Similarly as the rodent cells are non-permissive for HIV infection, attempts to engineer rodents have been partially successful (Shultz, et al., 2007). Recently, several groups have reported different humanized rodent models which are surgically implanted with fetal thymic and liver organoids, as in the SCID–hu system, that partially address these issue (Melkus, et al., 2006). Further animal models do not realistically represent the human disease and the results obtained using animal models often fail to hold true in human settings. In such a scenario use of human cell culture system as an *in vitro* system is a reasonable approach. Primary human brain cell culture system is an excellent experimental tool to be used for studying cellular and molecular basis of HIV-1 related neurological complications.

Several investigators, including us, are employing primary cell culture system of human brain cells. Few laboratories around the world have been successful in isolating and characterizing human neural stem/precursor cells for their use in neuroAIDS studies. In India, Seth and his colleagues at National Brain Research Centre, Manesar have successfully isolated and characterized several lines of human neural stem cells and are using them for neuroAIDS studies. Briefly, the human brain tissue are collected from completely aborted fetus of early gestation of around 8-16 weeks and samples are processed by strictly following the guidelines of Institutional Human Ethics Committee. Brain samples are processed to isolate neural precursor cells. These precursor cells are passaged several times to get the purified populations of human neural precursor cells (hNPCs). These cells are highly proliferative and express nestin, the neural precursor cell marker. This robust cell culture system has a potential to be used as a model for studying neuron-glia crosstalk and *in vitro* model for other neurodegenerative diseases.

5.13. Effect of HIV-1 Tat on Human Brain Cells

Neural and Glial Cells -To understand the neurological complications of HIV-1 or its protein, it is essential to focus on the complex interactions between the transactivating protein Tat, toxic cellular proteins and the immune response among various cell types in the CNS.

Although productive infection of HIV-1 is lacking in neurons, neuronal loss in distinct regions of the brain has been demonstrated in HIV/AIDS patients. HIV-1 Tat causes loss of selective populations of neurons *in vivo* as well as *in vitro* models (Hayman, et al., 1993; Jones, et al., 1998; Maragos, et al., 2003). The neurotoxiciy of Tat is distinct in particular regions which included striatum (Hayman, Arbuthnott et al., 1993), dentate gyrus and CA3 region of hippocampus (Maragos, et al., 2003) in several *in vivo* studies. There is a significant decline of dopamine in the striatum of Tat- treated animals which indicates dysfunction of nerve terminals among the dopamine producing neurons (Ferris, et al., 2009). HIV-1 Tat can cause neurotoxicity by releasing neurotoxic substances, stimulating cytokines and chemokines in macrophages and astrocytes in the brain (Kutsch, et al., 2000; D'Aversa, et al., 2004). Among these the most significant are MCP1/CCL2 and CXCL-10 (McManus et al., 2000; Eugenin, et al., 2005; Mishra, et al., 2008) and TNF alpha (New, et al., 1998; Shi, et al., 1998; Sui, et al., 2007). HIV Tat stimulates pro-inflammatory cytokines and neurotoxins in brain cells (Chen, et al., 1997; Pulliam, et al., 2007). Tat can also induce endothelin-1 in astrocyte (Chauhan, Hahn et al., 2007) and when synergized with gamma interferon it produces CXCL-10 (Dhillon, Zhu et al., 2008). Presence of MCP-1/CCL2 and CXCL10 in CSF has been correlated to severity of HIV dementia in HIV/AIDS patients.

Neural Precursor Cells - Despite advent of highly active anti-retroviral therapy the prevalence of HAND continues to be at alarming levels. Adult mammalian nervous system retains a limited capacity for self-renewal that is important for its normal functions, like learning and memory. Differentiation of neural stem cell or adult neurogenesis is reported to be impaired in several psychiatric disorders (Kempermann, Krebs et al., 2008) and neurodegenerative diseases including Alzheimer's disease and HIV associated dementia (Ming and Song, 2005). Until recently, HIV-1 was believed to be affecting only the astrocytes and microglial cells in the brain, however infection of neurons is rare (Gonzalez-Scarano and Martin-Garcia, 2005). Interestingly, it has been revealed that HIV-1 virus can also infect neural precursor cells in culture (Lawrence et al., 2004; Rothenaigner et al., 2007) as well as in archived brain sections from pediatric neuroAIDS patients.

The presence of HIV-1 in neural precursor cells (NPCs) has raised concern about the consequence of harboring of HIV-1 in NPCs. Furthermore number of adult neural precursor cells is far fewer in HIV patients with neurological deficits as compared to their age matched non-infected controls, or from HIV-1 infected but non-demented individuals (Krathwohl and Kaiser, 2004). However, such studies have been restricted to limited number of cases, as autopsies of AIDS patients are rare. Detailed investigation into the effect of HIV-1 and its proteins on properties of neural stem cells, particularly neurogenesis is lacking. According to another recent study, number of proliferating neural precursor cells are low in HIV/gp120 transgenic mice compared to the wild types, and gp120 protein seem to affect proliferation of neural precursor cells without affecting their viability (Okamoto et al., 2007). However, the details of such effects are poorly defined. The modulation by HIV-1 transactivating protein Tat, which actually affects multiple genes in the infected as well as neighbouring uninfected cells and has a profound effect on functions of neuron and glial cells, had not been studied in detail in human neural precursor cells. To fulfill this lacuna, Seth and colleagues carried out detailed investigations to understand the effect of HIV-1 transactivating protein Tat on proliferation and differentiation of hNPCs, the two most important properties of human neural stem/precursor cells. Their observations provide convincing evidence, that HIV-1 Tat attenuates the growth, proliferation as well as differentiation capabilities of human fetal brain derived neural stem/precursor cells. The study also defines possible cellular and molecular mechanisms for this phenomenon. The author and his colleagues are currently studying the effect of live virus infections of hNPCs with HIV-1 B and HIV-1 C to confirm their observations of the clade specific neurotoxicity of Tat live virus infections. Recent advances suggest that the HIV-1 infections in CNS are not limited to glial cells but also to hNPCs. These observations make the neuroAIDS field more challenging as the virus not only damages the neurons, but also the neuron precursor cells. The virus hence diminishes chances of replenishment with new neurons, or neurogenesis, this may also have serious implications in the cognitive deficits in HIV-1 infected individuals.

Improved understanding of clade specific effects of HIV-1 on human brain cells, particularly neurons, and their correlation with the clinical data of different levels of neurocognitive deficits in HIV-1 B and C in Western countries and India respectively, suggest that there is definitely a relationship between HIV-1 associated neurological disorders and clade / strain of the virus.

CONCLUSION AND FUTURE RESEARCH PROSPECTIVE

Further investigations to understand the intricacies of HIV-1 neuropathogenesis are being made worldwide by basic as well as clinical investigators and would remain a challenge in near future as it is a complex phenomenon. Combinatorial antiretroviral therapy has offered new hope for the AIDS patients, but as it appears from the preliminary observations, the therapy although successful in considerably reducing the viral load from the circulatory system, has failed till date in eradicating the virus from brain of the HIV/AIDS patients.

The paradoxical aspects of HAART have a profound impact on severe CNS disease complicating HIV-1 infection. On one hand it reduces the incidence of major CNS opportunistic infections and ADC, while on the other this success brings a question whether it might invite more indolent, sub-clinical brain injury that may have long term consequences. It is speculated that CNS infection, associated local inflammation and immune-activation may begin to damage the brain during the long period before treatment is initiated and may even continue in the presence of effective systemic viral suppression. The most conspicuous and severe neurological complications of HIV-1 infection can be largely managed. The effects of therapy on this less severe and more subtle form of brain injury must be carefully observed and explored. A number of important treatment issues need to be addressed. There is an urgent need for improved understanding of the interaction between HIV and its human host which may provide the hope for adjunctive therapies to antiretroviral treatment to improve the cognitive impairments that remains the major morbidity in HIV/AIDS individuals. Improved therapeutic regimens may be developed in future for reducing the burden of HIV-1 associated neurocognitive disorders.

ACKNOWLEDGMENTS

The research work of Dr. Pankaj Seth at NBRC is supported by research grants (BT/PR6615/MED/14/857/2005) from Department of Biotechnology (DBT), New Delhi,

India, National Institutes of Health, USA (5 R01 NS055628-02), and institutional core funding.

REFERENCES

Abubakar, A., Van Baar, A., Van de Vijver, F. J. R., Holding, P., and Newton, C. R. J. C. (2008). Paediatric HIV and neurodevelopment in sub-Saharan Africa: a systematic review. *Tropical Medicine and International Health, 13*, 880–887.

Aggoun-Zouaoui, D., Charriaut-Marlangue, C., Rivera, S., and Jorquera, I. (1996). The HIV-1 envelope protein GP120 induces neuronal apoptosis in hippocampal slices. *Neuroreport, 7* 433-436.

Aksenov, M. Y., Hasselrot, U., Bansal, A. K., Wu, G., Nath, A., Anderson, C., Mactutus, C. F., Booze, R. M. (2001). Oxidative damage induced by the injection of HIV-1 Tat protein in the rat striatum. *Neuroscience Letters, 305*, 5-8.

Albini, A., Ferrini, S., Benelli, R., Sforzini, S., Giunciuglio, D., Aluigi, M. G., Proudfoot, A. E. I., and Noonan, D. M. (1998). *HIV-1 Tat Protein Mimicry of Chemokines. Proceedings of the National Academy of Sciences of the United States of America, 95*, 13153-13158.

Ances, B. M., and Clifford, D. B. (2008). HIV-associated neurocognitive disorders and the impact of combination antiretroviral therapies. *Current Neurology and Neuroscience Reports, 8*, 455-461.

András, I. E., Pu, H., Tian, J., Deli, M. A., Nath, A., Hennig, B., and Toborek, M. (2005). Signaling mechanisms of HIV-1 Tat-induced alterations of claudin-5 expression in brain endothelial cells. *Journal of Cerebral Blood Flow and Metabolism, 25*, 1159-1170.

Annunziata, P. (2003). Blood-brain barrier changes during invasion of the central nervous system by HIV-1. Old and new insights into the mechanism. *Journal of Neurolology, 250*, 901-906.

Anthony, I. C., and Bell, J. E. (2008). The Neuropathology of HIV/AIDS. *International Review of Psychiatry, 20*, 15-24.

Aylward, E. H., Henderer, J. D., McArthur, J. C., Brettschneider, P. D., and Harris, G. J. (1993). Reduced basal ganglia volume in HIV-1-associated dementia, results from quantitative neuroimaging. *Neurology, 43*, 2099-2104.

Banks, W. A., Robinson, S. M., and Nath, A. (2005). Permeability of the blood-brain barrier to HIV-1 Tat. *Experimental Neurology, 193*, 218-227.

Bell, J., Arango, J.-C., and Anthony, I. (2006). Neurobiology of multiple insults, HIV-1-associated brain disorders in those who use illicit drugs. *Journal of NeuroImmune Pharmacology, 1*, 182-191.

Berger, J. R., and Arendt, G. (2000). HIV dementia, the role of the basal ganglia and dopaminergic systems. *Journal of Psychopharmacology, 14*, 214-221.

Bezzi, P., Domercq, M., Vesce, S., and Volterra, A. (2001). Neuron-astrocyte cross-talk during synaptic transmission, physiological and neuropathological implications. *Progress in Brain Research, 132*, 255-265.

Boissé, L., Gill, M. J., and Power, C. (2008). HIV infection of the central nervous system: clinical features and neuropathogenesis. *Neurologic Clinics, 26*, 799-819.

Brew, B. J., Crowe, S. M., Landay, A., Cysique, L. A., and Guillemin, G. (2009). Neurodegeneration and ageing in the HAART era. *Journal of NeuroImmune Pharmacology, 4,* 163-174.

Bruck, I., Tahan, T. T., Cruz, C. R., Martins, L. T., Antoniuk, S. A., Rodrigues, M., Souza, S. M., and Bruyn, L. R. (2001). Developmental milestones of vertically HIV infected and seroreverters children, follow up of 83 children. *Arquivos de Neuro-psiquiatria, 59*(3-B), 691-695.

Buzy, J., Brenneman, D. E., Pert, C. B., Martin, A., and Salazar, A. (1992). Potent gp120-like neurotoxic activity in the cerebrospinal fluid of HIV-infected individuals is blocked by peptide T., *Brain research, 598,* 10-18.

Chang, H. C., Samaniego, F., Nair, B. C., Buonaguro, L., and Ensoli, B. (1997). HIV-1 Tat protein exits from cells via a leaderless secretory pathway and binds to extracellular matrix-associated heparan sulfate proteoglycans through its basic region. *AIDS, 11,* 1421-1431.

Chauhan, A., Hahn, S., Gartner, S., Pardo, C. A., Netesan, S. K., Mcarthur, J., and Nath, A. (2007). Molecular programming of endothelin-1 in HIV-infected brain, role of Tat in up-regulation of ET-1 and its inhibition by statins. *Faseb Journal, 21,* 777-789.

Chen, P., Mayne, M., Power, C., and Nath, A. (1997). The Tat protein of HIV-1 induces tumor necrosis factor-alpha production. Implications for HIV-1-associated neurological diseases. *Journal of Biological Chemistry, 272,* 22385-22388.

Codazzi, F., Menegon, A., Zacchetti, D., and Ciardo, A. (1995). HIV-1 gp120 Glycoprotein Induces [Ca2^+^]i~ Responses not only in Type-2 but also Type-1 Astrocytes and Oligodendrocytes of the Rat Cerebellum. *European Journal of Neuroscience, 7,* 1333-1341.

Cohen, R. A., Boland, R., Paul, R., Tashima, K. T., Schoenbaum, E. E., Celentano, D. D., Schuman, P., Smith, D. K., and Carpenter, C. C. Neurocognitive performance enhanced by highly active antiretroviral therapy in HIV-infected women. *AIDS, 15,* 341-345.

Conant, K., Garzino-Demo, A., Nath, A., McArthur, J.C., Halliday, W., Power, C., Gallo, R.C., and Major, E.O. (1998). Induction of monocyte chemoattractant protein-1 in HIV-1 Tat-stimulated astrocytes and elevation in AIDS dementia. *Proceedings of the National Academy of Sciences of the United States of America, 95*(6), 3117-21.

Corasaniti, M. T., Nistico, R., Costa, A., Rotiroti, D., and Bagetta, G. (2001). The HIV-1 envelope protein, gp120, causes neuronal apoptosis in the neocortex of the adult rat, a useful experimental model to study neuroaids. *Functional Neurology, 16*(4 Suppl), 31-38.

Curtain, C. C., Separovic, F., Rivett, D., Kirkpatrick, A., Waring, A.J., Gordon, L.M., and Azad, A.A. (1994). Fusogenic activity of amino-terminal region of HIV type 1 Nef protein. *AIDS Research and Human Retroviruses, 10,* 1231-1240.

D'Aversa, T. G., Yu, K. O., and Berman, J. W. (2004). Expression of chemokines by human fetal microglia after treatment with the human immunodeficiency virus type 1 protein Tat. *Journal of Neurovirology, 10,* 86-97.

De Ronchi, D., Faranca, I., Berardi, D., Scudellari, P., Borderi, M., Manfredi, R., and Fratiglioni, L. (2002). Risk factors for cognitive impairment in HIV-1-infected persons with different risk behaviors. *Archives of Neurology, 59,* 812-818.

Del Valle, L., Croul, S., Morgello, S., Amini, S., Rappaport, J., and Khalili, K. (2000). Detection of HIV-1 Tat and JCV capsid protein, VP1, in AIDS brain with progressive multifocal leukoencephalopathy. *Journal of Neurovirology, 6,* 221-228.

Deng, H., Liu, R., Ellmeier, W., et al. (1996). Identification of a major co-receptor for primary isolates of HIV-1. *Nature, 381*(6584), 661-666.

Dhillon, N., Zhu, X., Peng, F., Yao, H., Williams, R., Qiu, J., Callen, S., Ladner, A.O., and Buch, S. (2008). Molecular mechanism(s) involved in the synergistic induction of CXCL10 by human immunodeficiency virus type 1 Tat and interferon-gamma in macrophages. *Journal of Neurovirology, 14,* 196-204.

Doranz, B. J., Rucker, J., Yi, Y., Smyth, R.J., Samson, M, Peiper SC, Parmentier M, Collman RG, Doms RW. (1996). A dual-tropic primary HIV-1 isolate that uses fusin and the beta-chemokine receptors CKR-5, CKR-3, and CKR-2b as fusion cofactors. *Cell, 85,* 1149-1158.

Ensoli, B., Buonaguro, L., Barillari, G., Fiorelli, V., Gendelman, R., Morgan, R.A., Wingfield, P., and Gallo, R.C. (1993). Release, uptake, and effects of extracellular human immunodeficiency virus type 1 Tat protein on cell growth and viral transactivation. *Journal of Virology, 67,* 277-287.

Eugenin, E. A., Dyer, G., Calderon, T. M., and Berman, J. W. (2005). HIV-1 tat protein induces a migratory phenotype in human fetal microglia by a CCL2 (MCP-1)-dependent mechanism, possible role in NeuroAIDS. *Glia, 49,* 501-510.

Eugenin, E. A., J. E. King, JE, Nath A, Calderon TM, Zukin RS, Bennett MV, and Berman JW. (2007). HIV-tat induces formation of an LRP-PSD-95- NMDAR-nNOS complex that promotes apoptosis in neurons and astrocytes. *Proceedings of the National Academy of Sciences of the United States of America, 104,* 3438-3443.

Eugenin, E. A., Osiecki, K., Lopez, L., Goldstein, H., Calderon, T. M., and Berman, J. W. (2006). CCL2/monocyte chemoattractant protein-1 mediates enhanced transmigration of human immunodeficiency virus (HIV)-infected leukocytes across the blood-brain barrier, a potential mechanism of HIV-CNS invasion and NeuroAIDS. *Journal of Neuroscience, 26,* 1098-1106.

Ferris, M. J., Frederick-Duus, D., Fadel, J., Mactutus, C.F., and Booze, R. M. (2009). In vivo microdialysis in awake, freely moving rats demonstrates HIV-1 Tat-induced alterations in dopamine transmission. *Synapse, 63,* 181-185.

Fischer-Smith, T., and Rappaport, J. (2005). Evolving paradigms in the pathogenesis of HIV-1-associated dementia. *Expert Reviews in Molecular Medicine, 7,* 1-26.

Geretti, A. M. (2006). HIV-1 subtypes, epidemiology and significance for HIV management. *Current Opinion in Infectious Diseases, 19,* 1-7.

Gharu, L., R. Ringe, R., Pandey, S., Paranjape, R., and Bhattacharya J. (2009). HIV-1 clade C env clones obtained from an Indian patient exhibiting expanded coreceptor tropism are presented with naturally occurring unusual amino acid substitutions in V3 loop. *Virus Research, 144,* 306-314.

Giunta, B., Hou, H., Zhu, Y., Rrapo, E., Tian, J., Takashi, M., Commins, D., Singer, E., He, J., Fernandez, F., and Tan, J. (2009). HIV-1 Tat Contributes to Alzheimer's Disease-like Pathology in PSAPP Mice. *International Journal of Clinical and Experimental Pathology, 2,* 433-443.

Glass, J. D., Fedor, H., Wesselingh, S. L., and McArthur, J. C. (1995). Immunocytochemical quantitation of human immunodeficiency virus in the brain, correlations with dementia. *Annals of Neurology, 38,* 755-762.

Gonzalez-Scarano, F., and Martin-Garcia, J. (2005). The neuropathogenesis of AIDS. *Nature Reviews Immunology, 5,* 69-81.

Goodkin, K., Wilkie, F. L., Baldewicz, T. T., Concha, M., Tyl, M. D., Lopiccolo, C. J., and Shapshak, P. (2000). HIV-1-associated cognitive-motor disorders: A research-based approach to diagnosis and treatment. *CNS Spectrums, 5,* 49-60.

Hall, M., Whaley, R., Robertson, K., Hamby, S., Wilkins, J., and Hall, C. (1996). The correlation between neuropsychological and neuroanatomic changes over time in asymptomatic and symptomatic HIV-1-infected individuals. *Neurology, 46,* 1697-1702.

Haughey, N. J., and Mattson, M. P. (2002). Calcium dysregulation and neuronal apoptosis by the HIV-1 proteins Tat and gp120. *Journal of Acquired Immune Deficiency Syndrome, 31 Suppl 2,* S55-61.

Hayman, M., Arbuthnott, G., Harkiss, G., Brace, H., Filippi, P., Philippon, V., Thomson, D., Vigne, R., and Wright, A. (1993). Neurotoxicity of peptide analogues of the transactivating protein tat from Maedi-Visna virus and human immunodeficiency virus. *Neuroscience, 53,* 1-6.

Hemelaar, J., Gouws, E., Ghys, P. D., and Osmanov, S. (2006). Global and regional distribution of HIV-1 genetic subtypes and recombinants in 2004. *AIDS, 20,* W13-23.

Ho, D. D., Bredesen, D. E., Vinters, H. V., and Daar, E. S. (1989). The acquired immunodeficiency syndrome (AIDS) dementia complex. *Annals of Internal Medicine, 111,* 400-410.

Hudson, L., Liu, J., Nath, A., Jones, M., Raghavan, R., Narayan, O., Male, D., and Everall, I. (2000). Detection of the human immunodeficiency virus regulatory protein tat in CNS tissues. *Journal of Neurovirology, 6,* 145-155.

Jayadev, S., and Garden, G. A. (2009). Host and viral factors influencing the pathogenesis of HIV-associated neurocognitive disorders. *Journal of NeuroImmune Pharmacology, 4,* 175-189.

Jones, G., and Power, C. (2006). Regulation of neural cell survival by HIV-1 infection. *Neurobiology of Diseases, 21,* 1-17.

Jones, M., Olafson, K., Del Bigio, M. R., Peeling, J, and Nath, A. (1998). Intraventricular injection of human immunodeficiency virus type 1 (HIV-1) tat protein causes inflammation, gliosis, apoptosis, and ventricular enlargement. *Journal of Neuropathology and Experimental Neurology, 57,* 563-570.

Jones, M. V., Bell, J. E., and Nath A. (2000). Immunolocalization of HIV envelope gp120 in HIV encephalitis with dementia. AIDS, 14, 2709-2713.

Ju, S. M., Song, H. Y., Lee, J. A., Lee, S. J., Choi, S. Y., and Park, J. (2009). Extracellular HIV-1 Tat up-regulates expression of matrix metalloproteinase-9 via a MAPK-NF-kappaB dependent pathway in human astrocytes. *Experimental and Molecular Medicine, 41,* 86-93.

Kaiser, P. K., Offermann, J. T., and Lipton, S. A. (1990). Neuronal injury due to HIV-1 envelope protein is blocked by anti-gp120 antibodies but not by anti-CD4 antibodies. *Neurology, 40,* 1757-1761.

Kandathil, A. J., Joseph, A. P., Kannangai, R., Srinivasan, N., Abraham, O. C., Pulimood, S. A., and Sridharan, G. (2009). Structural basis of drug resistance by genetic variants of HIV type 1 clade c protease from India. *AIDS Research and Human Retroviruses, 25,* 511-519.

Kaul, M. (2008). HIV's double strike at the brain, neuronal toxicity and compromised neurogenesis. *Frontiers in Bioscience, 13,* 2484-2494.

Kaul, M., Garden, G. A., and Lipton, S. A. (2001). Pathways to neuronal injury and apoptosis in HIV-associated dementia. *Nature, 410*(6831), 988-994.

Kaul, M., and Lipton, S. A. (1999). Chemokines and activated macrophages in HIV gp120-induced neuronal apoptosis. *Proceedings of the National Academy of Sciences of the United States of America, 96,* 8212-8216.

Kaul, M., and Lipton, S. A. (2006). Mechanisms of neuronal injury and death in HIV-1 associated dementia. *Current HIV Research, 4,* 307-318.

Kelder, W., McArthur, J. C., Nance-Sproson, T., McClernon, D., and Griffin, D. E. (1998). Beta-chemokines MCP-1 and RANTES are selectively increased in cerebrospinal fluid of patients with human immunodeficiency virus-associated dementia. *Annals of Neurology, 44,* 831-835.

Kempermann, G., Krebs, J., and Fabel, K. (2008). The contribution of failing adult hippocampal neurogenesis to psychiatric disorders. *Current Opinion in Psychiatry, 21,* 290-295.

Kramer-Hammerle, S., Rothenaigner, I., Wolff, H., Bell, J. E., and Brack-Werner, R. (2005). Cells of the central nervous system as targets and reservoirs of the human immunodeficiency virus. *Virus Research, 111,* 194-213.

Krathwohl, M. D., and Kaiser, J. L. (2004). HIV-1 promotes quiescence in human neural progenitor cells. *Journal of Infectious Diseases, 190,* 216-226.

Kumar, A. M., Fernandez, J. B., Waldrop-Valverde, D., Ownby, R. L., Kumar, M., Singer, E. J., and Commins, D. (2009). Human immunodeficiency virus type 1 in the central nervous system leads to decreased dopamine in different regions of postmortem human brains. *Journal of Neurovirology, 15,* 257-274.

Kutsch, O., Oh, J., Nath, A., and Benveniste, E. N. (2000). Induction of the chemokines interleukin-8 and IP-10 by human immunodeficiency virus type 1 tat in astrocytes. *Journal of Virology, 74,* 9214-9221.

Lackner, A. A., and Veazey, R. S. (2007). Current concepts in AIDS pathogenesis, insights from the SIV/macaque model. *Annual Review of Medicine, 58,* 461-476.

Lawrence, D. M., Durham, L. C., Schwartz, L., Seth, P., Maric, D., and Major, E. O. (2004). Human immunodeficiency virus type 1 infection of human brain-derived progenitor cells. *Journal of Virology, 78,* 7319-7328.

Letendre, S., Marquie-Beck, J., Capparelli, E., et al. (2008). Validation of the CNS Penetration-Effectiveness rank for quantifying antiretroviral penetration into the central nervous system. *Archives of Neurology, 65,* 65-70.

Levy, D. N., Refaeli, Y., MacGregor, R. R., and Weiner, D. B. (1994). Serum Vpr regulates productive infection and latency of human immunodeficiency virus type 1. *Proceedings of the National Academy of Sciences of the United States of America, 91,* 10873-10877.

Li, W., Huang, Y., Reid, R., Steiner, J., Malpica-Llanos, T., Darden, T. A., Shankar, S. K., Mahadevan, A., Satishchandra, P., and Nath, A. (2008). NMDA receptor activation by HIV-Tat protein is clade dependent. *Journal of Neuroscience, 28,* 12190-12198.

Li, W., Li, G., Steiner, J., and Nath, A. (2009). Role of Tat Protein in HIV Neuropathogenesis. *Neurotoxicity Research, 16,* 205-220.

Liner, K. J., 2nd, C. D. Hall, et al., (2007). Impact of human immunodeficiency virus (HIV) subtypes on HIV-associated neurological disease. *Journal of Neurovirology, 13,* 291-304.

Liner, K. J., 2nd, Hall, C. D., and Robertson, K. R. (2008). Effects of antiretroviral therapy on cognitive impairment. *Current HIV/AIDS Reports, 5,* 64-71.

Lipton, S. A. (1991). Calcium channel antagonists and human immunodeficiency virus coat protein-mediated neuronal injury. *Annals of Neurology, 30,* 110-114.

Lipton, S. A. (1993). Human immunodeficiency virus-infected macrophages, gp120, and N-methyl-D-aspartate receptor-mediated neurotoxicity. *Annals of Neurology, 33,* 227-228.

Lipton, S. A. (1994). AIDS-related dementia and calcium homeostasis. *Annals of the New York Academy of Sciences, 747,* 205-224.

Lipton, S. A., Sucher, N. J., Kaiser, P. K., and Dreyer, E. B. (1991). Synergistic effects of HIV coat protein and NMDA receptor-mediated neurotoxicity. *Neuron, 7,* 111-118.

Liu, Y., Jones, M., Hingtgen, C. M., Bu, G., Laribee, N., Tanzi, R. E., Moir, R. D., Nath, A., and He J. J. (2000). Uptake of HIV-1 tat protein mediated by low-density lipoprotein receptor-related protein disrupts the neuronal metabolic balance of the receptor ligands. *Nature Medicine, 6,* 1380-1387.

Lusti-Narasimhan, M., Power, C. A., Allet, B., Alouani, S., Bacon, K. B., Mermod, J. J., Proudfoot, A. E., and Wells, T. N. (1995). Mutation of Leu25 and Val27 introduces CC chemokine activity into interleukin-8. *Journal of Biological Chemistry, 270,* 2716-2721.

Madani, N., Kozak, S. L., Kavanaugh, M. P., and Kabat, D. (1998). gp120 envelope glycoproteins of human immunodeficiency viruses competitively antagonize signaling by coreceptors CXCR4 and CCR5. *Proceedings of the National Academy of Sciences of the United States of America, 95,* 8005-8010.

Magnuson, D. S., Knudsen, B. E., Geiger J. D., Brownstone R. M., and Nath, A. (1995). Human immunodeficiency virus type 1 tat activates non-N-methyl-D-aspartate excitatory amino acid receptors and causes neurotoxicity. *Annals of Neurology, 37,* 373-380.

Maragos, W. F., Tillman, P., Jones, M., Bruce-Keller, A.J., Roth, S., Bell, J.E., and Nath, A. (2003). Neuronal injury in hippocampus with human immunodeficiency virus transactivating protein, Tat. *Neuroscience, 117,* 43-53.

Masliah, E., Heaton, R. K., Marcotte, T.D., Ellis, R.J., Wiley, C.A., Mallory, M., Achim, C.L., McCutchan, J.A., Nelson, J.A., Atkinson, J.H., and Grant, I. (1997). Dendritic injury is a pathological substrate for human immunodeficiency virus-related cognitive disorders. HNRC Group. The HIV Neurobehavioral Research Center. *Annals of Neurology, 42,* 963-972.

McArthur, J. C. (2004). HIV dementia, an evolving disease. *Journal of Neuroimmunology, 157,* 3-10.

McArthur, J. C., Hoover, D. R., Bacellar, H., et al., (1993). Dementia in AIDS patients, incidence and risk factors. Multicenter AIDS Cohort Study. *Neurology, 43,* 2245-2252.

McManus, C. M., Weidenheim, K., Woodman, S.E., Nunez, J., Hesselgesser, J., Nath, A., and Berman, J.W. (2000). Chemokine and chemokine-receptor expression in human glial elements, induction by the HIV protein, Tat, and chemokine autoregulation. *American Journal of Pathology, 156,* 1441-1453.

Melkus, M. W., Estes, J. D., Padgett-Thomas, A., Gatlin, J., Denton, P.W., Othieno, F.A., Wege, A.K., Haase, A.T., and Garcia, J.V. (2006). Humanized mice mount specific adaptive and innate immune responses to EBV and TSST-1. *Nature Medicine, 12,* 1316-1322.

Meucci, O., Fatatis, A., Simen, A.A., Bushell, T.J., Gray, P.W., and Miller, R.J. (1998). Chemokines regulate hippocampal neuronal signaling and gp120 neurotoxicity. *Proceedings of the National Academy of Sciences of the United States of America, 95,* 14500-14505.

Ming, G. L., and Song, H. (2005). Adult neurogenesis in the mammalian central nervous system. *Annual Review of Neuroscience, 28,* 223-250.

Miralles, P., Berenguer, J., García de Viedma, D., Padilla, B., Cosin, J., López-Bernaldo de Quirós, J.C., Muñoz, L., Moreno, S., and Bouza, E. (1998). Treatment of AIDS-associated progressive multifocal leukoencephalopathy with highly active antiretroviral therapy. *AIDS, 12,* 2467-2472.

Mishra, M., Vetrivel, S., Siddappa, N.B., Ranga, U., and Seth P. (2008). Clade-specific differences in neurotoxicity of human immunodeficiency virus-1 B and C Tat of human neurons, significance of dicysteine C30C31 motif. *Annals of Neurology, 63,* 366-376.

Mollace, V., Nottet, H. S., Clayette, P., Turco, M.C., Muscoli, C., Salvemini, D., and Perno, C.F. (2001). Oxidative stress and neuroAIDS, triggers, modulators and novel antioxidants. *Trends in Neurosciences, 24,* 411-416.

Muller, W. E., Schroder, H. C., Ushijima, H., Dapper, J., and Bormann, J. (1992). gp120 of HIV-1 induces apoptosis in rat cortical cell cultures, prevention by memantine. *European Journal of Pharmacology, 226,* 209-214.

Nath, A. (2009). HIV/AIDS and Indian youth--a review of the literature (1980-2008). *Journal of Social Aspects of HIV/AIDS Research Alliance, 6,* 2-8.

Nath, A., Conant, K., Chen, P., Scott, C., and Major, E.O. (1999). Transient exposure to HIV-1 Tat protein results in cytokine production in macrophages and astrocytes. A hit and run phenomenon. *Journal of Biological Chemistry, 274,* 17098-17102.

Nath, A., and Geiger, J. (1998). Neurobiological aspects of human immunodeficiency virus infection, neurotoxic mechanisms. *Progress in Neurobiology, 54,* 19-33.

Nath, A., Haughey, N. J., Jones, M., Anderson, C., Bell, J.E., and Geiger, J.D. (2000). Synergistic neurotoxicity by human immunodeficiency virus proteins Tat and gp120, protection by memantine. *Annals of Neurology, 47,* 186-194.

Neri, E., Musante, V., and Pittaluga, A. (2007). Effects of the HIV-1 viral protein TAT on central neurotransmission, role of group I metabotropic glutamate receptors. *International Review of Neurobiology, 82,* 339-356.

New, D. R., Maggirwar, S.B., Epstein, L.G., Dewhurst, S., and Gelbard, H.A. (1998). HIV-1 Tat induces neuronal death via tumor necrosis factor-alpha and activation of non-N-methyl-D-aspartate receptors by a NFkappaB-independent mechanism. *Journal of Biological Chemistry, 273,* 17852-17858.

Okamoto, S., Kang, Y. J., Brechtel, C.W., Siviglia, E., Russo, R., Clemente, A., Harrop, A., McKercher, S., Kaul, M., and Lipton, S.A. (2007). HIV/gp120 decreases adult neural progenitor cell proliferation via checkpoint kinase-mediated cell-cycle withdrawal and G1 arrest. *Cell Stem Cell, 1,* 230-236.

Petito, C. K., and Cash, K. S. (1992). Blood-brain barrier abnormalities in the acquired immunodeficiency syndrome, immunohistochemical localization of serum proteins in postmortem brain. *Annals of Neurology, 32,* 658-666.

Philippon, V., Vellutini, C., Gambarelli, D., Harkiss, G., Arbuthnott, G., Metzger, D., Roubin, R., and Filippi, P. (1994). The basic domain of the lentiviral Tat protein is responsible for damages in mouse brain, involvement of cytokines. *Virology, 205,* 519-529.

Pocernich, C. B., Sultana, R., Mohmmad-Abdul, H., Nath, A., and Butterfield, D.A. (2005). HIV-dementia, Tat-induced oxidative stress, and antioxidant therapeutic considerations. *Brain Research: Brain research Reviews, 50,* 14-26.

Pomerantz, R. J. (2004). Effects of HIV-1 Vpr on neuroinvasion and neuropathogenesis. *DNA and Cell Biology, 23,* 227-238.

Power, C., Boisse, L., Rourke, S., and Gill, M.J. (2009). NeuroAIDS, an evolving epidemic. *Canadian Journal of Neurological Sciences, 36,* 285-295.

Pu, H., Tian, J., Andras, I.E., Hayashi, K., Flora, G., Hennig, B., and Toborek, M.. (2005). HIV-1 Tat protein-induced alterations of ZO-1 expression are mediated by redox-regulated ERK 1/2 activation. *Journal of Cerebral Blood Flow and Metabolism, 25,* 1325-1335.

Pulliam, L. (2009). HIV regulation of amyloid beta production. *Journal of NeuroImmune Pharmacology, 4,* 213-217.

Pulliam, L., Sun, B., Rempel, H., Martinez, P.M., Hoekman, J.D., Rao, R.J., Frey, W.H. 2nd, and Hanson, L.R. (2007). Intranasal tat alters gene expression in the mouse brain. *Journal of NeuroImmune Pharmacology, 2,* 87-92.

Ranga, U., Shankarappa, R., Siddappa NB, Ramakrishna L, Nagendran R, Mahalingam M, Mahadevan A, Jayasuryan N, Satishchandra P, Shankar SK, Prasad VR. (2004). Tat protein of human immunodeficiency virus type 1 subtype C strains is a defective chemokine. *Journal of Virology, 78,* 2586-2590.

Ranki, A., Lassus, J., and Niemi, K. M. (1995). Relation of p53 tumor suppressor protein expression to human papillomavirus (HPV) DNA and to cellular atypia in male genital warts and in premalignant lesions. *Acta Dermato-Venereologica, 75,* 180-186.

Ranki, A., Nyberg, M., Ovod, V., Haltia, M., Elovaara, I., Raininko, R., Haapasalo, H., and Krohn, K. (1995). Abundant expression of HIV Nef and Rev proteins in brain astrocytes in vivo is associated with dementia. *AIDS, 9,* 1001-1008.

Rao, V. R., Sas, A. R., Eugenin, E. A., Siddappa, N. B., Bimonte-Nelson, H., Berman, J. W., Ranga, U., Tyor, W. R., and Prasad, V. R. (2008). HIV-1 clade-specific differences in the induction of neuropathogenesis. *Journal of Neuroscience, 28,* 10010-10016.

Robertson, D. L., Anderson, J. P., Bradac, J. A., et al., (2000). HIV-1 nomenclature proposal. *Science, 288,* 55-56.

Rothenaigner, I., Kramer, S., Ziegler, M., Wolff, H., Kleinschmidt, A., and Brack-Werner, R. (2007). Long-term HIV-1 infection of neural progenitor populations. *AIDS, 21,* 2271-2281.

Rumbaugh, J. A., and Nath, A. (2006). Developments in HIV neuropathogenesis. *Current Pharmaceutical Design, 12,* 1023-1044.

Sabatier, J. M., Vives, E., Mabrouk, K., Benjouad, A., Rochat, H., Duval, A., Hue, B., and Bahraoui, E. (1991). Evidence for neurotoxic activity of tat from human immunodeficiency virus type 1. *Journal of Virology, 65,* 961-967.

Sacktor, N., Nakasujja, N., Robertson, K., and Clifford, D. B. (2007). HIV-associated cognitive impairment in sub-Saharan Africa--the potential effect of clade diversity. *Nature Clinical Practice Neurology, 3,* 436-443.

Sacktor, N., Nakasujja, N., Skolasky, R., Robertson, K., Wong, M., Musisi, S., Ronald, A., and Katabira, E. (2006). Antiretroviral therapy improves cognitive impairment in HIV+ individuals in sub-Saharan Africa. *Neurology, 67,* 311-314.

Sacktor, N., Nakasujja, N., Skolasky R. L., et al., (2009). HIV subtype D is associated with dementia, compared with subtype A, in immunosuppressed individuals at risk of cognitive impairment in Kampala, Uganda. *Clinical Infectious Diseases, 49,* 780-786.

Schwartz, L., Civitello, L., Dunn-Pirio, A., Ryschkewitsch, S., Berry, E., Cavert, W., Kinzel, N., Lawrence, D. M., Hazra, R., and Major, E. O. (2007). Evidence of human immunodeficiency virus type 1 infection of nestin-positive neural progenitors in archival pediatric brain tissue. *Journal of Neurovirology, 13,* 274-283.

Seth, P., and Koul, N. (2008). Astrocyte, the star avatar, redefined. *Journal of Biosciences, 33,* 405-421.

Shacklett, B. L. (2008). Can the new humanized mouse model give HIV research a boost? *PLoS Medicine, 5*(1), e13.

Shankar, S. K., Mahadevan, A., Satishchandra, P., Kumar, R. U., Yasha, T. C., Santosh, V., Chandramuki, A., Ravi, V., and Nath, A. (2005). Neuropathology of HIV/AIDS with an overview of the Indian scene. *Indian Journal of Medical Research, 121,* 468-488.

Sharer, L. R., Cho, E. S., and Epstein L. G. (1985). Multinucleated giant cells and HTLV-III in AIDS encephalopathy. *Human Pathology, 16,* 760.

Shi, B., Raina, J., Lorenzo, A., Busciglio, J., and Gabuzda, D. (1998). Neuronal apoptosis induced by HIV-1 Tat protein and TNF-alpha, potentiation of neurotoxicity mediated by oxidative stress and implications for HIV-1 dementia. *Journal of Neurovirology, 4,* 281-290.

Shultz, L. D., Ishikawa, F., and Greiner D. L. (2007). Humanized mice in translational biomedical research. *Nature Reviews Immunology, 7,* 118-130.

Siddappa, N. B., Dash, P. K., Mahadevan, A., et al. (2004). Identification of subtype C human immunodeficiency virus type 1 by subtype-specific PCR and its use in the characterization of viruses circulating in the southern parts of India. *Journal of Clinical Microbiology, 42,* 2742-2751.

Sinha, S., Mathews, T., Arunodaya, G. R., Siddappa, N. B., Ranga, U., Desai, A., Ravi, V., and Taly, A. B. (2004). HIV-1 clade-C-associated ALS-like disorder, first report from India. *Journal of the Neurolological Sciences, 224,* 97-100.

Stern, R. A., Silva, S. G., Chaisson, N., and Evans, D. L. (1996). Influence of cognitive reserve on neuropsychological functioning in asymptomatic human immunodeficiency virus-1 infection. *Archives of Neurology, 53,* 148-153.

Strizki, J. M., Albright, A. V., Sheng, H., O'Connor, M., Perrin, L., and González-Scarano, F. (1996). Infection of primary human microglia and monocyte-derived macrophages with human immunodeficiency virus type 1 isolates, evidence of differential tropism. *Journal of Virology, 70*(11), 7654-62.

Sui, Z., Sniderhan, L. F., Schifitto, G., Phipps, R. P., Gelbard, H. A., Dewhurst, S., and Maggirwar, S. B. (2007). Functional synergy between CD40 ligand and HIV-1 Tat contributes to inflammation, implications in HIV type 1 dementia. *Journal of Immunology, 175,* 3226-3236.

Tebit, D. M., Nankya, I., Arts, E. J., and Gao, Y. (2007). HIV diversity, recombination and disease progression, how does fitness fit into the puzzle? *AIDS Reviews, 9,* 75-87.

Toborek, M., Lee, Y. W., Pu, H., Malecki, A., Flora, G., Garrido, R., Hennig, B., Bauer, H. C., and Nath, A. (2003). HIV-Tat protein induces oxidative and inflammatory pathways in brain endothelium. *Journal of Neurochemistry, 84,* 169-179.

Tornatore, C., Chandra, R,, Berger, J. R., and Major, E. O. (1994). HIV-1 infection of subcortical astrocytes in the pediatric central nervous system. *Neurology, 44*(3 Pt 1), 481-487.

Tornatore, C., Nath, A., Amemiya, K., and Major, E. O. (1991). Persistent human immunodeficiency virus type 1 infection in human fetal glial cells reactivated by T-cell factor(s) or by the cytokines tumor necrosis factor alpha and interleukin-1 beta. *Journal of Virology, 65,* 6094-100.

Tristem, M., Marshall, C., Karpas, A., and Hill, F. (1992). Evolution of the primate lentiviruses, evidence from vpx and vpr. *EMBO Journal, 11,* 3405-3412.

Trono, D. (1995). HIV accessory proteins, leading roles for the supporting cast. *Cell, 82,* 189-192.

Van Rie, A., Mupuala, A., and Dow, A. (2008). Impact of the HIV/AIDS epidemic on the neurodevelopment of preschool-aged children in Kinshasa, Democratic Republic of the Congo. *Pediatrics, 122,* e123-128.

Wallace, D. R. (2006). HIV Neurotoxicity, Potential Therapeutic Interventions. *Journal of Biomedicine and Biotechnology, 2006*(3), 65741.

Webb, K. M., Aksenov, M. Y., Mactutus, C. F., and Booze, R. M (2010). Evidence for developmental dopaminergic alterations in the human immunodeficiency virus-1 transgenic rat. *Journal of Neurovirology, 16,* 168-173.

Wiley, C. A., Schrier, R. D., Nelson, J. A., Lampert, P. W., and Oldstone, M. B. (1986). Cellular localization of human immunodeficiency virus infection within the brains of acquired immune deficiency syndrome patients. *Proceedings of the National Academy of Sciences of the United States of America, 83,* 7089-7093.

Williams, K. C., Corey, S., Westmoreland, S. V., Pauley, D., Knight, H., de Bakker, C., Alvarez, X., and Lackner, A. A. (2001). Perivascular macrophages are the primary cell type productively infected by simian immunodeficiency virus in the brains of macaques, implications for the neuropathogenesis of AIDS. *Journal of Experimental Medicine, 193,* 905-915.

Xiao, H., Neuveut, C., Tiffany, H. L., Benkirane, M., Rich, E. A., Murphy, P. M., and Jeang, K. T. (2000). Selective CXCR4 antagonism by Tat, implications for in vivo expansion of coreceptor use by HIV-1. *Proceedings of National Academy of Sciences USA, 97,* 11466-11471.

In: Expanding Horizions of the Mind Science(s)   ISBN: 978-1-62808-705-5
Editors: P.N. Tandon, R.C. Tripathi and N. Srinivasan   ©2013 Nova Science Publishers, Inc.

Chapter 22

PSYCHONEUROIMMUNOLOGY – WHERE MIND AND BODY MEET

Vivek H. Phutane and Prabha S. Chandra

Department of Psychiatry, National Institute of Mental Health and Neurosciences,
Bangalore, India

"The mental outlook of a patient has a huge effect on their immune system and their illnesses. Studies have shown that..., so being committed about remaining healthy and being hands-on about your health care can make all the difference in the world."

Aushtin Primus

ABSTRACT

Psychoneuroimmunology (PNI) is an emerging branch of medicine which bridges the gap between psychiatry, neurology, immunology, endocrinology, neuroscience, and internal medicine. Both physical and mental stress affect the mind and body through immune system. Stress activates the HPA axis and increases the release of glucocorticoids which inhibits immune function of the body. Stress leads to dysregulation of innate (NK cells, TNF- α) and acquired immunity (T-cells and B-cells). Stressful life events are associated with poor immune response viz. increase in white blood cells, NK cell activity, and decrease in number of T lymphocytes especially T helper cells and delayed wound healing. This may be associated reduction of host resistance to tumour growth and increases the risk of various cancers. There are 4 functions of blood brain barrier (BBB) which has relevance in psychoneuroimmunology - barrier, permeability, clearance and secretary functions. Sickness behaviour resembles symptoms of depression and is triggered by pro-inflammatory cytokines. Neurological as well as psychiatric illnesses are associated with impaired immune functioning. Cytokines used in the treatment of autoimmune disorders and cancers also cause symptoms similar to depression. Women have stronger cellular and humoral immune response due to presence of estrogen. Mind body intervention reduces the stress and enhances overall well being of persons by changing the immune responses to stress. Thus, psychoneuroimmunology has significant implications in the etiology, presentation and management of various neuropsychiatric illnesses.

1. WHAT IS PSYCHONEUROIMMUNOLOGY?

Psychoneuroimmunology is the trans-disciplinary branch of medicine which deals with interactions between brain (mind and behaviour), the immune system and the clinical implications of this interaction. Other names for it are Neuroimmunomodulation or Behavioural Immunology (Solomon, 2003).

This discipline deals with immunologically produced psychiatric symptoms as well as understanding various biological mechanisms underlying the influence of psychosocial factors on the onset and course of immunologically mediated diseases. It hence tries to bridge different branches of science such as psychiatry, neurology, immunology, endocrinology, internal medicine, surgery, neuroscience, and psychology.

The term 'Psychoneuroimmunology' was coined by Robert Ader in 1975 who believed that there was a link between what we think (our state of mind), our health and our ability to heal ourselves.

Both physical and mental stress can influence the mind and body respectively. For example, when a person is depressed, the body shows lethargy, fever, malaise and disturbed biological functions. Similarly, when the person is suffering from a serious disease i.e. cancer, a negative mental state may occur.

It has been proposed that based on this link, communication between immune system and CNS can be established by conditioning the immune system through mental processes. This provides patients with physical illnesses some feeling of control over their circumstances that creates a positive outlook and positive attitude. Thus it may act as a valuable supplement to conventional medical care (Irwin, 2008).

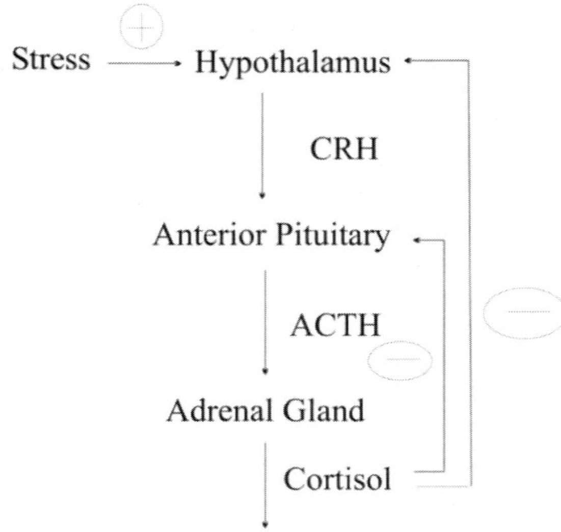

Figure 1. Interaction between stress and HPA axis.

Psychoneuroimmunology is hence considered as the scientific field which studies the communication between the nervous system, immune system and emotions and also focuses on the implications of these linkages for physical health (Ader, Cohen and Felten, 1995).

2. HOW DOES STRESS AFFECT THE IMMUNE SYSTEM?

Stress may be defined as a stimulus that challenges an organism to reach new homeostasis (Chrousos and Gold, 1992). The stimulus may be a *physical* one – such as injury, hunger, or cold – and / or *psychological* – such as bereavement, financial loss or social isolation. A stressor initiates a cascade of interaction between physiological and molecular events which starts with detection by immune or nervous system and ends with a new homeostasis.

In response to stress, corticotrophin releasing hormone (CRH) is secreted from the hypothalamus into hypophyseal – pituitary portal system. CRH acts on anterior pituitary which releases adrenocorticophic hormone (ACTH). ACTH acts on the adrenal cortex to stimulate the synthesis and release of glucocorticoids.

Glucocorticoids promote energy use, increase cardiovascular activity (increase flight or fight response), and *inhibit function* such as growth, reproduction, and *immunity*. This response also depends on frequency and duration of exposure.

Glucocorticoids influence the immune response by

1. Inhibiting cytokine production and function through interaction of glucocorticoid receptors with transcription factors (Auphan, DiDinato and Rosette, 1995) (Vacca, Felli and Farina, 1992).
2. Inhibit the inflammation by inhibiting the generation of products of arachidonic acid pathway (Goldstein, Bowen and Fauci, 1992).

The sympathetic nervous system (SNS) also plays an important role in immune changes induced by stressors as suggested by rapid onset and higher sensitivity to increased cardiovascular responses to stress (Herbert, Cohen and Marsland, 1994).

3. WHAT IS THE CLINICAL EVIDENCE THAT IMMUNE SYSTEM IS AFFECTED BY STRESS?

The clinical evidence for stress affecting the immune system comes from studies involving stressors such as care giving for Alzheimer's disease patients, during period of academic examinations, viral infections and wound healing. In a study of caregivers of patients with Alzheimer's, subjects (caregivers) were compared with controls for wound healing. The study showed that wound healing took significantly longer in caregivers than in controls. Peripheral blood leucocytes from caregivers produced significantly less IL-1β mRNA in response to lipopolysaccharide stimulation. These findings are important in the context of recovery from surgery where stress can prolong wound healing (Kiecolt-Glaser, Marucha, Malarkey, et al., 1995).

Antibody response to influenza vaccines has also been found to be reduced in the caregivers of Alzheimer's disease patients (Kiecolt-Glaser, Glaser, Cranestein, et al., 1996). This finding was replicated with hepatitis B virus vaccine in medical students experiencing acute examination stress (Glaser, Kiecolt-Glaser, Bonneau, et al., 1992). Acute respiratory infections caused by respiratory viruses such as rhinovirus, respiratory syncytial virus, and corona virus have also been found to be increased in direct proportion to the degree of psychological stress (Cohen, Tyrrell and Smith, 1991).

Stressful life events such as serious work related problems as well as change of residence are associated with occurrence of colorectal cancer (Courtney, Longnecker, Theorell, et al., 1993). Stressful life events reduce host resistance to tumour growth and are associated with recurrence in breast cancer (Palesh, Butler, Koopman, et al., 2007). Chronic depression and stressful life events have also been associated with HIV disease progression (Leserman, 2003).

Thus, naturalistic stressors lead to dysregulation of immune system with reduced responsiveness of immune system, increased susceptibility to infections as well as impaired wound healing. Stressful life events reduce the immune functioning of the host and are important factors in the recurrence of cancer as well as HIV disease progression.

4. WHAT ARE THE CELLULAR MECHANISMS BY WHICH STRESS AFFECTS THE IMMUNE SYSTEM?

In a study where eighteen elderly female caregivers of Alzheimer's disease patients were compared with 18 age and sex matched controls. It was found that caregivers had a significantly lower number of T cells especially of T-helper cells, higher T-suppressor cells as well as higher antibody titres for cytomegalovirus infection. Severity of stress correlated positively with T suppressor / cytotoxic cells and negatively with T helper / suppressor ratio (Pariante, Carpiniello, Orru, et al., 1997). Another study from a similar population reported reduced Natural Killer (NK) cell function in response to recombinant interferon-gamma (rIFN-gamma) and recombinant IL-2 (rIL-2) (Esterling, Kiecolt-Glaser and Glaser, 1996).

In a comparison between female caregivers of dementia sufferers with controls, female caregivers were found to have significantly higher antibody titres to the total viral antigen and a poorer HSV-1 specific T cell response than controls (Glaser and Kiecolt-Glaser, 1997). Examination taking students have been found to have lower levels of TNF- α which correlated with high anxiety scores (Chandrashekara, Jayashree, Veeranna, et al., 2007). Immune parameters of first year medical students were studied during their examination period. This study found significantly higher level of total T cells, reduced number of CD4+ and CD8+ cells, NK cells and IL-1, IL-2 producing cells (Uchakin, Tobin, Cubbage, et al., 2001). The above studies indicate that stress of care giving leads to dysregulation of innate (NK cells, TNF-α) as well as acquired immunity (T-cells and B-cells).

In a meta-analysis, Herbert and Cohen (Herbert and Cohen, 1993) reported that long term natural stressors lead to changes in the immune system such, as increased white blood cells, reduced lymphocytes in peripheral blood, reduced cytotoxic/suppressor T cells ratio, reduced lymphocyte response to phytohemagglutinin and concanavalin A and increase in antibody titres to herpes simplex virus type 1 as well as Epstein Barr virus. Social (interpersonal) stressors have more profound effects than non-social stressors.

Table. 1. Stress induced changes in immune function

Immune responses to acute stress:
↑ white blood cell count
↑ NK cell number
↑ NK cell activity
↓ number of T Lymphocytes
↓ proliferative responses to mitogens
↑ antibody titres to viruses
↓ antibody response to vaccine
Immune responses to chronic stress:-
↑ white blood cell count
↓ number of lymphocytes
↓ T helper : T suppressor ratio
↓ lymphocyte proliferative responses to mitogens
↓ lymphocyte production of $IL-2$ by stimulated lymphocytes
↓ NK cell activity
↑ antibody titres to viruses
↓ antibody response to vaccine
Delayed wound healing

(From Gumnick JF, Evans DL, Miller AH. Nervous, Endocrine, and Immune System interactions in Psychiatry. APA Textbook of Neuropsychiatry and clinical neurosciences).

5. WHERE DOES THE BLOOD BRAIN BARRIER COME INTO PICTURE?

Blood brain barrier acts as a barrier for free exchange of the circulating immune elements with the CNS as with the peripheral tissues. It regulates the exchange of immune substances between the blood and CNS and thus the interaction between the immune system and the peripheral tissues (Banks, 2009).

The BBB is the physical and physiologic dividing line between the immune system and the CNS and is the locale for interaction between them. The interactions of the BBB with events on its luminal (blood) and abluminal (brain) sides led the concept of Neurovascular Unit (NVU). There are four functions of the BBB which are fundamental to psychoneuroimmunology include- (1) Barrier function, (2) Permeability functions, (3) Clearance functions, (4) Secretory functions.

5.1. Barrier Function

Tight junctions of the endothelial cells of the capillaries in the brain and spinal cord and a lack of intracellular fenestrations with a decreased rate of pinocytosis help in performing the barrier function of the BBB. For restricted proteins, the serum: CSF ratio is about 200:1 (Davson and Segal, 1996).

At the same time, BBB has selective saturable transporters for nearly every substance needed for the homeostatic and nutritional maintenance of the CNS e.g. transporters for glucose, amino acids, vitamins, minerals etc. The most significant class of neuroimmune substances excluded by the BBB are *adrenergic compounds* (Sanders, 1995).

Disruption of BBB is a highly pathologic event and is associated with many neuro-immune phenomena e.g. lipopolysaccharide (LPS) disrupt the BBB. Release of cytokines such as tumor necrosis factor – α (TNF- α) is the mechanism for such disruption which is prostaglandin dependent. Most substances that disrupt the BBB do so by enhancing pinocytosis and other vesicular dependant mechanisms (transcytotic pathway).

5.2. Permeability Function

Substances crossing the blood brain barrier are vital for survival of the CNS. Steroid hormones, many peptides, regulatory proteins cross the BBB and influence brain functions. Cytokines can cross BBB and influence the neuroimmune system which is involved not only in pathologic but also physiologic states.

Other substances include interleukin-1 (IL-1), IL-6, TNF- α, nerve growth factor, neurotensin-3, brain derived neurotrophic factor, fibroblast growth factor, epidermal growth factor, leukemia inhibitory factor, ciliary neurotrophic factor, gamma interferon, and granulocyte macrophage colony stimulating factor.

Expression of intercellular adhesion molecule-1 by brain endothelial cells has been a key step in immune cell trafficking in the CNS. A prominent route by which immune cells cross the BBB is the process of diapedesis where immune cell tunnels through the brain endothelial cells.

5.3. Clearance Function

This occurs by two mechanisms viz the efflux of substances from brain interstitial fluid into the blood and the reabsorption of CSF which is a very important aspect in psychoneuroimmunology (Davson and Segal, 1996). The CNS can be a tremendous source of cytokine production, especially during meningitis. Except IL-2, remaining cytokines are reabsorbed at the arachnoid villi. About 30-40% of the circulating methionine- enkephalin is derived from the CNS (Banks and Kastin, 1997).

These substances are absorbed with the CSF and travel to the cervical lymphatics, where they affect immune cell function (Knopf, Cserr, and Nolan, 1995). Corticotrophin releasing hormone (CRH) is transported from brain to blood, where it circulates to the spleen and affects splenic production of the opiate neuroimmune modulator β-endorphin (Martins, Banks, and Kastin, 1997).

5.4. Secretion Function

Cytokines alter the permeability of the BBB by altering the receptor levels on brain endothelial cells as well as the ability of the cells to that comprise the BBB to secrete cytokines. Brain endothelial cells can secrete IL-1 and IL-6. Exposure to HIV-1 protein Tat induces release of IL-8 from brain endothelial cells (Hofman, Chen, and Incardona, 1999). Other substances that are important to neuroimmune system, such as prostaglandins and nitric oxide, also are released from the BBB. A brain endothelial cell might receive an immune signal at its blood side and respond by secreting a cytokine from its brain side (Varma, Nakaoke, and Dohgu, 2006).

In this manner the blood brain barrier plays an important role in the link between brain and immune mechanisms.

6. WHAT IS THE ROLE OF CYTOKINES IN SICKNESS BEHAVIOUR?

Sickness behaviour is the psychological and behavioural aspect of any sickness (Dantzer, O'Connor, Fruend, et al., 2008). It is triggered by pro-inflammatory cytokines such as IL-1, IL-6 and TNF-α produced by activated cells of the innate immune system. These cytokines act on the brain either via a neural or a humoral pathway (Dantzer, 2009). IL-1β is one of the most important cytokines for the induction of the symptoms of sickness behaviour such as fever, low energy levels, reduced social behaviour, anorexia, HPA axis activation (Anforth, Bluthe, Bristow, et al., 1998).

IL-6 is responsible for induction of the HPA as well as fever; but it has no effect on behavioural responses. It potentiates the behavioural responses of IL-1β. The evidence for these functions predominantly comes from the finding that injecting cytokines such as IL-1, IL-6 or TNF- α leads to symptoms of sickness behaviour while blocking them reduces the above symptoms. The imbalance between pro-inflammatory (e.g. IL-1, IL-6 or TNF- α) and anti-inflammatory (IL-10, TGF-β, IL-1Receptor antagonist) cytokines results in the ultimate response i.e. Sickness behaviour (Lenczowski et al., 1999).

Sickness behaviour is the initial stage of developing depression. In vulnerable patients [e.g. with functional polymorphisms in the promoter region of IL-6 and serotonin transporter gene (5-HTTLPR) (Bull, et al., 2008), presence of depressive symptoms at the start of cytokine therapy (Capuron and Ravaud, 1999), sickness behaviour leads to depression especially when the amount of pro-inflammatory cytokines exceeds the anti-inflammatory cytokines.

7. WHAT IS THE RELEVANCE OF PNI IN THE ONSET AND PROGRESSION OF DISEASES?

7.1. Cancer

Stressful life events such as serious work related problems as well as change of residence have been associated with occurrence of colorectal cancer (Courtney, Longnecker, Theorell, et al., 1993). Patients with breast cancer had significantly more life events, important losses, and difficult life situations before the onset of breast cancer than age, sex, number of child birth and language matched controls (Forsen, 1991).

Stressful life events reduce host resistance to tumour growth and are associated with recurrence in breast cancer. Women who had reported no traumatic or stressful life events compared to those who had experienced it had longer disease free interval (Palesh, Butler, Koopman, et al., 2007). Cytokines released by tumour cells lead to fatigue, depression, anorexia, cachexia and pain which significantly affect the quality of life of patients (Seruga, Zhang, Bernstein, et al., 2008).

7.1.1. Psycho-Oncology and Survival in Cancer - The State of the Evidence

The Mind Cancer Survival Question has dogged researchers for several years now. Some important issues that have been raised in the last 30 years of research in this area include, *remarkable patients* who survive against all odds with a positive outlook; the finding that stressful life experiences and low social support are associated with rapid progress; Personality style –the Type C personality being more prone to cancer i.e. people who have difficulty expressing negative emotions; helplessness and hopelessness, difficulty in asserting and handling one's needs and despite all this having a stoic façade (Temshok and Dreher, 1992).

One of the earliest studies described the relationship between survival in breast cancer and different types of Mental Adjustment to Cancer. The various types of coping and adjustment included-

> ➤ Fighting Spirit
> ➤ Hopelessness/Helplessness
> ➤ Fatalism
> ➤ Anxious Preoccupation
> ➤ Avoidance

It was found that the psychological response to breast cancer had an effect on outcome i.e. a cognitive coping style of a fighting spirit was associated with recurrence free survival (Greer, Morris and Pettingale, 1979).

However, a subsequent study did not replicate this finding. Influence of psychological response on breast cancer survival was studied in a 10 year follow up of a population based cohort of 578 women with stage 1 and 2 diagnosis of breast cancer. Overall and disease free survival were studied. 53% were alive at the end of 10 years without relapse and there was an increased risk of death in women with hopelessness and helplessness. The influence of depression was small and fighting spirit conferred no survival advantage (Watson, Homewood, Haviland, et al., 2005).

Steel et al. (Steel, Geller and Gamblin, 2007) studied depression and survival in hepatobiliary carcinoma. The mean survival times were as follows-

> ➤ With Depression and vascular invasion : 5.2 months
> ➤ With vascular invasion alone: 11.2 months
> ➤ Depression alone: 17 months
> ➤ No depression or vascular invasion: 26.6 months

What do these studies indicate about psychological factors and survival in cancer?

> ➢ Depression hastens progression of disease when associated with other risk factors
> ➢ Hopelessness and helplessness are linked to survival
> ➢ Absence of negative emotions appears to confer some advantage
> ➢ Presence of positive emotions alone however does not influence survival

Some of the possible modes of action of psychosocial well being in increasing survival include - better treatment adherence, better communication with doctors, lifestyle changes, improved social support and as a consequence low stress levels which have an influence on NK cell activity and TNF alpha levels.

The effects of biological stressors on various parts of immunological function and the association with cancer have been investigated in transverse and longitudinal prospective studies. At the cellular level, stressed and depressed patients had an overall leucocytosis, mild reduction in absolute NK-cell counts, and relative T-cell proportions, marginal increases in the ratio of CD4 to CD8, higher concentrations of circulating neutrophils, reduced mitogen-stimulated lymphocyte proliferation and neutrophil phagocytosis, moderate decreases in T-cell and NK-cell functions, and reduced and changed monocyte activity. At the molecular level, serum and plasma concentrations of basal cortisol, complement components C3 and C4, specific antibodies against herpes simplex virus type 1 and Epstein Barr virus, and acute-phase proteins were higher in depressed patients than in healthy controls (Reiche, Nunes and Morimoto, 2004).

7.2. HIV

The relationship of mental health specifically, depression and anxiety on HIV disease progression has been a topic of research for several years. While the findings have been mixed, studies have found low CD4 levels and high viral loads in HIV infected subjects with depression (Vedhara, Nott, Bradbeer, et al., 1997) (Kalichman, Difonzo, Austin, et al., 2002).

Longitudinal studies have also revealed that depression assessed over repeated measurements is associated with HIV progression (Leserman, Petitto, Gu, et al., 2002). Leserman (Leserman, 2003) in a review on this topic summarizes the studies done in the area and concludes that depression is indeed associated with disease progression, though the cause effect relationship has not been clearly delineated because of methodological issues. The review also concludes that the evidence that depression is related to HIV disease progression will be better demonstrated by longitudinal studies rather than cross sectional studies and by those studying the effects of chronic depression.

Several researchers have tried to understand the biological mechanisms of how depression, stress and anxiety might influence HIV disease progression. Two major mediating mechanisms have been implicated which include the Hypothalamic Pituitary axis (HPA) and the sympathetic nervous system pathway (SNS).

The HPA axis has been mainly evaluated through cortisol levels. Studies have reported mixed results about the association of depression and high cortisol levels in HIV infected populations. Goodkin et al. (Goodkin, Feaster, Tuttle, et al., 1996) demonstrated higher cortisol levels in bereaved compared to the non bereaved HIV infected population. However,

Leserman (Leserman, Petitto, Gu, et al., 2002) did not find a relationship of cortisol to depressed mood in a longitudinal study. While some studies have shown associations between cortisol and mental health parameters, researchers have also cautioned to keep in mind the fact that disease progression in HIV may also influence the HPA axis. Depression is also known to increase as disease progresses, hence it is important to control for disease progression, in order to find an independent association between depression, anxiety and the HPA axis parameters.

Cruess et al. (Cruess, Antoni, Kumar, et al., 1999) have demonstrated altered cortisol levels in depressed HIV positive individuals even after controlling for disease progression. In addition, several studies have now demonstrated decrease in cortisol levels with behavioural interventions, indicating that mental health parameters have an impact on the HPA axis (Antoni, Cruess, Klimas, et al., 2005).

7.3. Multiple Sclerosis

Patients with multiple sclerosis (MS) have reported marked difficulties a few months preceding the onset or exacerbation of illness (Grant, Brown, Harris, et al., 1989). One two year prospective study reported that the number of acute stressors predicted the relapse of multiple sclerosis (Brown, Tennant, Sharrock, et al., 2006).

Core features of pathogenesis in MS are local production of cytokines and breakdown of BBB which leads to entry of inflammatory cells in CNS. Cytokines cause damage to neurons and oligodendroglial cells which leads to behavioural symptoms (Gold and Irwin, 2006). These cytokines (IL-1, IL-6 and TNF-α) cause fever, malaise, headache, myalgia, anxiety, dysphoria, anhedonia, fatigue and affect quality of life (Flachenecker, Bihler and Weber, 2004). IFN-β used in the treatment of MS may cause depression and suicidal ideation especially in those who have past history of depression (Goeb, Even, Nicolas, et al., 2006). Improvement in depressive symptoms with antidepressants is associated with reduction in level of IFN-γ (Mohr, Goodkin, Islar, et al., 2001).

8. Methodological Issues in PNI Research in Clinical Conditions Like HIV and Cancer

Data in the PNI literature about the relationship between psychological factors, survival and the role of interventions is variable and often inconsistent. This appears to be related to heterogeneity of the populations, differing nature of stress and immune factors studied.

The stressor exposures assessed in the reviewed studies often differ according to the acute and chronic dimension and their intensity. The timing and duration of stress might substantially affect the nature of the effects of stress on immune function. During acute stress, stress hormones can help enhance immune function by informing the immune system about impending challenges that may be imposed by a stressor. However, chronicity has been shown to have an adverse effect on health. In addition to the direct effects of psychological states on physiological function, individuals who are stressed and depressed are more likely to have health habits that put them at great risk, including worse sleep, a greater propensity for alcohol and drug abuse, worse nutrition, and less exercise—health behaviours that have immunological and endocrinological consequences.

9. IS DEPRESSION AN IMMUNE DISORDER AS WELL?

Depression is associated with activation of HPA axis and elevated levels of cortisol which has significant implications on immune system as well as on metabolic changes in the body. Hypercortisolemia in depression is associated with significant increase in total leucocytes and neutrophils and decrease in lymphocytes (Kronfol, Turner, Nasrallah, et al., 1984). In a meta-analysis, Zorrilla et al. (Zorilla, Luborsky, McKay, et al., 2001) reported the heterogeneity in various studies in this area raising questions about the strength of this observation. Their meta-analysis showed the following immunological correlates of depression - leucocytosis, neutrophilia, lymphopenia, increased CD4/CD8 ratios, increased PGE2 and IL-6 levels, decreased in NK cell activity, and reduced lymphocyte proliferative response to mitogen.

Human studies of stimulated cytokines have however reported inconsistent findings about their levels in depression. There are increased levels of IL-1 and IL-6 but not of IL-2 in response to lipopolysaccharides (Kronfol, 2003). Another study showed an increase in production of IL-2 in both melancholic as well as non-melancholic depressed patients (Schlatter, Ortuno and Cervera-Enquix, 2004). In melancholic depressed patients, there was significant elevation in ACTH and cortisol levels which reduced partially on remission. The same study reported that, non-melancholic depressed patients produced significantly higher levels of IL-1β with lower ratio of IL-1ra / IL-1β as compared to melancholic depressed patients (Kastner, Hettich, Peters, et al., 2005). Fluoxetine has been found to increase NK cell activity and enhance the immune functioning in depressed patients (Frank, Hendricks, Johnson, et al., 1999).

Thus depression is associated with activation of HPA axis which leads to impairment in immune function that is associated with changes in cytokine levels. These return to normal levels with treatment and /or disease remission.

9.1. Immune Changes in Depression

Table 2. Immune changes in depression

↑ Total leucocytes blood count
↑ Neutrophil number
↓ Lymphocyte number
↑ CD4 / CD8 ratio
↑ PGE2 level
↑ IL-1 level (especially IL-1β)
↓ IL-1ra / IL-1β ratio
↑ IL-6 level
↑ IL-2 level
↓ NK cell activity
↓ Lymphocyte proliferative response to mitogen
↑ ACTH and cortisol levels

9.2. What are the Immune System Changes in Schizophrenia?

In patients with schizophrenia, prenatal infection causes early sensitization of immune system where there is partial inhibition of type 1 as well as hyperactivation of type 2 immune systems. This leads to inhibition of indoleamine dioxygenase (IDO) and activation of tryptophan 2,3-dioxygenase. This enhances breakdown of tryptophan to kynurenic acid (KYN-A) which is the only endogenous N-methyl-D-aspartate (NMDA)- receptor antagonist (Muller and Schwartz, 2008). High levels of kynurenic acid inhibit NMDA-receptors which leads to a hypo-glutamatergic state. This to an extent explains positive symptoms of schizophrenia. Lower levels of kynurenic acid blocks nicotinergic acetyl choline receptor which explains cognitive dysfunction in schizophrenia patients (Murphy, Sun, Murphy, et al., 2004).

9.2.1. Some Emerging Issues in the Field

Some of the emerging issues that have direct relevance to applications include the following -

9.2.1.1. What are the Psychiatric Adverse Effects of Therapeutic Cytokines?

Interferon-α (IFN- α) is used in the treatment of malignancies e.g. malignant melanoma, Kaposi's sarcoma, hairy cell leukaemia as well as in hepatitis B and C. It causes fatigue, malaise, psychomotor slowing, and cognitive changes such as decreased memory. Prolonged use of Interferon- β (IFN- β) used to treat multiple sclerosis and malignancies also may lead to fatigue, malaise and psychomotor slowing. Interleukin-2 (IL-2) which is used in the treatment of malignancies e.g. renal cell carcinoma, malignant melanoma also causes fatigue, malaise, depression, neurotoxicity, somnolence and disorientation (Patten, 2006). These side effects of various therapeutically used cytokines also point to a strong association between mood and the immune system.

9.2.1.2. Do Positive Emotions Improve Immune System Response?

The different emotional states affect the immune system. Positive emotions such as optimism and generalized positive affect enhance the immune response to stress by reducing the activity of HPA axis which is measured by salivary cortisol (Lai, Evans, Ng, et al., 2005). Dispositional optimism is the overall expectation that good rather than bad things will occur, especially during times of heightened stress in one's life (Scheir and Carver, 1992).Optimism improves the survival rates in conditions where immunity is lowered e.g. HIV (Ironson and Hayward, 2008), cancer (de Moor, de Moor, Basen Engquist, et al., 2006) as well as elderly people (Giltay, Zitman and Kromhout, 2006).

During the period of academic stress, students have lowered lymphocyte levels as well as function. Those with high score on optimism show increased numbers of cytotoxic T cells, enhanced natural killer cell (NKC) activity and larger antigen-stimulated delayed-type hypersensitivity responses than students with low optimism scores (Segerstrom SC, 2005). Optimism is associated with lowered IL-6 levels i.e. inflammatory response during acute psychological stress and thus promotes health of a person (Brydon, Walker, Wawrzyniak, et al., 2009).

9.2.1.3. Does Psychological Stress affect Immune Response to Vaccination?

A meta-analysis of 13 studies of 1158 patients which examined the effects of psychological stressors on the response of antibody following influenza vaccine reported a significant negative association between psychological stress and antibody response to influenza vaccine [pooled effect size: -0.18; p<0.0001] (Pedersen, Zachariae and Bovbjerg, 2009). This effect size did not vary according to type of stress. Antibody response to the A/H1N1 and B-influenza types appeared to be more sensitive to stress than A/H3N2 influenza type (Phillips, Burns, Caroll, et al., 2005). Both young and elderly age groups had significant negative associations between stress and peak antibody titres, but the association appeared larger for old than young patients. Timing of collection of blood after the stress also affects antibody response to stress. It appears from these studies and some reported earlier in the chapter that vaccination response appears to be influenced by psychological states and stress. This is an important field of study considering new vaccines being considered for large scale use and has relevance to the findings of vaccine trials.

10. WHAT IS THE ROLE OF PNI IN PLANNING MIND BODY INTERVENTIONS?

Mind body interventions are developed with the aim of reducing stress and enhancing overall well being of persons. These interventions increase the control over responses to stress and focus on better ways of handling stress, person's attitudes and their behaviours (Lorentz, 2006).

Meditation enhances relaxation of the body parts and helps the person in calming his mind. Relaxation response has been found to improve the physiological responses occurring during stress which include reduction in heart rate and blood pressure, decrease in blood lactate levels and increase in alpha brain waves on EEG.

The overall effect is a decrease in stress induced activation of sympathetic nervous system and increase in activity of parasympathetic nervous system. Mind body interventions improve the responses to stress which may enhance the immune system (Jacobs, 2001).

CONCLUSION

Psychoneuroimmunology is an emerging branch of medicine which bridges the gap between psychiatry, neurology and immunology, endocrinology, neurosciences, and internal medicine. It helps in better understanding of the interrelationship between these systems. Stress leads to activation of HPA axis and changes the functioning of the immune system which affects the onset and progression of various neurological as well as psychiatric illnesses. Thus, stress modulates the activity of CNS, endocrine and immune systems.

Sickness behavior is the psychological and behavioral responses to the sickness which are triggered by pro-inflammatory cytokines. The symptoms resemble depressive symptoms with mild variety. Depression is associated with hypercortisolemia and impaired immune functioning. Cytokines used in treatment of various cancers and autoimmune disorders causes symptoms suggestive of sickness behaviour and depression.

Mind body interventions may have a significant application as a cost benefit add-on treatment in stress related disorders by reducing the stress and enhancing overall quality of life.

> Health and disease are the cumulative and compound sequelae of interactions across biological, psychological, and social systems
>
> George Engel

REFERENCES

Adams, P.B., Lawson, S., and Sinclair, A.J. (1996). Arachidonic acid to eicosapentaenoic acid ratio in blood correlates positively with clinical symptoms of depression. *Lipids, 31*, 157-161.

Ader, R., Cohen, N. and Felten, W. (1995). Psychoneuroimmunology: An interaction between the nervous system and the immune system. *Lancet, 345*, 99-103.

Anforth, R., Bluthe, R.M., Bristow, A. et al. (1998). Biological activity and brain actions of recombinant rat interleukin-1 alpha and interleukin-beta. *European Cytokine Network, 9*, 279-288.

Antoni, M.H., Cruess, D.G., Klimas, N. et al. (2005). Increases in a marker of immune system reconstitution are predated by decreases in 24-h urinary cortisol output and depressed mood during a 10-week stress management invervenion in symptomatic HIV infected men. *Journal of Psychosomatic Research, 58*, 3-13.

Auphan, N., DiDinato, J.A., and Rosette, C. (1995). Immunosuppression by gluco-corticoids: inhibition of NK-kappa B activity through inducitn of I-kappa B sysntesis. *Science, 286*, 286.

Banks, W.A. (2009). The blood brain barrier in psychoneuroimmunology. *Immunology and Allergy Clinics in North America, 29*, 223-228.

Banks, W.A., and Kastin, A.J. (1997). The role of blood brain barrier transporter PTS-1 in regulating concentrations of methionine enkephalin in blood and brain. *Alcohol, 14*, 237-245.

Brown, R.F., Tennant, C.C., Sharrock, M. et al. (2006). Relationship between stress and relapse in multiple sclerosis: Part I. Important featurs. *Multiple Sclerosis. 12*, 453-464.

Brydon, L., Walker, C., Wawrzyniak, A.J. et al. (2009). Dispositional optimism and stress induced changes in immunity and negative mood. *Brain, Behavior and Immunity, 23*, 810-816.

Bull, S.J., Huezo-Diaz, P., Binder. E.B. et al. (2008). Functional polymorphisms in the interleukin-6 and serotonin transporter genes and depression and fatigue induced by interferon-alpha and ribavarin treatment. *Molecular Psychiatry, 14*, 1095-1104.

Capuron, L. and Ravaud, A. (1999). Prediction of depressive effects of interferon alpha therapy by the patient's initial affective state. *New England Journal of Medicine, 340*, 1370.

Chandrashekara, S., Jayashree, K., Veeranna, H.B. et al. (2007). Effects of anxiety on TNF-alpha levels during psychological stress. *Journal of Psychosomatic Research, 63*, 65-69.

Chrousos, G.P., and Gold, P.W. (1992). The concepts of stress and stress system disorders. Overview of physical and behavioral momeostasis. *Journal of the American Medical Association, 267,* 1244-1252.

Cohen, S., Tyrrell, D.A., and Smith, A.P. (1991). Psychological stress and susceptibility to the common cold. *New England Journal of Medicine, 325,* 606-612.

Courtney, J.G., Longnecker, M.P., Theorell, T. et al. (1993). Stressful life events and the risk of colorectal cancer. *Epidemiology, 4,* 407-414.

Cruess, D.G., Antoni, M.H., Kumar, M. et al. (1999). Cognitive behavioral stress management buffers decreases in dehydroepiandrosterone sulphate (DHEA-s) and increses in the cortisol / DHEA-S ratio and reduces mood disturbances and perceived stress among HIV seropositive men. *Psychoneuroendocrinology, 24,* 537-549.

Dantzer, R. (2009). Cytokine, sickness behavior, and depression. *Immunology and Allergy Clinics in North America, 29,* 247-264.

Dantzer, R., O'Connor, J.C., Fruend, G.G. et al. (2008). From inflammation to sickness and depression: when the immune system subjugates the brain. *Nature Review Neuroscience, 9,* 46-56.

Davson, H. and Segal, M.B. (1996). *The return of the cerebrospinal fluid to the blood: the drainage mechanism.* CRC Press, Boca Raton (FL).

de Moor, J.S., de Moor, C.A., Basen-Engquist, K. et al. (2008). Optimism, distress health-related quality of life, and change in cancer antigen 125 among patients with ovarian cancer undergoing chemotherapy. *Psychosomatic Medicine, 68,* 555–562.

Esterling, B.A., Kiecolt-Glaser, J.K., and Glaser, R. (1996). Psychosocial modulation of cytokine induced natural killer cell activity. *Psychosomatic Medicine,* 58: 264-272.

Flachenecker, P., Bihler, I., and Weber, F. (2004). Cytokine mRNA expression in patients with multiple sclerosis and fatigue. *Multiple Sclerosis, 10,* 34-39.

Forsen, A. (1991) Psychological stress as a risk for breast cancer. *Psychotherapy and Psychosomatics, 55,* 176-185.

Frank, M.G., Hendricks, S.E., Johnson, D.R. et al. (1999). Antidepressants augment natural killer cell activity: in vivo and in vitro. *Neuropsychobiology, 39,* 18-24.

Giltay, E.J., Zitman, F.G., and Kromhout, D. (2006). Dispositional optimism and the risk of depressive symptoms during 15 years of follow-up: The Zutphen Elderly Study. *Journal of Affective Disorders, 91,* 45–52.

Glaser, R., and Kiecolt-Glaser, J.K. (1997). Chronic stress moldulates the virus specific immune response to latent erpes simplex virusv type-1. *Annals of Behavioral Medicine, 19,* 78-82.

Glaser, R., Kiecolt-Glaser, J.K., Bonneau, R.H. et al. (1992). Stress induced modulation of the immune response to recobinant hepatitis B vaccine. *Psychosomatic Medicine,* 54: 22-29.

Goeb, J.L., Even, C., Nicolas, G. et al. (2006). Psychiatric side effects of interferon-beta in multiple sclerosis. *European Psychiatry, 21,* 186-193.

Gold, S.M., and Irwin, M.R. (2006). Depression and immunity: Inflammation and depressive symptoms in multiple sclerosis. *Neurologic Clinics, 24,* 507-519.

Goldstein, R.A., Bowen, D., and Fauci, A.S. (1992). *Adrenal Corticosteroids.* Raven Press, New York.

Goodkin, K., Feaster, D.J., Tuttle, R. et al. (1996). Bereavement is associated with time-dependent decrements in cellular immune function in asymptomatic human immunodeficiency virus type 1 seropositive sexual men. *Clinicial and Diagnostic Laboratory Immunology, 3*, 109-118.

Grant, I., Brown, G.W., Harris, T. et al. (1989). Severely threatening events and marked life difficulties preceding onset or exacerbation of multiple sclerosis. *Journal of Neurology, Neurosurgery and Psychiatry, 52*, 8-13.

Greer, S., Morris, T. and Pettingale, K.W. (1979). Psychological response to breast caner: Effect on outcome. *Lancet, 13*, 785-787.

Herbert, B.B., Cohen, S. and Marsland, A.L. (1994). Cardiovascular reactivity and the course of immune response to an acute psychological stressor. *Psychosomatic Medicine, 56*: 337.

Herbert, T.B. and Cohen, S. (1993). Stress and immunity in humans; a meta-analytic review. *Psychosomatic Medicine, 55*, 364-379.

Hofman, F., Chen, P., and Incardona, F. (1999). HIV-tat protein induces the production of interleukin-8 by human brain derived endothelial cells. *Journal of Neuroimmunology, 94*, 28-39.

Ironson, G., and Hayward, H. (2008). Do positive psychosocial factors predict disease progression in HIV-1? A review of the evidence. *Psychosomatic Medicine, 70*, 546-554.

Irwin, M.R. (2008). Human Psychoneurology: 20 years of discovery. *Brain, Behavior and Immunity, 22*, 129-139.

Jacobs, G.D. (2001). Clinical applications of the relaxation response and mind body interventions. *Journal of Alternative and Complementary Medicine, 7*, 93-101.

Kalichman, S.C., Difonzo, K., Austin, J. et al. (2002). Prospective study of emotional reactions to changes in HIV viral load. *AIDS Patience Care and STDs, 16*, 113-120.

Kastner, F., Hettich, M., Peters, M. et al. (2005). Different activation patterns of pro-inflammatory cytokines in melancholic and non-melancholic major deprssion are associated with HPA axis activity. *Journal of Affective Disorders, 87*, 305-311.

Kiecolt-Glaser, J.K., Glaser, R., Cranestein, S. et al. (1996). Chronic stress alters the immune system to influenza virus vaccine in older adults. *Proceedings National Academy of Sciences USA, 93*, 3043-3047.

Kiecolt-Glaser, J.K., Marucha, P.T., Malarkey, W.B. et al. (1995). Slowing of wound healing by psychological stress. *Lancet, 346*, 1194-1196.

Knopf, P.M., Cserr, H.F., and Nolan, S.C. (1995). Physiology and immunology of lymphatic drainage of interstitial and cerebrospinal fluid from the brain. *Neuropathology and Applied Neurobiology, 21*, 175-180.

Kronfol, Z. (2003). *Cytokine regulation in major depression*. Kluwer Academic Publishers, Boston.

Kronfol, Z., Turner, R., Nasrallah, H. et al. (1984). Leucocyte regulation in depression and schizophrenia. *Psychiatry Research, 13*, 13-18.

Lai, J.C., Evans, P.D., Ng, S.H. et al. (2005). Optimism, positive affectivity, and salivary cortisol. *British Journal of Health Psychology, 10*, 467-484.

Lenczowski, M.J., Bluthe, R.M., Roth, J. et al. (1999). Central administration of rat IL-6 induces HPA activation and fever but not sickness behavior in rats. *American Journal of Physiology, 276*, 652-658.

Leserman, J. (2003). HIV disease progression: depression, stress, and possible mechanisms. *Biological Psychiatry, 54*, 295-306.

Leserman, J., Petitto, J.M., Gu, H. et al. (2002). Progression to AIDS, a clinical AIDS condition and mortality: psychosocial and physiological predictors. *Psychosomatic Medicine*, 32: 1059-1073.

Lorentz, M.M. (2006). Stress and Psychoneuroimmunology Revisited: Using mind body interventions to reduce stress. *Alt Journal of Nursing, 11*, 1-11.

Martins, J.M., Banks, W.A. and Kastin, A.J. (1997). Transport of CRH from mouse brain directly affect peripheral production of beta-endorphin by the spleen. *American Journal of Physiology, 273*, 1083-1089.

Mohr, D.C., Goodkin, D., Islar, J. et al. (2001). Treatment of depression is associated with suppression o non-specific and antigen specific t(H)1 responses in multiple sclerosis. *Archives of Neurology, 58*, 1081-1086.

Muller, N. and Schwartz, M.J. (2008). A psychoneuroimmunological perspective to Emil Kraepelinsm dichotomy: schizophrenia and major depression as inflammatory CNS disorders. *European Archives of Psychiatry and Clinical Neuroscience, 258*, 97-106.

Palesh, O., Butler, L.D., Koopman, C. et al. (2007). Stress history and breast cancer recurrence. *Journal of Psychosomatic Research, 63*, 233-239.

Pariante, C.M., Carpiniello, B., Orru, M.G. et al. (1997). Chronic caregiving stress alters peripheral blood immune parameters; the role of age and severity of stress. *Psychotherapy and Psychosomatics, 66*, 199-207.

Patten, S.B. (2006). Psychiatric side effects of interferon treatment. *Current Drug Safety, 1*, 143-150.

Pedersen, A.F., Zachariae, R. and Bovbjerg, D.H. (2009). Psychological stress and antibody response to influenza vaccination: A meta-analysis. *Brain, Behavior and Immunity*, 23: 427-433.

Phillips, A.C., Burns, V.E., Caroll, D. et al. (2005). The association between life events social support, and antibody status following thymus-dependent and thymus-independent vaccinations in healthy young adults. *Brain, Behavior and Immunity*, 19: 325-333.

Reiche, E.M., Nunes, S.V., and Morimoto, H.K. (2004). Stress, depression, the immune system, and cancer. *Lancet Oncology, 5*, 617.

Sanders, V.M. (1995). The role of adrenoceptor-mediated signals in the modulation of lymphocyte function. *Advances in Neuroimmunology, 5*, 283-298.

Scheier, M.F., and Carver, C.S. (1992). Effects of optimism on psychological and physiological well being: theoretical overview and empirical update. *Cognitive Therapy and Research, 16*, 201-228.

Schlatter, J., Ortuno, F., and Cervera-Enquix, S. (2004). Lymphocyte subsets and lymphokine production in patients with melancholic versus non melancholic depression. *Psychiatry Research, 128*, 259-265.

Segerstrom, S.C. (2005). Optimism and immunity: do positive thoughts always lead to positive effects? *Brain, Behavior and Immunity*, 19:195–200.

Seruga, B., Zhang, H., Bernstein, L.J. et al. (2008). Cytokines and their relationship to the symptoms and outcome of cancer. *Nature Reviews Cancer, 8*, 887-899.

Solomon, G.F. (2003). Psychoneuroimmunology. In: Gelder MG, Lopez-Ibor JJ, Andreasen N (eds). *New Oxford Textbook of Psychiatry*, 1st ed. Oxford University Press, New York.

Steel, J.L., Geller, D.A., and Gamblin, T.C. (2007). Depression, immunity, and survival in patients with hepatobiliary carcinoma. *Journal of Clinical Oncology, 25*, 2397-2405.

Su, K.P. (2008). Mind body interface: the role of omega-3 fatty acids in psychoneuroimmunology, somatic presenation, and medical illness co-morbidity of depression. *Asia Pacific Journal of Clinical Nutrition, 17*, 151-157.

Temshok, L.R., and Dreher, H. (1992). *The type of C-connection: The behavioral links to cancer and your health.* Random House, New York.

Uchakin, P.N., Tobin, B., Cubbage, M. et al. (2001). Immune responsiveness following academic stress in first year medical students. *Journal of Interferon and Cytokine Research, 21*, 687-694.

Vacca, A., Felli, M.P., and Farina, A.R. (1992). Glucocorticoid receptor mediated suppression of the interleukin 2 genes expression through impairment of the co-operativity betwen nuclear factor of activated T-cells and AP-1 enhancer elements. *The Journal of Experimental Medicine, 175*, 637.

Varma, S., Nakaoke, R., and Dohgu, S. (2006). Release of cytokines by brain endothelial cells: a polarized response to lipopolysaccharide. *Brain, Behavior and Immunity, 20*, 449-455.

Vedhara, K., Nott, K., Bradbeer, C.S. et al. (1997). Greater emotional distress is associated with accelerated CD4+ cell decline in HIV infection. *Journal of Psychosomatic Research, 42*, 379-390.

Watson, M., Homewood, J., Haviland, J. et al. (2005). Influence of psychological response on breast cancer survival: 10 years follow up of a population based cohort. *European Journal of Cancer, 41*, 1710-1714.

Zorilla, E.P., Luborsky, L., McKay, J.R., Rosenthal, R., Houldin, A., Tax, A., McCorkle, R., Seligman, D.A., and Schmidt, K. (2001). The relationship of depression and stressors to immunological assays: a meta-analytic review. *Brain, Behavior and Immunity, 15*, 199-126.

Section 4. Philosophical Issues

In: Expanding Horizions of the Mind Science(s)   ISBN: 978-1-62808-705-5
Editors: P.N. Tandon, R.C. Tripathi and N. Srinivasan   ©2013 Nova Science Publishers, Inc.

Chapter 23

A SKETCH OF AN INDIAN PSYCHOLOGY

P.K. Mukhopadhyay

Associate Professor and Head, Department of Condensed Matter Physics and Material
Sciences, Salt Lake, Kolkata

ABSTRACT

The possibility and success of interdisciplinary dialogue and research demand that we carefully choose the right type of disciplines such as, say, psychology and neuroscience; further we should choose the right kind or variety of such disciplines. Among all Indian psychologies, Nyaya psychology offers us immediate and best opportunity of useful dialogue with a number of emerging disciplines of knowledge. In view of this it is intriguing that whether in India or outside scholars, generally speaking, know little to nothing about Nyaya or Nyaya system of psychology. In consideration of all these attempt has been made here to give a brief exposition of Nyaya Psychology by way of mainly listing a few general theses as well as some specific theses of it. Some of these theses may be seen as challenge or invitation to designing appropriate experiment and testing the theses in question for their tenability or otherwise as the case may be.

1. PRELIMINARIES

If one so likes one may skip this section in the first reading. For the best result it would be ideal, in that case, to follow up the first selective reading with a second reading from the beginning to the end. There are many important theses which should have been but are not included in the penultimate section. Those that have been listed in this section include many controversial theses and some in urgent need of explanation. But where it is incomplete the paper is an invitation to start dialogue and discussion, preferably interdisciplinary ones.

1.1. Why Choose Nyāya Psychology?

What follows is merely a sketch, an incomplete sketch, of a single system of Indian psychology (IPsy) – namely Nyāya psychology (NPsy) to be precise. It is to be admitted that there are other schools of Indian psychology (IPsy). We can still justifiably choose to begin our account with that of NPsy. One good reason for such choice is that NPsy presents the core of our thinking in the matter. It is closest to the commonsense beliefs and usages which constitute both the natural starting point of any (empirical) theory and one good testing ground of a theory in respect of its acceptability. NPsy starts with the sort of beliefs and usages in question and also remains most loyal to them throughout. It is arguably the best among the theories of its kind and provides us with the most comprehensive and coherent interpretation of the large body of commons sense beliefs and usages in the matter and of the most familiar experiences of the relevant phenomena. Closeness to or distance from, as the case may be, NPsy can be used as one of the measures of the comparative merit and acceptability of a psychological theory.

There may be and are theories which begin not from ordinary experience or commonsense beliefs and usages in the matter. Instead they begin with some mystic experience of some extra ordinary men. Later on these theories find it rather very difficult if not impossible to make sense of our ordinary experience and language which, they know, cannot be wished away. The explanation these theories are eventually forced to provide is very similar to the explanation we find in NPsy. For that very reason and to that extent even the advocates of these theories find that their explanation conflicts with their basic mystical position, their starting point. In a bid to effect some sort of reconciliation between their avowed mystical stand and their actual theory or explanation it is said that the explanation provided as well as the facts explained are not ultimately true or real. This amounts to saying, on the part of these advocates, that NPsy is all right as far as it goes but it does not go very far in that though it explains well our empirical beliefs and facts in the matter yet these are not true from the ultimate and transcendental standpoint; that like the phenomena it explains NPsy lacks ultimate truth or value. If it is an objection, Naiyāyikas would say it is a welcome objection. For the standpoint of NPsy is avowedly phenomenal rather than transcendental. However, it is to be admitted that the inverted procedure of theorizing adopted by the transcendentalists is informed by certain compulsions; but the compulsions are neither empirical nor ordinary or factual. NPsy so far represents the common core of psychological thinking which all the different systems of psychology are to reckon with. It is to be recognized therefore that NPsy represents the natural starting point (even if some may find it to be ultimately unacceptable due to its incompatibility with the truths of some mystical experience) if one wants to study and understand different psychological phenomena as well as theories of them.

Another reason why we should choose NPsy is that contrary to what one would expect, from what is said above, majority of scholars in or outside (modern) India knows little to nothing about NPsy. The systems that are widely popular are, besides Yoga, Advaita Vedānta (henceforward Vedānta would mean Advaita Vedānta unless stated otherwise) and Buddhism. We will presently see that Nyāya philosophy and psychology have some more features which scientific study of psychological phenomena presupposes. To put it differently, these features facilitate scientific study of the phenomena in question. A related point is that NPsy could more naturally, effectively and meaningfully enter into interdisciplinary dialogue with such emerging studies as Cognitive Neuroscience and Consciousness Studies. Therefore there seems to be an urgency to make the philosophical and psychological thoughts of Nyāya available to wider section of scholars. It will not seem an easy job if we remember that NPsy uses for theorizing highly developed technical (Sanskrit) language and in addition it conducts very deep, rich and intricate analyses. However, these are all the more reasons why we should make this system of thought available to wider section of scholars and sensitize them about the real value of its doctrines and methods. Further, what are deterrent to ordinary men – the rich vocabulary, high precision, economy of style, subtle analysis etc. – are likely to prove great attractions for the scholars. Today's world of scholarship will remain lamentably poorer if it does not include or cover NP and NPsy.

The value of the interdisciplinary and intercultural dialogue in question cannot be missed today. If it is undertaken in right earnest then it will benefit in different ways old classical systems of thought like Nyāya and also such emerging disciplines as "evolutionary neuro psychiatry". To put it more precisely and definitely, the members of the Nyāya school of thought will have a chance to be fairly sure if and how much of modern scientific and experimental findings (which, given its general standpoint, Nyāya is committed to accepting) could be incorporated into the frame work of classical Indian psychology of Nyāya. This in its turn will enable us to determine if and what vitality Nyāya (philosophy and) psychology (still) has. Once that is decided, we would be sure as to if and how Nyāya could be further developed or advanced and in which direction. There is the other side of the picture also. Today it is generally admitted that though enormous amount of important data (believed to be important for psychological study also) have been made available to us by improved technology and new techniques of say brain scaning and imaging, yet we do not seem to have any adequate theory which could coherently organize or satisfactorily explain them all. In the present circumstance and in the context of the quest for an adequate theory it might be found useful to be acquainted with the largely unknown but otherwise very rich theory of mind which was discovered and formulated by the Naiyāyika-s of the classical period and which is being uninterruptedly studied and examined since then. In a bid to making our points more concrete we will expose below the general nature and form of the psychological theory developed in the school of Nyāya. We would state and interpret in addition a number of specific truths or theses of NPsy. Some of these theses may prove to be a challenge for the experimentalist to devise experiment to test them and turn interdisciplinary dialogue and research into concrete work agenda rather than mere wishful thinking. These theses will also give us an idea how the psychological theory of Nyāya accommodates, coherently organizes and satisfactorily explains diverse facts and experiences of our internal life. If it turns out to be both logically consistent and empirically confirmed then we would be sure that NPsy might prove to be quite useful in the present state of scholarship in psychology and related disciplines of knowledge. For all these reasons we think that we need to know or know more about NPsy which deserves to be given a fair trial by various groups of scholars in different related disciplines of same or different cultures.

1.2. Some Initial Difficulties: Translating Technical Terms

We have spoken about the prospect of interdisciplinary study and dialogue. We also need to be explicitly aware of some of the difficulties and problems. Such study or dialogue could hardly begin or yield useful results of the kind stated above unless we could ensure intelligibility and precision. One of the things that stand in the way is terminological difficulty. It is very difficult to inter translate technical vocabularies of different developed theories. The difficulty increases enormously if the disciplines belong to different linguistic community and culture. It is extremely difficult to find English equivalent of key expressions of the technical Sanskrit language which the psychological and other theories of Nyāya use. In later day works of medicine or even Yoga and advaita Vedānta technical language of Nyāya is found to be extensively used. Similarly the technical expressions of Neurosciences or physics are not easy to translate in Indian languages including Sanskrit, nor even in the technical Sanskrit of say NPsy. We however do not propose to follow the more familiar method of *beginning* by giving a list of technical terms of a certain theory or language and a corresponding list containing equivalent technical expressions in another theory or language. Matching the key expressions of systems like psychology and those of whatever is the Indian counterpart of it is a task as much necessary as it is difficult. The question of using expressions of ordinary or literary language does not arise. Even within the same language the problem of translating and understanding technical expressions is sometimes felt to be quite acute. Usually rich metalanguage along with natural language is resorted to. But deeper understanding results only when we closely follow the technical expressions in their actual use within the theory in different natural contexts. The other alternative – of each person constructing a list of new technical expressions each time – does not serve the desired purpose, rather it makes mutual understanding and communication all the more difficult if not impossible. The best course is to explain technical terms of related disciplines and languages when they first occur in their natural context in course of a certain discourse like the present one. There is certainly some risk of introducing many unwelcome breaks and digressions in our exposition (of NPsy). But the advantage is that we explain the terms in question more easily and briefly, and definitely and convincingly. Thus we do not propose to introduce and explain technical vocabularies in one go.

1.3. Some Initial Difficulties: Style of Presentation

We do not envisage presenting here a textual or historical account of Nyāya psychology by way of summarizing or translating full length in English, portions of standard texts in Sanskrit. Such dead translation turns out to be not only not illuminating but also uninteresting. Sometimes people claim in favour of such translation the virtue of authenticity. But they mistake literality as authenticity. To be authentic it is not enough to follow the letter of the text when translating it in another language. We need to follow the spirit of it. For interdisciplinary and intercultural dialogue we need to express and highlight the deeper intention than the surface and literal meaning. We do not minimize the value of authenticity but we are insisting on the right understanding of authenticity. Creative interpretation, not dead translation, by people having gasped the deeper and analytical content of a text can alone authentically communicate the thoughts of the text. We are fully aware that philosophical texts, in Sanskrit, of psychology or epistemology or whatever have been written in such a style that it is not at all easy to translate them. Further it is not at all easy to understand such translations unless they are accompanied by studies or glosses. In short the very purpose of making Nyāya available to the greater world of scholars would be defeated if we chose to remain too bookish and present dead translation of portions of often elementary texts in Sanskrit. We therefore envisage that our presentation of part of Nyāya psychology would be a sort of creative reconstruction. We foresee that it would be necessary to take some liberty in our presentation. However, we will avoid violating the requirement of authenticity unless it means mere literality. If sometimes we sacrifice literality that would be done precisely to preserve authenticity in terms of intended meaning and significance of thought. Within this limit our reconstruction would amount to adding, amending, reinterpreting, even sometimes rejecting parts of what may be called literal Nyāya thought. We thereby want to achieve something valuable for Nyāya itself, though not for it only. We want to facilitate optimally accurate understanding of NPsy on the part of whoever is not a scholar of the Nyāya system of thought which is available in Sanskrit. Further we want to facilitate academically useful dialogue at a level of optimal complexity and depth between NPsy on the one hand and on the other hand such systems as psychology and emerging branches of knowledge such as Consciousness Studies and Cognitive neuroscience which sometime partially overlap.

Incidentally it may be said that here we have followed the convention of not italicizing those Sanskrit expressions in their transliterated form which involve diacritical mark. Thus we write Nyāya but *manas*.

2. NYĀYA, PSYCHOLOGY AND NYĀYA PSYCHOLOGY

2.1. The Word Nyāya is the Name of One of the Major Schools of Indian Philosophy

There is a Sanskrit word dars′ana which is often translated as philosophy. It is to be noted that the idea of dars′ana is different in many important respects from that of philosophy. They have some comparable features also. In the respective cultures they are taken alike as intellectual and analytical enterprise, which studies issues of epistemology, ethics, logic, metaphysics, psychology and so on. Nyāya may be found to place greater emphasis on and specializes in the study of logic (in the broad sense) and epistemology. However it is in its own way a complete system of philosophy. Till at least early modern period it was the age of system philosophy, as it is called, in Europe. In India in the classical period at least philosophy developed as so many systems each fully comprehensive and complete in itself. Nyāya (as also each one of the other systems) therefore naturally contains many theses about man – his mind, morals and religion. Some of these are here collectively called NPsy.

More precisely Psychology may be understood here in two different but related senses first as a particular theoretical *inquiry*, distinguished by its subject matter or method or outlook, and secondly as a *theory*, i.e., an interconnected body of finite number of truths and theses about the subject matter in question. In our presentation or reconstruction of Nyāya psychology here we have given prominence to the second sense. The theses in question may be and are collected in three broad groups. The first group consists of truths or theses that specify the general nature and form of the Nyāya system of philosophy as also of psychology and its distinctive perspective. The second group includes those theses that directly relate to and specify the Nyāya view of man whose internal life is the broad field, scope or subject matter of psychology. The third group contains specific psychological theses or truths about various mental states and their relationships. Groups and the truths they contain are too closely related to be kept strictly separate. For example the thesis Gth1.00 below can legitimately claim a place in the first as well as the third group. But the thesis that the psychology under reference is dualistic and realistic can hardly claim a place in any one of the last two groups of truths.

Be that as it may, our proposal to take a certain particular body of truths of NP as constituting Nyāya psychology will be deemed legitimate if these truths are comparable with some of those which figure in the system of knowledge called psychology. Both in Europe and India psychology was at least in the beginning a part of philosophy (dars'ana) which also contained important theses about mind and consciousness. Further in each of these cultures there are too many forms or kinds of psychology. This makes it very difficult to precisely define psychology. It is said that the word psychology does not have a single clear sense in English, in the literature of the subject or its history in English. Sometimes the word is taken to denote the study or empirical study of mind – the structure and function of it. This goes perfectly well with Cartesian dualism in which mind is admitted to be a substance which stands distinguished from the only other substance – the material substance or matter – by its exclusive property of thought or consciousness, which happens to be its *essential* quality also. But those who do not admit, for some reason or other, a substantive soul cannot accept this to be the sense of the expression psychology. When James Ward, a Kantian psychologist, revolutionized the subject by publishing his seminal paper in Encyclopedia Britannica, which is said to have inaugurated thereby a new age in the history of the subject, he viewed psychology as the study not of mind but of experience. (I doubt whether the word should be spelt with a capital E). Though apparently this view of psychology (that it is a study of experience rather than substantive soul) is also the view of the associationists, like Hume say, yet as a matter of fact Ward's position was consciously offered as a rejection of the associationistic psychology in some other aspect of it. According the associationists, experience, such as the cognitive experience of an adult human being, a perception of a table on the floor, say, is the result of associative combination of some isolated fundamental bits of experience called ideas. According to them therefore experience comes to us first in the form of isolated small bits derived from sense impression. These are, so to say the mental atoms or, in the language of another celebrated psychologist, William James, mind stuff. Adult human experience is built up by connecting these isolated bits according to laws of association. The Kantian picture presented by James Ward is just the opposite. Experience, according to him comes to us or is produced in us in lump so to say. The expression Ward uses is that we are presented with a totum objectivum, an undistinguished mass experience or totality which then gets discriminated by the operation of attention. Normal experience of an adult human being is thus the discriminative work of attention focused on the undistinguished mass presentation we first have. Some again tell us that psychology is the study of human behaviour. The scope of this psychology is further extended by some to include animal behaviour or experience of abnormal men. As opposed to psychology of the associationists or classical atomistic behaviourists, Gestalt psychology emphasizes on the point that experience is always a whole patterned response; this is not the same as though similar to the view of Ward in some respect. We need not enumerate more of such familiar facts and information to impress our readers that there is not only no single meaning of the term psychology but also it is nearly impossible to unify the widely diverse senses in which the term has been used in the history of the subject. As a last word in the matter for the present I would like to recall that Brunswick produced his small but beautiful work, *Conceptual Framework of Psychology*, as a part of the project called Encyclopedia of Unified Sciences. The project was launched by the logical positivists but it had to be abandoned by them very soon. To come back to the situation in psychology just described, one wonders if we have any way out. For without working consensus about the meaning of the term or the nature of the subject we can hardly hope to have a confident start in our effort to present a fairly comprehensive, though not complete or accurate in all respect, account of Indian psychology or Nyāya psychology. We may recall the way out, if it can be said to be so, which was suggested by Quine in a similar situation. He found the same problem with the expression philosophy and suggested that it was better not to ask the question what is philosophy or seek any common feature in all that is called philosophy. He advised us on the other hand to take into account what the philosophers actually do and allow the word philosophy to fall wherever it suits. Adapting this strategy and slightly changing the language we would like to say that in spite of all the difficulties shown above there is some sort of agreement or consensus as to who to call a psychologist or which work is to call a piece of psychological writing. Our situation is somewhat like that of a common man who unmistakably identifies a man and tells it from say a tiger. But he finds himself completely lost if he is asked to precisely formulate a *definition* of man. We begin almost like a common man of this description.

The term sometimes used to denote the Indian counterpart of Psychology namely manovidyā, has been recently coined. It is an example of using ordinary language to translate a technical expression of some advanced theory. Anyway, the expression manovidyā seems to be more confusing than illuminating. Taking psychology to mean the study of the mind and taking *manas* or mānasa in manovidyā to be synonym of mind, psychology is translated as manovidyā. Too many systems of psychology are there which do not admit the existence of mind in its standard sense of a spiritual *substance*. On the other hand almost nobody in India admits *manas* or manaḥ to be a spiritual substance. It is a material substance and we call it here psychic matter which seems to us more accurate a translation of *manas* or manaḥ. More importantly manas is not the seat of consciousness, that is, experience or thought is not a quality, let alone essential quality, of *manas*. This is more than enough to convince us that mind is sharply different from *manas* and vice versa. Further *manas* is regarded, in many systems of Indian philosophy and psychology such as Nyāya, as a sensory organ or internal sensory organ, to be precise, which enables us to have perception of such internal states as pleasure, pain, desire etc.

Anticipating some of the specific theses of NPsy to be presented later we may conclude this section by proposing to mean and understand by the expression Indian psychology or at least Nyāya psychology (NPsy), the study of the internal life – the internal sates and processes – of man; primarily and mostly normal, adult, average human individual. It is concerned with the internal, and, in this sense, the mental. The word mental may be translated as mānasa, which strictly means internal, and then it should be contrasted with vāhya which means external. Psychology is the study of the mānasa phenomena, internal states and processes. Mānasa vidyā (study of internal phenomena) conveys the sense of psychology more accurately (even if it sounds somewhat less pleasing or elegant) than manovidyā which is more familiar now.

2.2. Perspective of Nyaya and General Form of Nyaya Psychology

In this section we will present a selection of theses or truths, of the first and the second group, of NPsy. Strictly speaking the first group of truths relate to the philosophy of Nyāya rather than to Nyāya psychology. However, an account of NPsy must incorporate these as they determine the general form of NPsy as much as or because they specify the general perspective of Nyāya philosophy as a whole.

Even if we confine ourselves to the conformist (āstika) schools of Indian philosophy, there are six different schools to consider. For our purpose it is important that these philosophies stand divided into two major groups: dualist and monist. The spiritualistic monistic systems of Indian philosophy like advaita Vedānta and Bauddha (the anglicized form of Bauddha is Buddhist or Buddhism. Here we will use Baudha to mean only the idealistic Bauddha thought of the Yogācāra or Vijñānavāda Bauddha-s) are basically reductionistic. In fact all monistic systems of the world, including the materialism of scientific Europe, are reductionist. It is quite normal to expect and easy to understand that materialism or material monism would deny the fundamental reality to mind. Only reality is matter. All our beliefs and talks about mind and the mental are, in so far as they are to be at all accepted as true and meaningful, to be reduced to talks about matter and the material. This is part at least of what is meant by materialist reduction of mind to matter. What is likely to strike one, who is uninitiated in Vedānta or Bauddha philosophy, as strange is that each of the two forms of Indian spiritual monism noted above, denies the (ultimate) reality of mind and hence of the mental (though they admit the reality of consciouisness). This (apparent) problem or anomaly is largely, if not wholly, due to the terminological difficulty. The English expression mind is indiscriminately used as synonym of both *manas* and ātmā both of which are internal substance according to Nyaya. It would be less confusing if we use here the expression Mind with a capital M for ātmā or self and mind with a small m for *manas*. What in these two Indian systems of thought, particularly in Vedānta, is taken to correspond to mind, namely, *manas* or *antāhkarana*, is material. The only non-material thing in the Vedānta is Consciousness (*Caitanya* or as, in Buddhism, Vijñāna). In view of all this it is better to translate *antāhkarana* or *manas* as psychic matter (psyM). If a piece of stone or certain water body is a physical matter (substance) (pM) and a human body is a biological matter (bM), then *manas* is a psychic matter (psyM) and ātmā is a spiritual matter (sM). Though otherwise material, *manas* is called psyM because it is responsible for many, if not all, of the states and functions that are called mental or internal and form the subject matter of psychology.

For the ease of reading we will use here the expressions like psychology, mind, soul and so on. If we keep the explanation given above in mind then we can avoid confusion and understand the text more easily. Most difficult to understand are the conventions adopted here about mind and Mind. The reason is that the idea of an internal sense organ, for which we will frequently use the expression mind here, is not very familiar in European culture or psychology. Some psychologists even said the internal sense organ is a misnomer. But in India idea of internal sense organ (*antarendriya*) as different from self (ātmā) is very familiar even if it is not accepted by philosophers of all the schools. To strike a balance between popular practice (of translating *manas* as mind) and accuracy in theory we have been forced *manas* to adopt the convention of using the two expressions Mind and mind for self and internal sense organ respectively. The latter is a material principle and will be called psychic matter.

2.3. General Perspective or Framework of Nyāya Thought which Determines the Form of Npsy

The source of Nyāya psychology is basically the classical literature of Nyāya philosophy (NP). In the classical period at least, psychology in India was a part of general philosophy. The same was the case in Europe for a long time. Nonetheless we can easily identify in any comprehensive *system* (and earlier it was, as it is called, the age of system philosophy) of philosophy good many number of theses or doctrines which collectively constitute the respective psychological theory. In a sense therefore there are as many different psychologies as there are systems of philosophy. Psychologies embedded in different philosophies expectedly differ among themselves, sometimes in important ways.

Both forms of Indian spiritual monism noted above, deny the reality of mind and, so far, of the mental. If we attribute to them a belief in the mental then we should remember that what they understand by mental, if they understand anything by it at all, has little or nothing to do with mind or even Mind as the Naiyayika-s understand them and as we have explained them above. In fact they deny both mind and Mind as anything real whereas Nyaya admits both as real. For Nyāya *manas* which is otherwise a matter is distinguished from other kinds of matter and is called psychic matter or mānas *dravya* because we owe to *manas* (though not to *manas* alone) the possibility or being as well as knowledge of the mānas, i.e.,the internal, states and process, in a word, our internal life, which therefore is called mānas avasthā, āntara jīvana etc.

Realism is strongly opposed to reductionism. This is one of the root causes of opposition between various forms and aspects of reductionism on the one hand and realism on the other. Nyāya is a major school of realistic philosophy of India. Two major manifestations of its realism and anti-reductionism in ontology are ontological dualism and anti-phenomenalism. Nyāya believes in the first place that there are in the world two fundamentally different sorts of reality: matter and mind (jaḍa and *cetana*, or some say, more strictly speaking, jaḍa and *chaitanya*). Both are equally real; none of them is reducible to the other. This may be called dualism of substance (or *dravya-dvaya*-vāda or jaḍa *caitanya bheda* vāda). It is opposed to ontological monism which admits either matter or mind as the sole fundamental reality. Though not strictly correct yet let us translate this view as dravyādvaitavāda to contrast it (ontological monism in its usual sense) with phenomenalism. Phenomenalism goes a step further. Within ontological monism of substance one may admit dualism of quality and substratum (*dharma-dharmi bheda*). For example one may say that matter is the only reality (monism) but a material substance is not just the qualities we associate with it. It is something more than its qualities. This is called physicalism which is a form of realism and dualism; it is not dualism of substance but dualism of substance and quality. Phenomenalists say a substance is not different from the qualities we associate with it. In this sense a substance, whether material or mental, is reduced not to a substance of another kind but to its own qualities or qualities we associate with it. Nyāya is strongly opposed to both spiritualistic or materialistic reductionism and phenomenalistic reductionism. Of particular interest for us here is the truth that Nyāya admits two distinct kinds of internal substance called ātmā (soul or Mind) and *manas* (an internal substance which is also the internal sense organ or *antarendriya* or, as we call it here, mind). Besides, Nyāya admits that the internal phenomena like cognition, conation, feeling, desire etc which we associate with the Mind or the soul are qualities of which Mind or soul or ātmā is the substratum. However, the production as well as perception of these qualities is made possible by certain operations of *manas* or mind. Mind or ātmā is not reducible to these mānasa or ābhyantara guṇ-s i.e., internal qualities. So the two general theses of Nyāya philosophy are:

- *Gth.1 Matter and Mind (sometimes, deha and ātmā, say) are two fundamentally real and irreducible substances (dravya-s) that are there in the world.*
- *Gth.2 A substance is different from the qualities which it has and or which we associate with it; it is something more than these qualities, it is in fact their substratum.*

These two theses determine the perspective of Nyāya philosophy and ontology as realistic and dualistic. In so far as this general perspective of the Nyāya philosophy determines the general nature and form of the psychology embedded in it, NPsy may be deemed to be on the one hand a form of Cartesian psychology and on the other it is a psychology which believes in the interaction between Mind and body. Nyāya is thus committed to accepting independently existing Mind or spiritual substance and holding the following general thesis.

- *Gth.3 Mind can causally affect the course of material substances like human body (and some of its qualitie and actions).*

The theses Gth.1 and Gth.3 are very much unacceptable to a large section of working scholars in the field of psychology and such other related studies as cognitive neuroscience. So the objection runs

- Obj.1 Contrary to what has been claimed above, NP and NPsy are unsuitable for interdisciplinary studies of consciousness for their general framework or perspective is a sort of Cartesian dualism whereas majority of related scientific disciplines are inclined, if not committed, to materialism or material monism .

In our answer to this objection we need to discuss very briefly two things. First the days of simple minded rejection of dualism by naïve materialists are over. Secondly Nyāya dualism is different in important ways from the Cartesian dualism.

It is true that many of those who are actively engaged today in the field of consciousness studies, including the cognitive psychologists and cognitive neuroscientists, are materialist in their beliefs and practices. But they hardly can deny outright the existence of, or problems related to, qualia or of subjective experience. They generally acknowledge that though they are working for it, and are hopeful of eventual success, yet the problem of subjective experience has not been solved along fully materialist line and that this phenomenon cannot be wished away. They feel that it is not as easy to solve or ignore the problem of qualia as the older issue of whether there is in the universe a Mind in addition to material substance. These scholars, even when they have strong inclination in favour of material monism, maintain in practice, if not in belief also, a sort of methodological neutrality in respect of the metaphysical issue of dualism – the issue of the existence of Mind and the mental over and above matter and the material. They are more interested in, and almost solely concentrate on, the empirical, scientific (piece meal) and experimental study of the functioning of the brain and their correlation with the human responses – the responses that we are accustomed to describing by using mental words. Not so hidden an agenda which they have is to convincingly show that advances of science have made it possible today to give a fully scientific account (that is an account in terms of the structure and function of the brain matter and the like) of what is *called* the mind, Mind or the mental. Many, however, believe that subjective experience marks a limit to which, according to the current conception of science, we call science and scientific explanation.

The second point we need to note is that one of the important ways in which Nyāya dualism differs from the Cartesian dualism is that the former is, as we will soon see, much more liberal and weaker a position than the latter. According to Nyāya conception of it the Mind – ātmā or the soul – is not as different from matter or body as it is in Cartesian dualism. In Nyāya, not in Cartesian system, thought or consciousness is *not a strongly essential* quality of Mind. And further, whereas according to Vedanta, not so clearly in Cartesian system, consciousness is a thing of the transcendental realm, in Nyāya, however, consciousness is essentially a kind or form of phenomenal experience.(Even divine consciousness is closer to phenomenal conscioiusness).

Another very important difference of Nyāya from current science relates to Nyāya theory of causality. An important thesis or truth of this theory which is an integral part of NP as well as NPsy is

- *Gth.4 Causality operates across different orders of thing.*

Current science believes that causality operates among things of one single order, the material order. Two or more physical systems can causally interact. Sometimes when forced, the followers of this science grudgingly extend the scope of causality to epiphenomena. But it strictly remains a form of upward or bottom up causality. Matter can causally affect the mental, which is an epiphenomenal product of matter. But this quasi mental cannot effect changes in the matter. From the Nyāya point of view this restricted functioning of causality or the narrow conception of it which the current science has, is more a matter of assumption than a fully reasoned or experienced truth. Enough ordinary usages, beliefs and experiences are there to strongly suggest that mind interacts with matter as much as matter does with mind; and the nature of the interaction is also the same or strongly similar.

We propose to note only one more feature of the perspective or general form of NPsy. Though it can hardly be found in this form in the literature of Nyāya yet for convenience of reference we formulate this feature in the form of a general thesis or truth of or about NPsy.

- *Gth.5 Nyāya psychology is a form of philosophical psychology.*

This point needs to be understood properly and with due emphasis. It should not be over interpreted to mean that the truths of NPsy are purely speculative and have no relevance or bearing on the different related branches of studies mentioned above. Nor should it be interpreted to mean that if certain truth of NPsy cannot be verified by the means used in these branches then that is necessarily a falsification of that truth or thesis. Again Gth.5 should not be taken to imply that NPsy is speculative. Nyāya psychology is neither speculative nor experimental. It is not even a priori or conceptual. It is empirical and factual. This is what makes its dialogue with the many modern branches of studies in the related field possible, easy and fruitful. We would also enumerate some specific theses of NPsy for the verification of which it is quite pertinent to devise appropriate laboratory experiment. Anyway, NPsy is to be distinguished as much from experimental studies like cognitive neuroscience as from contemporary philosophy of mind. The latter is avowedly a conceptual study or even analysis of mental *words*. We say NPsy is philosophical psychology first because it never sharply dissociated itself from the general philosophy of Nyāya and established itself as an autonomous discipline of knowledge. Secondly and more importantly the methods used to uncover or justify the truths and theses of NPsy are exactly the same as those used in general philosophy.

3. ON UNDERSTANDING SOME SPECIFIC THESES OF NYĀYA PSYCHOLOGY

We will group under this head theses and truths about human experience and the structure and role of three sorts of substance that go to make up a concrete human individual which constitutes the central part of the study of the theory we call here NPsy.

- *Sth.1 Psychology is the study of human experience – the experience of average,adult and normal human being.*

Man, according to Indian thinking and hence Nyāya, belongs to the broad class of sentient beings which includes on the one hand plants and animals and on the other such beings as gods. Though we will not attempt it here yet it appears to be an interesting and feasible a task task to extend the current NPsy to the study of animal and abnormal human behaviour.

- *Sth.2 Human experience is broadly of three kinds (this closely corresponds to the familiar tripartite division of mind in European psychology) cognitive, affective and conative – anubhava, icchā and prayatna.*
- *Sth.3 A living human individual is a sort of complex unity of three sorts of substance or matter, a living body (s'arīra or deha), a conscious soul (ātmā) and an internal sensory organ (antarendriya or manas).*

A living body is otherwise just a physical or material substance like a stone but for the presence in it of the principle called life. This principle, an essential component of the structure and function of a human individual, is conceived of as a form of air (vāyu) which is manifested in our act of breathing. The act of inhaling and exhaling, maintenance of body temperature etc. are made possible due to the active presence of this kind of air called life air, that is say, life preserving air (prāna vāyu). Physical death occurs when this air leaves the body; normal life or living is conceived of as the active presence of this air in the body. The mobility of *manas* or psychical matter is closely related to the movement of this life air.

When an individual succeeds, normally through certain practices, in quieting this air the *manas* also becomes immobile (*sthira*). Anyway, in view of what has been just said it seems proper to describe a (living) human body as a biological matter. A dead body is just a physical matter but for its history which shows that it was born out of biological bodies, lived for sometimes, that is, till thetime life air leaved the body. During its life span a living body is seen to grow, decay and die. Unlike a stone or a water boidy, a human body and other living bodies whether of a plant or an animal manifest, in some sense, such functions as metabolism, digestion, procreation and so on. This way of looking at a human individual or worldly sentient creatures, is quite pervasive in Indian culture. It broadly tallies with the account we find in Indian medical treatises as well as in philosophical and religious literature of India.

The soul or ātmā may legitimately be called a spiritual *matter*. This is the important difference of Nyāya from Cartesian thought. Both in Cartesianism and in Nyāya consciousness is admitted to be an owned state or quality of the soul substance (ātmā). In both these systems consciousness is an essential quality and mark of the soul. However in Cartesian system it is an essential quality in the *strong sense* of essential. A quality F of a substance x is an essential quality in the strong sense if x cannot exist without F. Wherever x is found it is found as Fx. Nyāya contends that consciousness is an essential quality of the self in the *weaker sense*. According to this sense the necessity is more from the side of the quality than from the side of the substance. Whenever there is consciousness (jñāna or *caitanya*) it exists as the quality of some soul substance. It cannot exist all by itself (even when, as in the case of supreme God head or Is'vara, it is a permanent or eternal quality). However, a soul substance can exist, temporarily though, without consciousness. The difference between a physical matter like a stone and a soul substance is that the former lacks consciousness *absolutely* (always and unconditionally) but there is not absolute lack of consciousness (jñānātyantābhāva) in the latter. Since it does not possess consciousness as its strongly essential quality or always, a soul can be seen and described as a matter; at least when it exists without consciousness. But it is a spiritual matter, for even in the weak sense of essential, no other substance or matter has consciousness as its essential quality. Answer to the question why Nyāya prefers weak sense of essential to strong sense it may be had in the following thesis of Nyāya.

- *Sth.4 The life of a human individual is marked as much by the presence of consciousness as by occasional arrest of consciousness or destruction of states of consciousness.*

This thesis or truth is a characteristically empirical truth if we have in mind such occasion of arrest of consciousness as deep dreamless sleep, (perhaps deep coma also), or a state of swoon. Vedānta offers sort of a priori argument against the possibility of arrest of consciousness or deep sleep in our sense of the term. We may be misled to thinking that the Vedānta thesis that consciousness does not suffer arrest or destruction is an empirical thesis. For they refer to certain common experience and usage in support. On waking from alleged state of deep dreamless sleep, we *remember*, Vedanta says, that we slept happily or peacefully. Since the content of memory must have been the content of some previous non-recollective cognition, Vedanta concludes that we not only had happiness but had experience of happiness during the so-called deep sleep. In other words even during deep sleep there was consciousness; it did not suffer arrest. But if one was not already *non-empirically* convinced that consciousness was permanent and could suffer no change or variation, one would easily view the situation in question in this way. The usage in question attributes the pastness to the state of sleep and not to the alleged experience of happiness. For who says "I slept well" is evidently in a waking state. And the sate of sleep being talked about is a past state of sleep. So it is not (at least not necessarily) the past (pastness of) happiness but past (pastness of) state of sleep that is being conveyed by the utterance. Secondly the happiness spoken about is nothing more than absence of experience of any suffering. So all that is conveyed by the common utterance like "I slept well or slept happily" is that during the past sate of sleep I *did not have any experience* of pain or suffering. The Vedānta interpretation of the utterance in terms of positive state of experience of pleasure violates the norm of logical economy. And Vedānta would not have risked this if it was not already committed, on the strength of some a priori or non-empirical considerations, to accepting that consciousness suffers no change like arrest or disappearance. We also know what non-empirical consideration they bank on.

The arrest of consciousness is not, however, an isolated empirical truth. It is one of a body of empirical truths which is consistent with other truths contained in that body. This body of truths also contains some central truths of Nyāya philosophy. One related empirical truth in the matter, for example, is the following thesis or doctrine.

- *Sth.5 Consciousness is produced only when specified conditions of it obtain.*

Under observable and experimental conditions, that are specifiable and specified, consciousness is produced. When those conditions do not obtain consciousness is not produced. Non-production of consciousness and the destruction or loss of it, are not one and the same thing. In NP and NPsy conditions of these are separately specified. Be that as it may, Nyāya contends, what is a common truth, that we can, so to say, force a certain consciousness to be produced. A man may be in a totally absorbed sate of listening to a certain particular music, quite oblivious of the moving hands of the watch. But if someone sitting next to him tells his friend loudly that it is getting late then who was so long absorbed in listening music, looks at his watch and a particular perception is produced in him. Similar facts of common experience are found mentioned in standard text books of Nyāya philosophy.It has been cited as an example that if one resolves before going to sleep that he would wake up at a certain particular hour in the morning, then his waking and hence experience that marks his waking are instances of production of states of consciousness. Neuroscientists know there is any number of examples of production of particular states of consciousness under normal and unusual conditions.

There are some important philosophical truths of Nyāya which lend further support to otherwise empirically supported belief that consciousness suffers birth, arrest and destruction.

- *Sth.6 Soul (ātmā) has consciousness (caitanya or jñāna) only in its embodied state.*

This is only partly an empirical truth. Every normal and adult human individual knows that he has a body. Commonsense experience and ordinary usage testifies to the truth that we *have* a body and not that we *are* body. That we have body is given, so to say, to us. But it does not also seem to be given to a man what else he is. Ordinary use of language and experience justify our giving the name soul to this second component of a human individual. Some hold that soul is given to us only in our mystical experience, that is, if and when we have this sort of experience. But Nyāya is convinced that some of the ordinary means or sources of knowledge that make us aware of the existence of body also give us the knowledge of the self.

It is also empirically known to us that in his embodied state a man, that is, his soul has consciousness. But what is the ground for the stronger statement that *only* in such a state soul has consciousness? Do we or could we know that self had a disembodied state and that in that state it lacked consciousness? Instead, we know men or souls are always conscious, but for occasional arrest of consciousness during states of deep sleep or swoon. So Sth.6 seems to be more a philosophical truth rather than an empirical one. The point is that Sth.6 is not an isolated truth but is a member of the connected body of truths which constitute the theory called NP or NPsy. Some other truths of the theory with which Sth.6 is closely connected are

- *Sth.7 Consciousness is an impermanent quality of the self.*

We can hardly adduce empirical evidence for the creation or production of the first ever state of consciousness of a human individual. But from familiar experience of phenomena like death of a man and occasional arrest of consciousness in a living man we know that consciousness is impermanent. Further, ordinary experience and common usage show that consciousness is an attribute of whatever the personal pronoun I (*aham*) stand for. And it stands for self. So far Sth.7 is an empirical truth.

It is the normal practice with the Naiyāyika-s to offer, in their theory we call philosophy or psychology, causal explanation of the phenomena they study. For this reason also, and to that extent NP as well as NPsy may be regarded as an empirical study. Nyāya argues that we do not have (empirical) evidence for any consciousness which does not have any origin. The related point is that we do not have evidence to show either that there is a consciousness which is not owned by some self as its conditional quality. It follows

- *Sth.8 Consciousness, which Psychology studies, is the phenomenal consciousness which is a certain episodic and internal state or quality owned by some self.*

Nyāya knows of no transcendental consciousness which is permanent and without an owner. In spiritual monism of Vedānta such a consciousness is not only admitted but admitted as the sole single reality. But quite naturally it has been also said there that any study of this consciousness (at least in the ordinary sense of the term) is theoretically impossible. Therefore psychology can have nothing to do with such a consciousness. A word of caution is necessary. Nyāya does not contend that all qualities are impermanent or that all impermanent things are quality. But consciousness is an impermanent quality of the soul. As impermanent it is there in the soul only when it is produced or occurs in the soul and so long it is not destroyed). In other words when conditions for the occurrence of a consciousness state obtains, a conscious experience is produced in the self as its quality. Among the causal conditions of consciousness Nyāya distinguishes two types. Condition (or conditions) that accounts for the *very possibility* of consciousness and condition (or conditions) which in combination with the former accounts for where when and what state of consciousness is produced.

- *Sth.9 Soul's relation of effective conjunction with the psychic matter called manas is the condition of the very possibility of consciousness and;*

- *Sth.10 A soul (ātmā) enters into effective conjunction with a manas only in its embodied state.*

We should note that two among the three substances which constitute the complex unity called a human individual are permanent. Only component substance of that complex which is impermanent is the body. However, when all these substances are effectively combined and formed the complex in question – ({Mind^1, m^1, b^1}) that is a particular living human individual – consciousness is produced and manifested. If soul and *manas* were the only or sufficient conditions of consciousness then we could hardly account for the impermanence of consciousness. As permanent substance soul and manas are always there. What is not always there, and what is therefore impermanent, is the effective conjunction of these two substances. And this effective conjunction of self and *manas* is, according to Nyāya the condition of the very origin or existence of consciousness. If this conjunction is not effected (as, in case of an individual man, when he is not born) or is somehow sort of obstructed (as perhaps when a man is in deep sleep) there is total absence of all experience whatsoever and not of this or that particular experience.

Nyaya gives argument to show why there could never be a time when soul is not in conjunction with some mind or *manas*. But it is effectively conjoined with a *manas* only in its embodied state. In other words effective conjunction between a self and a *manas* means such conjunction between them (a soul and a *manas*) provided they are, at the time, the components of the same single complex which is a single human individual. In a sense this self as well as this *manas* is equally embodied and are *in* the same body.

- *Sth.11 There are at least as many selves and as many manas as there are human individuals (dead or alive).*

Except in some very rare and sort of esoteric situation no two bodies can have the same self embodied in them. That is no self can be a member of the two complex unities we call two human individuals. No two bodies can share same *manas* either. Each human individual is a unique whole of three sorts of substance. Within such unique and complex unity, a self stands effectively conjoined with a *manas* and this makes it possible for experience to occur or manifest. But no two bodies can have in common the same soul or same manas.

Naiyāyikas raise and answer a question here. Since there are at least as many different selves and *manas* as there are human bodies the question is do we have any law that can explain which particular self combines, or enters into effective conjunction, with which particular body or which particular *manas*. We discussed the Nyāya answer to this question elsewhere.

- *Sth.12 Consciousness is a conditional or adventitious quality of self.*

The quality (*guṇa*) of a substance (say, magnitude or *parimāṇa*) is called unconditional *(svābhāvika)* if it is not conditional (āgantuka) . A quality of a substance is conditional or adventitious if does not accrue to the substance unless it stands related to another substance. Consciousness, however, is a conditional or adventitious (āgantuka) quality of the self. A condition under which a soul has this quality is its effective conjunction with a particular other substance, *manas*. Such a conjunction obtains between the two when both of them are the components of the same single complex unity called a (living) human being. So it is said that a soul does not have the quality of consciousness all by itself; it has it only in so far as or as long as it stands effectively conjoined with certain other substance (*dravya*). [So consciousness is called (dravyāntara)*saṃyogaja* guṇa of ātmā. Dravyāntara saṃyoga is the condition and the dravyāntara is *manas*.]

- *Sth.13 In its totally disembodied state a self ceases to have its impermanent and adventitious quality of consciousness.*
- *Sth14 Such permanent loss (ātyantika nivṛtti) of consciousness is distinguished from ordinary and familiar states of arrest of consciousness.*

The familiar human body we are acquainted with normally is not the only kind of body that can embody a self or form suitable complex, with self and *manas,* that can be identified as an individual sentient being. But all human bodies are impermanent and are normally made of elements (bhūta-s) of one kind only. There are altogether five elements or primordial matter (*nitya* bhūta *draya* or bhūta paramāṇu). Some bodies like worldly human body are perceptible. There are also imperceptible kinds of body in which a self can be embodied. Be that as it may, since all bodies are impermanent and self is permanent it is at least theoretically possible that a particular self (we will soon discuss how the identity of a self in such a state is decided) may be there which is not attached to or embodied in any particular body whatsoever. If such a state of detachment with body as such never ceases, that is if having lost its connection with previous body or bodies the self in question never gets attached to any other body then it will remain without its quality of consciousness or experience for all time to come.

Such a soul (and also *manas* that is not attached to any particular body whatsoever) is a matter of no concern or interest to psychology or related studies. Further such a soul will never again experience any pain or suffering nor will it be embodied ever. This is called the state of spiritual freedom (or mokṣa).

- *Sth.15 In a state of spiritual freedom a self ceases to have any experience including experience of pain or suffering. Alternatively when a self transcends the very condition of the possibility of experience or suffering it attains spiritual freedom.*

Nyāya has some explanation, into which we will not go, about why at all a self enters into the relation of embodiment with a body and with which body or which kind of body it enters into such relation.

We have already seen the role the vital body or biological matter plays in the overall explanation of consciousness which psychology studies. We now need to note a few more important functions of the human body.

- *Sth.16 A body is that "in" which a self experiences pleasure and pain*
- *Sth.17 Consciousness "is" where the body is.*

Both the truths are somehow included in the Sanskrit utterance s'arīram bhogāyatanam. It literally means body is the place where consciousness and immediate experience of pleasure and pain occurs. The point, however, is that we have said above that the seat of consciousness or experience is the self and not body (or the brain) or *manas*. Self is said to be the seat of consciousness because when (a state of) consciousness is produced it inheres, that is, exists in the relation of inherence, in the self. All qualities inhere in the substances of which they are the qualities. However, existing or existence *in inherence* has some distinguishing features which are not there in the existing or existence *in some other relation* like conjunction. When we say a table is or exists on the floor it is an instance of existing in the relation of conjunction. And we know the table does not exist in whole of the floor but only in some part of it. It makes sense therefore to ask, when it is a case of existing in conjunction, in which part of the locus what exists in conjunction exists. But in case of quality existing in a substance or generic property (jāti) existing in the individual (*vyakti*) it characterizes, we cannot meaningfully ask the same question. We cannot ask, for example, where in the milk or in which part of it white colour exists. Similarly no competent user of the concerned language asks or is permitted to ask where in man, in which part of a man, does manhood exist. Ocean is ocean all over, man is man all over. In such cases oceanhood or manhood exists in the ocean or man in the relation of inherence. Similarly, it makes no sense to ask where in the self or in which part it there is consciousness. Self is conscious all over and not in any particular part or parts of it. The reason why we cannot ask those questions or speak in that way is not that self has no 'part' or extension. For Nyāya holds that

- *Sth.18 Self is ubiquitous or infinitely extended; it is vibhu.*

But if self is infinitely large or ubiquitous and it is conscious all over then how is it that we do not experience the presence of consciousness outside our body? The body has finite extension and covers a small part of the infinitely extended space. Self which is ubiquitous extends far beyond the body; it is also infinitely extended. Most of the self is outside the body and consciousness is in this portion of the self also. Why do we not then experience the presence of consciousness outside the body? Before we answer the question let us note that it is admitted to be a fact of experience that we notice the presence of consciousness only where there is body. And this fact shows that body is a necessary condition for the *manifestation* of consciousness. Thus body not only helps the origin of consciousness through facilitating effective conjunction between a self and a *manas*, but it also *individuates* consciousness, so to say, by determining it to be manifested only where there is body. (For the phenomenon of localization of sensation see Sth.31 below). Even if there was no difference, as some people say, between consciousness of Peter and consciousness of Paul (as consciousness), Peter's consciousness is to be found (manifested) only where Peter's body is there and not where the body of Paul or someone else is there. It is in this special sense that an individual's body is said to be the place (āyatana) of his consciousness. And if it is the "place" of consciousness, it is also the place of our *bhoga* (it is bhogāyatana) that is, the place of the immediate experience of pleasure and pain. This point needs a lot more explanation for which we have no scope here.

In NPsy due importance is given to body as a factor which plays important roles in the origin and individuation of consciousness. Further this body itself is a (another) complex whole besides being a necessary component of the whole which is a living human individual.

- *Sth.19 A human body has parts and has some other bodies connected to it.*

Hands and feet, the brain and the kidney are proper parts of the composite biological matter or substance called a human body. The network of nerves is to be regarded also as proper part of the body. It is also so treated in the medical treatises. But the sensory organs, five of them, which are there in the body in the sense of being attached to it are independent substances and are created, in some sense or other. They are not proper parts of the body but are somehow attached to it.

- *Sth.20 The self is the seat of consciousness.*

The difference between a *seat* of consciousness and *place* of consciousness is, briefly speaking, that place here means place of *manifestation* and seat means *locus* in the relation of inherence.

Since NPsy admits the importance of body (having as its parts brain and nervous system) in the origin and individuation of consciousness and in the overall being and functioning of the human individual viewed in the way described, most of the recent findings of neurosciences can, in principle, be accommodated in the Nyāya theory of consciousness. A Naiyāyika can even hold that it is theoretically possible that we could find some sense of the expression 'seat' in which one could say with some justification that brain was the seat of consciousness. He further says that already there is (has been found) another sense in which brain is not the seat of consciousness but the self is.

4. A FEW MORE SPECIFIC THESES OF NYĀYA PSYCHOLOGY

In the rest of the paper we would like to just formulate some of the important and interesting theses of NPsy with the hope that interdisciplinary discussion and dialogue on them would be profitably initiated. Some of these theses may be found to put up the challenge to experts of the relevant fields to devise or design laboratory experiments with a view to confirming or disconfirming, as the case may be, them. It would however be too premature to arrive at any final evaluative judgments, on the basis of such experimental findings, about the theses of the philosophical psychology of Nyāya. This is no dogmatic or a priori ruling. It is rather the recognition or better reiteration of widely acknowledged truth that even the latest machines are incredibly slow compared to the speed of the mind or *manas*. Some of the theses of the classical NPsy emphasize this aspect of the *manas*. Anyway, many of the following theses are proposed by the Naiyāyika-s as empirical or they are held by them, at least so they appear to claim, on empirical evidence.

However let us first begin with an objection. Nyāya seems to define self as what has consciousness as its essential quality, though it is essential in a weaker sense. On the other hand consciousness is defined as an exclusive quality of self. This is clearly circular.

The following theses of NPsy provide reply to the objection.

- *Sth.21 The self is what has, as one of its marks, consciousness which it owns as a quality.*

- *Sth.22 Consciousness or better cognition (jñana) is that quality which is the condition of our use of language.*
- *Sth. 23 Manas has two aspects which determine it to function in two different ways. As a substance it facilitates the origin of internal states including consciousness. As an internal sense organ it enables us to perceive such internal phenomena as particular cognitive states or states of desires, fears etc.*
- *Sth.23a Manas as a substance further facilitates the occurrence of sensory perceptions by external sensory organs by assisting the latter. Thus no perception, whether external or internal, is produced without the help of manas.*
- *Sth.24 The realm of the mental consists of mainly two internal substances (self and manas) and many internal properties mainly of the substance self. Quality of subjective and private experience is one such property.*

Nyāya is a form of soft dualism. In it mind and matter are more alike than they are in the Cartesian system. First consciousness is only a weekly essential quality of the self. Secondly the self is admitted to have spatial extension; it is ubiquitous or infinitely extended. But in the Cartesian system as well as in Europe and elsewhere spatial extension is regarded generally as a mark of matter or of the material. For this reason we prefer to describe NPsy not as the study of the mental but the study of the human experience which is an *internal* and phenomenal state. The distinction between the internal and the external is clearer than the distinction between the material and the mental.

- Sth.25 What is not external is internal. And something is external if it is an object of external perception or is the seat of something that is externally perceived.
- Sth.26 All sense organs are substances and are imperceptible.
- Sth.27 *Manas* is the only internal sense organ. (Internal sense organ has been described by some European psychologist as misnomer) There are five external sense organs in addition. These are organs respectively of visual perception, tactual perception, auditory perception, gustatory perception and olfactory perception.
- Sth.28 *Manas* is described as psychic matter because it plays an important role in the origin and perception of all that are internal, perceptible and impermanent. It however enables us to perceive even those internal things that are permanent.
- Sth. 28a Though these things are permanent, various experiences we have of them are impermanent and need causal explanation. Major task of psychology is to find such explanation.
- Sth.29 NPsy thus includes such sub theories as theory of perception, inference, language etc. on the one hand and on the other theory of memory, emotion, feeling, volition and so on.
- Sth.30 Ubiquitous soul accounts for the unity of experience. Neither *manas* or mind nor body can do that.
- Sth.31 *Manas,* because it is atomic in dimension (paramāṇuparimāṇa) and is extremely mobile, accounts for such phenomena as shift of attention, localization of sensation, succession of episodes of experience etc.
- Sth.31a *Manas* also explains what is known in psychology as the phenomenon of localization of sensation.

- Sth.32 No two states of experience occur simultaneously. The feeling that they do is the result of our misperception.
- Sth.33 Generally speaking no state of experience exists for more than two moments (only some special experiences exist for longer period).
- Sth.34 We have empirical evidence that attention or shift of attention can be caused as much by physical conditions as by mental resolve.
- Sth.35 There can be and are internal states of consciousness or cognition which remain (or can remain for unspecified period of time) unknown. There may be produced certain internal state in us without our being aware of them.
- Sth.36 A state of consciousness and the corresponding state of self consciousness are two different acts of consciousness which are separated in time or are only successive internal states.
- Sth.37 Nyāya has answer to the objection that the immediately previous thesis leads to infinite regress.
- Sth.38 Occurrence, in a soul, of an internal state annihilates or removes any already existing states provided the latter perceptible.
- Sth39 Apart from the said existential or factual opposition between perceptible internal states of one and the same soul there obtains among different cognitions a sort of logical and causal opposition. Doubt about a thing is removed by or is prevented from occurring again by the definitive knowledge of it, if and when we have that (and that kind of) knowledge.
- Sth.40 Though we cannot perceive the internal states of other Minds or individuals we can infer such states. As such we can attribute to an individual certain belief or emotions.
- Sth.41 Such third person(inferential or indirect) account finds steady corroboration by first person account.
- Sth.42 Human individuals who are structurally similar also behave similarly. New born babies not only suck mother they also respond alike by rejecting bitter testing substance and welcoming the taste of honey, say. Such unlearned uniformity of behavior including life saving ones (jīvana *yoni prayatna*) is more similar than not to our learned behavior and are explained in the same causal way. In a bid to provide uniform causal explanation to both so-called instinctive and non-instinctive behavior NPsy refers to some imperceptible factor or adṛṣta.
- Sth.43 There are some individual differences in the experience of pleasure and pain which are not explainable causally in terms of observed or observable causal condition. Instead of saying and believing that they are not causally explainable NPsy admits that some imperceptible, though identifiable, causal factors are responsible.
- Sth. 44 Nyāya and Nyāya psychology advocate, consistently with the above thesis, a form of spiritual pluralism.
- Sth.45. Some successive states of cognition, affection, conation and physical behavior form a single causal chain. (This shows one side of, the more controversial aspect of, mind body interaction – interaction between the physical and the mental).
- Sth. 46 Memory, unity of experience, perfect coordination between, so to say, the independent workings of different sense organs or better the data made available by

the independently working sensory organs, are facts of experience. These phenomena are made possible by an enduring self (*sthira* ātmā).Which remains unchanged amidst almost constant bodily changes.

5. PSYCHOLOGY AND YOGA

We stop here arbitrarily with one last remark. Psychology qua psychology is a descriptive account and theoretical study of the workings of the *manas* and the properties of internal states of self. Existence and some features at least of these states are made possible by the atomic structure and extremely mobile nature of the *manas*. With the detachment of a scientific theory or inquiry NPsy studies normal phenomenal human experience and attempts empirical causal explanation of it. But there is another sort of study called Yoga which is informed by the necessity of certain useful practical results. Yoga conducts its study of the *manas* not so much to yield merely *theoretical* understanding of the workings of it or the properties of human experience but to discover means of controlling or disciplining *manas* and the results thereof. Yoga unlike NPsy is not a merely descriptive and theoretical study but a normative study related to certain spiritual praxis (or mokṣa sādhana) which is of immense help if man is to attain spiritual freedom. Mano-eva manuṣyānam kāranam *bandha* mokṣayoḥ: Manas, the internal mobile substance which binds man also frees him. Since spiritual freedom is a common goal of all men Nyāya also accepts and recommends the measures of quieting *manas* which Yoga teaches. So far Yoga and Nyaya are complementary studies.

Psychology, however, does not interfere with the normal functioning of the *manas*. It studies as an observer the workings or the contributions of the *manas* along with those of the ātmā and s'arīra which make possible the origin, growth, decay, change etc of internal states of ordinary, normal adult human life. It can also observe and explain the way techniques of yoga introduce changes in our experience and self-consciousness. Psychology also notes that it is through a sort of untrained *manas* that the mind quieting techniques of yoga are put to use. To put it differently through *manas*, *manas* is to be trained. Be that as it may, the objectives of NPsy and Yoga are different. Yoga aims at the stopping of the workings and activities of the mind or *manas*. NPsy seeks to study them as they are, and as they naturally flow, so to say.

In: Expanding Horizions of the Mind Science(s)   ISBN: 978-1-62808-705-5
Editors: P.N. Tandon, R.C. Tripathi and N. Srinivasan   ©2013 Nova Science Publishers, Inc.

Chapter 24

THE QUEST AND DENIALS

Kireet Joshi

Almora Area, India

ABSTRACT

The chapter discusses the materialistic views of the Universe in the context of the uncertainty associated with processes leading to questioning the ultimate efficacy of explaining the mysteries of the Universe. The problem of consciousness and theory of evolution are discussed in the context of modern developments in science. Possible ways to synthesize science and spirituality are also discussed. The chapter concludes with philosophical views related to consciousness and spiritual experience based on the works from Indian philosophy with special emphasis on Aurobindo's thought.

A pilgrim of our times, earnest in his quest and scrupulous in his methods, burning with a single aspiration to search for truth and to consecrate himself to the tasks that would flow from that search for the fulfillment of himself and for the eventual fulfillment of the human journey, finds himself arrested by various forms of denial and skepticism that pronounce the message that the present organization of consciousness is the last limit and that the course of rationality or reasonableness is to limit himself to the boundaries of that organization and be content with limited certainties and probabilities and ever crumbling ideas of tolerable existence for oneself and the world.

There is, indeed, a layer of human intellectuality which is tied to senses so exclusively that it finds its station and resting place in what these instruments perceive and cognize, and even if it reflects, it finds nothing more satisfying than the argument that senses are our sole means of knowledge and that nothing can be known and nothing can be valid if it does not refer back to its origin in sense experience.

Not long ago, materialism held the central field of inquiry, and it had predominant influence over the seekers of knowledge and persuaded them to limit the aim of life to a feverish effort of the individual to snatch what he may from a transient existence of to dispassionate and objectless service of the race and of the individual, knowing well that the latter is a transient fiction of the nervous mentality and the former only a little more long lived collective form of the same nervous spasm of Matter.

It is true that materialism is no more being advanced so overwhelmingly, and it has been largely conceded that materialism cannot be defended as a metaphysical philosophy. This is because, in the first place, those who are inclined psychologically to favor materialism have largely come to the logically sustainable It has also come to be recognized increasingly that it is impossible to argue that they are the only means of knowledge and conclude that from the premise that matter alone exists.

Nonetheless, materialistic bias has continued to preponderate in the field of philosophy, epistemology, science and philosophy of science. And this preponderance can be seen in the way in which concepts such as those of infinity, eternity, universality, essence, explanation, causality and others have come to be dealt with during the last hundred years. All these concepts have been scrutinized with the microscopic lens which permits only those deliverances which are ultimately warranted by physical senses.

At one time, the knowledge of the phenomena was sought to be understood in the light of 'noumena', but this no more is considered to be a necessary requirement of rational understanding. At one time, there seemed in the world, an iron insistence on order, on a law basing the possibilities. But today, even while granting some apparent operations of laws of nature, what is predominantly emphasized is the unaccountability and freak and fantasy and random action. It increasingly is being recognized that the theory of Mechanical Necessity by itself does not elucidate the free play of the endless unaccountable variations which are visible in the universe, as also, in the evolutionary processes. Perceiving, however, that there is a good deal of freak and fantasy combined with a good deal of order, there is a trend to explain the world by pointing to a self-organizing dynamic chance or to give up the idea altogether of explanation in terms of any universal law of causation. It has even been argued that one should not aim at explanation of the phenomena of the world, but one should remain content with their descriptions.

It is true that science still aims at explaining phenomena apart from describing them in terms of the how and why. But considering that induction is necessary method of science, and since the assumptions of induction in regard to the law of uniformity of nature and the law of the world-phenomenon, various theories have been developed to explain induction empirically without the need to acknowledge non-empirical belief in laws of universality and causality. As a result, philosophers of science have largely come to various shades of skepticism or fallibility.

It is against this background that science is being looked upon as a body of knowledge, the certainty of which can never be guaranteed, and which will constantly be subject to corrigibility and fallibility. It is being admitted that on account of fallibility of scientific knowledge, while the boundaries of knowledge can be indefinitely expanded in course of time, questions such as those of the origin of the universe, as also questions pertaining to the aim of life, and those relating to fulfillment of the individual and the collectivity, need not be raised because they cannot be answered. It is also acknowledged that this conclusion and consequent attitude may be found unsatisfactory, but it is argued that there are no ascertainable facilities other than those of physical senses and those of anthropological rationality by the help of which any higher knowledge can be attained.

Bertrand Russell, speaking on his own behalf and like-minded philosophers, stated towards the end of his 'History of Western Philosophy', "They confess frankly that the human intellect is unable to find conclusive answers to many questions of profound importance to mankind, but they refuse to believe that there is some 'higher' way of knowing by which we can discover truths hidden from science and the intellect." (p.789)

D.P. Chattopadhyay in his book, *Induction, Probability and Skepticism* has stated:

> "The real world and the objects in it are not revealed to us at once. We know them gradually, historically, and can never be cognitively sure of their absolute certainty. The knowing self and the known (and knowable) world are differently interlocked, inter-animated, and interactive, both biologically and epistemologically. In our knowledge of the world we do not, rather cannot, stand totally apart from it. From the evolutionary point of view what is called survival value is in a way akin to truth value, truth of what our organism is informed about the world around us. Situated in the world, we have to know it, in spite of our ability to transcend our situation, in a limited way; we cannot not totally get out of it. It is somewhat like our inability to jump out of our own skins and schemes." (p. xxxi)

Science and philosophy of science have, thus, come to advise us to accept in all humility our limitations in respect of knowledge, in respect of truth and in respect of certainty. But is this the end of the journey? Human aspiration refuses to accept it, and a question is raised: Are there any avenues which we can justifiability pursue, such as those proposed by religion, occultism, rationalistic philosophy and spirituality and persuade science and philosophy of science to join what can become a combined or synthetic quest, not only for the purposes of expanding conquest of the realms of truth but also those of the realization of the highest possible ideals of human welfare, human solidarity and human fulfillment?

1. PHENOMENON OF CONSCIOUSNESS: PROBLEM OF SCIENCE AND SPIRITUALITY

In an important development in the recent march of science, the phenomenon of consciousness has begun to command increasing attention. As a matter of fact, as far back as the close of the 19th century, the great Indian scientist, Jagdish Chandra Bose, had demonstrated the presence of consciousness in plants and even in metals. But consequent upon the latest developments in Quantum Mechanics, such as Bell's Theorem, scientists have begun to inquire into the mystery of consciousness.

Roger Penrose has in his celebrated book 'Shadows of the Mind: A Search for the Missing Science of Consciousness', attempted to address the question of consciousness from a scientific standpoint. He has strongly contended that an essential ingredient is missing from the contemporary scientific picture. He acknowledges the precision and scope of physical laws but he complains that they contain no hint of any action that cannot be simulated computationally. He points out that there are, however, reasons for believing that there is in them a hidden non-computational action that the functioning of our brains must somehow be taking advantage of. He further argues that questions should be asked as to why the scientists have failed to recognize the fact that a non-computational phenomenon like consciousness should be inherent at least potentially in all material things. According to him, it is quite possible in physics, to have a fundamentally new property completely different from any contemplated hitherto, hidden, in the behaviour of ordinary matter.

David Bohm in his book, 'Wholeness and the Implicate Order', argues that the Quantum Theory presents a serious challenge to the mechanistic order. He points out the key features of the quantum theory that challenges mechanistic order are:

1. Movement is in general discontinuous, in the sense that action is constituted of indivisible quanta implying also that an electron, for example, can go from one state to another, without passing through any status in between.
2. Entities, such as electrons can show different properties (e.g., particle-like, wave-like or something in between), depending on the environment or context within which they exist and are subject to observation.
3. Two entities , such as electrons, which initially combine to form a molecule and then separate, show a peculiar non-local relationship, which can best be described as a non-causal connection of elements that are far apart(as demonstrated in the experiment of Einstein, Podolsky and Rosen).

These three key features of quantum theory, according to Bohm, clearly show the inadequacy of mechanistic notion. Thus, if all actions are in the form of discrete quanta, interactions between different entities (e.g. electrons) constitute a single structure of indivisible links, so that the entire universe had to be thought of as an unbroken whole. In this whole, each element that we can abstract in though shows basic properties (wavelike and particle like, etc.) that depend on its overall environment, in a way that is much more reminiscent of how the organs constituting living being are related, than it is of how parts of a machine interact.

Bohm (ibid) discusses the relationship of matter and consciousness on the basis of some common ground and propounds and idea of a new world-view based on the concept of unbroken wholeness. He states:

"To obtain an understanding of the relationship of matter and consciousness has, however, thus far proved to be extremely difficult, and this difficulty has its root in the very great difference in their basic qualities as they present themselves in experience. This difference has been expressed with particularly great clarity by Descartes, who described matter as 'extended substance' and consciousness as 'thinking substance'. Evidently, by 'extended substance' Descartes meant something made up of distinct forms existing in space in an order of extension and separation basically similar to the one that we have been calling explicate. By using the term 'thinking substance' in such sharp distinct forms appearing in thought do not have their existence in such an order of extension and separation (i.e. some kind of space), but rather in so in a certain sense Descartes was perhaps anticipating that consciousness has to be understood in terms of an order that is closer to the implicate than it is to the explicate."

According to Bohm, matter as a whole can be understood in terms of the notion that the implicate order is the immediate and primary actuality (while the explicate order can be derived as a particular, distinct case of whether or not the actual substance of consciousness can be understood in immediate actuality. He contends that if matter and consciousness could open for comprehending their relationship on the basis of some common grounds. He concludes that in this way one could come to the term of a new notion of unbroken wholeness, in which consciousness is no longer to be fundamentally separated from matter.

2. SCIENTIFIC THEORY OF EVOLUTION VS. THEORY OF INTELLIGENT DESIGN

There has been a debate between the scientific theory of evolution and the special creation theory. This debate is most acute in the United States, where Christians faithful to the creation account in the Book of Geneses have seized upon "intelligent design" (I.D.) to show that the blanks in the evolutionary narrative are meaningful. There is a debate between Richard Dawkins, who holds the view that on the basis of the physical evidence that we possess today of the universe, evolution should lead towards atheism, and Francis Collins, who holds the view that material science not only points to Immanent God but also to Transcendental God, who exists outside of space and time, - the account following statement of Roger Sperry, who made the following important statement on 1995:

> "**With a scientist's faith** in empirically verified truth and a long commitment to research in the brain, behavioral and life sciences, I spent most of my working years accepting the scientific account of the nature and origins of life and the universe. If science said that human life is lacking in any ultimate purpose, value, or higher meaning – that we and our world are driven merely by mindless, indifferent physical forces – I was prepared to face this. Like many scientists, I preferred to seek out and confront the truth, however harsh, rather than live by false premises and illusory values. The more I learn about the workings of the brain and how it processes information, the stronger becomes my allegiance to the type of truth that receives consistent empirical validation in the outside world.
>
> "**Nevertheless, without abandoning or compromising scientific** principles, I have come around almost full circle today to reject the type of truth science traditionally has stood for, along with its dominant central tenet that everything in our universe, including our human psyche, can be accounted for in terms entirely physical – that science has a absolutely no need for recourse to conscious mental or spiritual forces. As a brain scientist, I have come to believe in the reality and power of conscious mental/cognitive entities of the mind or spirit and the indispensability of their causal control for both brain function and its evolution – and that science has been wrong all along with categorical denial of this."

We may also refer to Shewin Nuland, a clinical professor of surgery at Yale University who, in an interview on his book published in 1997 "*The Wisdom of the Body*", stated as follows:

> "**I think there is an evolutionary accomplishment of the human** cortex, the cortex of the brain, and the way it accepts and synthesizes the information, uses information from the environment, from the deepest recesses of the body, the way it recognizes dangers to its continued integrity. And I think this is precisely what the human organism because of our cortex and our ability to process information, I think that there is an awareness of the closeness of chaos.
>
> "**And I think that there is a lot of evidence** for that including cultural **evidence** ..."
> (Vide American Public Media, interview of Ms. Tippett with Dr. Nuland, 2007)

It can be said that, as the field of research is expanding, both microcosmically and microcosmically, fresh data are pouring upon us to suggest that the materialistic view of the universe and the materialistic interpretation of evolution should lose force and give way to a new way of looking at the universe and evolution where consciousness may come to be regarded as more primary and fundamentally determine power.

3. SYNTHESES OF SCIENCE AND SPIRITUALITY: NOT YET

With the increasing evidence of the presence of consciousness or of the possibility of presence of consciousness in the physical, we seem to have entered into a new of the development of scientific knowledge and of the philosophy of scientific knowledge. We seem to be arriving at a meeting point of matter and consciousness, of the physical and the super-physical. But it would be premature to conclude that the identity of matter and consciousness and of the physical and the supra-physical has been proved, and that, therefore we have arrived or that we are soon going to arrive at the syntheses of science and spirituality.

The presence and operation of consciousness in the physical world has been admitted by a number of philosophical systems, and these philosophical systems, even though they contain valuable insights, do not provide as yet any firm ground for the relevant verifiable knowledge. It has been found necessary, in order to know with greater certitude, to follow the curve of evolving consciousness until it arrive at a height and largeness of self-enlightenment in which highest subjective consciousness is discovered to be capable of the generation of objective universe.

4. QUEST OF SUPRA-PHYSICAL REALITIES: FOUR AVENUES

A crucial problem of thought as also of life is centred on the distinction between the physical and the supra-physical, on the possibility of transition from the physical to the supra-physical, and on establishing, if possible, the relation between the physical realities and supra-physical realities. It is, indeed, possible to question the existence of the supra-physical, and several systems of thought and philosophy have been built in the course of human history to convince the human that there is no such thing as the supra-physical, and all that can be considered to be supra-physical is nothing else than an epiphenomenon of the physical. Even subjective experience which can be regarded as supra-physical is sought to be shown to be a twitch of the physical. But the human quest refuses to be permanently tied up to the physical and to the sensuous. The data of experience so vast, so complex and so intriguing that persistent efforts have been made through ought the history of humankind to transcend the limitations of the physical realities. The human tendencies to concepts and theories are a minimum starting point of the west for the supra-physical. If we do look at the world with fresh eyes and with wonder that accompanies every impartial look at the world and events, we seem to be surrounded by something that is not physical and it betrays itself to be other than what it appears to be.

That the physical can be transcended and that there can be transition from the physical to the supra-physical, and even there can be built up relations between the physical and supra-physical has been constantly affirmed and religion, occultism, philosophy, and spiritual experience. These four lines of human quest have some kind of interdependence, but they have been pursued from time to time independently of each other or with predominance of one line of development in comparison with others. Of these four, the first three are approaches to the spiritual; evolution of the human consciousness and being, but the last is the central avenue of entry.

5. RELIGION

It is hard to define religion, considering that it is seen to have been manifested in various forms and at various levels. It has been conjectured that some kind of animism was the beginning if religion and that it developed in spiritism, totemism and in beliefs and experiences of gods and goodness, and finally in the beliefs and experiences of one universal God or one universal principle of Being or Non-Being. At one stage, tribal religions with tribal gods and goodness fought among themselves; at another stage, regional or national religions flourished; it was only at advanced levels that there have grown up universal religions, and even there, there have been claims and counter-claims regarding their universality. This complex history of the development of religion makes it difficult for us to define religion in some precise terms. But it may be said that religion is that avenue of human quest which is essentially supra-physical or invisible, which have super human qualities and powers, and which can be invoked for aid in the growth of human life and in transgressing obstacles that lie in the process of what comes to be conceived as human fulfilment or human salvation or human perfection. But religions develop doctrines that tend to a character that demands unquestioning acceptance in the ground that the truth that it claims cannot be proved rationally and can only rarely be verified, if at all, in experiences that lay beyond our senses and reason. Religion also tends to expand its domain of influence and tends to build up edifices, structures, forms by the help of those realities which bridges can be built between humanity and that reality or those realities which are celebrated as if supreme importance. Religion is, therefore, conceived as something that that binds humanity with God or gods or those beings or states of consciousness which are worthy of worship and celebration. In the process of building up these bridges, a prominent part is played by the growth of rituals and ceremonies as also of prescribed acts, of laws of conduct and of institutions by the aid of which human life can be molded in forms that are considered helpful to the growth of human life on the lines that can eventually be harmonized with the sacredness that is attributed to the highest reality of realities that are prescribed to be objects of faith, reverence, admiration, and worship. Religions tend to trace their origin in scriptures. Words of Revolution or inspired avenues of approach to the Ineffable, Incommunicable and Unknowable.

6. RELIGION AND OCCULTISM

Religion has an element of occultism - sometimes this element is predominant, sometimes preserved only in rituals, sometimes even largely exiled or condemned. In modern times, as a result of excessive rationalization, there has come about a discrediting condemnation of most of the occult elements which seek to establish communication with what is invisible. In a sense, occultism aims at knowledge of secret truths and potentialities of Nature which enable man to lift himself out of slavery to his physical limits of being. It is an attempt in particular to posses and organize the mysterious, occult and subliminal power of Mind upon Life and of both Mind and Life over Matter. Occultism also strives to know the secret of physical things, and in this striving the development of several sciences aim to be promoted, particularly, astronomy chemistry, geometry and the sciences aim to be promoted, particularly, astronomy, chemistry, geometry and the science of numbers. In fact, the knowledge developed by physical sciences can also be regarded as a part of occultism, since that knowledge reveals the secret knowledge and application of secret knowledge so as to make secret knowledge and utilizable for enlargement of discoveries and utilization of forces of Nature. It is true that occultism is popularly associated with magic and magical formulae and supposed mechanism of forces which are supra-physical of supernatural. Often occultism is looked upon as an aspect of Nature-Force experimented with their possibilities; occultism is not altogether a superstition. Just as in physical science, when the forces of Life and Mind are studied, one can discover secret formulas and methods of their application, and one can even develop secret formulas and methods of their application, and one can even develop a domain of mechanization of latent forces - mental and vertical in a restricted and specialized sense.

7. ROLE OF PHILOSOPHY

Nonetheless, in the multi-sided effort of humanity, the role of philosophy as a means of effecting a transition from the physical to supra-physical has made contributions which can be considered to be of capital importance. Reason is the central instrument of philosophy, and it enables the formation of perceptions and conceptions which can be obtained by penetrating appearances of phenomena. It is true that in its mixed action. Reason confines itself to the circle of sensible experience and to the appearances of things in their relations, processes and utilities. But Reason asserts its pure action , when even while accepting human sensible experiences as a starting-point, it refuses to be limited by them; it goes behind, judges, works in its own right and strives to arrive at generalizations and even at unalterable concepts which attach themselves not to appearances of things, but to that which stands behind their appearances. The most important function of pure Reason is to correct the errors of the mind which is normally tied up both the senses. It is by the complete use of the pure Reason that one can arrive from physical to metaphysical knowledge, which is the highest contribution of philosophy.

Philosophy aims at embracing vastest possible ranges of facts, physical and psychological and, in an attempt to ascertain truths behind appearances, develops criteria of validity of knowledge. In this process, philosophy tends to criticize and control experience and it criticizes itself and strives to question all that is dogmatically assumed in all domains of experience, knowledge and utilities. The formulations of philosophy follow standards of intellectual knowledge, and philosophy may be regarded as only a way of formulating to ourselves intellectually in their essential significance the psychological and physical facts of existence and their relation to any ultimate reality that may exist.

The basic problem of philosophy is not only related to its role as a critique of experience and as a critique of Reason, but in collecting data of existence, and thus one of the main reasons why philosophical systems are riddles with claims and counter-claims, and battles of logomachy.

In search of comprehensive data, philosophy is obliged to consider the claims of occultism, religion and spiritual experience. Occultism, if admitted, provides a vaster field of experience, but at the best, occultism, points of spiritual experience which lies beyond its proper domains of enquiry. Religion tends to claim the knowledge of spiritual realities; it pronounces its judgments based on intuitions and revelations that are proper to spiritual experience. But religion tends to set itself in opposition to the demands of Reason and rational enquiry, and tends to claim even that intuitions and revelations or inspirations of spiritual experiences are rather rare and cannot be and ought not to be subjected to strict demands that science and philosophy make in regard to the validity of knowledge. Religion prescribes faith as a method of holding in to the truths which have been claimed to have been received in rare states or in rare individuals where revelations, institutions or inspirations have flowered. In regard to those rare occurrences of illuminations, religion tends to formulate doctrines and puts them forward as dogmas, which are not defensible in terms of rationality but which are yet declared to be unquestionably valid. Religion and philosophy, religion and science stand in conflict in this very important issue. Philosophy and science often tend to reject the claims of religion, or while admitting their own limitations in arriving at conclusive statements of truths create double standards of truth: those relevant to science and philosophy and which provides justification for acceptance of the dogmas of religion and yet apply rational methods in explaining those phenomena of the world which can be explained in the light of supra-natural revelations and dogmatic doctrines.

This entire field of philosophy in its field of philosophy in its interaction with religion and spiritual experience has been a field of interminable debates and even violent disagreements. Philosophy has therefore often tended to end these debates and disagreements by excluding altogether the realms of religion and spiritual experience.

Nonetheless, philosophy, even when obliged to reject religion and spiritual experience, has attained greater clarities, subtleties and complexities, and provides to the human experience those questions and led to admit as important ot else as important and worthy of pursuit by methods of intellect to their farthest possible capabilities or by opening to some other or higher methods of knowledge and experience.

Philosophy has, sometimes striven to elevate Reason to such a degree of refinement that it opens up to its deepest recesses where it discovers incorrigible intuitions, and it is capable of such catholicity and flexibility that the truths of intuitions and revelations can even be grasped by a logic that can formulate truths of spiritual experiences, not by resorting to dogmas, but by the cultivation of a larger Reason or by Reason where concepts of Reason and derived from spiritual experience can meet together in harmony. There are, indeed, problems in arriving at that harmony, and these problems need to be confronted in order that humanity can eventually find data of rational thought of spiritual experience in such wide totality that it can resolve conflicts between religion, science and philosophy, conflicts among religions spiritual experiences and conflict in the field of direct spiritual experiences. In the meantime it is to be admitted that philosophy has played a role of a bridge between the spirit and the intellectual Reason, and it has shown that the light of a spiritual or at least a spiritualized intelligence is necessary for the fullness of our total inner evolution, and without it, if another deeper guidance is lacking, the inner movement may be erratic or one-sided or incomplete in its catholicity.

At the highest summit if philosophical thought, pure reason comes to conceive infinity and eternity. All end and beginning presuppose something beyond the end or beginning. In the first place, infinity is conceives in terms of Time and Space, an eternal duration, interminable extension. In the philosophy of Immanuel Kant, Time and Space are conceived as conditions of consciousness under which we arrange our perceptions of phenomena. But existence-in- itself has to be conceived as something beyond Time and Space, if the antinomies that attend on Time and Space, as also those in respect of other categories of quantity, quality, relation and modality, are to be avoided. In the philosophy of Bradley, which describes the movement of Pure Reason in the analysis of appearances, we find self-containing ever-new movement, and infinity comes to be conceived as the same all-containing all-pervading point without magnitude. It is admitted that there is here a conflict of terms, even a violent conflict, and yet these terms express quite accurately what is perceived or conceived by Pure Reason of Space and Time, and this perception or conception ends in conceiving Reality as Other than Thought, as something incorrigibly Real, transcending the limitations of Space and Time and all other appearances that are found to be in the universe. It is, indeed , possible to question the need of posting the conception of Infinity and Eternity and the Pure Reason they are incorrigible concepts carrying with them rational certainty. But it is not possible to go farther and to deny the very concepts of Pure Reason formulated by relations and to conceive of Reason in empirical terms. The empirical concept of Reason, however, is a fluctuating concept, considering that empiricism itself has verifying forms. Nonetheless the strength of empiricism lies in the fact that the concepts of the Pure Reason do not in themselves fully satisfy the demand of our integral being. Just as our physical body sees things through two eyes always, even so the integrality of our nature demands integral seeing as consisting of ideative conception and factuality verified in experience. Mere ideative conception or mere ideative certainty of the Reality caught in conception as actually existing beyond thought remains to the demand of our integral nature incomplete and at least to a part of our nature almost unreal until it becomes an experience. This is the standing ground on which empiricism finds a perennial source of strength.

But empiricism, as formulated in the materialistic or allied systems of thought, confines itself to the deliverances of sensuous experience; it avoids self-criticism and tends to develop some kind of dogmatism in regard to the incorrigibility if sensuous experience. Or else, it develops skepticism. Or at a more reasonable and at a more irrefutable level, it acknowledges the deliverances of scientific knowledge but pronounces that knowledge to be valid until it comes to be contradicted or modified in the light if further advances of scientific knowledge. This attitude is truly skeptical in character, and although it avoids any extreme position of skepticism, it refuses to grant the possibility of human Reason to arrive at any irrefutable certainty.

But human cannot for ever rest in a state in a uncertainty. Even the needs of human survival demand the quest of certainty, and realizing that philosophy is unable to provide that kind of certainty, unless our ordinary limits of consciousness are transcended in supra-sensuous experiences, there arises, at the borders of philosophical thought, an irresistible urge to expand the capacities of human experience.

8. PHILOSOPHY AND SPIRITUAL EXPERIENCE

This is where philosophy tends to admit the relevance of religion and occultism. But realizing the limitations of these two avenues of the expansion of human experience, it pushes the borders of quest into the realm of direct spiritual experience, and revelatory experience. There are, however, great difficulties in understanding this realm of spiritual and hardly in its purity, since it is overwhelmingly mixed up with religion or gets confined to the limits of occultism or to those occasional experiences of flashes which fail to illumine in some steady light of verifiability and comprehensiveness. And yet, it is precisely this field of direct spiritual experience and consciousness to which we are obliged to turn, if we are earnest in our inquiry and if we have patience to endure the persistence that is demanded by the inescapability if our need to know and our need to apply our knowledge to the problems of practical life in the midst of which we find ourselves poised uncomfortably and even painfully.

An object of inquiry reveals its truth in some kind of fullness only when it is examined in its outermost appearance as also in its loftiest or highest possibilities of being and manifestation. Thus, if consciousness is our object of inquiry we need to study the nature of consciousness at various levels of its appearance and reality, and we have to explore those methods, by means of which various layers of consciousness can be uncovered, observed and experimented upon with a view to arrive at the true of consciousness.

It is in this context that the study and practice of yoga can be perceived as not only relevant but indispensable.

For the body of knowledge that we call yoga has been developed as a science and technology of consciousness and its history in India extends right from the Vedic times to the present day. If we study the Veda seers had already attained to the loftiest secrets of consciousness. As Sri Aurobindo points out:

> "They may not have yoked the lightening to their chariots, nor weighed sun and star, nor materialized al the destructive forces in Nature to aid them in massacre and domination, but they had measured and fathomed all the heavens and earths within us, they had cast their plummet into the inconscient and the subconscint and the superconscient; they had read the riddle of death and found the secret of immortality; they had sought for and discovered the One and known and worshiped Him in the glories of His light and purity and wisdom and power."[1]

The sea of the superconscient, the sea of the subconscient, anf the sea of the conscient, of the light of the life of the living being between the two, -- this is the Vedic idea of existence which consists of three oceans. Vamadev in Rigveda IV.58.11 describes these three oceans of consciousness as follows."

- Dhaman te visvam bhuvanam adhisritam[1]
- Antah samudre hryantar ayusi[2]

---

[1] Sri Aurobindo: The Secret of the Veda, Centenary Edition, Volume 10, p. 439.

[2] The entire creation extends from Thy seat (which is in the superconscient) to the lowest sea the interior of the heart of living beings.

This concept of threefold consciousness may be regarded as the basis of the explorations of the Vedic seers as also of the methods which they developed for affecting the journey of the search of the relations that can be secured in the plane of the superconscient. It is these methods which constituted what may be called the technology of consciousness or the practical psychology of yoga as it developed in India throughout the ages. From the Veda to Sri Aurobindo is a long and unbroken journey and story of the development of the science of Yoga.

Whole giving the meaning of the term "Consciousness", Sri Aurobindo states:

"Ordinary we mean by it our first obvious idea of a mental waking consciousness such as is possessed by the human being during the major part if his bodily existence, when he is not asleep, stunned or otherwise deprived of his physical and superficial methods of sensation In this sense it is plain enough that consciousness' id the exception and not the rule in the order of the material universe. We ourselves do not always possess it. But this vulgar and shallow idea of the nature of consciousness, thought still colors our ordinary thought and associations, must now definitely disappear out of philosophical thinking. For we know that there is something in us which is conscious when we asleep, when we are stunned or drugged or in a swoon, in all apparently unconscious states of our physical being. Not only so, but we may now be sure that the old thinkers were right when they declared that even in our waking state what we call then our consciousness is only a small selection from our entire conscious being. It is a superficies; it is not even the whole of our mentality. Behind it, much vaster than it, there is a subliminal of subconscient mind which is the greater part of ourselves and contains heights and profundities which no man has yet measured of fathomed. This knowledge gives us definitely a starting-point for the true science of Force and its workings; it delivers us definitely circumscription by the material and from the illusion of the obvious."[3]

In Sri Aurobindo's writings of Yoga we have a systematic and detailed statement if the realms of consciousness discovered thorough the yogic effort, and we may describe here the secrets of consciousness that have been developed in their subtleties, profundities and applications, -- again through the ancient or new methods of yoga.

One of the most important counsels that Sri Aurobindo gives us is to avoid a plunge into the subconscient when we are not sufficiently ready. He also points out that our first concern must be with all that we are conscious of, and it is only when there has been already a good deal of harmonization of our conscious being and ascent to higher levels of consciousness that it becomes easier and safer to deal with the subconscient. An early plunge into the subconscient, he warns us, is unsafe and it would lead us into incoherence of into sleep or dull trance or a comatose torper. The higher we rise, he points out, the greater is the capacity that we achieve to deal with the lower. The lowest layer of the subconscient, which Sri Aurobindo calls the incoscient, can effectively be dealt with transformed only by the highest powers of the supramental superconscient.

The most important question is related to the methods by which one can rise into the higher levels of consciousness. Right from the Vedic times, it was discovered that the path of ascent lies in the utilization of normal functionings of our normal psychological facilities, namely, those of cognition, affection and conation. The experimentations and discoveries made by the yogins have also revealed that even the functionings of our physical being, and

---

[3] Sri Aurobindo: Archives and Research, Volume 10, Number 1, pp.10-12.

the postures of the body and processes of breathing, if properly purified, combined, strengthened and heightened in their capacities, can produce results which might seem to us extraordinary and even miraculous. The entire systems of Hatha yoga and the system of Raja yoga are based of badid of these discoveries. That there can be life-process without heartbeats and there can be thoughts independent of the operation of the brain have been demonstrated by the Hatha yoga and Raja yogic practices. The realizations, which are grouped under the concept of *asha siddhi* are verifiable and can be attained by the methodized effort that we call yoga.

Just as by the dexterous and scientific handling, the natural forces of electricity or of steam can be enhanced in their powers and in their possibilities of utilizations, similarly, the customary psychological workings can be scientifically and dexterously handled and their utilization can be facilitated in various ways. Yogic methods depend on the perception and experience that our inner elements, combinations, functions, forces, can be supported of dissolved, can be combined and set to novel and formerly the impossible workings can be combined and resolved into a new general synthesis by fixed internal processes. The discovery of the power of concentration of contemplation of the power of witnessing if the movements of prakriti in the state of purusha consciousness are extraordinary achievements of the yogic science. Raja yoga, yoga of devotion, yoga of knowledge, and yoga of divine works, start from use of some faculty in us by ways and for ends not contemplated in their everyday spontaneous workings. It can be said that all methods grouped under the common name yoga are special psychological processes founded on a fixed truth of Nature and developing out of normal functions, powers and results which were latent to which her ordinary movements do not easily or do not often manifest.

In one of the statements on *ashtha siddhi,* Sri Aurobindo states:

> "All siddhis exist already in nature. They exist in you. Only owing to habitual limitations, you make a use of them which is mechanical and limited. By breaking these limitations, one is able to get the conscious and voluntary use of them."

The eight sidhis of which yogic science speaks are obtained in varying degrees as one begins to utilize various methods of concentration. These siddhis are, first of all, those of *mahima* (including *garima*), *laghima* and *anima*. These three siddhis are siddhis of being as distinguished from siddhis of knowledge and siddhis of power. *Mahima* is unhampered force in the mental power or in the physical power. In the physical it shows itself by and abnormal strength which is not muscular and may even develop into the power of increasing the size and the weight if the body, etc.

Laghima is a similar power of lightness, that is to say, of freedom from all pressure or weighing sown in the mental vital or physical being. Laghima is the basis of the power to overcome gravitation, and thus it is the basis of *uthapana.*

Anima is the power of freeting the atoms of subtle or gross matter from their ordinary limitations. It is by this power that yogins were supposed to make themselves invisible and invulnerable or to free the body from decay and death.

Apart from these three powers, there are three siddhis of the powers in the consciousness of acting upon other conscious being or even upon things without physical means or persuasive compulsion. Great men are said to make others do their will by assort of magnetism, by a force in their words, in their action, or even in their silent will or mere presence. These three powers are *aishwarya, ishita,* and *vashita.Aishwarya* is when one merely uses the will and things happen or people act according what is willed. *Ishita* is when one does not even need to have a will but when one has a want or need or a sense that something ought to be and that thing comes about or happens. *Vashita* is when one concentrates one's will on a person or object so as to control it. The yogic science gives warning that these powers can only be entirely acquired or safely used when we have got rid of egoism and identified ourselves with infinite will and infinite consciousness.

There are other two *siddhis,* namely *vyapti* and *prakamya.* These are *siddhis* of knowledge. *Vyapti* is obtained when the thoughts, feelings, etc. of others or another kind of things etc. are felt rising from those things. There is also a power of communicative *vyapti* when one can send or put one's own thoughts, feelings, etc., into someone else. *Prakamya* is when one looks mentally or physically at somebody or something and perceives what that of the mind but of the senses is in it. It is the power of perceiving smells, sounds, contacts, tastes, lights, colours, and other objects of sense which are either not all perceptible to ordinary persons or beyond the range of one's own ordinary senses.

The most important aim of the yoga is to arrive at higher states of consciousness and the powers of union, unity or identity with universality and transcendence. These states are basically the states of wideness, or friendliness, of intense will power of askesis and of inner delight. In the Veda, these four powers were symbolically called the powers of *Varuna, Mitra, Aryaman,* and *Bhaga,* and it was ;aid sown that without the attainment of these combined powers, one cannot attain to the state of Truth-Consciousness or Sri Aurobindo calls the supermind which lies above the states of the higher mind, illumined mind, intuitive mind overmind.

The yogic science recognises four faculties which begin to develop when by the processes of meditation or by the renunciation of desire or by the utter purification of emotions or by the combination of all these three; one is able to transcend the limitations of the mental consciousness. These four faculities are those of revelation, inspiration, intuition, and sponataneous discrimination, the four faculties which were known to the Vedic seers as the powers of *Illa, Sarswati, Sarama* and *Daksha.*

What is meditation and how the process of meditation can lead to the states of silence in which higher faculties begin to operate is a vast field of exploration and long and arduous practice in Jnana Yoga. Similarly, how desire can be eliminated form our psychological complex and how desire can be by that elimination, the universal and transcendental will can begin to operate in us is also a field of vast askesis, control and mastery. Again, how various stages of the elimination of desire and along with the elimination of annihilation or egoism can occur is a field of difficult process of Karma Yoga. In the same way, how emotions in our consciousness which are attached to small and narrow aims and which are related to immediate and binding relationships can be purified and transformed so as to become a huge reservoir of universal love and transcendental compassion is also a vast field in Bhakti Yoga of untiring process of purification and of generation of secret delight in everything and perception of beauty and joy in all contacts of consciousness, and it is a field also that takes us into the supreme mysterious of consciousness and existence.

These are only introductory perceptions and conceptions, and can only faintly indicate wonder of consciousness and the object that lies beyond consciousness, although seizeable by consciousness. All these attainments and processes by which our entire physical life can be transformed into divine life on the earth – all this and much more is the subject matter of the theme of yoga and consciousness.

An extremely important concept that has been developed in the yogic science of consciousness is that of the human bondage and liberation. And underlying this concept of Ignorance, which is symbolically described as *ratri* in the *aghamarshana* mantra of the Rigveda.

A central knot of ignorance is the knot of the ego, *ahankara*. The analysis of ego in one of the profoundest in the yogic science. The ego presents itself as though it is an entity, separate, distinct and independent of everything else. As a result, the egoistic consciousness is constantly engaged in the processes of divisions; divisions between self and others, divisions between oneself and the world, and divisions even within oneself. This egoist consciousness is also found to be at the root of what yogic science calls dualities or dvandvas, -- dualities of joy and sorrow, of success and failure, of the superior and inferior, and the dualities which whirl round in cycles and circles of fluctuating and unequal rotations that stifle and afflict the individual consciousness imprisoning it constantly within narrow bounds from which escape or liberation is found to be extremely difficult.

At a deeper level of analysis, it is found that there is really no such thing as an entity which can be called the entity of the ego. It is found, on the other hand, that ego is simply a construction, a sense or a mere idea corresponding to which there is really nothing excepting a narrow construction, a sense of *"I-ness"* that prevents the larger perception of a true entity within, which is not egoistic in character.

That entity is variously understood and realized in the different disciplines. In the *jnana yoga*, that larger entity is the Substance, the self, the Atman which is same and the only universal and even transcendental Reality, which is termed the Brahmin, and which when experienced is found to be ever free Sachchidanand, the pure Existence-consciousness reveals itself as the Instrument that is inalienably related to the supreme Master, who is also experiences as Sachchidananda. In the Yoga of Divine Love, the entity experiences behind egoistic consciousness are entity that can be entitled the Child of the divine Mother, who is also Sachchidanad. And, in several other synthetic systems of Yoga, these three states are experienced in some kind of combination or integrality. And in all these experiences, the identification if the inner entity with the outer body, life, mind and ego through the process of *adhyasa* is lost; it disappears and there is annihilation of the ego. It is this annihilation of this ego which marks a momentous stage in yoga, and which is known as *mukti* of *moksha*.

Indian tradition makes a distinction between three states of *mukti*,-- *salokya mukti*, *sayuiya mukti and sadharmya mukti*. In Salokya mukti, the of consciousness that is attained is the same as the state of transcendental reality; In sayujya mukti there is a union of the individual and the supreme reality, and there is a relationship of constant nearness and constant self-giving on the part of the individual to the supreme reality. In *sadharmya mukti*, not only is the individual liberated from the limitation of the *prakriti* but even the nature, prakriti, becomes transformed and manifests the divine nature. This transformation of the lower nature of *apara prakriti* into the higher nature, divine nature, *para prakriti*, is conceived to be a state not only of liberation but also of perfection.

Consciousness as revealed through various processed of yoga can be viewed in three distinct but interrelated standpoints. In the first place, higher states of consciousness developed through gradual process of widening of consciousness. In different systems of yoga, these three processes are aided by certain specific prescriptions regarding the development of attitudes, moral and spiritual control and mastery over impulses, desires, preferences and inclinations of will, action, and rational or aesthetic perceptions and tastes. One of the highest attainments is that of equality, *samattvam*, -- equality of both inner sates of quietude, silences and peace as also equality in dynamic states of will and action.

In the second place, highest states of consciousness are sought to be attained by a conscious application of the secret knowledge in regard to the relationship between two statuses of consciousness, the state of *purusha* and the state of *prakriti*. In the ordinary movement of consciousness in which human beings dind themselves, there is normally predominance or rule of movements of prakriti, or dynamic nature, -- analysed in Indian psychology as a triple thread, of *tamas, rajas* and *sattva,--* and *purusha* which is secretly present behind the movement of *prakriti* is found to be normally dormant or under constant subjection of *prakriti*. According to the secret knowledge of the relationship between *purusha* and *prakriti,* purusha is the originator of the movement of *prakriti*, and *prakriti* works out or executes the will of the *purusha.*

From this point of view, if *purusha* is found to be subject to *prakriti*, it can be traced to an inner consent on the part of the *purusha* to become ao subjected. In the same way, it is up to the *purusha* to withdraw its consent so as to become a ruler of *prakriti*. Yoga reveals that this secret knowledge is variable, and if one begins to will freedom from subjection to *prakriti's* movements, *prakriti* itself will collaborate and create conditions, under which purusha can recover its position as the originator and therefore the ruler of the movements of *prakriti*. It is in this process that one experiences the status of the witness, *sakshi bhava*, the state of *bharta*, the state of the giver of consent or withholder of the consent, the state of *anumanta*, the state of *bharta*, the state of the master in which the *prakriti* obeys the will of the *pursha*, and even the state of *ishwarva* in which *purusha* experiences the lordship over *prakriti*. And change in the ordinary nature and transmuted into divine nature. This is the process that we find elaborated in the lasr six chapters of Bhagvadgita, where Sri. Krishna speaks of *trigunatita* and *envisage*s the transmulation of ordinary nature, *sarva dharma*n *partajya*, so as to attain oneness with the nature of the divine consciousness.

Thirdly, the process of development of consciousness can be seen as a gradual rising of states of consciousness which are normally fixed in physicality and which can be raised by various process of the awakening of *kundalini*, which has been discovered by the yogic sciences to be coiled at the lowest physical base or the centre of chakra called *muladhara*. On this line of development, there are elaborate centres of chakras until it is raised up to sahasrar, thousand petalled lotus, on the top of skull or beyond. Here, consciousness, as power, power of vision and power of thought and light. It is with the development of power of consciousness that is lower as to attain a permanent station in the higher states of consciousness. The entire *shastra* of *Tantra* is based upon the discovery of this yogic science, and we have in this exposition, revaltions of various states of consciousness which ultimately can be used for attaining a complete mastery and extraordinary *siddhis* of consciousness and power.

These three basic lines of development are, in a sense, interrelated. The higher is the state of consciousness, the greater is the predominance of *purusha* over *prakriti*, and the greater is also the rise of *kundalini* in the scale of centres of *chakras*. And this is true also *vive-versa*. In the ultimate analysis, one can combine all the three development of consciousness. The result, too, could be richer and more fruitful.

The entire process of yogic development of consciousness is intricate and it requires, on the part of the individual, not only a steady and unfailing aspiration, *utsaha*, but also uplifting guidance of the teacher, *guru*, who has the true knowledge of the principles and states of consciousness attainable by the process of yoga. And, finally, the entire eternity to arrive at realization and yet a constant effort which insists on the development of capacities which ultimately at their can bring about instantaneous realisation.

B

E

I

J

K

L

O

P

Q

R

S

sadness, 271, 272
SAIM, 120
saliva, 75
sample survey, 325
Sanskrit, 332, 429, 430, 431, 444
saturation, 58
sawdust, 254
scatter, 366
schizophrenia, 367, 418, 422, 423
scholarship, 429
school, 5, 19, 60, 207, 217, 219, 223, 225, 226, 288, 289, 296, 299, 310, 311, 312, 314, 315, 316, 317, 318, 319, 320, 321, 322, 323, 324, 325, 327, 329, 332, 333, 334, 428, 429, 434, 435
schooling, 296, 303, 305, 310, 311, 322
science, vii, ix, 3, 4, 5, 7, 13, 14, 15, 16, 18, 19, 21, 30, 33, 54, 122, 136, 159, 223, 224, 226, 228, 240, 247, 314, 323, 324, 364, 379, 408, 437, 449, 450, 451, 453, 454, 456, 457, 459, 460, 461, 462, 463, 464
scientific knowledge, 450, 454, 458
scientific method, 3
scientific theory, 448, 453
scientific understanding, 239
sclerosis, 341, 416
scope, 90, 117, 118, 119, 122, 124, 159, 322, 431, 432, 437, 444, 451
scope of attention, 117, 118, 119, 122, 124
script dependent hypothesis, 222
scripts, 218, 219, 220, 221, 222, 223, 315, 320, 323, 324, 327, 328, 329, 330, 331, 332, 333
sea level, 51
second language, 202, 203, 204, 205, 206, 211, 212, 215, 217, 220, 222, 224, 225, 228, 229, 233, 314, 319, 335, 336
secrete, 412
secretion, 144, 154, 156, 389
segregation, 65, 293, 298
selective attention, 60, 64, 105, 106, 107, 108, 109, 112, 120, 121, 122, 123, 124, 125, 126, 128
selectivity, 110, 118, 259, 370
self consciousness, 447
self esteem, 317
self-consciousness, 448
self-consistency, 294
self-doubt, 241
self-enhancement, 284
self-regulation, 380
semantic analysis, 107, 109
semantic association, 107
semantic information, 108, 211

semantic memory, 27, 37, 42, 130, 190
semantic processing, 107, 109, 230, 334
semantic storage, 328
semantics, 168, 214, 221, 229, 310, 314, 333
semisyllabic, 332
senile dementia, 360
sensation(s), 22, 70, 71, 73, 75, 76, 91, 93, 94, 100, 101, 366, 444, 446, 460
senses, 22, 369, 431, 432, 449, 450, 455, 456, 462
sensing, 29, 90
sensitivity, 63, 107, 174, 218, 219, 315, 323, 343, 349, 352, 367, 409
sensitization, 130, 133, 134, 138, 139, 418
sensors, 16, 39, 40, 168
sensory modalities, 370
sensory nerves, 22
sensory perceptions, 446
sensory systems, 69, 70, 89, 91
sentence processing, 56, 211
separatism, 19
septum, 374
sequencing, 343, 345
serotonin, 11, 13, 17, 133, 134, 135, 144, 147, 153, 242, 375, 413, 420
serum, 134, 390, 402, 412, 415
services, 14
SES, 204
severe stress, 142
sex, 195, 229, 294, 295, 373, 385, 410, 413
sex difference(s), 229, 294, 295
sex role, 295
shade, 260
shape, 25, 27, 36, 39, 50, 83, 85, 112, 113, 127, 153, 288, 368
shock, 131, 143
short term memory, 27, 116, 218
shortage, ix
short-term memory, 130, 131, 133, 193, 227, 329, 370
showing, 57, 83, 95, 97, 98, 112, 148, 204, 211, 251, 299, 368
siblings, 373
side effects, 355, 418, 421, 423
signal transduction, 85
signaling pathways, 130, 134
signalling, 154
signals, 24, 25, 26, 28, 39, 40, 43, 44, 51, 52, 53, 57, 58, 59, 69, 70, 72, 73, 74, 75, 76, 78, 81, 82, 90, 110, 117, 127, 160, 170, 171, 182, 184, 189, 269, 347, 423
signal-to-noise ratio, 58
signs, 275, 332, 345, 346, 350, 353, 354, 358, 366, 367, 368

T